Intelligence and Evolutionary Biology

NATO ASI Series

Advanced Science Institutes Series

A series presenting the results of activities sponsored by the NATO Science Committee, which aims at the dissemination of advanced scientific and technological knowledge, with a view to strengthening links between scientific communities.

The Series is published by an international board of publishers in conjunction with the NATO Scientific Affairs Division

A Life Sciences B Physics	Plenum Publishing Corporation London and New York
C Mathematical and Physical Sciences D Behavioural and Social Sciences E Applied Sciences	Kluwer Academic Publishers Dordrecht, Boston and London
F Computer and Systems Sciences G Ecological Sciences H Cell Biology	Springer-Verlag Berlin Heidelberg New York London Paris Tokyo

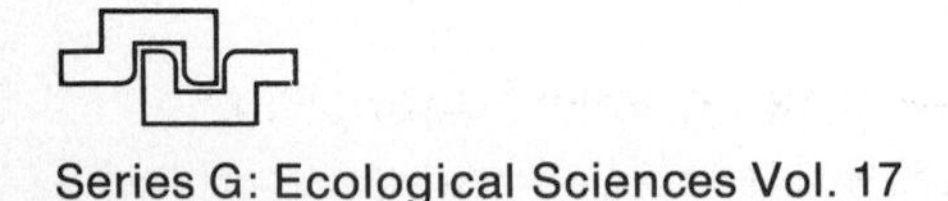

Series G: Ecological Sciences Vol. 17

Intelligence and Evolutionary Biology

Edited by

Harry J. Jerison and Irene Jerison

Department of Psychiatry and Biobehavioral Sciences
UCLA School of Medicine
University of California, Los Angeles
760 Westwood Plaza
Los Angeles, California 90024, USA

Springer-Verlag
Berlin Heidelberg New York London Paris Tokyo
Published in cooperation with NATO Scientific Affairs Division

Proceedings of the NATO Advanced Study Institute on The Evolutionary Biology of Intelligence held at Poppi, Italy, July 8–19, 1986

ISBN 3-540-16085-X Springer-Verlag Berlin Heidelberg New York
ISBN 0-387-16085-X Springer-Verlag New York Berlin Heidelberg

Library of Congress Cataloging-in-Publication Data. NATO Advanced Study Institute on the Evolutionary Biology of Intelligence (1986 : Poppi, Italy) Intelligence and evolutionary biology / edited by Harry J. Jerison and Irene Jerison. p. cm.—(NATO ASI series. Series G, Ecological sciences ; vol. 17)
"Proceedings of the NATO Advanced Study Institute on then Evolutionary Biology of Intelligence, held in Poppi, Italy, July 8–19, 1986"—T.p. verso. Includes index.
ISBN 0-387-16085-X (U.S.)
1. Animal intelligence—Congresses. 2. Intellect–Congresses. 3. Evolution—Congresses. I. Jerison, Harry J. II. Jerison, Irene. III. Title. IV. Series: NATO ASI series. Series G. Ecological science ; vol. 17. [DNLM: 1. Evolution—congresses. 2. Intelligence—congresses. BF 431 N2785i 1986] QL785.N39 1986
591.51—dc 19 DNLM/DLC 87-37660

Printed in Germany

Printing: Druckhaus Beltz, Hemsbach; Binding: J. Schäffer GmbH & Co. KG, Grünstadt
2131/3140-543210

To Professor B. A. Chiarelli

PREFACE

Sponsored by the North Atlantic Treaty Organization (NATO), the Advanced Study Institute (ASI) on the Evolutionary Biology of Intelligence was held in Poppi, Italy, July 8 - 19, 1986. Support was provided by other agencies, including the National Science Foundation (USA) and agencies in Italy. Most of the participants received some support either from their universities or other sources, independently of the ASI's funding. NATO ASI's are intended primarily for the dissemination of knowledge in rapidly growing scientific areas. The director is responsible for organizing the course of study, recruiting lecturers and choosing participants. Like most of the NATO ASI's, this one inevitably had some of the ambience of a symposium or conference, because of the advanced level of many of the participants. A sense of the mood of the meeting is presented in the final chapter.

The book consists of rewritten and edited versions of most of the main lectures and several of the contributions by "participants." To suggest something about the high level of the participants, note that Dr. Terrence Deacon of Harvard was in that category, although he was recruited for the role as a result of strong recommendations from other anthropologists. His outstanding contributions were the subject of several "rump" sessions in addition to his formal presentation to the Institute. Revisions of many of the chapters (including Deacon's) were received as long as a year after the Institute, and the book's rather than the Insititute's dates are the correct indicators of the recency of the contributions.

The general organization of the book has analytic and philosophical chapters at the beginning, followed by chapters emphasizing neurobiology and language. There follow chapters on evolutionary issues and morphometrics as these are relevant for intelligence. There were then several chapters on behavioral evidence, including concerns with human as well as animal intelligence. Beginning with Chapter 19, there is a reprise of the earlier organization, emphasizing neurobiology and philosophy, with a final chapter on the director's afterthoughts about the Institute.

The production of the book is, perhaps, unusual but likely to become less so as the technology of word processing and "desk-top publishing" develops. To take advantage of new technology and to speed publication, most contributors provided their texts on computer diskettes as well as in typed form. Another purpose was to limit copy-editing and the introduction of type-setters' and proofreaders' errors, by having each contributor responsible for these mechanical features of the text. These were then reviewed by the editors, who had undertaken to provide "camera-ready" copy for the publisher. A few words on problems and opportunities that arose here are in order, because publication is so central a part of the scientific enterprise and others may learn from our experience.

We were, of course, unsuccessful in eliminating proofreaders' errors, because scientists are no better than anyone else in catching these. We were helped by computer methods for checking spelling, which can catch many typographical error, but at one point the entire production of this book was delayed by a failure in our computer's hard disk on which the files of the text were stored. The failure was a strange one that we did not recognize until our spell-checking program began to catch errors such as tzpe or txpe for type. This is a "one-bit" error in the computer (ASCII) code for the letter "y," which might be inserted by a human typist switching from typewriters designed for one country to those of another. In our case it was clearly a computer error; the main problem was to diagnose it. When we did this the correction was relatively easy because of the availability of the spell-checking programs.

The most useful technological advances in the preparation of the book were the international computer networks that have been established and the advent of desk-top publishing. We received edited versions of two of the contributions via the network. Ian Deary sent his chapter from Edinburgh to Florence in March of 1987, and it arrived without errors (we think). In this case transmission was from Edinburgh to Switzerland to Wisconsin (USA) to Florence (again, we think; it was fascinating to receive messages about the march of information through the network -- a model for the distributed systems in the brain discussed in the "Afterthoughts" chapter). We received the final text of Martin Pickford's contribution in a transmission from Florence to Los Angeles in April or May of 1987, this time with some garbled lines that we think were corrected by printed copies mailed in a standard way. For readers who may wish to communicate (gentle) messages to the editors, note that we can be reached on BITNET (see Jennings et al., "Computer Networking for Scientists," *Science*, 28 February 1986, Vol. 231, pp. 943-950); as this book goes to press, our BITNET is: IJC1HJJ@UCLAMVS.

Desk-top publishing made it possible to rush through the final printing of Deacon's chapters. The critical feature was that the book was being printed in one type font of a LaserJet printer (B Font: Times Roman and Helvetica), but we were under severe time pressure to meet a schedule that made it impossible to prepare Deacon's illustrations properly. Deacon himself undertook to print his chapters, but in order to incorporate his illustrations into his text he had to use a different printing system. His chapters were not prepared on the same printer as the others in this book, and the reader who enjoys typography can compare the chapters to recognize how good a type-matching can be. (The minor differences between Deacon's and other chapters will also inform the reader about the limits of our ability communicate with one another about layout.)

These not entirely minor matters of form had one unfortunate consequence that required an impossible editorial decision. The format of the book was as uniform as we could make it within the limits of our printer, a Hewlett-Packard LaserJet. Unfortunately, the type font cartridge that produced the most attractive page was limited in its characters, and we could not produce proper diacritical marks. We chose to eliminate most of them, including *Umlauten*. Replacing the *Umlaut* in the standard way (e.g., "fuer" for "für" -- note our printer's rendering of "ü") produces some spellings that are jarring and unfamiliar to many native English readers, and our decision was to defer to the requirement for readability in that group. The

Umlaut was entered by hand in a few instances, where it would have been especially offensive to neglect proper German spelling. We apologize for our inability to resolve this problem satisfactorily.

The work of many people was involved in the production of this book. Professor B. A. Chiarelli of the Department of Anthropology, University of Florence, first proposed organizing the NATO Advanced Study Institute that led to this book. He was co-director of the ASI and accepted responsibility for its administration. Unfortunately, he could not participate in editing this book because of other commitments. It is a pleasure to dedicate this book to him, to recognize his inspired leadership.

The NATO ASI could not have succeeded without the devoted efforts of Chiara Bullo, executive secretary of Professor Chiarelli's group, and the Ph. D. students in that group. Among these, Dr. Andrea Camperio Ciani was especially helpful, both at the ASI and during the preparation of this book, supporting most of the computer communications between Florence and other stations in this international effort.

For a period of about six weeks the editorial effort was centered at the University of Hawaii, where we were guests of the University, and we acknowledge the gracious support of Professor L. M. Herman, who provided access to his computer facilities and editorial review for one of the papers. The review contributed balance to these published proceedings; it is quoted extensively in the "Afterthoughts" chapter.

The actual printing of the manuscript as "camera ready" copy for publication was done at the Department of Biology, University of California at Los Angeles. We thank June Baumer, director of computer facilities in that department for technical advice and for access to the facility. Although most of the artwork was provided by the contributors, Sharon Belkin, artist at the Department of Psychiatry and Biobehavioral Sciences redrew several figures (in the contributions by Pickford and by Schusterman and Gisiner) and, with photographer Maria Karras of that department, helped in the final preparation of camera-ready copy, working long hours in this effort, beyond the normal duties of the departmental staff.

Harry J. Jerison and Irene Jerison
November 6, 1987

CONTENTS

EVOLUTIONARY BIOLOGY OF INTELLIGENCE: THE NATURE OF THE PROBLEM
Harry J. Jerison 1

INTELLIGENCE AND NATURAL SELECTION
Michael Ruse 13

THE CONCEPTUAL ROLE OF INTELLIGENCE IN HUMAN SOCIOBIOLOGY
Paul Thompson 35

ARTIFICIAL INTELLIGENCE AND BIOLOGICAL INTELLIGENCE
Margaret A. Boden 45

AN EVOLUTIONARY EPISTEMOLOGICAL APPROACH TO THE EVOLUTION OF INTELLIGENCE
H.C. Plotkin 73

COMPARATIVE NEUROANATOMY AND THE EVOLUTION OF INTELLIGENCE
William Hodos 93

THE FOREBRAIN AS A PLAYGROUND OF MAMMALIAN EVOLUTION
H.-P. Lipp 109

COMPARING THE STRUCTURE OF BRAINS: IMPLICATIONS FOR BEHAVIORAL HOMOLOGIES
A. Fasolo and G. Malacarne 119

LANGUAGE, INTELLIGENCE, AND RULE-GOVERNED BEHAVIOR
Philip Lieberman 143

THE EVOLUTION OF HUMAN CEREBRAL ASYMMETRY
Jerre Levy 157

THE EVOLUTION OF INTELLIGENCE: A PALAEONTOLOGICAL PERSPECTIVE
Martin Pickford 175

ALLOMETRIC ANALYSIS AND BRAIN SIZE
Paul H. Harvey 199

MAMMALIAN DOMESTICATION AND ITS EFFECT ON BRAIN STRUCTURE AND BEHAVIOR
D. Kruska 211

VERTEBRATE-INVERTEBRATE COMPARISONS
M. E. Bitterman 251

SPECIES-SPECIFIC DIFFERENCES IN LEARNING
Marco D. Poli 277

CONTRIBUTION OF THE GENETICAL AND NEURAL MEMORY TO ANIMAL INTELLIGENCE
V. Csanyi 299

ANIMAL LANGUAGE RESEARCH: MARINE MAMMALS RE-ENTER THE CONTROVERSY
Ronald J. Schusterman and Robert Gisiner 319

BASIC PROCESSES IN HUMAN INTELLIGENCE
Ian J. Deary 351

HUMAN BRAIN EVOLUTION: I. EVOLUTION OF LANGUAGE CIRCUITS
Terrence W. Deacon 363

HUMAN BRAIN EVOLUTION: II. EMBRYOLOGY AND BRAIN ALLOMETRY
Terrence W. Deacon 383

THE FUNCTION OF INFORMATION PROCESSING IN NATURE
J. M. H. Vossen 417

A NEUROPHYSIOLOGICAL BASIS FOR THE HERITABILITY OF VERTEBRATE INTELLIGENCE
T. Edward Reed 429

BRAIN, MIND AND REALITY: AN EVOLUTIONARY APPROACH TO BIOLOGICAL INTELLIGENCE
Michel A. Hofman 437

THE EVOLUTIONARY BIOLOGY OF INTELLIGENCE: AFTERTHOUGHTS
Harry J. Jerison (with contributions by L.M. Herman and W. Hodos) 447

AUTHOR INDEX 467

EVOLUTIONARY BIOLOGY OF INTELLIGENCE: THE NATURE OF THE PROBLEM

Harry J. Jerison
Department of Psychiatry and Biobehavioral Sciences
University of California, Los Angeles, Medical School
Los Angeles CA 90024 USA

If our topic presents a problem, it is one of the good ones: like working a puzzle with a known solution. Most of us enjoy a tough puzzle, especially when we finally solve it. The challenge of the game of solving (or resolving) the "problem" of the evolution of intelligence has the added attraction that success will deepen our understanding of ourselves and of our world. We know enough about both intelligence and evolutionary biology to think sensibly about the intersection of these topics. Our problem is really our opportunity to enjoy the problem.

During the course of these meetings [of the NATO Advanced Study Institute, Poppi, Italy, July, 1986] the two general fields, intelligence and evolution, will be reviewed from a number of perspectives, and although I am obviously interested in all of the approaches I am not eclectic enough to view them as equally useful. However, the utility of a perspective depends on how it is to be used, and my own view is, if anything, more idiosyncratic than most. In these remarks I emphasize directions of work that have been most relevant for me.

Intelligence as Encephalization

My objective has been to place intelligence in as broad a biological framework as is consistent with its identification with a specific phenotypic expression. My scientific tactic has been to examine the evolution of that "expression" as a concrete problem, and I solved (or more or less solved) that problem by analyzing measurements of the "expression" in fossil and living animals. For this reason, I have studied encephalization, a morphological trait, because there was evidence (Dubois, 1897; Lashley, 1949) that it could be used as a measure of animal intelligence for evolutionary purposes. Encephalization is not a behavioral measure, of course. It is a simple transformation of a relationship between brain and body size. The measure works (as far as I can tell) because the numbers that it yields are fairly directly related to the capacities of animals to handle information. As a first approximation, information handling capacity is related to body size according to the allometric brain:body relationship. Encephalization is the residual capacity that remains after basic allometric requirements are satisfied, and it represents additional information that an animal handles, beyond that used in routine control of body functions. Encephalization is my phenotypic "expression" of intelligence.

NATO ASI Series, Vol. G17
Intelligence and Evolutionary Biology
Edited by H. J. Jerison and I. Jerison

As you will appreciate during the coming days, not everyone is satisfied with this approach to intelligence, which is broad in some ways yet quite limited in others. I am not especially satisfied with it myself. I present it, because I think most of us would agree that at least in some ways, intelligence in a biological sense is equivalent to residual processing capacity.

The evolution of residual processing capacity (i.e., encephalization) can be measured, and we know many details of its history and distribution in nature. This knowledge has important implications for our view as evolutionists about the kind of trait (or set of traits) with which we are dealing and the problems that are likely to arise in its evolutionary analysis. For example, encephalization occurs in many animal species, and comparable grades of encephalization occur in very distantly related species. Man, dolphin, and killer whale (*Homo, Tursiops, Orcinus*) are approximately equal in encephalization, despite their evolutionary separation by more than 60 million years of history and by their niches in terrestrial versus marine adaptive zones. The harbor porpoise (*Phocoena*) and the great apes are another example of distantly related species that are at one comparable grade. At the other extreme, the marsupial Virginia opossum (*Didelphis*) and the European hedgehog (*Erinaceus*) are at the same grade of encephalization, and that grade is the same as in the earliest mammals, at least those that lived 150 million years ago. Let us see what these few facts imply.

Evolutionary Models

If we equate encephalization with intelligence, these facts imply that similar grades of intelligence have been achieved in widely disparate species. They point us to an anagenetic analysis, above the species level. Some of the most important results for a broad evolutionary perspective have been on such anagenetic (i.e., progressive) changes in encephalization. Not all modern evolutionary models are relevant for such an analysis, even of phenotypic effects. One contribution of the simplifying assumption that intelligence and encephalization are the same is that it suggests the relevant models.

The analysis of intelligence can proceed equally correctly with a cladistic analysis of its distribution in related species, or with an anagenetic analysis of the "topography" of grades of intelligence as they are represented among animal species over space and time. Cladistic analysis of encephalization generally demonstrates that this trait has remained stable and unchanging in related species and in phyletic lineages. Stability, or conservation, is one of the most important facts about the evolution of encephalization: encephalization is a conservative trait. Having said that, not much more remains to be said. So cladism is not a very interesting approach. The one exception to the boredom of stability saves the day. In the face of stability that occurs in almost all lineages, the hominids were unique. Encephalization in hominids from the near-pongid australopithecine grade achieved about 5 million years ago to the present sapient grade of the past quarter million years, has been one of the most remarkable features of our history, inevitably emphasized by all students of primate evolution.

We would expect to restrict our evolutionary models to those concerned with phenotypic as opposed to genotypic variation. And we would also emphasize models that can work above the species level. Let me review some analyses that I have seen recently, which propose models important for the understanding of the evolution of intelligence.

One of the most interesting models is Maynard Smith's (1982), which treats evolution as a game that is resolved for its players (genotypically mutant "entities" that can adopt behavioral strategies for their social interactions) either by extinction or by the appearance of an evolutionary stable strategy in one of the players -- which becomes the universal phenotype. I was especially interested in Maynard Smith's analysis, because of the important role of differences among strategies with respect to their load on information processing capacity. Information processing capacity was shown to be a "trait" on which natural selection could act. The demonstration was elegant and precise and did not depend in any way on vague intuitions.

A few days ago I finished reading Dawkins (1982) on the "extended phenotype," an unusual gene's eye view of evolution (cf. Dawkins, 1976). Even more than Maynard Smith's work, it forced me to do some serious rethinking, which I will share with you now. The problem that Dawkins raised can be fundamental for cognitive psychologists and for our view of the nature of intelligence. It is essentially one of how to define the phenotype on which natural selection acts. Classically, the answer is that natural selection acts on the individual. Dawkins makes a strong case that this is too limited a view, for example, for the evolution of parasitism and symbiosis. He makes the case for many evolutionary phenomena but is most challenging in his ethological examples of social parental and kin behavior.

Dawkins argues that rather than restricting it to the individual animal, the phenotype carrying the genetic material can more profitably be viewed as extended to include all the interacting individuals that carry the genetic system evolving with respect to a particular inter-individual relationship. I have just published a paper (Jerison,1986) that includes a wild speculation on the nature of the individual as a self, and without presenting my reasoning in detail here, I note that I proposed that bottlenosed dolphins (*Tursiops*) could fuse mentally, as it were, so that several animals would perceive and know the external world as if they were a single individual in the conventional sense. After reading Dawkins, this speculation did not seem quite as wild as when I first presented it. Without committing the sin of group selectionism I was creating a group that could also be an individual, an extended phenotype responsive to natural selection, though not quite as extended as some that Dawkins discusses.

(To explain a bit more, because this story has a personal angle that I'd like to share with you, I first presented the speculation to titillate colleagues in marine biology and warn them that we know both less and more than we think about the way animals know the external world. We know less, because we almost always assume, incorrectly, that other animals experience, or know, the same reality that we do. But we know more, because we have discovered many unusual dimensions of information that can be processed. Echolocation was the critical dimension for dolphins, and I contrasted the possibilities available for small-brained bats and highly encephalized dolphins in the construction of "reality" by their brains. My audience was so outraged by the speculation that when I prepared it for the inevitable publication I decided, in defense of my scientific reputation, such as it is, that I had better justify my position. Almost the entire paper had to be devoted to setting the stage for the odd view of *Umwelt* in dolphins. The point in a nutshell is as follows. If brains work to construct an experienced external world as a model of a possible world -- a position that is easy to defend -- one of the features of the human construction of reality is surely the self, or ego. I take the self as one construction, and knowledge of the self as another, quite comparable to the experience we have of anything external to ourselves and our knowledge of that experience, and the construction can be fairly arbitrary with respect to the distinction between self and not-self in different species.)

In introducing the Institute and its program with this lecture I vowed to try not to anticipate the lecturers and their contributions. I cannot avoid commenting on Henry Plotkin's work, however, which I was reading at the same time as Dawkins. I became involved in an imaginary dialogue (trialogue?) with Dawkins and Plotkin, which I could not disentangle. Plotkin's, it seemed to me, was a view even broader than that of the extended phenotype. It was much broader than my view of encephalization as an "it" and intelligence as an "it" and the idea that these two "its" might be one. The very unusual and important contribution of evolutionary epistemology seemed to be that, in its evolution, intelligence may represent an aspect of nature, a kind of entropy that should somehow be incorporated into evolutionary thinking. A symposium on our topic inevitably conjures up the image that I presented: a trait or set of traits, perhaps a morphological trait, that is or are at least analogous in its operation to such traits. The alternative view puts intelligence on a different level in the analysis.

I must stop this digression now. Plotkin may see it as a transgression and he would very likely be right.

There are important contributions from quantitative population biology and theoretical evolutionary biology to our analysis. Some of these, on allometry and encephalization, will be considered in depth here. We may find ourselves spending less time on other topics, because of the vagaries of scheduling, the ability of people to meet their commitments to contribute to our proceedings, and the inevitable limitations of organizing an institute. Not all topics can be given the time that they deserve, and often by chance a scheduled topic is not covered. I have tried to emphasize evolutionary biology rather than intelligence when I was planning the Institute. The idea was to indicate to students of intelligence something about the form of data that can be analyzed easily by present methods.

With respect to some other well known developments in evolutionary biology, we may be thankful (in my view) that we can probably avoid participating in the recent battle between gradualists and punctuationalists (Gould, 1980; Lande, 1980). As far as I can tell, intelligence (as encephalization) could have evolved in ways consistent with either scheme. There have clearly been long periods of equilibrium of grade of encephalization in large numbers of species. But the fossil evidence (as I read it) has never implied that changes of grade occurred so rapidly that a gradualist model (e.g., Lande, 1976, 1979) could not account for them.

I have calculated some conditions associated with evolutionary rates and found that a gradualist model works quite well for the evolution of brain size in hominids. Surprisingly, the model (Lande, 1976) indicated that the rate of evolution of brain size in hominids was slow enough to make it impossible to reject genetic drift as its determinant. The idea was to set selection pressure at zero, and to determine the values of the free parameters. With no natural selection, these should have had impossible values, such as a breeding population size of 1.5 individuals. Instead the parameters that were computed were not only possible but plausible. Population size, for example, was 50 - 150. This result does not necessarily mean that we hominids became so brainy as a result of genetic drift. It does mean that either the evidence or the evolutionary model is not good enough to discriminate between natural selection and drift in the evolution of our brain.

In any event, a gradualist model of the kind used by Lande has no difficulty with data on encephalization (cf. Lande, 1979). It is hard to make the punctuationalist point if the case believed to represent the most rapid phenotypic evolution displayed in the fossil record, namely, that of the evolution of our brain, was slow enough to be accounted for without even requiring significant selection pressures. Some of our participants may disagree with that conclusion, and I look forward to discussions on points like these.

The final models developed by evolutionary biologists in recent years include Van Valen's (1974) "Red Queen's Hypothesis" which I will describe briefly, and the elaborate system proposed by Lumsden and Wilson (1981) on which I will pass for the present because I don't know it well enough. Van Valen's model is essentially that evolution as a whole is a zero sum game played by populations of species. Gains by some species are offset by losses by other species. Progress is an illusion that results from concentrating on the gains, because in the long run species maintain a kind of balance with respect to one another; change is a tracking of a changing environment.

This account fits in well with the evidence on encephalization, in that cases of encephalization can almost always be explained as responses to unusual environmental challenges or opportunities -- the invasion of empty niches, as it were, which can be occupied only at the expense of investing in increased processing capacity. When such an adaptive zone is first occupied, all species that invade it have to measure up to the high processing requirements, and a balanced trend toward encephalization should appear over a range of species. That is exactly what one finds in a broad look at the fossil record. The events in the hominid lineage are, therefore, odd in evolutionary perspective, because we were the only species displaying progressive encephalization during the Plio-Pleistocene.

Although I will refrain from discussing Lumsden and Wilson (1981) I can do this safely, first, because several of our contributors will be concerned with the perspective that is associated with Wilson's synthesis of sociobiology. Second, I have already described several contributions from a sociobiological "school," although the contributors, Dawkins and Maynard Smith, might reject the labels. Theirs is just good biology, good genetics, and good thinking. That it relates remote phenotypic behavioral phenomena, including intelligent behavior, to genotypic events is not surprising, since discovering such relationships is a major goal of biology.

Encephalization and the Meaning of Intelligence

To return to the evolution of encephalization and its implication for the evolution of intelligence, I know that the facts of encephalization are at least once removed from the facts of intelligence. Yet it is easy to show that encephalization means more than a pound (or ounce or gram) of flesh on a scale, that it tells us something about the amount of neural information that is handled by nervous systems. In addition, the evidence on encephalization tells us about the kind of model of an external (and internal) world that may be used by nervous systems to handle the information. It summarizes facts about information processing capacity. These are also facts about evolutionary biology when they are collected across species and ecosystems and geological time. They demonstrate that processing capacity evolved to different extents in different species, presumably because of different adaptations to different environments.

From the facts about the evolution and present diversity of encephalized species, and from a neurological understanding of the diverse ways in which encephalization occurs, we can infer one of the fundamental aspects of intelligence from an evolutionary perspective. It is impossible to believe that intelligence represents a singular direction of specialized adaptation. Rather we must assume that excess processing capacity is used in different ways in different species. We must assume that when intelligence evolved it was not a single type of intelligence. Rather that a variety of intelligences in different species evolved and the evolution of intelligence in the human species represents just one kind of intelligence.

What can such intelligence represent? Here I prefer a neurological to a psychological or behaviorist orientation. The basic question is about how and why brains are organized to handle large amounts of information. The question can be put in energetic terms. If natural selection minimizes the expenditure of energy relative to a particular biological goal, why must so much energy be invested in the operation of large nervous systems as part of adaptation? The answer is at least partly evident in the actual structure of nervous systems. We must look at how neural activity is apportioned among neural structures and how much material in the brain is devoted to the various activities that the brain performs.

Vertebrate brains, in particular mammalian brains, appear to be large primarily to support the neural structures that are identifiable projection systems, and these are conventionally assigned sensory and motor functions. The projection systems are demonstrably association systems as well. That is, classically defined association functions, such as those of attention or goal-directedness, are disrupted by lesions in sensory "projection" systems, and the systems as a whole may be thought of as supporting higher order functions distributed throughout the brain. The brain itself works as a "distributed system" (Mountcastle, 1978), with specialized functions distributed through much of its structure. Structurally, the largest organ in the mammalian brain, the neocortex, is now known to be surprisingly uniform throughout its extent (Rockel, Hiorns and Powell, 1980), a consistency related to the organization of the neocortex (and probably many other regions in the brain) into columnar modules for processing information). These modules characterize all of the projection systems, and association functions are distributed among the modules as part of their operational characteristics. This is the consensus among neuroscientists today, and it is entirely consistent with the position on intelligence that I would defend, namely, that intelligence should be equated with cognition, that is, with the way an animal knows its external world.

My Problems with Learning as Intelligence

I will not deny the validity of approaches to animal (and human) intelligence that begin by examining learning ability. But my judgment of the results of that approach is that it is as consistent with the perceptual/cognitive model that I prefer as it is with classical learning-theory approaches that produced associationistic models for estimating response probabilities in the face of various environmental antecedents and consequences of the responses. It may be possible to develop equivalent models for the same data, but the evidence of the brain favors the idea that it is sensory-perceptual activity that requires large amounts of neural tissue, and I would add cognitive, analysis as the brain-work that can only be done by large brains.

I am suggesting that the fundamental properties of learning are inherent in the structure of nervous systems. Everything now seems to point to miniature nervous systems as plastic rather than hard-wired with respect to the way they function, even if their general structure is fixed. In that sense all nervous systems can and do "learn." Computer modelers have shown us that relatively small information processing systems can learn in that sense and should be designed to learn if they are to work efficiently. Learning ability, per se, should not normally require large systems of processing machinery such as large brains. Selection pressures to be able to learn should lead to differently organized brains, perhaps, but not to especially enlarged brains.

The literature on learning does show that encephalized species may perform better on some tasks than do less encephalized species. This evidence is clearest within reasonably closely related species, such as the set of all primate species (Passingham, 1982). When some tasks are harder to learn than others, it is not because they require more elaborate learning mechanisms but because they depend on more elaborate encoding of the external world. The encoding is perceptual/cognitive work, supportable only by an enlarged brain.

The intelligence that evolved when encephalization evolved is, therefore, a way of knowing. I have said elsewhere (e.g., Jerison, 1985a) that the work of the brain is correctly thought of as *creating* a real world. That act of creation is essentially the same as an act of cognition.

Within this framework an enormous variety of specialized adaptations can be analyzed. As humans, we are inevitably drawn to the adaptation that we call language, since that is so human a characteristic. The evolution of language and the evolution of the structural specialization to support it are major themes of our Institute, and I must defer to our expert lecturers on this subject. But allow me to present one idea that I have presented before in discussing language as an adaptation.

Human Language

Like learning, language is paradoxically difficult for the understanding of encephalization from a neurological perspective. Learning is a feature of all nervous tissue, I argued, hence learning ability is unlikely to explain the evolution of enlarged brains. It is what is learned that is likely to be correlated with that evolution rather than how (at a basic level) the learning takes place. The obvious and outstanding aspect of language has always been that this is the human way of communicating. The great ethological discoveries of elaborate systems for symbolic communication among individuals of various species, and especially von Frisch's (1950) discovery of the "language" of bees suggested a continuity among animal species with respect to communication and, perhaps, language. The word, "language," was used to describe such communication with symbols. The discoveries of capacities for sign language in great apes, now almost too familiar to cite (cf. various contributions in Weiskrantz, 1985), reinforced the view of continuity with respect to language and communication.

The difficulty is not so much about whether language requires a large brain. It is that communicational functions in most species do not appear to encumber very much brain tissue. It does not take a large brain to handle elaborate communication in bees, for example, and selection pressures for superior communication would not necessarily be expected to elicit an evolutionary "response" of brain enlargement.

There is no question that a large fraction of the human brain -- much more than just the language areas -- is involved in some way in human language. But if we look at the evolution of language as a problem in economical evolution it becomes clear that its early evolution was unlikely to have been as a method of communication. Elaborate and efficient communication of social information is handled by bee-brains, and if one insists on mammalian (or vertebrate) models for brain function one can make the same argument for social communication among wolves which is handled by brains that are at an average mammalian grade of encephalization. Communication per se does not normally require a large brain. If communication is at issue it must be a matter of what is communicated rather than how communication is effected. The situation is exactly analogous to the learning problem, of what is learned rather than the elementary mechanisms of learning.

The difficulty is resolved if we recognize that human language has an important cognitive role, that our knowledge of reality has linguistic as well as more conventional sensory "elements." If human language first evolved in response to selection pressures for cognitive capacity the paradox of the large size of the language system of the brain is resolved. Cognitive systems require large brains, and language as a cognitive system would have begun as a system from an enlarged brain.

This argument on the nature of language as a feature of human intelligence, a specialized adaptation in our species, has one unusual conclusion, and it is on that note that I also want to conclude this lecture.

As a cognitive adaptation, human language must contribute to the reality constructed by each of our brains. In this respect it may be like vision or olfaction or other senses. But language is also a major adaptation for human communication. The implication is that we communicate by sharing parts of our real worlds, sharing consciousness with one another, because we share part of the information used by an individual brain in constructing its reality. We can verify this by the often repeated "experiment" in which we have all participated. When we read and become engrossed in a realistic novel, we experience the characters as real, and the experience may be almost as vivid as if we were living their lives. This verifies the fact that when we communicate with language we share experiences, in this case the experience of the author -- and the experiences of many other readers at the time that they read. It is an unusual way for animals to communicate. It is as if we could communicate by having others see what we see and hear what we hear. The normal role of language in communication is very close to fictional accounts of communication by extrasensory means and may explain the attractiveness of ideas of such psychic powers. These imagined powers are not far removed from what we do in everyday life when we use ordinary language.

Normal animal communication, we have learned from the ethologists, is direct and certain. It is with commands that are usually obeyed: *sign stimuli* and *releasers* and *fixed action patterns.* We can use elements of human language to train other animals, and there is some suggestion that when this has been done with great apes and dolphins, cognitive activity becomes involved in the communication (Premack and Woodruff, 1978; Herman, 1986). But there is as yet no evidence that the communication is in any sense like human communication with shared consciousness. Rather the homologue for animal communication in the human species may be the important communication with 'natural' gestures, such as unrehearsed facial expressions, which might be described as communicating directly without the intervention of consciousness or a sense of identification. It is also sometimes described as limbic language (Myers, 1976), and it has some of the characteristics of animal communication in its universality and lack of ambiguity.

Why would as remarkable an adaptation as human language evolve? We cannot accept the idea there would have been selection for this remarkable, but peculiar communicational skill when language began to evolve. It is too uncertain a method of communication, too costly and unreliable compared to the system of commands that is normal animal communication. Rather, if we accept the view that the system is neurologically a cognitive one (that is, if we accept the idea that large systems are almost always cognitive), it is more reasonable to assume that our hominid ancestors at a very early point in their divergence from other large hominoids would have been under selection for augmented cognitive capacities -- beyond those of the great apes, for example. The cognitive capacities about which I have speculated are comparable to those of social predators such as wolves and are related to the capacity to generate elaborate cognitive maps. Animals generally are good at this sort of thing (Olton, 1985), and this is one of the more remarkable skills demonstrated by Kanzi, the pygmy chimpanzee with which Savage-Rumbaugh (personal communication; see Levy, this volume) has been working.

The problem of the early hominids, which appear to have invaded a hunter-gatherer niche in which a group's territorial hegemony could cover several hundred square kilometers, would be to generate cognitive maps adequate to enable them to navigate so extensive a range. The analogy is to the adaptation of wolves (Peters and Mech, 1975) for their predatory role. Wolves do the job with an elaborately organized olfactory system supported by limbic structures. Although the central part of this system is essentially normal in primates the peripheral part has been degenerate for 20 million years according to the fossil record by the time the first hominids appeared. I have suggested that the central portion of this limbic mapping system was available to the early hominids, and that alternative peripheral input to this polysensory system could have evolved from the auditory and auditory-vocal systems. The system would have had to be elaborated to handle the mapping problem, and in its early evolution the involvement of an auditory-vocal dimension could have been entirely accidental. But involving that system would have made it possible for that system to evolve into a communication system based on vocal transmission and auditory reception of information. The unusual element in the analysis is that the system would then be involved in both cognition and communication.

The rest of the story is easy to tell. If the same data contribute to the construction of a "privately" experienced reality and to communication with others, the experiences themselves could be communicated. That was the burden of the example presented earlier of reading as representing a way of sharing consciousness or awareness. Although the unique auditory-vocal system evolved into an inefficient (in the sense of error-prone) system compared to typical animal communication systems, it is useful despite its uncertainty. It was evidently useful enough in the hominids to evolve into an extraordinarily flexible communication system, perhaps because it could be used to communicate cognitive "data" directly. The argument can be developed in more detail, but this is enough on the subject for this occasion.

Conclusions

Intelligence evolved, I have argued, as an aspect of the evolution of encephalization. The "excess" capacity represented by encephalization is used primarily for the construction of reality -- of the representation of a world that is the reality of each species. Human reality, according to this view, is deeply associated with human language. It is reasonable to extend this implication to the specialized correlates of encephalization in other species, and to suggest that their adaptations may be as unusual as language.

The correct view is surely that in the evolution of excess processing capacity, that is, in the evolution of encephalization, a variety of intelligences evolved. Human intelligence, deeply associated with human language, is one kind of intelligence. The evolutionary message about intelligence, like the message about so many other dimensions in biology, is a message about pluralism and diversity, about the variety of intelligences in the biological world.

Recognize, now, that this "intelligence" is a trait that can be selected for in a species. For the present I am happiest with a conclusion at this level and with the cognitive baggage that it implies. Whether I am obliged to go further as an evolutionist may be settled in the conferences that will be part of our Institute.

References

Dawkins R (1976) The Selfish Gene. Oxford: Oxford University Press

Dawkins R (1982) The Extended Phenotype. Oxford and San Francisco: W H Freeman and Co

Dubois E (1897) Sur le rapport du poids de l'encephale avec la grandeur du corps chez mammiferes. Bulletins de la Societe d'Anthropologie de Paris 8:337-376

Gould SJ (1980) Is a new general theory of evolution emerging? Paleobiology 6:119-130

Herman L (1986) Cognition and language competencies of bottlenosed dolphins. In Schusterman RJ , Thomas J , and Wood FG (Eds.) Dolphin cognition and behavior: a comparative approach. 221-252 Hillsdale, N J Erlbaum

Jerison HJ (1985a) Animal intelligence as encephalization. Philosophical Transactions of the Royal Society (London) B308:21-35

Jerison, HJ (1985b) Issues in brain evolution In Oxford Surveys in Evolutionary Biology 2:102-134

Jerison HJ (1986) The perceptual worlds of dolphins. In Schusterman RJ , Thomas J, Wood FG (Eds.) Dolphin cognition and behavior: a comparative approach 141-166 Hillsdale, N J Erlbaum

Lande R (1976) Natural selection and random genetic drift in phenotypic evolution Evolution 30:314-334

Lande R (1979) Quantitative genetic analysis of multivariate evolution, applied to brain:body size allometry Evolution 33:402-416

Lande R (1980) Microevolution in relation to macroevolution. (Review of Macroevolution: Pattern and Process by S S Stanley) Paleobiology 6:233-238

Lashley KS (1949) Persistent problems in the evolution of mind. Quarterly Review of Biology 24:28-42

Lumsden CJ, Wilson EO (1981) Genes, Mind, and Culture: The Coevolutionary Process. Cambridge, Mass.: Harvard University Press xii+428

Maynard Smith J (1982) Evolution and the Theory of Games. London and New York: Cambridge University Press

Mountcastle VB (1978) An organizing principle for cerebral function: The unit module and the distributed system. In Edelman GM and Mountcastle VB The Mindful Brain 7-50 Cambridge, Mass.: MIT Press

Myers RE (1976) Comparative neurology of vocalization and speech: proof of a dichotomy. New York Academy of Sciences 180:745-757

Oakley DA (ed.) (1985) Brain and Mind. London Methuen

Olton DS (1985) The temporal context of spatial memory. Philosophical Transactions of the Royal Society (London) B308:79-86

Passingham RE (1981) Primate specialization in brain and intelligence. Symposia Zoological Society London 46:361-388

Peters RP, Mech LD (1975) Scent-marking in wolves. American Scientist 63:628-637

Premack D, Woodruff G (1978) Does the chimpanzee have a theory of mind? Behavioral and Brain Sciences 4:515-526

Rockel AJ, Hiorns, RW, Powell TPS (1980) The basic uniformity in structure of the neocortex. Brain 103:221-244

Van Valen L (1974) Two modes of evolution. Nature 252:298-300

von Frisch K (1950) Bees: their chemical senses, vision, and language. Ithaca N.Y. Cornell Univ Press

INTELLIGENCE AND NATURAL SELECTION

Michael Ruse
Department of Philosophy
University of Guelph
Guelph, Ontario N1G 2W1
Canada

We humans are modified monkeys, not the favoured miraculous creation of a good god some six-thousand years ago. This fact has been known for well over a hundred years, and on the surface it has been accepted by educated people for almost as long. And yet, in many respects, I believe that the acceptance is rather superficial. I do not mean to suggest that people harbour deep preferences for the literal story of Genesis - endorsing a tale of human origins akin to that espoused by today's so-called "creation scientists" - but rather that very many people still make pre-evolutionary assumptions about human uniqueness, a uniqueness that supposedly separates us in kind as well as degree from our simian ancestors.

For this reason, it is argued (or assumed) by people on all sides of the scientific spectrum, on all sides of the religious spectrum, on all sides of the political spectrum - particularly those on the extreme left and on the extreme right - that there is something special about *Homo sapiens's* power of thought and action. As Aristotle pointed out 2500 years ago (it is surely significant that an arch-anti-evolutionist like Aristotle is taken as a guide), although humans may be contingently featherless bipeds, we (and we alone) are essentially rational animals. It is in this rationality - in our powers of *intelligence* - that we find the heart of the distinctively and exclusively human. There are, of course, other moderately clever mammals, but none with anything like the abilities possessed by us - abilities which enable us to create our own environment and to transform our lives from that of brutes to that of cultural beings or persons. (For the rightist position, see Scruton, 1986; for the leftist, see Lewontin, Rose, and Kamin, 1984; for a good, middle-of-the-road position, taken by a well-known philosopher, see Rorty, 1979).

Moreover - in a move reminiscent of the nineteenth century - it is suggested that there is something morally and politically unsavoury about the ideas (if not the very personalities) of those who would claim that our evolutionary past is relevant to an understanding of our cultural present. Terms like "reductionism" and "genetic determinism" are bandied about, and although it is not always made quite clear precisely what these notions mean, it is made quite clear that they represent traps - traps into which those who argue that evolution is pertinent have fallen. It is hinted - and if the hints are not taken, it is affirmed explicitly - that denials of human uniqueness lead, first to racism and sexism, and if they are not

NATO ASI Series, Vol. G17
Intelligence and Evolutionary Biology
Edited by H. J. Jerison and I. Jerison

then barred, ultimately to the gas ovens of Auschwitz (Kitcher, 1985).

It is surely foolhardy to oppose such a chorus of nay-sayers, yet I believe that the chorus is as fundamentally mistaken (and for many of the same reasons) as was the church at the time of Galileo. I do not want to preach some slavish rule of science, but I do believe that a full appreciation of the truths of evolution will be as revolutionary for our understanding of ourselves and our surroundings as was the appreciation of the great revolution in astronomy in the 16th and 17th centuries. To see ourselves fully in the light of evolution is as transforming for us as were Copernicus' revelations for the medieval mind.

What precisely is it that I want to argue, and why do I - who am a philosopher and not an empirical scientist - make such grandiose claims? Let me lay out my case starkly, and then go on to argue for it. First, I want to suggest that we have now, within our grasp, a good tough theory of organic evolution, and can rightly claim that this theory tells us much about human origins. Second, I want to suggest that on the basis of our biological understanding, we can go far towards unpacking our distinctive human abilities - those abilities that cluster around the general term "intelligence", understood in some loose sort of sense. Third, with the human intellect thus laid bare, we can become more introspective and can approach some of the deeper questions about human understanding, *per se*.

You will find, if you do not realize from the just-given brief description, that I move from the empirical and scientific to the *a priori* and philosophical. To me, as a deeply committed naturalist, this is as it should be, for I see both scientists and humanists as working on the same problems, with the hope that their hands might grasp in the middle (Ruse, 1986). However, since I am a philosopher and not a scientist, I must warn you that although this essay may be both good and original, those parts which are good (the early parts) will not be very original, and those parts which are original (the latter parts) may not be very good. However, because the topic is so important, I am willing to be one of those people who are - in the words of the social psychologist Donald Campbell (1977) - "marginal scholars prepared to be incompetent in several fields at once" (1).

The Darwinian Background

In thinking about evolution, I find it useful to make a three-fold division: evolution as *fact*; evolution as *path*; and evolution as *theory* (Ruse, 1984). This is a somewhat artificial trichotomy and certainly does not point to real barriers - after all, you could hardly have either path or theory of evolution were it not already a fact - but, it is a useful classification for all that.

I take it that, in the minds of reasonable, informed people, the fact of evolution is beyond question (Ruse, 1982; Ayala, 1985). Although he was certainly not the first evolutionist, thanks to the seminal work of Charles Darwin - most particularly in his *On the Origin of Species*, first published in 1859 - the claim that the organic world, including all plants and animals (for the moment, let us put our own species on one side), came about through some continuous, law-bound ("natural") process, from primitive forms (ultimately presumably from inorganic substances), is beyond reasonable doubt. In the minds of many people, the fact of evolution rests primarily on the fossil record; but, as Darwin showed, this need not (and should not) be so. The whole of the organic world - embryology, systematics, anatomy, biogeography, paleontology, and more - combines in one massive "consilience" to prove that evolution is a solid fact (Ruse, 1975a, b, 1979).

The path of evolution is much less well established. Darwin himself said little on the subject. However, today, thanks in large part to an ever-growing fossil record - but thanks also to many other biological sub-disciplines, not the least important of which are those at the molecular end of the spectrum - we now have a firm grasp of the major events in the organic world's history. Life on this planet has been here for (at least) 3 1/2 billion years - that is, since the globe was cool enough to support any life as we know it. Significant multi-cellular life forms exploded upwards in number and variety some 600 million years ago. Mammals (or mammal-like reptiles) have existed for 200 million years, but the triumph of these warm-blooded creatures had to wait on the collapse of the dinosaurs. This happened around 65 million years ago. Since then, there has been much diversification, although not necessarily more than occurred at other points in Earth's history (Cloud, 1974; Valentine, 1978).

Which brings me to the third part of the evolutionary picture: the theory of mechanism. Here we enter a realm of much controversy; but, partly that I might get quickly to my main theme and partly because I have dealt with these matters in great depth and at much length elsewhere, I shall state categorically (or, if you prefer, dogmatically) my own position - one which, I believe, is still held by the great majority of professionals in the field. The causal key to evolutionary change is that discovered by Darwin, and expounded and defended at length in the *Origin: natural selection.* More organisms are born than earth and its supplies can support or hold. This leads to an ongoing race - "struggle" - for existence and (more importantly) reproduction. But not all organisms are born equal. Some have features not possessed by others, and these give their owners an edge in life's struggles. There is, thus, a differential reproduction, as the "fitter" win out over the rest. Given enough time, this "natural selection" leads to full-blown evolution (Ruse 1982; Dobzhansky et al, 1977; Ayala, 1985).

I shall not now dwell on the additions to and refinements of Darwin's ideas that have occurred in the past century. As is well-known, these have been chiefly in the area of heredity - a topic about which Darwin knew little and about which he guessed correctly even less. I shall, rather, make three general points which will help guide subsequent discussion.

First, evolution through natural selection does not lead simply to change. It leads to change of a particular kind, namely towards greater "adaptedness". The features that organisms have - hands, eyes, teeth, wings, bark - are used to survive and reproduce. In pre-Darwinian days, such "adaptations" were thought to be evidence of God's design. Now, more prosaically, they are thought the result of natural selection (Lewontin, 1978; Mayr, 1982). But the Darwinian, no less than the natural theologian, looks at the features of organisms, inquiring into their function or purpose or value in the endless competition which is the lot of every living creature.

Second, notwithstanding the point that selection does tend towards adaptive excellence, one must never suppose that the path hewn out by the selective mechanism is in any sense progressive. The old picture of a chain of development from monad to man - still implicitly endorsed by so many (including good scientists) - is alien to the Darwinian perspective. Evolution (to the selectionist) is opportunistic. It is (I like to say) a "string and sealing wax" operation. It does not set goals, and then try to achieve them (as did the hero of *Pilgrim's Progress*). It works with what is at hand, pushing, shoving, twisting, taking from one purpose and using for another, and all of the time trying to stay one step ahead of the competition - and of extinction. Sometimes, in human eyes, evolution goes forward. As often as not, it goes backwards. Sometimes, in human eyes, evolution goes quickly. As often as not, it goes slowly. The point is, it goes nowhere (Williams, 1966).

Third, selection demands variation. In a family of clones, there can be no Darwinian evolution. However, thanks to our post-Darwin discoveries about heredity, there is today little doubt that within populations of organisms lie vast reserves of variation, waiting on selective demands. Indeed, there are reasons to think that selection itself can produce, or at least maintain, variation. Thus, you need not fear either that evolution will run out of steam, or that so "blind" a process is, almost a priori, incapable of producing the intricate organic adaptations we see about us on all sides (Ruse, 1982).

There is much, much more that could be said about the theory of evolution through natural selection. I shall say nothing, except to repeat my previously stated conviction that it is a good tough theory, with plenty of strong evidence.

Human Evolution

Again, let me invoke the trichotomy: fact, path, and theory. The definitive case for the fact of human evolution was made by Darwin's great supporter Thomas Henry Huxley, in his *Man's Place in Nature* (1863). The argument therein was endorsed by Darwin himself, a few years later, in the *Descent of Man* (1871), and in essence it is simply a specific application of the general argument for the overall fact of organic evolution. Through the various branches of biology, we find voice after voice speaking to the evolutionary origins of humankind. The person who is not persuaded by the skeletal similarities (homologies) between human and ape will never be persuaded by anything.

When we turn to the path of human evolution, we find ourselves in more difficult waters - deep and extremely fast moving. Summarizing, what I candidly admit I know only through second-hand report, it is growing increasingly clear that the crucial time-span for human evolution extends at most a mere 10 million years. Probably, indeed, the really critical period was little more than half of that, for it was then that the prospectively human line split from the prospectively chimpanzee line, the prospective gorillas having gone sometime earlier (Pilbeam, 1984).

We now know, thanks to the discoveries of the paleoanthropologists like Donald Johanson, that around 5 million years ago our ancestors got up off their haunches and walked. It was when, and only when, this vital evolutionary move had occurred that the really significant evolution of the brain occurred - or, rather, the really significant evolution of the size of the brain occurred. We go from *Australopithe-*

cus afarensis, some 3 - 4 million years ago, with an ape-sized brain of around 500 - cc (not necessarily an apelike brain) to *Homo sapiens*, with a brain of around 1300 cc or a little more (Johanson and Edey, 1981; Isaac, 1983).

There is much that could be said in detail about the course of human evolution, but I will rush on to my central concerns. I pause only to note that for some time there were extant two species of quasi or proto intelligent ape, of which only one (our ancestors) survived. There was, apparently, no more of a uni-directional climb to the top in the human world than there was elsewhere in the organic world.

I come now to the question of theory. What mechanism(s) produced *Homo sapiens*? At the risk of sounding circular, I want to make an important point. We do not, need not, and should not, pretend that we approach this question with a blank mind (Ruse, 1986). The presumption must be that natural selection figures largely in the answer. Physiologically, paleontologically, and in every other way, humans are animals. Unless good reasons are given to the contrary, we expect natural selection to be important. This point sounds circular, because the fear is that - having declared the importance of selection - any discovered evidence will at once be christened support for the supposed selection. However, obviously, there is vicious circularity only if all evidence is treated as support, without any independent evaluation of its true nature. The presumption is that selection is important. But, this is a presumption that must be tested against the real world.

As it happens, there is little need for genuine concern. The fossil and archaeological record needs no special interpretation before it can be brought to bear favourably on the claim that the most distinctive aspect of human evolution - brain-size growth - was a function of natural selection. Without too deeply anticipating discussion which is to come, the overwhelming evidence is that those proto-humans with bigger brains were brighter than those with smaller brains - and they used this brightness to advantage in tool use and manufacture, in cooperation, and the like. Whatever may be the case today, natural selection was important in our past, specifically in the evolution of our intelligence (Isaac, 1983; Jerison, 1982).

One final point, and then I am done with this first part of my discussion - a part which I have admitted candidly is a review and not at all original. Articulate, spoken language is certainly not a necessary condition for any kind of intelligence whatsoever (although perhaps intelligence presupposes at least some ability to communicate). However, human intelligence as we know it is very much bound up with our language abilities. Indeed, today many philosophers would say that language-use is the very essence of rationality and intelligence (Black, 1968).

I am only too aware that there are few areas in the biological sciences which cause more controversy than the origin of language. Even to an outsider, the reasons for this are obvious. There are simply horrendous methodological and associated problems that any hypothesis about language's origins must face. What is a language? How would its existence get reflected in the fossil record? What would be its adaptive value? And so forth. I suppose that any extensive study of the subject will, sooner or later, have to confront those powerful ideas about the nature of language which stem from Noam Chomsky and his followers (e.g., 1957, 1965, 1966, 1980). Hence, speaking now less as a philosopher and more as a Darwinian, let me say two things.

On the one hand, Chomsky's belief that language (all languages) is rooted in a shared biologically based "deep structure" is just what one would expect from a process of evolution through selection. Biologically speaking, language cannot just exist in the abstract - and, given that during our evolution we would have been a much much smaller group than today, we would expect that there would be fundamental similarities in languages, stemming from a shared mechanism or capacity. On the other hand, Chomsky's own belief that language-powers came through a one-step evolutionary jump is just plain unacceptable. Evolution, especially Darwinian evolution, simply does not happen that way.

Fortunately, there are other accounts, preserving Chomsky's insights, which use and confirm a selection-based perspective. Philip Lieberman's (1975, 1984) hypothesis that language ability depended on certain crucial changes in throat structure is highly controversial; but, it does have a ring of authenticity to a Darwinian - not the least of which is Lieberman's claim that our speaking abilities were achieved at some cost (the dangers from choking) and that other humans (Neanderthals) successfully pursued an alternative strategy for quite some time.

Lieberman may or may not be right. He does bring the origins of language down into the biological arena, showing that language (like the rest of human intellectual abilities) does not seem to put up insuperable barriers to a Darwinian account of origins and development. And with this, I can conclude the section. There is no negative reason to think that *Homo sapiens* was produced by factors beyond natural selection, and growing positive reason to think that natural selection was the crucial causal factor.

The Paradox of Culture

Deliberately, thus far, I have been keeping the discussion in a biological vacuum. But, we must try to break out, and as soon as we do we come right up against the notorious barrier of culture - a barrier cherished and strengthened by the efforts of social scientists and humanists. As beings with intellects - as thinking persons - we humans transcend our biology. We exist in a world (or dimension) of artifacts, of customs, of ceremonies, of literature and art and music, of technology, of clothing and housing. In short, we live in a world of culture.

Supposedly, or so say the critics, no part of this world could possibly be subject to biological causes. It is usually admitted, at least it is usually admitted by its more tolerant devotees, that it is not so much that culture denies biology, as escapes it: somehow sitting on top. But, beyond biology culture undoubtedly is. Apart from anything else, the variability of cultures hints against the significance of biology, and the speed at which cultures can change shouts against any such connection. Natural selection can work very rapidly when conditions are right, but it still takes generations - comparatively, many generations - to effect major change. Yet, as is well-known, cultures can change drastically, even within the living memory of one generation of humans. Consider, for instance, the Japanese, who have moved firmly from fascism to democracy within 30 or 40 years (Runciman, 1986).

In other words, even though the intellectual abilities or capacities may have evolved, intellect has now entered a new, non-biological world. Natural selection is irrelevant to human understanding. Indeed, in a way, having used the ladder of biology to climb up into culture, we are now able to throw away that ladder, for it is clear that, in important respects, intelligence itself is beyond biology. If one wanted to alter someone's intelligence (or to change the intelligences of a group), one would not turn to biology for help, possibly simulating natural selection through artificial breeding programmes. Instead, one would turn to culture, for instance by improving educational opportunities. This is the way to alter intelligence - through culture, not biology. In short, intelligence (that is, human intelligence) and natural selection are now things apart. The marriage was long and mutually satisfying; but, the time came for a divorce, and this occurred.

To many - including most of my fellow philosophers - this is the end of matters. There is no "paradox" here, as I have suggested in the title to this section. Biology is irrelevant. And yet, to one like myself (even a philosopher!) who takes seriously the story that has been told thus-far in this essay, this cannot be the end of matters. Grant the importance of culture. There is still the question of biology. There really is a paradox.

Why do I speak so doggedly, if not abrasively? Because everything we have learnt suggests that biology cannot be treated in the dismissive way taken by most social scientists and students of the humanities. Evolution does not proceed by sharp breaks; but, rather, runs gradually and continuously. Parts are modified and used for other ends - rarely (if ever) is something scrapped entirely and a whole new design incorporated from scratch. There are remnants and vestiges of the past in just about everything, no matter how "up to date" and efficient it may be. Natural selection is not about to be cast rudely aside, after 3 1/2 billion years of crucial importance. More directly, the biological elements (i.e. elements produced by natural selection) of human intelligence will be used - possibly in a transformed way - in the present cultural milieu in which we now exist.

Of course, this is just a feeling or suspicion - even though backed by evolutionary biology. [It is also a feeling or suspicion made uneasy by the curious coincidence that it just so happens that we humans - who hardly want to see ourselves as mere brutes - have uniquely escaped those forces which make the animals mere brutes]. (2) The culture-as-autonomous supporter might agree with all that I have said, but simply claim that in this one case - the present-day realm of human intelligence - an exception must be made. A committed Darwinian like myself must therefore launch a positive counter-argument. To this end, there are, I think, two separate matters which must be teased apart.

On the one hand, there is the question of whether biology - more specifically, the causal effect of natural selection - is important in the understanding of the functioning of human intelligence today. This question is irrespective of whether natural selection still operates on (human) intelligence. It is akin to a question about whether human two-limbedness is a result of our aquatic past or of some more recent set of events (Maynard Smith, 1981). In asking (and answering) such a question, no one thinks we now live in the sea, even though this is not to deny the great significance of two-limbedness. Similarly, even though natural selection may no longer operate, its legacy may be very important in an understanding of intelligence.

Then, on the other hand, there is the question of the putative on-going effects of natural selection on intelligence. Does natural selection matter? Does intelligence help in the struggle for survival and reproduction? Is intelligence now something which is really under the exclusive control of the environment (and hence of culture), or does biology really have a role to play here?

Here, then, we have two questions, and it is to these (in the order given) that I turn now.

Underlying Dispositions of Thought

I have spoken of the need for the sciences to stretch across and clasp hands with the humanities, stretching the other way. Let me make my image more specific, speaking of biology (the centre of my discussion so far) clasping hands with philosophy, which is not only my own discipline but one whose practitioners tend to fall firmly in the culture-as-autonomous camp. Additionally, let me move across to philosophy - taking its content and claims as representative of the human cultural realm - and work backwards (or forwards, if you prefer the metaphor the other way) to biology. Is there that in the content of philosophy (or in that which is the subject matter of philosophy) which might incline you to seek an origin rooted, not just in biology, but in the causal effects of natural selection? (3)

Think, for a moment, about philosophy - or rather about its subject matter. There are two basic questions that the philosopher must answer. What can I know? What should I do? These take us into the areas of knowledge or "epistemology", and of morality, or "ethics". They were the foci of Kant's two masterpieces: *The Critique of Pure Reason* (1929) and *The Critique of Practical Reason* (1949). In epistemology, we deal with knowledge claims, paradigmatic of which (and that on which I shall concentrate) are scientific claims about the world around us. In ethics, we deal with moral and social theories and systems, enterprises which attempt to guide us (primarily) in our dealings with our fellow humans.

Now, it is clear that a scientific theory, per se, is not simply a function of natural selection (as recorded in our biology) - it is not even a function of the results of natural selection. People change their minds about the worth of scientific theories at far too rapid a pace for biological causes to be paramount. In our own lifetimes, for instance, active geologists have changed their allegiances from a static earth picture to the dynamic theory of continental drift through plate tectonics (Hallam, 1973; Ruse, 1981). Analogously with morality. It is true that, for instance, Christian ethics has been around a long time, but it changes (and goes on changing) at a high speed. For instance, the past thirty years have seen a significant revolution in thinking about the morality of sexuality, by Christians as well as by others.

The question, therefore, has to be whether there is that in science and in moral systems which, in some sense, counters or persists through all of the change. If there is not, the quest for biological importance must cease. If there is, then the quest is still open (although, far from complete). Fortunately, however, as soon as one puts things in this way, the prospects start to seem most hopeful. The very essence of philosophical inquiry is to seek order and regularity within the change and diversity. Does one scientific theory, in important respects, resemble another? Does one proper moral action share features with any other such action? And the answer is, resoundingly, that common patterns and themes do emerge. The products of human thought, of human intelligence, are not just random, one-off, unique phenomena. Rather, they conform to deep-seated rules of inference and reasoning (Nagel, 1961; Hempel, 1966).

Thus, on the one hand, we find that scientific theories - all scientific theories - must be compatible with laws of mathematics and of logic, they must relate to the physical evidence in certain acceptable ways (as, for instance, in the Popperian demand that science be "falsifiable"), and they must illuminate our understanding of reality according to certain conventions (Ruse, 1982 a). I mentioned, in discussing Darwin's ideas, that the great merit of his theory was that it was "consilient" - it gathered together many disparate areas of knowledge into one whole. It is clear that this consilience is a mark of all good science. Successful theories are judged successful for the same reasons - and one of the main reasons is the successful incorporation of a "consilience of inductions" - to use the full name, as introduced by the philosopher William Whewell (1840).

On the other hand, in morality we likewise find common themes or standards emerging. The criteria for judging actions or deeds good in one case are the same criteria used for judging actions or deeds good in other cases. It is true that philosophers frequently differ over the exact nature of the criteria - Kantians, for instance, endorse the "Categorical Imperative" ("Treat others as ends rather than as means") whereas Utilitarians endorse the "Greatest Happiness Principle" ("Actions are good inasmuch as they promote happiness"). But, do not let the differences mislead you. Almost always, different criteria yield the same results (pointing to their essential similarities). Kantians and utilitarians come together on actual decisions about good and bad (Kant, 1949; Mill, 1910; Taylor, 1978).

Given that human thought does seem to follow shared rules or patterns, you will know now what I am going to suggest. I claim - at least, I hypothesize - that these patterns are rooted in our biology. Intelligence is structured according to innate tendencies or propensities, presumably with a material correlate in the brain. We think in the ways that we do - both about the world and about our fellow humans - because we are so constrained (admittedly liberatingly and creatively constrained) by our biological nature. And the reason why our nature is as it is, is because of natural selection. The innate propensities exist because those of our would-be ancestors who had them - and thus had their thought and action guided - out-survived and out-reproduced those of our would-be ancestors who did not have them.

It is as simple as that. The paradox of culture has dissolved. The products of the mind escape the tight controls of traditional biology. However, they exist only by courtesy of structures devised by selection.

Evidence

Simple, the case I advocate may be. Unproven, it certainly is. Let me now make five points, briefly, in its favour.

Firstly, everything we know about biology - particularly, everything we know about the evolution of intelligence - suggests that my perspective on intelligence is empirically plausible. If you deny innate constraints, then on the one hand you have to concede that the (human) brain is like a large, all-purpose computer - lots of hardware, but no software built in. Since, apparently, we do think in certain patterns - I take this fact now as a phenomenological given, whatever the cause - presumably these were fed in by culture at some point. But, this means there is all sorts of surplus capacity, which may or may not get used. To say the least, this is highly inefficient - not only going against the workings of selection in general, but also against all we know about the innate programming which is so characteristic of virtually all other animals in the world (Ruse and Wilson, 1986).

On the other hand, the all-purpose computer/brain suggestion lays itself wide open to misuse - we might learn and believe all kinds of counter-adaptive rules: "2 + 2 = 5". "Parents, eat your children". "The quickest way to warm yourself is to sit right in the fire, rather than next to it". It is highly improbable that selection would have gone to all of the effort of creating humans and allowed the appalling possibility of beliefs like these.

Second, other researchers - working independently from the side of biology - have reached conclusions similar to mine, which have come from the side of philosophy. For instance, the sociobiologists Charles Lumsden and Edward O. Wilson hypothesize that the mind is constrained by innate propensities, which they call "epigenetic rules".

> Any regularity during epigenesis that channels the development of an anatomical, physiological, cognitive, or behavioral trait in a particular direction. Epigenetic rules are ultimately genetic in basis, in the sense that their particular nature depends on the DNA developmental blueprint... In cognitive development, the epigenetic rules are expressed in any one of the many processes of perception and cognition to influence the form of learning and the transmission of [units of culture]. (Lumsden and Wilson, 1981, p.370)

Arguments from authority are, of course, no more than that. Even the greatest authorities can go astray, and should never be taken totally on trust. However, because these particular authorities come from the other end of the spectrum from mine, they back their case with physiological, psychological, genetic, sociological, and other kinds of empirical data. Hence, I would like to think it is not too immodest to suggest that we have here a consilience between humanities and sciences.

Third, although world-wide conformity, with respect to some particular human characteristic, does not definitively prove a biological base, it is supportive - after all, world-wide diversity if taken as counter-supportive! As it happens, both in epistemology and in ethics, the cultural-anthropological evidence is that peoples of widely different surface beliefs and practices share linking underlying propensities. The situation is much the same as with language, which of course - given the basic truth of claims about deep structure - is merely a special case of the general position on human intelligence that I advocate.

Thus, for instance, we find that Indian and Chinese logics - although formulated quite separately from those in the West - bear remarkable similarities to those that we know and use (Staal, 1967; Bochenski, 1961). The basic Aristotelian premises of logic, like the law of excluded middle, are common to all peoples. (This is not to say that all make such premises explicit, or even that they all use them fully in every possible situation). The same is true also of morality, despite cultural differences. The Love Commandment, for instance, ("Thou shalt love thy neighbour as thyself") appears again and again in widely varying cultures (Ginsberg). Humans really do think alike.

Fourth, the kinds of innate dispositions I have in mind are just those you might have expected natural selection to have cherished. They really are adaptive advantages to thinking that "1 + 1 = 2" and not that "1 + 1 = 3". Similarly, the proto-human who took seriously consiliences would have had a reproductive edge. Pulling different things together and making informed inferences needs no defence from a Darwinian. Analogously, in morality. Until recently, many people might have doubted this second point. There was the belief that selection promoted selfishness - that only thus would one succeed, given "nature red in tooth and claw". But now, thanks especially to the work of the sociobiologists, we know that this is not so. Caring and altruism - *genuine* caring and altruism - can pay far greater reproductive dividends than can all-out hostility and self-regard. At least, they can for social animals, which is what we humans pre-eminently are. There is, in short, every reason to think that our morality was conferred by biological causes (Ruse, 1979, 1985, 1986; Wilson, 1978; Ruse and Wilson, 1985, 1986; Murphy, 1982).

Fifth and finally, the evidence from our nearest neighbours supports the innate disposition hypothesis. I do not assume that all intelligences are the same - indeed, I shall later be raising and deeply questioning such an assumption. However, we have seen that we were with the apes until very recently, and we know furthermore that selection is a fairly conservative mechanism. If something is working well, then why change it? One would expect, therefore, to find that the chimpanzees, at the very least, will think in patterns very similar to ours. And, indeed, such does seem to be the case. An increasing number of studies show that chimpanzees reason in ways much as we do - using analogies, making simple calculations, and the like (Woodruff, Premack and Kennel, 1978; Gillan, Premack and Woodruff, 1981; Gillan, 1981). Moreover, socially, the chimpanzees relate in ways which certainly qualify as quasi- or proto-moral. They think things through, show concern, help one another - and do so in repeatable and repeated ways (de Waal, 1982).

I would not want to say that the case for innate propensities is proven beyond all doubt. But, there are enough straws in the wind to make some fairly solid bricks.

Selection Today?

For the sake of argument, grant all that has been said. Natural selection was crucially important in making our thinking capacity what it is. Is there now any reason to think that natural selection is still an important, on-going influence on human intelligence? Are we still evolving, perhaps in the direction of super-brains, or perhaps slipping back to a pre-human-like phase or simply going side-ways into a new kind of intelligence? The answer I suspect is yes - and no!

Note as a preliminary, that natural selection does not necessarily have to be changing things to be effective. It can be working flat-out, keeping things much as they are at present. Note also, that there is nothing in Darwinian evolutionary theorizing which suggests that, because human intelligence seems to have evolved up, fairly steadily, from the early Australopithecines, such evolution must go on. This is to incorporate within Darwinism an unwarranted progressivism. Any change must be because there is a good reason - and not otherwise.

With these points as background, it is obvious, on the one side, that intelligence is hardly something which has run its biological course - a vestigial organ like the appendix, which may once have had an adaptive value, but which has one no longer. Culture - the product of intelligence - is obviously that which makes humans such a successful organism. Our numbers multiply up explosively, suggesting that we are an extremely well-adapted animal, and it is obvious that it is intelligence that is at work here. I might add, forestalling one possible objection that understanding (as I am) "intelligence" in the broad sense of including social and moral abilities (Kant's "practical reason"), it seems to have given us the ability to live with each other, as well as to conquer nature. Despite this century's dreadful wars, comparatively speaking, we humans are a remarkably peaceable animal. We kill each other, but not at the rate most other mammals do (Wilson, 1975).

On the other side, however, it is surely true that the success of intelligence has in great measure shielded it from the rigours of selection - indeed, for many of us, it has shielded, more than just itself from the rigors of selection. Certainly in Western society, there seem to be few reproductive penalties for those less blessed with intelligence - subject, of course, to the qualification that the really badly handicapped tend to be low reproducers. Indeed, folklore (at least) has it that the less intelligent out-produce the more intelligent. How true this is, on a long-term basis, may be questioned - but, selection has indeed lost a lot of its former power (Dobzhansky, 1961).

Nevertheless, having agreed that intelligence may now have the power to shield itself from selection, one would have to be insensitive to the point of callousness not to recognize that for many people on this globe - perhaps indeed a majority - natural selection, in the form of disease, starvation, limited living resources, and the like, is still a major factor. Africa, India, South America and more are still places where culture fails to mask the forces of biology. Here, one might surely have some actual ongoing selective pressures in favour of intellectual capacities. Although, in line with the qualifications made just above, the pressures might be directed more towards appreciation of innate dispositions that we have already, rather than the creation of anything new. The aim of evolution is to be better than your competitor. Not to be top, on some absolute scale.

At this point, one matter must not be passed over in silence. If selection is presumed still to be an effective causal factor molding or maintaining human intelligence, then it must be supposed that (in this respect) there are biological, heritable differences between humans. This is, to say the least, a highly controversial claim, but it cannot be glossed over. Actually, indeed, if one thinks that intelligence is the product of selection (and that selection has left its mark), then in line with general points made at the beginning of this essay, one would expect biological variation in intellectual capacities. Not necessarily, I hasten to add, between races or between sexes, but with groups. However, one can argue (although how plausibly is another matter) that any such differences are masked by environmental factors. It is not that there are no differences, or that biology is not a factor in intelligence, but rather that upbringing and the like are clearly crucially important in adult intellectual abilities, and they may be so important as to swamp all else.

However, if selection is actually impinging on intelligence, then the biology must be (relatively) at the fore. One assumes that this is not too implausible a possibility, given that education and the like will undoubtedly be (to say the least) inadequate in those parts of the world where the struggle for existence is most severe.

To go deeply, in this discussion, into the matter of biological factors behind the unarguable fact of different intellectual abilities would lead us too far astray. Let me, therefore, say simply and categorically that, for all the problems and difficulties (including bad and fraudulent studies), I do believe the case has now been made showing that biology matters. Intelligence (understood in some broad sense as an ability to follow the kinds of rules I have highlighted) is just as much a function of biology as other physical human characteristics, like height and weight. This is not a statement of prejudice - I emphasize again I am not linking this conclusion with racial or other like factors - but, it does seem to be a statement of fact. In this respect, there is no reason why selection should not occur (Ruse, 1982b; Scarr and Carter-Saltzman, 1982). (4)

One last point, in this sub-discussion about the possible ongoing effects of selection on intelligence. It must be emphasized that natural selection is an opportunistic mechanism. It works for the present. It does not think about the long-term benefits or dangers. It pays dividends for the successful individual here and now. It says nothing of the needs of the group, which may well suffer as the individual benefits (Brandon and Burian, 1984). These, and like factors and qualifications, apply particularly to intelligence. Our powers of reason evolved because they were useful to our ancestors. They exist now - we exist now and flourish - because they are useful to us. It is possible that our intelligence will turn against us, as a species in the long run.

In fact, to speak pessimistically for a moment - but, I fear not unreasonably - I fully expect our intelligence to be the ruin of *Homo Sapiens*. In an age of Hiroshima - not to mention Chernobyl - I cannot believe that we shall not poison ourselves with radioactivity in, say, the next 10,000 years. Either that, or we shall collapse beneath the burden of overpopulation - brought about, entirely, through technology produced by our intelligence. Unfortunately, these gloomy thoughts in no way deny the truth of what I have been saying about the evolution of intelligence. That ability evolved to get us out of the jungle and to the point where we are now. There is no presumption of - and much to deny - the power of human intelligence adequately to control the Pandora's box it has helped us to open.

The Status of Human Knowledge

I move now to the third and final part of my discussion. As I have warned you, I now become almost entirely theoretical and philosophical - although, I hope to be able to circle right back to a fundamental empirical point.

Grant, as I keep asking you to grant, everything that has gone before. Grant, that is, that our thinking is informed and constrained by innate dispositions, put in place by natural selection for their adaptive value. What kind of product emerges from such an intellect? What is the nature of the knowledge which we claim to have, both of the world around us and of the way in which we should treat our fellow humans?

I suspect that most people are (what we philosophers would call) "realists" - indeed, they are probably (what we rather condescendingly refer to as) "naive realists"! By this is meant that, for most people the aim of intelligence is to discover and record a real world, "out there" - and, in the opinion of most people, this is what we truly do. There are chairs and tables, molecules and electrons, laws and morals, existing independently of us. The task and the ability of human intelligence is to find and use this independent reality.

I hardly need say, by now, that this will not do. It is, at best, excused thinking, where we see ourselves simplistically as discovering God's good creation, which was made exclusively for us. (5) [I hasten to note that I do not see Darwinism as incompatible with religion as such - but, as might be expected, it is not compatible with a religion based on the Genesis story]. Apart from anything else, our powers of sense crucially affect the way we see the world - were we, for instance, to use pheromones rather than sight as the principle medium of observation, it would be a very different "real world" that would surround us. Note (to pick up on an old conundrum), I am not saying that the apple before me is not really red - as opposed to blue. What I am saying is that the redness does not exist in splendid isolation - in the mind of God, as it were. The redness is as crucially dependent upon us as observing entities, as it is on the apple's molecules and on our nature as observing entities, as given to us by evolution. (6)

The evidence of our senses is a function of our evolution, and the way we think about the world and ourselves is likewise a function of our evolution. All of this may start to sound as if human thought is very subjective, so let me hasten to add that nothing so far supports what is often taken to be an immediate consequence of subjectivism, namely relativism. We humans share a common gene pool, and our similarities much outweigh our differences. Moreover, intelligence - particularly at the social level - only really works, if people think more or less alike. If I think killing is wrong, and you disagree, someone is going to suffer. (There are some fairly persuasive models showing that the would-be killer would not necessarily be the one to succeed (Dawkins, 1976). People trying to work together can be quite effective at dealing with dissenters). Hence, within obvious bounds, you expect people to think in more or less the same ways - at least, inasmuch as biology has any say in the matter.

But, what about the subjectivity? Prima facie, one might expect that biologists would be more worried about this than philosophers; but, paradoxically, I suspect that the worry might be more the other way. As Darwinians, we must keep always in mind the non-progressive nature of the evolutionary process. There is no ultimate goal or resting point - or absolutely better or right solution. Our intelligence has evolved and works very well, but it is not correct or right or perfect in some absolute or ontological sense. We are not made in God's Image. Thus, we must take seriously the possibility of intelligence having evolved in ways quite different from ours - and being just as successful at coping. (7) (Would it properly be called "intelligence"? More on this point in a moment).

Philosophers - like most folk - will find this radical answer uncomfortable, if not implausible. They may agree that there is no absolute "real world", out there, waiting to be discovered, but the radical subjectivity of the Darwinian will trouble them. The position towards which our biology has pushed us bears many similarities to the philosophy of Immanuel Kant (1929), and his response (to similar sorts of arguments) can be taken as typical. He argued that the mind plays a crucial role in understanding. Nevertheless, Kant preserved objectivity of a kind by arguing that the way we think is a necessary condition of thought by any rational creature *whatsoever*. You cannot be a rational thinking being and think in any other way. Analogously, today's philosopher will argue that the truths of the innate dispositions - "2 + 2 = 4", "Parents, love your children" - transcend the contingently biological, and take on a strengthened necessity of their own.

Now, at one level, Kant and his modern-day supporters are surely right. From our perspective, you cannot imagine - that is, there is nothing in evolutionary theory suggesting - that beings are going to evolve thinking that 2 + 2 = 5 and that you should kill your neighbour. There is much against this, for such beings would be less fit than we. However, the Darwinian challenges the claim at two other vital levels. First, were circumstances greatly different from the globe on which we live (especially from the part of Africa where we evolved), then the way we think would be affected by a much different biology. Our intelligences would not be the same. Suppose we had evolved (for instance) in an aquatic environment, breathing under water, like fish. Suppose also that there was not much light. Fairly obviously, our organs of sense would be much altered and our perceptions of the world and ourselves would be most unlike those which we have now. We would have abilities we do not now possess; but, other things, which we quite take for granted, would be quite alien. (Jerison, 1982, points to the fascinating fact that cetaceans are so used to things jiggling about in the water, that they cannot respond properly to stationary objects! See also Herman, 1980).

In a sense, this is no more than has been touched on already (although no less important for that). Let me now go on to the second level at which Darwinism challenges the Kantian. I believe that we might have evolved in such a way that we would be intelligent, rational beings, but subject to quite different innate dispositions - to quite different "epigenetic rules". (Given the arguments already offered, I will assume that there will be epigenetic rules, nevertheless). Why would one speak of "intelligence" at this point? (to pick up on a question asked earlier). Because such beings would go through the motions (some, if not all) that we associate with intelligence. However, the reasons why they would go through the motions would not be the reasons why we would go through the motions. Their rules would not be our rules.

Can one prove this, other than by emphasizing yet again the non-progressive nature of Darwinian evolution? Let me try a couple of thought examples. (8) First, why do we not put our hands in the fire? Because we have learnt not to, associating fire with pain. Suppose that there was no such association - suppose that we could not feel pain - but that we worshipped fire and felt too unclean to draw near to it. Damage which occurs after fire exposure is taken to be a mark of God's displeasure, and was not caused directly by the fire. We would do exactly what we do now, but for quite different reasons. Second, why do we not kill each other? Because we are filled with sentiments about love and kindness. Suppose that there were no such sentiments, that we hated each other, but knowing that others likewise hate us, settle for uneasy armistice. Again, we would do exactly what we do now, but for quite different reasons. (Remembering how America dealt with Russia during the Cold War of the 1950s, I call this the "John Foster Dulles systems of morality". For related arguments, see Ruse, 1984, 1985, 1986; Ruse and Wilson, 1986).

I suggest that we would have here quite intelligent beings, but beings that would think in ways radically different from us. Yet, they are beings which are as likely as we, given the nature of Darwinian natural selection. In short, as its very deepest level, our intelligence has no ontological or epistemological - let alone theological - superiority. [Am I not arguing this, assuming that my theory of natural selection has a privileged position, and is true whatever system of rationality one has? I am assuming natural selection, because it is true from the only perspective I have. But, I do not conclude that other intelligences will know of selection - nor that I know of their ways of coping with the world].

Down to Earth

Which brings me to the final question I shall ask in this discussion, and as I promised we come full circle, back from philosophy and into empirical science again. Strange intelligences, beings which are not subject to our innate dispositions, are (I presume) flourishing today elsewhere in the universe. Andromedans go about their own business, according to their own rules and laws. But, what about down here on earth? Does every being think in the same way as we? I ask the question now about non-human organisms, recognizing that they are not as clever as we but recognizing also that in their domains other organisms may be highly intelligent.

Clearly, at the surface level there are many differences, due to different powers of sensation, different needs, and so forth. But, what about the epigenetic rules - the innate constraints? You might hazard the guess that there may not be much difference, in part because we have all evolved together for a long time (especially those organisms you would want to call "intelligent") and in part because many of the challenges we all face are similar.

I suspect there is truth in this, but I doubt it is the whole tale. Apart from anything else, in nature (as in business) you usually find that, although success calls forth imitation, it also creates other alternative niches and possibilities. It might be that (speaking broadly) primate intelligence set up pressures and opportunities on and for other animals to evolve intelligences of radically different forms, even to the epigenetic rules. (And, conversely, on the primates).

Is this possible? Is this so? I do not know. Obviously, at this point I am indulging in armchair theorizing, whereas what is needed is empirical study. So, I will hand the quest over to others, better suited than I. (10)

Notes

1. Although I am aware of the futility of the attempt, I really feel that I - especially since I am a philosopher - ought to attempt some definition of the notion of "intelligence". The trouble, I find, is that psychologists (i.e. those most concerned professionally with intelligence) tend to put too narrow an interpretation on the notion - at least, considered either biologically or philosophically. Problem solving (the focus of things like IQ tests) is important, but I want "intelligence" to cover such things as "ability to get on in society". This might not require a huge amount of IQ, but is crucial biologically and philosophically. However, broader definitions usually strike me as too broad. For instance, "goal-directed adaptive behavior" (Sternberg and Salter, 1982) covers plants turning to the sun, which hardly strikes me as very intelligent. In this discussion, I shall compromise, and focus on the following of rules - which certainly seems important as intelligent humans are concerned. After all, it is just plain stupid to put your hand in the fire, when you have already been burnt once.
2. It is also a feeling or suspicion made uneasy by the curious coincidence that it just so happens that we humans - who hardly want to see ourselves as mere brutes - have uniquely escaped those forces which make the animals mere brutes.
3. The ideas I am about to explore have been dealt with, more fully (although from more of a philosophical perspective) in my recent book, *Taking Darwin Seriously: A Naturalistic Approach to Philosophy.* (1986) My scientific debts are to the sociobiologist Edward O. Wilson, especially to his co-authored Lumsden and Wilson, 1981, and more distantly to the ethologist Konrad Lorenz (1940). However, I doubt they would follow me in everything I argue.
4. Criticisms are usually levelled against IQ tests, and indeed some telling points are made (Kamin, 1974; Block and Dworkin, 1974; Lewontin, Rose, and Kamin, 1984). However, no one seems to want to deny that there is something like "intelligence" in existence - the question is, rather, whether IQ tests spot and measure it. My point is simply that it is bad biology to assume (as, for instance, does Daniels, 1976) that all intelligences are biologically identical. He writes: "Like the abilities to acquire spoken and written language, to exercise memory, to coordinate tactual and visual sensory inputs, [many mental capacities and abilities] are part of the normal functioning of humans". (p. 154) That is true. But some people can see better than others, and no one would doubt that the genes play a role in such differential abilities.
5. I hasten to note that I do not see Darwinism as incompatible with religion as such - but, as might be expected, it is not compatible with a religion based on the Genesis story.
6. In Ruse (1986), I try to show why my claims do not plunge us into idealism, with everything "simply in the mind of the observer". Essentially, my claim is that ide-

alism can be avoided if you adopt a coherence theory of truth rather than a correspondence theory. In my thinking, I have been much influenced by Rorty (1979) and Putnam (1981). I argue strongly against "hypothetical realists" like Popper (1970), who believes in the existence of an independent albeit unknown real world. I believe that the world is not independent (of us) but that it is known.

7. This point is a major theme in the writings of Harry Jerison. For instance, in a recent review, he writes: "Our experienced real world...is a possible world that is constructed in a way that makes the neural information reasonably self-consistent. Animals with different kinds of neural information, for example, ones lacking color vision or having limbs that functioned as wings or flippers, would construct at least slightly different realities" (1982, p. 725). The problem Jerison does not solve - Why should he? He is a biologist, not a philosopher - is why we should take Jerison's reality seriously. Even agreeing that Jerison's reality is our reality - and you will see from the text that I do agree to this - the problem of subjectivity remains. In the above passage Jerison writes as if he were a disinterested external observer, but a porpoise could complain that this is just a *Homo sapiens* perspective. Pushing the attack to the limit, a non-human critic who cares little for natural selection might argue that the whole evolutionary approach is itself subjectively held, and thus its conclusions about subjectivity collapse in contradiction (or paradox like "This Statement is false").

As noted in the last footnote, I believe the adoption of a coherence theory of truth shows the way out of these problems. See Ruse (1986); also Ruse (1985) and (1985).

8. My thought examples cannot be as radical as I would like them to be, because the whole point is that I am talking of beings that think in ways different from me - beings whose thinking is "incommeasurable" from mine, to use a term made trendy by Thomas Kuhn (1962).

9. Am I not arguing this, assuming that my theory of natural selection has a privileged position, and is true whatever system of rationality one has? I am assuming natural selection, because it is true from the only perspective I have. But, I do not conclude that other intelligences will know of selection - not that I know of their ways of coping with the world. (See also the comments in footnotes 6 and 7).

10. In his classic paper, Lorenz (1940), rather assumes that other animals can and will think in ways fundamentally different from us. I believe this is Jerison's (1982) position also, although his rather different interests from mine make a definite decision difficult (see also Jerison, 1973).

References

Ayala FJ (1985) The theory of evolution: recent successes and challenges In E McMullin (ed) Evolution and Creation. Gary, Indiana: University of Notre Dame Press 59-90

Black M (1968) The Labyrinth of Language. New York: Praeger

Block N, G Dworkin (1974) IQ heritability, and equality. Philosophy and Public Affairs 3:331-409; 4:40-99

Bochenski I (1961) History of Formal Logic. Trans. I Thomas, Gary, Indiana: Notre Dame University Press

Brandon R, R Burian Eds (1984) Genes, Organisms, Populations: Controversies Over the Units of Selection. Cambridge Mass: MIT Press

Campbell DT (1977) Descriptive Epistemology: Psychological, Sociological, and Evolutionary. Unpublished, William James Lectures (given at Harvard University)

Chomsky N (1957) Syntactic Structures. The Hague: Mouton

Chomsky N (1965) Aspects of the Theory of Syntax. Cambridge, Mass: MIT Press

Chomsky N (1966) Cartesian Linguistics. New York: Harper and Row

Chomsky N (1980) Rules and representations. Behavioral and Brain Sciences 31-61

Cloud P (1974) Evolution of ecosystems. American Scientist 62:54-66

Daniels N (1976) IQ heritability, and human nature. R S Cohen et al eds PSA 1974 Dordrecht: Reidel 143-180

Darwin C (1859) On the Origin of Species. London: John Murray

Darwin C (1871) The Descent of Man. London: John Murray

Dawkins R (1976) The Selfish Gene. Oxford: Oxford University Press

de Waal F (1982) Chimpanzee Politics. London: Collins

Dobzhansky T (1961) Mankind Evolving. New Haven: Yale University Press

Dobzhansky T, F Ayala, G Stebbins, J Valentine (1977) Evolution. San Francisco: Freeman

Gillan D (1981) Reasoning in the Chimpanzee: 2, Transitive inference. J Exp Psych: An Beh Processes 7:150-164

Gillan D, D Premack, C Woodruff (1981) Reasoning in the chimpanzee: 1 Analogical reasoning. J Exp Psych: An Beh Processes 7:1-17

Ginsberg M Evolution and Progress London: Heinemann

Hallam A (1973) A Revolution in the Earth Sciences. Oxford: Clarendon Press

Hempel C G (1966) The Philosophy of Natural Science. Englewood Cliffs: Prentice-Hall

Herman L (ed) (1980) Cetacean Behavior: Mechanisms and Functions. New York: Wiley

Huxley TH (1863) Man's Place in Nature. London: Williams and Norgate

Isaac G (1983) Aspects of human evolution. In D S Bendall ed Evolution from Molecules to Men. Cambridge: Cambridge University Press 509-543

Jerison HJ (1973) Evolution of the Brain and Intelligence. New York: Academic Press

Jerison HJ (1982) The evolution of biological intelligence. In R Sternberg ed Handbook of Human Intelligence Cambridge: Cambridge University Press 723-791

Johanson D, M Edey (1981) Lucy: The Beginnings of Humankind. New York: Simon and Schuster

Kamin L (1974) Science and Politics of IQ. Potomac Md: Erlbaum

Kant I (1929) The Critique of Pure Reason. Trans NK Smith London: Macmillan

Kant I (1949) The Critique of Practical Reason. Trans L W Beck Chicago: University of Chicago Press

Kitcher P (1985) Vaulting Ambition. Cambridge, Mass : MIT Press

Kuhn TS (1962) The Structure of Scientific Revolutions. Chicago: Univ Chicago Press

Lewontin RC (1978) Adaptation. Scientific American 239 3 212-230

Lewontin RC, S Rose, and L Kamin (1984) Not In Our Genes. New York: Pantheon

Lieberman P (1975) On the Origins of language. New York: Macmillan
Lieberman P (1984) The Biology and Evolution of Language Cambridge, Mass: Harvard University Press
Lorenz K (1940) Kant's Lehre von apriorischen im Lichte geganwartiger Biologie. Blatter fur Deutsche Philosophie 15 94-125
Lumsden C and EO Wilson (1981) Genes, Mind and Culture. Cambridge Mass: Harvard University Press
Mayr E (1982) The Growth of Biological Thought. Cambridge Mass: Harvard University Press
Maynard Smith J (1981) Did Darwin get it Right? London Review of Books 3 (11) 10-11
Mill JS (1910) Utilitarianism. London: Dent
Murphy J (1982) Evolution, Morality, and the Meaning of Life. Totowa, N.J : Rowman and Littlefield
Nagel E (1961) The Structure of Science. London: Routledge and Kegan Paul
Pilbeam D (1984) The Descent of Hominoids and Hominids. Scientific American 250/3 84-97
Putnam H (1981) Reason, Truth and History. Cambridge: Cambridge University Press
Rorty R (1979) Philosophy and the Mirror of Nature. Princeton: Princeton University Press
Runciman WG (1986) On the tendency of human societies to form varieties. London Review of Books 8 (10) 16-19
Ruse M (1975) Charles Darwin's theory of evolution: an analysis. J Hist Bio 8 219-241
Ruse M (1975) Darwin's debt to philosophy. Studies in History and Philosophy of Science 6 159-181
Ruse M (1979) Sociobiology: Sense or Nonsense? Dordrecht: Reidel
Ruse M (1979) The Darwinian Revolution: Science Red in Tooth and Claw. Chicago: University of Chicago Press
Ruse M (1981) What kind of revolution occurred in geology? P. Asquith and I Hacking Eds PSA 1978 East Lansing: PSA 2 240-273
Ruse M (1982a) Creation-science is not science. Science Technology, and Human Values 40 72-78
Ruse M (1982b) Darwinism Defended: A Guide to the Evolution Controversies. Reading, Mass: Addison-Wesley
Ruse M (1984a) The morality of the gene. Monist 67 167-199
Ruse M (1984b) Is there a limit to our knowledge of evolution? BioScience 34 (2) 100-104
Ruse M (1985) Is rape wrong on Andromeda? In E Regis ed Extraterrestrials Cambridge: Cambridge University Press 43-78
Ruse M (1986a) Evolutionary ethics: a Phoenix arisen. Zygon
Ruse M (1986b) Taking Darwin Seriously. Oxford: Blackwell
Ruse M, EO Wilson (1985) The evolution of ethics. New Scient. 1478 (17 Oct) 50-52
Ruse M, EO Wilson (1986) Ethics as applied science. Philosophy
Scarr S, L Carter-Saltzman (1982) Genetics and intelligence In R Sternberg ed Handbook of Human Intelligence. Cambridge: Cambridge University Press 792-896
Scruton R (1986) Sexual Desire. London: Weidenfeld and Nicolson

Staal JF (1967) Indian Logic. In P Edwards ed Encyclopedia of Philosophy New York: Collier Macmillan 4 520-523

Sternberg R W Salter (1982) Conceptions of intelligence. In R Sternberg ed Handbook of Human Intelligence Cambridge: Cambridge University Press 3-28

Taylor P (1978) Problems of Moral Philosophy. Belmont, Calif : Wadsworth

Valentine J W (1978) The evolution of multi-cellular plants and animals. Scientific American 239 September 140-158

Whewell W (1840) The Philosophy of the Inductive Sciences. London: Parker

Williams GC (1966) Adaptation and Natural Selection. Princeton: Princeton University Press

Wilson EO (1975) Sociobiology: The New Synthesis. Cambridge, Mass: Harvard University Press.

Wilson EO (1978) On Human Nature. Cambridge, Mass: Harvard University Press

Woodruff G, D Premack, K Lennel (1978) Conservation of liquid and solid quantity by the chimpanzee. Science 202 991-994

THE CONCEPTUAL ROLE OF INTELLIGENCE IN HUMAN SOCIOBIOLOGY

Paul Thompson
Department of Philosophy
University of Toronto
Toronto, Ontario
Canada. M5S 1A1

In this paper I discuss some of the logical and methodological features of evolutionary explanations of intelligence-based human behavior. Hence, the main thrust of the paper will be theoretical attempting to set out the conceptual role and importance of theories of cognition and neurobiology in applying evolutionary theory to the explanation of human behavior. My main thesis is that evolutionary explanations of human behavior involve complex causal chains the links of which are justified by reference to numerous quite distinct theoretical frameworks. Among these theoretical frameworks are theories of cognitive psychology and neurobiology that connect cognition to behavior, and theories of cultural transmission of information and behavior patterns. My central goal is to provide a theoretical framework for understanding the need for, and the logic and methodology underlying, such causal chains.

To this end, I shall contrast two alternative views of theory structure and the respective accounts each gives of the relationship of theories to phenomena. And, I shall argue that many of the methodological problems with human sociobiology and the abundance of ad hoc hypotheses in sociobiological explanations of human behavior are a result of inadequate attention to the need for theories of cognitive psychology and neurobiology as part of causal chains which link evolutionary theory to phenomena. Indeed, while I shall not concentrate on them, theories from various social sciences are usually required in adequate causal chains. I suggest that the inadequate attention to these causal chains in theoretical discussions is, in large part, due to wide acceptance of a view of theories in which these causal chains play no role in relating a principle theory to phenomena. I shall set out an alternative view which provides a richer account of the relation of theories to phenomena including the need for causal chains of imported theories.

I

I do not intend to spend a great deal of time expounding the two views of theories to which I shall be referring. A brief sketch of the two views will suffice for the purposes at hand. My main concern is with the implications of accepting one

NATO ASI Series, Vol. G17
Intelligence and Evolutionary Biology
Edited by H.J. Jerison and I. Jerison

or the other of them for relating evolutionary theory to human behavior.

The view of theories which still dominates philosophy of science is a view that has its roots in logical positivism (see, fact the 1977 for an excellent exposition and historical account of this view; see also Braithwaite 1953; Carnap 1936, 1937; Hempel 1965, 1967). It is this view that underlies the accounts of evolutionary theory given by a number of philosophers of biology (see, for example, Ruse 1973, Hull 1974, Rosenberg 1985) and which most biologists will produce, if pressed, as the correct account of the structure of evolutionary theory. It is clearly the view that underlies most discussions of reductionism.

On this view, dubbed by one of its critics - Hilary Putnam (Putnam 1962) - as the Received view and by which I shall hereafter refer to it, a scientific theory is an axiomatic deductive structure which is partially interpreted in terms of definitions called correspondence rules. Correspondence rules define the theoretical terms of the theory by reference to observation terms. That is, a term like 'fragile' is partially defined by reference to 'breaking when struck by a specific force'. And, a theoretical term like 'voltage' is partially defined by reference to 'readings on a calibrated meter such as a galvanometer'. The definition is only partial because a term can be defined by reference to an open ended number of observational situations and procedures. New technologies, for example, will make possible new observations and new empirical operations and if the term was explicitly, rather than only partially defined by a particular operation, the new technological procedure would in effect amount to a redefinition rather than simply an expansion of the definition of the term. For some theoretical terms empirical meaning is indirect. That is, some theoretical terms are defined by reference to one or more other theoretical terms. Ultimately, any chain of such definitions must end in theoretical terms that are defined by reference to observations. Because of this complex interconnection of theoretical terms, the meaning of any one term is seldom independent of the meaning of many if not all of the other terms of the theory. Theories, hence, have a global meaning structure - changes to the meaning of one term will have consequences for the meaning of many and usually all the other terms of the theory.

Hence, on the received view, a theory consists of a set of deductively related statements, the structure of which is provided by mathematical logic and the empirical meaning of which is provided by definitions in terms of correspondence rules which link theoretical terms like 'mass', 'electron', 'spin', etc. to other theoretical terms or to observations. Ultimately these definitions provide direct or indirect partial empirical meaning to all the terms of the theory. In this way the theory is, as a whole, given empirical meaning. The statements of a theory are generalizations (laws), a small subset of which are taken as the axioms of the theory. The axioms are laws of the highest generality within the theory. They constitute a consistent set no one of which can be derived from any subset of the others. All laws of a theory, including the axioms, describe the behavior of phenomena. All laws except the axioms, in principle, can be derived from the axioms. Usually such deductions require numerous subsidiary assumptions. Explanation and prediction of phenomenon consists in demonstrating that the phenomenon can be deduced (or that it follows logically with a high probability) from some subset of laws conjoined with a relevant description of the prior state of the system.

The important features of this view for this paper are: (1) that the definitions, in terms of correspondence rules, which provide meaning to (i.e., semantics for) the formal structure are part of theory without which the deductive structure would be empirically meaningless; (2) that the definitions provide the link between theory and phenomena by virtue of their defining the terms of the theory by reference to phenomenal observations; (3) that because of (2) the theory specifies the way in which it relates to the world.

According to a recently developed alternative to this view which has become known as the semantic conception of theory structure, a theory is an extralinguistic mathematical entity which consists in the specification - in mathematical English - of a physical system (see, Beatty, 1980, 1981; Lloyd, 1984; Sneed, 1971; Stegmuller, 1976; Suppe, 1972, 1976, 1977; Suppes 1957, 1967; Thompson 1983, 1985, 1986, 1987; van Fraassen 1970, 1974, 1980). Theories understood in this way can be regarded as models in the sense of a model *interpretation* of a formal system. On this view laws do not describe the behavior of phenomena but specify the behavior of a system. The relationship of the theory to the world is one of isomorphism. That is, the behavior of a particular phenomenal system is claimed to be isomorphic to the physical system specified by the theory. The theory does not specify either the domain of its application or the methodology involved in establishing the isomorphism.

The important features of this view for this paper are: (1) theories are semantic structures which specify the behavior of systems not the behavior of phenomena; (2) phenomena are not explained by deducing them from laws but by asserting that the phenomenal system of which they are a part is isomorphic to the physical system specified by the theory and, hence, has the same causal structure; (3) the existence and nature of the isomorphic relationship between theory and phenomena are not specified by the theory, and its establishment is complex, requiring reference to theories of experimental design, goodness of fit, analysis and standardization of data, etc., and to causal chains based on other scientific theories.

For some this account of the two views may be too brief to be clear. I think, however, that considerable clarity will be forthcoming from the discussion of the significance of these views for understanding evolutionary explanations of human behavior. And it is to this that I now turn.

II

During the last decade there has been a storm of controversy over sociobiology - the application of evolutionary theory to animal behavior (see Allen 1976; Burian 1978; Caplan 1978; Gould 1976, 1978; Lewontin 1976; Montague 1980; Sahlins 1976). Much of the controversy has been over the application of evolutionary theory to human behavior. This is what I am calling human sociobiology though I do not ascribe to the view that there are two distinct sociobiological theories (see Thompson, 1980, 1985). Critics of human sociobiology claim that it is methodologically flawed and consists of unbridled speculation. Although I am a strong supporter of the theoretical principles of sociobiology - a point I wish to emphasis at the outset - there is, I think, considerable truth in many of the criticisms leveled against it.

Sociobiologists *have* offered rather grand speculations, involving numerous ad hoc hypotheses, about the evolutionary explanation of human behavior (consider, for example, the accounts of the persistence of homosexual behavior). In my view there are undeniable methodological problems (see Burian 1978; Thompson 1985). I suggest, however, that these problems are a function of adherence to the received view of theories. I shall argue that adherence to this view has resulted in insufficient attention being given to the methodological role of theories other than evolutionary theory in relating evolutionary theory to human behavior (e.g., theories of cognitive psychology, physiology, neurobiology, and various social science). Sociobiology, far from pre-empting the social sciences, brings them within a more unified framework within which they play a crucial role.

Evolutionary theory involves: heredity, variability and selection (see Lewontin 1974; Sober 1981). All three components are necessary for evolutionary change to occur. Without variation there is no possibility of directional selection. Without selection there is no exploitation of variability and, hence, no non-random alteration of the phenotype and, consequently, of the genetic structure of populations. Though there are other mechanisms of population change (random drift, for example,) selection remains of prime importance. However while selection will alter the phenotypic structure of a population, unless characteristics are heritable there will be no effect of this selection on subsequent generations. Heredity, therefore, is of fundamental importance. Hence, any biological evolutionary account of phenomena will involve an establishment of a genetic basis for the phenomena. This is one place where human sociobiology runs into methodological trouble. I have argued elsewhere (Thompson 1986) that since there is no unified theory describing the mechanisms of all three of these components, no received view account of the structure of evolutionary theory is possible. This has resulted in severe constraints on the nature and scope of the available responses to conceptual and methodological problems with evolutionary theory and sociobiology.

Cases of 'fixed action patterns' of behavior as found in arthropods and cases of threat and courtship displays in vertebrates are cases where evolutionary explanations are widely accepted and uncontroversial. The reason for this is that these behaviors are known to have a genetic basis and are, in the main, unaffected by learning, practice, problem solving abilities, etc. For example, the success of the sociobiological explanation of the ostensibly altruistic behavior of hymenoptera (honey bees) in terms of heterozygote superiority results from the fact that the behavior and its persistence are a function of genes and not of cognitive processes (i.e., intelligence) or of culture which requires the presence of specific cognitive process. In effect, it is a 'fixed action pattern'. Given this feature Hamilton's coefficient of genetic relationship allows the application of the laws of modern population genetics and selection theory to be employed in a deduction of the apparently altruistic behavior.

Problems arise, however, in cases of behaviors that are not known to have a genetic basis or are substantially affected by learning, practice, problem solving abilities, etc. - in short, by intelligence (cognitive abilities). These cognitive capacities allow a significant cultural component and personality component to affect behavior and also allow the cultural transmission of behavior patterns from generation to generation. Almost all human behavior is affected by cognitive capacities. Consequently, evolutionary explanations of human behavior will be complex and simple explanations patterned after explanations of 'fixed action patterns' will be methodologically inadequate. Not surprisingly, a common tactic of critics who see sociobiology as offering explanations of human behavior patterned on explanations of 'fixed action patterns' is to argue that human sociobiology is, in effect, a biological determinist view within which culture plays no significant role. Frequently, a presupposition of this criticism is that if the role of culture is taken seriously, human sociobiology, with its need to establish a genetic basis for the behaviors it is to explain, will simply cease to be tenable.

This presupposition, I shall argue, is incorrect. If, however, one is constrained by adherence to a received view of theory structure, the position can appear to have considerable force. This is due to the fact that in order to explain a particular human behavior it must be *deducible* from the laws of evolutionary theory. However, behaviors that are significantly affected by learning, practice, problem solving, etc. are not linked to genes in the direct way required in order for deductive explanation using evolutionary theory *alone* to be given. Instead, cognitive capacities intervene to modify or radically alter whatever genetic links there might be.

The central problem is that evolutionary theory does not contain either the concepts or the laws of cognitive psychology, neurobiology or any of the social sciences which are required in order to deduce behaviors that are significantly affected by cognitive abilities. For example, in the altruistic case discussed above were there a significant component of learning involved in the altruistic behavior of hymenoptera, explanation in terms of evolutionary theory alone would be impossible since no deduction from the laws of population genetics, ecology, and selection theory (the component parts of evolutionary theory) would be possible. What is absent from evolutionary theory but is needed in order to provide, by the deductive method of explanation of the received view, an account of learned behavior is a theory of the neuropsychology or neurobiology of learning which will specify the effects, if any, of certain patterns and structures of learning on behavior.

This significant role of cognitive abilities in human behavior (as well as much animal behavior) and the impossibility of deducing behaviors so affected from the laws of evolutionary theory has caused many biologists, anthropologists and philosophers to reject human sociobiology as irredeemably methodological flawed. Its usefulness and acceptability are at best considered to be in the realm of 'fixed action patterns' of behavior and behaviors which are largely hormonally driven like threat and courtship patterns. That is, in cases where a direct genetic basis for the behavior can be found or where the behavior is not significantly affected by cognitive abilities. In an attempt to meet these criticisms Charles Lumsden and Edward O. Wilson (Lumsden and Wilson 1981, 1983), and Michael Ruse (Ruse 1986b; Ruse and Wilson 1986) have attempted to introduce the concept of epigenetic rules which constrain but do not dictate behavior. While pointing in the right direction this move still fails adequately to take account of the role of intelligence because it does not provide a theoretical basis for understanding the nature and mechanics of the epigenetic rules. Its fundamental failure is its apparent adherence to the received view while not providing even a sketch of the deductive character of explanations that the view requires. This is not surprising since such a deduction will involve reference to many disparate theories and on the received view this is not logically possible because different theories will have unique correspondence rules which will result in unique global meaning for the theory. What is needed is a view of theories that provides a logical framework within which a number of theories can all contribute to the explanation of a phenomenon.

What emerges, then, is that the natural way to understand the relationship between evolutionary theory and human behavior is in terms of the evolution of neurological structures which *determine* cognitive abilities. Behavior while not determined by the neurological structures is constrained and shaped by them. A methodologically adequate explanation will have to account for behavior within these constraints but also in terms of theories about cognition and culture (including cultural transmission of information and modes of behavior). Hence, theories of cognitive psychology and theories of neurobiology are important to our understanding of the relationship of evolutionary theory to behavior. So are theories about the formation and transmission of information and patterns of behavior through cultural mechanisms.

The schema of the relationship is that evolutionary theory is related to human behavior through a causal chain of theoretical frameworks. Hence, the content and nature of the relationship of evolutionary theory to human behavior is integrally dependent on theories of cognitive psychology, neurobiology, and cultural transmission and formation of information and patterns of behavior. This kind of multiple theory type explanation is precisely what the promising accounts of Robert Boyd and Peter J. Richerson (Boyd and Richerson 1978, 1985), and Henry Plotkin and F. J. Odling-Smee (Plotkin and Odling-Smee 1981) attempt to provide. The major problem for explanations of this kind, on a received view of theories, is that the relationship of a theory to phenomena is deductive and is specified by the theory itself and there is, therefore, no method of employing other theories in relating the theory to phenomena. What is needed is a theoretical framework within which a multiple number of theories can be be employed in conjunction in explaining phenomena. This is exactly what the semantic account of theories provides.

On a semantic view, the relationship of a theory to phenomena is not specified by the theory and a multitude of other theories can be employed. For example, on a semantic view, other theories can be employed in order to account for the fact that the asserted isomorphism of the physical system specified by the theory and a phenomenal system fails to obtain. This will sometimes be the role played by theories of cognitive psychology, neurobiology and culture in the sociobiological application of evolutionary theory to human behavior. The effect of cognitive capacities and cultural transmission of information and patterns of behavior is radically to alter the effects that genes have on behavior. And, theories of cognitive psychology, neurobiology and culture provide a basis for explaining the nature and mechanisms of these alterations.

Consider two simple examples of this kind of role for imported theories. Population genetical theory describes a breeding system, the principal laws of which are Mendel's law of independent assortment and his law of segregation and the related

Hardy-Weinberg equilibrium. A wide array of populations of organisms including humans are held to be systems of this kind. However, because of crossing over, linkage, inversion, translocation and a whole host of other factors almost no population behaves exactly the way a Mendelian system, as specified by the theory, behaves. What is needed here is a theory about linkage, inversions, trans-locations, etc. that specifies the effects these phenomena will have on the structure of actual populations. In this case the theoretical grounding comes from cytological theory and molecular genetics.

The relationship between population genetics and cytological theory is best understood as follows. Cytological theory, neither the terms of which nor the laws of which are part of population genetics, is employed in applying (relating) population genetics to phenomenal systems (populations of organisms). By employing cytological theory, population genetics can be related in a mediated way to populations and this is possible even though no direct deduction, as required by the received view, of certain phenomena from population genetics alone is possible since phenomena are affected by cytological factors which are not described by population genetics. What cytological theory does, in effect, is to provide a basis for asserting that actual populations are systems of the kind described by population genetics even though there are clear differences in the behavior of the two systems. It does this by providing accounts of linkage, translocation, etc. which make clear that the differences between the systems are not a result of the inapplicability of the theory to the particular phenomena but of other causal factors described by other theories.

In a similar way, the relation between evolutionary theory and human behavior is mediated by theories of cognitive psychology, neurobiology and culture. These theories provide a theoretical grounding for the differences in the way the system specified by evolutionary theory behaves and the way phenomenal systems (i.e., human individuals and societies) behave. The differences in this case are a result of the effects of learning, problem solving abilities, etc. the nature and effects of which are the subject of theories of cognitive psychology, neurobiology and culture.

Though there are differences in content, the logic of the genetic/cytology case and the sociobiological case are identical. In the genetic/cytology case, the behavior of chromosomes is genetically based and, hence, amenable to evolutionary explanation. However, cytological theory is an important part of the explanatory application of population genetics to actual populations of organisms. In the case of sociobiology, the nature and extent of cognitive abilities is a product of evolution and, hence, amenable to evolutionary explanation. However, and this is the crucial point, theories of cognitive psychology and neurobiology about the nature of human cognitive abilities and about their effects on human behavior are important to applying evolutionary theory to human behavior. Without theories of cognitive psychology the application of evolutionary theory to human behavior is as methodologically and logically flawed as the attempt to apply population genetics to the behavior of actual populations without reference to cytology.

The upshot of this is that, on the received view, failure of the phenomena to be deducible from the theory has but two remedies. Rejection of the theory - a drastic and seldom employed remedy if no alternative theory is available - or the development of ad hoc hypotheses which will permit the deduction of the phenomena. This was the route taken by FitzGerald and Lorentz when faced by the null result of the Michelson-Morley experiment. Since the results indicated a discrepancy between the actual phenomena under investigation and what could be derived from Newtonian mechanics, FitzGerald and Lorentz (independently) proposed the ad hoc hypothesis that a body contracted in the direction of its travel through the ether. Indeed, this technique is resorted to frequently when the phenomena cannot be deduced from the laws of a theory as understood on a received view account.

What has happened in the case of sociobiology is that its strongest critics have, in the face of the obvious difficulties in deducing human behavior from evolutionary theory, advocated the reject-the-theory resolution. Sociobiologists, on the other hand, have chosen to resolve this problem by developing ad hoc hypotheses (e.g., the multiplier effect) - some of which are more plausible than others. These hypotheses are ad hoc because they are not derivable from the axioms of the theory nor are they axioms of the theory. They are non-integrated "add-ons" to the theory, their sole purpose being to allow the deduction of phenomena.

What I have been pressing for is a rejection not of sociobiology, but of the view of theories that makes it appear as if there are only these two options in favor of the richer and more descriptively accurate semantic conception of theories. What one has, on a semantic view, is a situation where evolutionary theory encompasses a number of specifications of physical systems, the laws of which are those of population genetics, selection, molecular genetics, and ecology. Actual populations are held to be systems of the kind jointly specified by these theories. Faced with discrepancies between the two systems one can opt for a different, I think preferable, methodological solution, namely, the employment of other theories from different but relevant domains that provide a theoretical, and not ad hoc, account of the nature of the discrepancies and provide a method of resolving them. And that is the solution I have been pressing for in this paper.

This solution is possible on a semantic view because the relationship between theories and phenomena is not deductive. The deductive component is one of deducing states of the system on the basis of laws which specify the behavior of the system. The relationship between the theory and phenomena is one of asserted structural and behavioral identity. An assertion which is to be empirically tested. This testing will involve careful applications of theories of experimental design, data analysis as well as other theories which will often indicate why the phenomenal system seems to behave differently than the system specified by the theory.

Therefore, it is, I submit, adherence to a view of theory structure that fails to provide a role for theories of cognitive psychology, neurobiology and culture in evolutionary explanations of human behavior that had resulted in the methodological and logical heavy weather in which sociobiology has sailed. It also, I suggest, has resulted in a level of speculation and ad hoc hypothesis formation that exceeds acceptable bounds. The speculation and ad hoc hypothesis formation has been a substitute for a logical role for theories of cognitive psychology, neurobiology and culture which have not been employed where they are clearly required. And, the failure to see the need and nature of their employment is a function of adherence to a received view of theory structure.

III

I have been arguing that the methodological and logical problems with human sociobiology can be clearly identified and remedied by changing the view of theory structure being employed to formalize the theory and its relationship to phenomena. The alternative view of theory structure that I am promoting makes clear the methodological and logical role of other theories in applying evolutionary theory to phenomena. In particular, it makes clear the logical and methodological role (as well as necessity) of theories of cognitive psychology, neurobiology and cultural transmission and formation of information and patterns of behavior in applying evolutionary theory to human behavior. This is something that is entirely absent from the received view conception of theory structure but is crucial to a logically and methodologically sound human sociobiology (as well as many other theoretical domains). Far from pre-empting the social sciences, sociobiology must make significant use of them.

References

Allen E, et al., (1976) Sociobiology: another biological determinism. Bioscience 26:182-186

Beatty J (1980) Optimal-design models and the stategy of model building in evolutionary biology. Phil of Sci 47:532-561

Beatty J (1981) What's wrong with the received view of evolutionary theory? In Asquith, PD and Giere, RN (eds.) PSA 1980. vol 2, Philosophy of Science Association, East Lansing, pp. 397-426

Braithwaite RB (1953) Scientific Explanation Cambridge University Press, Cambridge

Burian R (1978) A methodological critique of sociobiology. In Caplan AL (ed) The Sociobiology Debate Harper & Row, New York, pp. 376-395

Caplan AL (1978) The Sociobiological Debate Harper and Row, New York

Carnap R (1936) Testability and meaning 1. Phil of Sci 3:420-468

Carnap R (1937) Testability and meaning 2. Phil of Sci 4:1-40

Gould SJ (1976) Biological potential vs biological determinism. Nat Hist Mag May

Gould SJ (1978) Sociobiology: the art of storytelling. New Scientist 80:530-533

Hempel CG (1965) Aspects of scientific explanation The Free Press, New York

Hempel CG (1967) Philosophy of natural science. Prentice-Hall, Englewood Cliffs

Hull D (1974) Philosophy of biological science. Prentice-Hall, Englewood Cliffs

Lewontin R (1974) The genetic basis of evolutionary change. Columbia University Press, Columbia

Lewontin R (1977) Sociobiology: a caricature of darwinism. In Suppe F, Asquith PD (eds) PSA 1976, vol. 2. Philosophy of Science Association, East Lansing

Lloyd E (1984) A semantic approach to the structure of population genetics. Phil of Sci 51:242-264

Lumsden C, Wilson EO (1981) Genes, mind and culture. Harvard University Press, Harvard

Lumsden C, Wilson EO (1983) Promethean fire. Harvard University Press, Cambridge

Montague A (ed) (1980) Sociobiology examined. Oxford University Press, Oxford

Plotkin HC, Odling-Smee FJ (1981) A multiple-level model of evolution and its implications for sociobiology. Behav Brain Sci 4:225-268

Putnam H (1962) What theories are not. In Nagel E, et al. (eds) Logic, methodology and philosophy of science. Stanford University Press, Stanford, pp. 240-251

Richerson PJ, Boyd R (1978) A dual inheritance model of the human evolutionary process: I basic postulates and a simple model. J of Soc Bio Struct 1:127-154

Richerson PJ, Boyd R (1985) Culture and the evolutionary process. University of Chicago Press, Chicago

Rosenberg A (1985) The structure of biological science. Cambridge University Press, New York

Ruse M (1973) The philosophy of biology. Hutchinson, London

Ruse M (1986) Taking Darwin seriously. Basil Blackwell, Oxford

Ruse M, Wilson EO (1986) Ethics as applied science. Philosophy

Sahlins MD (1976) The uses and abuses of biology. University of Michigan Press, Ann Arbor

Sneed J (1971) The logical structure of mathematical physics. D. Reidel, Dordrecht

Sober E (1981) Holism, individualism, and the units of selection. In Asquith PD, Giere RN (eds) PSA 1980, vol 2. Philosophy of Science Association, East Lansing, pp. 93-121

Stegmuller W (1976) The structure and dynamics of theories. Springer Verlag, New York

Suppe F (1972) What's wrong with the received view on the structure of scientific theories? Phil of Sci. 39:1-19

Suppe F (1976) The structure of scientific theories. University of Illinois Press, Chicago

Suppe F (1977) Theoretical laws. In Prezlecki M, et al., (eds) Formal methods in the methodology of the empirical sciences. Ossolineum, Wroclaw, pp. 247-267

Suppes P (1957) Introduction to logic. Van Nostrand, Princeton

Suppes P (1967) What is a scientific theory? In Morgenbesser S (ed) Philosophy of science today. Basic Books, New York, pp. 55-67

Thompson P (1983) The structure of evolutionary theory: a semantic approach. Stud in Hist and Phil of Sci 14:215-229

Thompson P (1985) Sociobiological explanation and the testability of sociobiological theory. In Fetzer JH (ed) Sociobiology and epistemology. D. Reidel, Dordrecht, pp. 201-215.

Thompson P (1986) The interaction of theories and the semantic account of evolutionary theory. Philosophica 37:73-86

Thompson P (1987) A defence of the semantic conception of theories Bio and Phil 2:26-32

van Fraassen BC (1970) On the extension of Beth's semantics of physical theories. Phil of Sci 37:325-339

van Fraassen BC (1974) A formal approach to the philosophy of science. In Colodney RE (ed) Paradigms and Paradoxes. Univ. of Pittsburgh Press, Pittsburgh, 303-366

van Fraassen BC (1980) The scientific image. Oxford University Press, New York

ARTIFICIAL INTELLIGENCE AND BIOLOGICAL INTELLIGENCE

Margaret A. Boden
Cognitive Studies Programme, University of Sussex

It is often said that there can be no such thing as artificial intelligence. For certain, there is no such thing as biological intelligence. There are, however, biological *intelligences*. Animals have evolved many different motor and perceptual strategies to cope with the world -- indeed, these strategies are so different that one might rather say "their worlds", as we shall see. The "space" of biological intelligences is not yet fully charted, nor have all the relevant dimensions even been identified. I shall argue that this exploration and theoretical mapping may benefit from some of the insights of artificial intelligence (AI).

Biologists often doubt that AI could possibly throw light on animal intelligence. For computers are very unlike brains. They are made of quite different stuff, and most are digital, serial-processing (and general-purpose) devices. Furthermore, AI-workers typically ignore neuroscience. The intellectual divide seems deep: how can AI be of interest to biologists?

The accusation that AI ignores biology is broadly correct. But, as we shall see, ideas from neuroscience are playing a larger role in AI now than in the past. During the 1960's and 1970's, research in AI-vision (for example) had abandoned the early attempts of the 1950's to mimic the parallel embodiment of the brain. Instead, it focussed on discovering the knowledge (and the ways of using it) involved in vision, on the computational functions required for the basic task of sight: 2-D to 3-D mapping. (Comparable work was being done in other areas of AI, with respect to the computational constraints on different information-processing tasks.) It was because the computational functions necessary for sight had not yet been properly identified that the early parallel-processing models of vision had to be abandoned. Recently, however, some of the special features of brains -- their parallelism and dedicated machinery -- are being taken seriously again in AI-vision (and in the modelling of language and memory too).

Even those computer-modellers who work with parallel ("connectionist") systems still live in happy ignorance of biochemistry. For those properties of brains which enable them to function as organs of intelligence almost certainly have less to do with what they are made of than with how they are organized and -- above all -- what it is that they do. AI-workers concentrate on the latter two questions: on the identification, in abstract terms, of specific computational functions, and on the way in which computational units can be organized so as to carry out those functions.

NATO ASI Series, Vol. G17
Intelligence and Evolutionary Biology
Edited by H.J. Jerison and I. Jerison

This process may cast light on how we, and other animals, do things. For at least some of the computational processes in computers (whether parallel or not) may be significantly similar to the processes going on in human brains. So psychology and ethology can expect to profit from AI. (And AI can profit from psychology and ethology, which provide a rich source of examples of drawn from the space of intelligent phenomena.)

Even neuroscience, the study of "wetware", might profit from AI -- indeed, it already has. It was the abstract computational arguments about perception put forward in the late 1950's by the programmer of *Pandemonium* (an early computer model of vision) which first suggested that feature-detectors might exist in the nervous system, and which prompted neurophysiologists to search for them (initially, in the frog's retina (Lettvin, Maturana, Pitts and McCullough, 1959); Selfridge, 1959). Computer scientists can tell neuroscientists nothing about the material nature of the brain. But they may (as in this case) be able to suggest what sort of *functional* unit neuroscientists might fruitfully look for. Correlatively, neuroscience may suggest that certain sorts of organization can be put to good computational effect in generating intelligent behaviour.

I shall argue that an explanation of biological intelligence must credit many animals besides man with the ability to form representations -- and the concepts of AI are specifically concerned with representations, and their transformations. "Representation" is not a concept of physiology (the term "cerebral model" is a psychophysiological hybrid), so the physiological differences between brains and computers do not make AI biologically irrelevant. The foundations of an adequate explanation of "intelligence" would consist of abstract analyses of the information-processing tasks carried out by various species. Resting on these foundations would be an account of the computational processes that perform those tasks, and (ideally) of how they are embodied in the neurophysiology of organisms.

Like the ancient city of Romulus and Remus, this explanatory edifice will not be built in a day. We shall see that the problems of formulating plausible hypotheses about representations in animal (and human) mind are even more complex than is generally believed. The moral of this paper, then, is not that AI-workers have achieved solutions to ethologists' problems, but that AI raises theoretical questions about intelligence which ethologists should not ignore.

In Section II, I say something in general terms about AI as the study of representation, and explain why it suggests that we must attribute representations even to non-communicating animals. Next, I relate some problems about motor action (Section III) and perception (Section IV) in animals to examples of work in AI. These problems are typical of those raised within "cognitive ethology", a term recently coined to cover studies of the psychological competence of animals (Griffin, 1978), such as the work on chimps directed by D. Premack or D. M. Rumbaugh. Finally, in Section V, I discuss three common objections to computer models of intelligence: the difficulty of validating such models by experiments; the architectural differences between brains and digital computers; and the problem of ascribing conscious phenomenology to animals.

II. AI as the Study of Representation

Years ago, I saw in the pages of *Punch* a cartoon more memorable than most (I have redrawn it in Figure 4.1). It showed a kingfisher sitting on a willow-branch, staring at fish in the river below, and thinking to itself, "u = sin a /sin b." This cartoon is no mere triviality, for it is a reminder of some deeply puzzling questions.

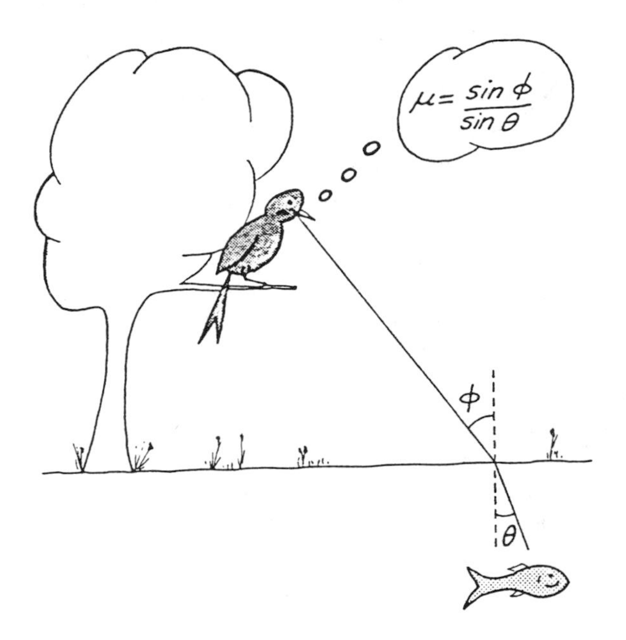

Figure 4.1

How does the kingfisher manage to catch the fish, no matter (within limits) where it is in the water? Unlike some birds, it does not dive vertically into the water, nor does it pursue the fish while under water. Kingfishers are plunge-divers who go rapidly straight to the target. Given that the kingfisher has never heard of Snell's law, does it have to go through some alternative process of computation to adjust its angle-of-dive appropriately -- and if so, what? A less obvious puzzle is how the bird manages to identify part of the scene as a fish, or as food, in the first place, and how it is thereupon led to take appropriate action (that is, how does it know that it should dive, irrespective of how steep the dive should be)?

All these puzzles concern the information being used by the animal, and the way in which the animal is using it. One might expect, then, that AI should be somehow relevant. AI is a recent branch of information-science that is suited to the needs of ethology or theoretical psychology because it defines a wide range of qualitatively distinct and structurally complex symbolic representations and interpretative procedures. The computational concepts used in AI are concerned with the reception, storage, transformation, interpretation, and use of information by information-processing systems which employ and construct symbols -- and symbol-manipulation procedures -- of many distinct kinds.

It is often regarded as problematic whether or not animals have mental representations, or use symbolic systems or languages. Sometimes it is even stated categorically that they do not. Piaget, for example, says that animals (like newborn babies) do not have representations, and Chomsky denies that even chimps enjoy language. To some degree, these disputes turn on terminological differences in the use of terms such as "representation" and "language". But even setting aside such terminological differences, it remains true that whether or not any animals employ symbolic representations is widely regarded as doubtful.

The lesson of AI is that many animals must have representations enabling them to interpret stimulus-information sensibly in widely differing contexts and to take appropriate action accordingly. The more flexible the action, the more complex must be the computational resources for monitoring, planning, and scheduling different types of activity. In particular, when the creature has to take account of a wide range of *structural* differences and similarities between distinct situations (as opposed to concentrating on only one or a few physical parameters), these structural features can only be represented *symbolically* -- for, by hypothesis, they have no *physical* features in common.

It does not follow, of course, that the representations in animals' brains are comparable in detail to the symbol-manipulating processes in current digital computers (in which representations are coded at distinct, and often physically separated, locations in the machine). Indeed, some of the most interesting recent work in AI, and in computationally-based psychology, focusses on *connectionist* systems, in which extensive parallel-processing allows for a representation to be *distributed* across an entire network of computational units (Rumelhart and McClelland, 1986). I shall say a little about such systems below, but -- since the vast majority of AI-work has not been of this type -- most of the computer-models I shall mention are conceptualized in terms of (as well as being implemented in) digital computers.

That animals need to employ (computational mechanisms functioning as) abstract, symbolic, representations is true *irrespective* of whether the animal is able also to communicate with its conspecifics, by warning-cries, mating-calls, and the like. And it is true whether or not the animal is able, like humans, to employ a syntactically structured public language, using units of meaning whose semantic import is determined by social conventions rather than by fixed genetic mechanisms. The point of present importance is that even much *noncommunicative* behaviour has to be understood in computational terms such that internal symbolic processes must be attributed to the creature. Indeed, the interpretation of audible or visible signs, words, or gestures *as* communications with a certain meaning presupposes the computational mechanisms involved in sensory perception in general. This is why one AI-worker has referred to "the primacy of non-communicative language" (Sloman, 1979).

This is not to say that all computations carried out by animals are effected symbolically. For example, hoverflies appear to compute their interception paths with conspecifics according to a simply specifiable rule, one which could plausibly be "hardwired" into the flies' brains (Collett and Land 1978). Although this rule *could* be represented and applied within a symbolic system, it is reasonable to suggest that it has been "learnt" by the evolutionary process and is embodied in the flies' neurophysiology. It is significant, however, that the computations concerned are relatively inflexible: the fly in effect assumes that the size and velocity of the target are always those corresponding to hoverflies, and on this rigid (and fallible) basis the creature determines its angle of turn, when initiating its flight, according to the variable approach angle subtended by the target. Moreover, the fly's path cannot be adjusted in mid-flight, there being no way in which the pursuer can be influenced by feedback from the (perhaps unpredictable) movement of the target animal.

This rigid behaviour is fairly common in insects, but the higher animals are capable of considerable flexibility in adjusting their behaviour to widely differing (and continuously changing) circumstances, where the relevant "parameters" are structural features rather than physical ones (such as angle-of-approach). The visual capacities of the higher animals are typically more flexible also: for instance, many species can discriminate the position, shape, texture, and surface-orientation of arbitrarily many physical objects -- even objects seen for the very first time. Recent computational work in "low-level" vision -- an example of which is discussed below -- shows that these capacities, even though they are largely hardwired (innate), cannot be understood unless they are conceptualized as involving (multi-levelled) representational processes (Mayhew and Frisby, 1984).

The term "cognitive ethology" is only recently coined, but the problems are not new. In his seminal paper on "the invisible worlds of animals and men," Jacob von Uexkull (1957) showed that the task of a cognitive ethology is to articulate the varied *Umwelten* of different species. To do this, we need to ask what a given species can perceive, and what it can do accordingly. Those aspects of environment or action which a creature does not have the epistemological resources to represent cannot form part of its cognitive world.

Von Uexkull illustrated these points by his unforgettable pictures of the living-room as seen by fly, dog, or man, and of the fish and the boat as seen by a sea-urchin. But, charming though they are, his pictures do not clearly articulate the similarities and differences between the invisible worlds of these species. Work in AI might help us to a richer understanding of such matters, as I shall now try to show by reference to examples concerning the planning of action and the perception of the physical world.

III. Computational Concepts in the Explanation of Action.

The concept of purposive action has often entered ethological discussion. Purposive action is behaviour controlled by a guiding representation of some desired state, whose overall plan allows for obstacles to be overcome by appropriate variations in the activities selected as means to the end (notice that this is an essentially *psychological* definition, so that purposive behaviour is not the same as behaviour controlled by feedback of the sort studied in classical cybernetics or control-theory) (Boden, 1972). To be sure, ethologists in the past have been more concerned to deny the relevance of this concept to animal behaviour than to insist on its applicability. Thus in 1937 Konrad Lorenz criticized anthropomorphic attributions of "instinct", saying that, "To assume a 'whole-producing,' directive instinct superior to all part reactions could evidently be justified only if the effects of a regulative factor, exceeding the experimentally demonstrable regulative faculty of the single reactions, could be observed" (Lorenz, 1957).

Since then, entomologists and ornithologists in particular have often identified independently controlled units of behaviour, which in normal circumstances combine to give the appearance of activity that is planned as a whole and dependent on a recognition of complex means-end relationships. D. S. Lehrman's studies of parental behaviour in ring-doves are an example of this sort of analysis (Lehrman, 1958). Various hormonal factors and several "social" releasing stimuli interact, so that the behaviour of birds and squabs is reciprocally determined in an appropriate manner. But if the normal sequence is upset the doves do not engage in variation of means, they do not adapt their behaviour intelligently so as to achieve the desirable (though evidently not *desired*) end-state of a nestful of happy, healthy chicks. (It does not follow, of course, that the independent parts of the "parental" sequence are not flexible to some degree according to circumstance: but the guiding goal of *rearing healthy chicks* cannot be posited as an explanation of this sequence.)

Recently, however, primatologists have begun to ask whether the behaviour of apes, at least, may sometimes be directed by plans or strategies guided by an idea of the goal. And some ethologists raise this question also about non-primate, and even non-mammalian, species such as beavers and bees (Premack and Woodruff, 1978). But it is generally agreed, even by those willing to consider such questions, that they are very difficult to answer. This difficulty rests partly in the fact that psychology has not provided a theoretical vocabulary for expressing the structure of purposive activity. Indeed, for many years Anglo-Saxon psychologists actively discouraged any such endeavour, because of the anti-mentalistic bias of behaviourism. AI may be helpful here, for there are already a number of AI programs concerned with planning, in which are defined procedures of varying complexity for comparing current with desired state and selecting activities accordingly (Boden, 1977; Charniak and McDermott). The computational concepts involved offer the beginning of a theoretical taxonomy of plans. Such a taxonomy could aid the behavioural analysis of those forms of animal activity that are apparently purposive, rather than being simply "automatic" or "mechanical" in nature. In a recent publication, Lorenz has cited examples, arguing that this is a continuous range rather than a bipolar distinction within animal behaviour (Lorenz, 1977). Indeed, it might better be described as a multidimensional space. Computational considerations could help distinguish the different points in the space of actual (and, ideally, of possible) behaviours.

Many actions of insects are sequential patterns of invariant order which, once started, are "automatically" executed to the bitter end even in inappropriate circumstances. Sometimes there is a degree of flexibility due to local conditions (such as the configurations of the terrain), but there is no feedback of information capable of altering the overall pattern; at best, it can be interrupted, cut short without the possibility of restarting later at the same point. And some other examples of animal action (such as the parental behaviour of Lehrman's ring-doves) are composed of units which follow "mechanically" in a fixed order provided that at each point the relevant releasing stimulus occurs.

AI-plans are not like this, although most people unfamiliar with AI assume that they are. They are hierarchically organized wholes, variable according to circumstance. Many programs have a "heterarchical" control-structure, in which control is widely distributed throughout the system: the sub-programs on various levels can communicate up and down *and sideways*, so that decisions can be taken at a local level relatively independently of the overall goal of the system as a whole. This type of control-structure (which is often compared to a human committee of experts) makes it easier to effect subtle variations according to context, so that what at a higher level is clearly "one and the same plan" can be interpreted in importantly different ways on different occasions. Various sorts of monitoring activity are employed to schedule different sorts of action and to make adjustments to ongoing action that is not proceeding (in its relation to the problem-environment) as well as it might be.

For example, some programs monitor and adjust the execution of their plans by reference to their internal representation of the preconditions and consequences of different actions. Thus the mobile robot SHAKEY, while executing a plan for moving blocks from one room to another, asks itself at each step whether the plan as executed so far has produced the expected results (which it may not have done if the environment has changed unexpectedly); what portion of the plan needs to be executed next (which may not be the portion initially foreseen, if the previous question was answered in the negative); and whether this next portion can indeed be executed in the current state of the world (if not, a sub-goal may be set up to realize the necessary preconditions). Other planning programs exist with a richer representational power and so a greater flexibility of action. Some can choose in a principled fashion whether or not to commit themselves to a specific ordering of subgoals ahead of time, and accordingly decide sensibly when the time comes to execute the plan. Some can generate an outline into detailed effective action when necessary. Some can anticipate unwanted side-effects and modify the plan accordingly, so as to avoid them or neutralize their unwanted aspects. Some can envisage different alternative strategies for achieving a goal, and use both reasoning and empirical enquiry in choosing between them. Some can recognize a *cul-de-sac* and re-enter a strategy at the precise point where it was previously abandoned, possibly generating a new mini-plan for overcoming the local obstacle which (as it remembers) led to its abandonment in the first place. And some can construct a representation of the goals and plan-following of another program, using it to guide interaction between the two systems (Sacerdoti, 1974).

Were one to apply the insights gained in the development of these programs to the experiments on chimps done by D. Premack and G. Woodruff, or by D. M. Rumbaugh et al., one would be led to ask a number of questions not mentioned by them. For example, how sensitive are chimps to constraints on the temporal ordering of certain units of behaviour in the context of an overall problem, such that *this* sub-unit has to be performed before *that* one? (They clearly are sensitive to such constraints in some degree, since they will often go to fetch a tool before attempting to do the task for which the tool is required.) Does a chimp have the representational complexity to gather together two or three tools, each of which will be needed in the ensuing task? Or must the chimp think about only one step at a time? If it sees another individual attempting the second step before trying the first (where this ordering is mandatory), can the chimp realize and communicate the information that the required step should be taken instead? If a chimp decides to abandon a task, what features influence its decision? Is it capable of coming back to that task at an appropriate moment, and, if so, can it remember where it was in the task previously, or must it begin again from scratch? Does a chimp ever engage in activity which looks as though it is a preparation for some later task, either in establishing necessary preconditions or in forestalling unwelcome consequences that would otherwise ensue on later performance of the task in question? If a chimp is interrupted during its problem-solving by some irrelevant occurrence, does it remember the unfinished task, and does it remember what stage it had reached at the time of interruption? And so on, and so on. We must not merely ask whether chimps generate representations of plans, but must distinguish the computationally different types of plan that they might be using in the control of their behaviour, and that they might be attributing to other individuals (whether chimp or human).

Some AI-workers would echo Lorenz here, objecting that one cannot assume that apparently integrated behaviour is controlled by some integrally organized plan, or that flexible, context-sensitive behaviour is guided by a representation of the desired overall result. They would refer to programs called "production systems," in which control rests in a number of largely independent rules, each of which may be acquired in isolation, and each of which expresses a Condition-Action pair. Each rule tests for a certain condition (in input or in short-term memory) and then carries out the relevant Action (either producing output or altering the contents of short term memory). This approach is quite different in spirit from the "planning" approach previously described. It is better able to represent the continual shifting of the focus of attention, and also the interruption of behaviour, whereby appropriate action can be instantly taken on the occurrence of an unexpected event. Yet it can model problem-solving behaviour which one might have thought to be controlled by a plan explicitly representing the structure of the task as a whole. However, this purposive structure has to be implicit in the system if it is to model hierarchically integrated behaviour. So, for example, constraints have to be written into the content of the rules, or the priority and/or temporal ordering of the rules have to be constrained, to decrease the independence of the several rules and so go against the spirit of the approach in its pure form (Davis and King, 1976).

Production systems can sometimes be matched to detailed behavioural protocols, and studied *pari passu* with experimental results. For example, a system of rules whose subsets generate different patterns of seriation (staircase-building) can be matched to children's motor and verbal behaviour so as to capture a wide range of detailed observations (Young, 1976); and a production system for subtraction can model the many commonly observed errors in subtraction sums that children make (Young, 1977). These examples show that even a small number of production rules can give rise to performance that is considerably more varied and flexible than the relentless formula-following common in insects or the successive behaviour-triggering seen in Lehrman's ring-doves. And large production systems, incorporating many hundreds of rules, can generate problem-solving performance comparable to subtle and complex human behaviour.

It may be that much animal behaviour, especially non-mammalian forms, could fruitfully be modelled in these terms. For the Condition may be an external environmental condition (temperature, sunrise, or the presence of a fish or a cat), a state of the animal's internal environment (hormonal concentration), or an inner psychological condition (such as the impulse or desire to catch a fish). And the Action may be motor behaviour (as in diving for the fish or fleeing the cat), or psychological processing (as in activating the desire to catch a fish, or checking to see whether it is on the surface or deep in the water). Ethologists might find it useful to try to write production systems modelling behaviour in different species, and to enquire into the acquisition (whether genetic or through learning) of individual rules. Running one's set of rules on a computer enables one to test not only their coherence but their implications, for one can systematically omit or alter individual rules and observe the performance so generated. In this way, one might enrich one's understanding of what Lorenz termed "the experimentally demonstrable regulative faculty of the single reactions".

The early robot SHAKEY, as its name implies, was not impressive considered as a motor system: its interest was in its capacity to plan and monitor its actions. Recent work in robotics recognizes that motor action involves not only trajectory planning, but detailed movement-control, maintenance of stability, calculation of forces, and visuomotor coordination. One might think that only the first of these (planning) involves artificial intelligence: the others, it seems, are mere engineering. However, the engineering is not "mere" at all: very difficult theoretical problems are involved in (for example) controlling jointed robot-manipulators with the degrees of freedom of the human hand (Hardy 1984). Likewise, enabling a camera-equipped robot to *avoid* the obstacles it sees in its (potential) path is no simple matter. Work in robotics is increasingly grounded in the detailed theory of mechanics and dynamics, in roboticists' attempts to develop computational procedures that can exploit this knowledge in performing practical tasks. The crucial theoretical problems have not all been identified, still less solved (today's commercial robots avoid these problems, but only by compromising their flexibility). But because the basic information-processing task involved in motor action is common to robotics and motor-psychology alike, theoretically-grounded work in robotics can be expected to move nearer to work in motor-physiology and motor-psychology.

IV. Computational Concepts in the Understanding of Perception.

Coordinated with its active aspect (whether this be regulated by an overall plan or by isolable rules), the *Umwelt* of any animal has a perceptual aspect. For example, many species are assumed by ethologists to enjoy motion-perception and object-concepts of some sort. Just what sort, however, is usually unclear. Even in the human case, the psychological processes underlying motion-perception are not fully understood.

Some recent computer-modelling work done by Shimon Ullman (1979) suggests computational questions and hypothetical answers that are relevant not only to human vision (Ullman's prime focus), but to animal vision also. Like the psychologist J. J. Gibson (1979), Ullman attempts to show that many perceptual features can be recognized by relatively low-level psychophysiological mechanisms, whose functioning relies on the information available in the ambient light rather than on high-level concepts or cerebral schemata. But unlike Gibson, who posits a "direct" unanalysable perceptual process of "information pick-up," Ullman views this functioning as a significantly complex process intelligible in computational terms.

Ullman reminds us that if two differing views or input-arrays are successively presented to the visual system, then one of several phenomenologically distinct perceptions may arise. We may see an object (visible in the earlier view) disappearing, and being *replaced* by another one--as in a game of "peekaboo"; we may see one and the same (rigid) object *moving*, perhaps involving a change in its appearance due to rotation; we may see one and the same object *changing* in shape so as to be transformed into something different -- as the baby that Alice was holding gradually turned into a pig before her eyes; finally, we may see an object moving and changing shape at the same time (as does a walking mammal). Using the experimental technique of "apparent movement" (which, interestingly, has been shown to occur in some animal species (Rock, Tauber and Heller, 1964 and 1966), the conditions under which these perceptions are elicited can be empirically investigated.

Ullman's project is to discover the series of computations that the visual system performs on the input-pairs so as to arrive at an interpretation of the (2-D) array in terms of (3-D) replacement, motion, or change. In particular, he asks whether (and how) these distinct percepts can be differentially generated without assuming reliance on high-level concepts of specific 3-D objects (such as fish or sticklebacks), and even without assuming the prior recognition of a specific overall shape (such as a sort of narrow pointed ellipse with sharp projections on its upper surface). As the ethologist might put it, Ullman attempts to follow Lloyd Morgan's Canon, in asking what are the *minimal* computational processes that need to be posited to explain motion-perception. As we shall see, Ullman is misled by his concentration on mathematically minimal computations into assuming a specific hypothesis which is ethologically implausible--but this does not destroy the general interest of his approach.

As regards the visual interpretation of each array considered in isolation, Ullman relies on the work of D. A. Marr. Marr (1980) studied the information potentially available in the ambient light falling on the retina, and the image-interpreting computations performed on it by peripheral levels of the visual system. He is largely responsible for the insight, mentioned above, that even low-level vision has to be understood as a *representational* process -- one in which the visual organism constructs a series of representations on successive levels, culminating in an object-centred (viewer-independent) 3D-description of the object in question. ("Low-level" vision involves the interpretation of the 2D input, or intensity array, in terms of 3D properties of physical surfaces/objects; "high-level" vision involves the recognition of specific categories of objects, and the use of expectations and concepts to guide the process of visual interpretation.)

The first stage of visual computation, according to Marr, is the formation of a "Primal Sketch," an image consisting of descriptions of the scene in terms of features like *shading-edge*, *extended-edge*, *line*, and *blob* (which vary as to *fuzziness*, *contrast*, *lightness*, *position*, *orientation*, *size*, and *termination* points). These epistemological primitives are the putative result of preprocessing of the original intensity array at the retinal level -- that is, they are not computations performed by the visual cortex (still less, the cerebral cortex). Marr defines further computations on these primitive descriptions, which group lines, points, and blobs together in various ways, resulting in the separation of figure and ground. He stresses that these perceptual computations *construct* the image, which is a symbolic description (or articulated representation) of the scene based on the initial stimulus-array. The computations are thus interpretative processes, carried out by the visual system considered as a symbol-manipulating system rather than simply as a physical transducer (though Marr attempts to ground his computational hypotheses in specific facts of visual psychophysiology).

Starting with Marr's basic meaningful units, Ullman defines further visual computations which would enable the system, presented with two differing views, to make a perceptual decision between replacement, motion, or change. Ullman divides the computational problem faced by the visual system into two logically distinct parts, which he calls the *correspondence* and the *interpretation* problems. (The latter term unfortunately obscures the fact that *all* these computations, including Marr's, are interpretative processes, carried out by the visual system in its role as a symbol-manipulating device.)

The correspondence problem is to identify portions of the changing image as representing the same object at different times. This identity-computation must succeed if the final perception is to be that of a single object, whether in motion or in change. Conversely, the perception of replacement presupposes that no such identity could be established at the correspondence stage. The interpretation problem is to identify parts of the input arrays as representing objects, with certain 3-D shapes, and moving through 3-space (if they are moving) in a specific way. In principle, correspondence- and interpretation-computations together can distinguish between the three types of perception in question. And, if specific hypothetical examples of such computations are to be of any interest to students of biological organisms, they should be able to distinguish reliably (though not necessarily infallibly) between equivalent changes in the real-world environment.

This last point is relevant to the way in which Ullman defines specific correspondence - and interpretation - algorithms. In principle, any part of one 2-D view could correspond with (be an appearance of the same object as) many different parts of another; similarly, any 2-D view has indefinitely many possible 3-D interpretations. (Anyone who doubts this should recall the images facing them in distorting mirrors at funfairs.) Faced with this difficulty, Ullman makes specific assumptions about normal viewing conditions, and takes into account certain physical and geometrical properties of the real world, as well as (human) psychological evidence based on studies of apparent motion. Accordingly, he formulates a hypothetical set of computational constraints which he claims will both assess the degree of match between two views so as to choose the better one, and typically force a 3-D interpretation which is both unique and veridical. For instance, for the correspondence stage he defines "affinity functions" that compute the degree of match between two points or short line-segments, depending on their distance, brightness, retinal position, inter-stimulus (time) interval, length, and orientation. And for the interpretation stage, he defines a way of computing the shape and motion of a rigid object from three views of it, making his system assume that if such a computation succeeds then it is indeed faced with a rigid body in motion (as opposed to two different objects or one object changing its shape). He justifies this by proving mathematically that, except in highly abnormal viewing conditions, three views of a rigid object can uniquely determine its shape and motion.

Given that Ullman's computations can indeed interpret correspondence, shape, and motion in a wide range of paired 2-D views (which can be tested by running his system in its programmed form on a computer provided with the relevant input), what are we to say about the ethological importance of his work?

The first thing to notice is that Ullman embodies implicit assumptions about the physics and geometry of the real world, and about biologically normal viewing conditions, into the computations carried out by the visual system. It is plausible that many species may have evolved such implicit computational constraints. That is, the animal's mind may implicitly embody knowledge about its external environment, which knowledge is used by it in its perceptual interpretations. Something of the sort seems to be true for migratory birds, who have some practical grasp of the earth's magnetic field or of stellar constellations; and, as I shall suggest presently, the kingfisher may have some practical grasp of the refractive properties of water.

What is ethologically implausible about Ullman's hypotheses is not that they embody some knowledge about material objects and normal viewing conditions, but rather that they assume the perception of rigid objects to be basic, while perception of non-rigid movement is taken to be a more complex special case. *Mathematically*, of course, the perception of non-rigid motion is more complex; but this does not prove that it is *biologically* secondary to the perception of rigid objects. At least in the higher animals, it is more likely that the visual perception of shape and motion has evolved in response to such biologically significant environmental features as the gait or stance of hunter or prey, or the facial grimaces and tail-waving of conspecifics. The fact that human beings do not always perceive the correct (rigid) structure when presented with a mathematically adequate though impoverished stimulus, may be due not (as Ullman suggests) to their failing to pick up all of the mathematically necessary information in the stimulus, but rather to their using computational strategies evolved for the perception of non-rigid objects which -- *even* when directed at rigid objects -- need more information than is present in the experimental stimulus concerned (Sloman, 1980). Admittedly, a robot could be provided with an Ullmanesque capacity to perceive rigid objects in motion; but whether any creature on the phylogenetic scale employs such visual mechanisms is another question.

Our friend the kingfisher apparently possesses computational mechanisms which can discover the real position of a fish at varying depths in the water. Ullman's general approach suggests that these could well be relatively low-level processes, not requiring cerebral computations (as puzzling out Snell's law presumably does). For the visual computations algorithmically defined by Ullman do not depend on high-level processes capable of identifying (recognizing) objects as members of a specific class: the system does not need to know that an object is a fish, or even that it has the 3-D shape that it has, in order to know that it is an object. Nor does it need any familiarity with the object; that is, it does not need to have experienced those two views in association beforehand. Ullman therefore suggests (*contra* empiricists and Piaget) that a baby--or, one might add, a kingfisher -- can see that two appearances are views of one and the same object even if it has never seen that sort of object before, and even if it has no tactile or manipulative evidence suggesting that they pertain to one and the same thing. These conclusions follow from the fact that all of the correspondence-computation, and much of the interpretation-computation, is via low-level, autonomous processes that do not depend on recognition of the input as a familiar 3-D object. The correspondence-computations match primitive elements (those defined by Marr) in successive views, and do not depend on computation of the overall shape as a whole.

It follows that creatures incapable of computing shape in any detail, or of recognizing different classes of physical object, may nonetheless be able to compute motion. As the example of von Uexkull's sea-urchin suggests, this is no news to ethologists, who often have behavioural evidence that an animal can perceive motion though they doubt its ability to be aware of detailed shapes. But Ullman's achievement is to have complemented this empirically-based intuition by a set of admirably clear hypotheses about precisely what visual computations may be involved, at least in the human case. That some of his hypotheses are biologically dubious does not destroy the ethological interest of his general approach.

Ullman's work also casts some light on our kingfisher-cartoon. For if the general shape, the location, and the motion of objects can be computed in a low-level, autonomous fashion, then it is not impossible that a kingfisher may possess comparable perceptual mechanisms capable of computing the depth of a fish in water. The refractive index of water would be implicitly embodied in these computational mechanisms, perhaps in an unalterable fashion. So a kingfisher experimentally required to dive into oil might starve to death, like newborn chicks provided with distorting goggles that shift the light five degrees to the right, who never learn to peck for grains of corn in the right place (Hess, 1956). This assumes (what is the case for the chicks) that the kingfisher utilizes an inborn visuomotor coordination, linking the perceptual and active aspects of its *Umwelt*, a coordination that is not only innate but unalterable. Psychological experiments on human beings, and comparable studies of chimps, show that these species by contrast can learn to adjust to some systematic distortions of the physics of the visual field (Stratton, 1896, 1897).

In their paper asking whether chimps are lay psychologists, Premack and Woodruff remark in an aside that one might also enquire whether they are lay physicists. Before being in a position to do this at any level of detail, one will need a clearer sense of what the content of a lay physics might possibly be. The foregoing discussion of Ullman's work suggests some part of the answer, in articulating assumptions about the physics and geometry of 3-D objects viewed in air (and, perhaps, in water) that may inform the *Umwelten* at least of some animals. But, presumably, human beings and many other species possess many more concepts and inferential structures that embody everyday knowledge of the material world, much of which knowledge may be acquired through learning. Some work in AI, which admittedly is programmatic rather than programmed, is an interesting preliminary attack on this problem (which is attracting increased attention: a recent issue of *Artificial Intelligence* (13, 1980; see also Bobrow, 1985; Hobbs and Moore, 1984) was entirely devoted to it.

In his "Naive Physics manifesto," P. J. Hayes (1979) asks how one might construct a formalisation of our everyday knowledge of the physical world. Ethologists may be tempted to dismiss such an enquiry as irrelevant to their problems: human beings have Newton and Einstein, whereas animals do not, so human knowledge of physics cannot be relevant to enquiries about chimps, beavers, or bees. That this would be an inappropriate objection is evident from the fact that the *Punch* cartoon (Figure 4.1) would have been almost as funny if it had figured a human fisherman rather than a kingfisher. Not only do we not usually think of Snell's law when we try to net a fish or tickle a trout, but we could not use it to help us do so even if we did. Similarly, we do not balance a bicycle by applying the formulae of mathematical dynamics. Our everyday intuitions of concepts such as *weight*, *support*, *velocity*, *height*, *inside/outside*, *next to*, *boundary*, *path*, *entrance*, *obstacle*, *fluid*, and *cause* (to name but a few) are pretheoretical. It is this pretheoretical knowledge which interests Hayes.

It is apparent that some animals share much of this pretheoretical knowledge with us -- often, as in the case of the kingfisher, also knowing things which we do not. (Though in some cases "pretheoretical knowledge" may be grounded in a small number of independent condition-action rules, corresponding broadly to Gibson's notion of perceptual "affordances," rather than in prelinguistic conceptual networks of the sort posited by Hayes.) A cat or monkey leaping from wall to wall, or branch to branch, needs some representation of *support* and *stability*, and diving animals need some grasp of the difference between *solids* and *fluids*, as well as of *depth*, *movement*, and *distance*. Chimps clearly have some grasp of notions such as *inside*, *obstacle*, *place* .... We will not be in a position to ascertain how much grasp, of which concepts, until we are clearer about the nature of these concepts in our own case. And this means knowing the perceptual evidence in which the concepts are anchored and the motor activities which test for them or which are carried out on the basis of conditional tests defined in terms of them. For example, newborn creatures who refuse to cross a "visual cliff" apparently have some innate procedure for recognizing the absence of *support*, where the object to be supported is their own body. It does not follow that they understand in any sense that the bottom bricks of a tower support the top ones -- although this is something which a leaping animal living in a jungle or an untidy house may have to learn. To understand a concept involves having some representation of the inferences that can usefully be drawn to link it with other concepts in the same general domain. *Support*, for instance, has something to do with *above* for leaping creatures who can recognize the potential for action in a pile of bricks. Hayes outlines some ways in which the core concepts of naive physics, and groups of cognate concepts, may be organized, so that inferential paths can be traced between them. His work is an intriguing beginning of a very important enterprise, which should help us understand *how* perceptual experience functions in the control of motor action.

A word of warning is in order here, however. Hayes is primarily interested in the human *Umwelt*, which is informed through and through by natural language. It is true that our earliest knowledge of naive physics is prelinguistic: the baby's sensorimotor understanding is prior to her acquisition of English or French. But it follows from Hayes' account of meaning that, once such natural languages are acquired, the meaning of the more primitive core concepts is altered -- not merely added to. In principle, then, even if we had a precise account of adult human knowledge of *inside*, *support*, and *behind*, we could not equate any part of this with the chimp's knowledge simply by jettisoning those parts of it influenced by our linguistic representations. Rather, we would need to be able to trace the development of our naive physical concepts, distinguishing their earlier, sensorimotor, forms from the later, linguistically-informed, semantic contents and inferential patterns. Hayes makes some relevant remarks, but even more apposite here is the computationally-informed work of the psycholinguists G. A. Miller and P. N. Johnson-Laird (1976), who have studied the basic perceptual procedures in which our linguistic abilities are grounded.

Miller and Johnson-Laird define a number of perceptual discriminations in detailed procedural terms, utilizing what is known about our sensorimotor equipment and development. They then show how these discriminatory procedures could come to function as the semantic anchoring of our lexicon. For example, perceptual predicates that can be procedurally defined include the following spatial descriptions: x is higher than y; the distance from x to y is zero; x is in front of the moving object y; y is between x and z; x has boundary y; x is convex; x is changing shape; x has the exterior surfaces y; x is included spatially in y; x, y, and z lie in a straight line; x travels along the path p. They give both psychological and physiological evidence for the primacy of these notions, and they use them to define object-recognizing routines of increasing power. Their sensitivity to computational issues leads them to ask not only *which* predicates are involved in a certain judgment, but *when* each predicate is applied in the judgmental process. (For example, the logically equivalent "y over x" and "x under y" are not psychologically equivalent: the first term in the relation should designate the thing whose location is to be determined, while the second should represent the immobile landmark that can be used to determine it.) The perceptual routines they define as the meaning of words such as "in," "on," "outside," and "at" are surprisingly complex.

Were a chimp to grasp the meaning of "in" or "on" in Ameslan, therefore, this would presuppose extremely complex perceptual computations on the chimp's part. And animals which, unlike chimps, have no great manipulative ability, would not be able to compute those perceptual discriminations requiring motor activities such as putting bananas inside boxes, so that their understanding of naive physics would be correspondingly impoverished. Von Holst's (1954) studies of reafference, and Held and Hein's (1963) experiments on visual development in kittens, suggested that many perceptual discriminations require active bodily movement: insofar as this is so, the creature could not substitute an understanding of "putting in " derived merely from watching others. (It is perhaps worth remarking that limbless thalidomide babies apparently reach a normal understanding of physical concepts: whether their natural language plays an essential part in enabling them to do so is not known.) Irrespective of chimps' potential mastery of Ameslan, the implication common to the work of Ullman, Marr, Hayes, and Miller and Johnson-Laird is that the perceptual and motor abilities of animals far lower in the phylogenetic scale than chimps must be based on representational competences of a highly complex kind. So an increased sensitivity to computational issues might help ethologists to investigate the symbol-manipulations carried out by different species, and to compare *Umwelten* in a systematic fashion.

In addition to empirical observations (about which, more in the following section), it may be that general results in the abstract theory of computation might help in this systematic comparison. If it could be shown, for example, that a given type of representation in principle could not express a certain type of information, or that it would be computationally enormously less efficient than some other type of representation, such insights might help guide the ethologist in attributing specific representational capacities to different animals.

For instance, abstract considerations show that computational mechanisms of a certain type (namely, "perceptrons," of which an example would be a nervous net with no significant prior structure) simply cannot achieve specific kinds of learning or spatial pattern recognition. Since it is abundantly clear that animal brains do have a significant prior structure, this result is somewhat academic from the point of view of the ethologist. But other results of this general type might be more relevant.

For example, in discussing what F. Rosenblatt (1958) had termed "perceptrons," M. L. Minsky and S. Papert (1969) claimed to show that certain mechanisms capable of performing some nontrivial computations are incapable of performing others which at first sight might appear to be within their range. Perceptrons are parallel-processing devices which make decisions on the basis of weighted evidence from many local operators, and various physiological examples have been suggested by cybernetically-inclined neurophysiologists interested in pattern-recognition and "self-organizing systems." Minsky and Papert sought to show, by way of abstract considerations alone, that no simple perceptron (without loops or feedback paths) could compute spatial connectedness, though it could compute convexity. Similarly, they claimed that no system without significant prior structure could in practice learn discriminations of high complexity, even given the existence of feedback paths.

Likewise, current work on "connectionist" systems (networks of parallel-processing units, wherein a representation is embodied by the pattern of excitation distributed across the network) has explored a number of different types of such systems: deterministic or stochastic, binary-valued or continuous. It has been proved, for example, that a certain sort of network (called a "Boltzmann machine" because it consists of stochastic units whose behaviour can be described by the equations of thermodynamics) can learn to represent any structure, given infinite time (Ackley, Hinton and Sejnowski, 1985). In practice, however, most real learning-problems (since they involve significant levels of noise) cannot be solved in a reasonable time by Boltzmann machines.

Clearly, results such as these are relevant to the representational capacity of nervous systems of different kinds, whether in the form of more or less complex nervous nets or of highly structured cerebral systems. Whether these abstract considerations can soon be brought into articulation with specific neurophysiological data is another question, since in only very few cases can we realistically hope to have an adequate (still less, complete) understanding of the neural connections within an entire nervous system.

Another suggestive example of abstract work that might throw light on issues of interest to ethologists is provided by John McCarthy (1969). McCarthy has long been interested in the representation of basic epistemological concepts (such as those discussed by his student, Hayes), and has recently embarked on what he terms "meta-epistemology," the attempt to define general representational or computational constraints on the sorts of mechanisms in principle capable of grasping particular notions. The account of perceptrons was in fact an early example of meta-epistemology, but was not conceived as part of an integrated research-programme directed to a wide range of representational systems.)

Again, A. Sloman (1978) has shown that "analogical" representations may be in various ways more computationally efficient than "Fregean" ones. He defines an analogical representation as one in which there is some significant correspondence between the structure of the representation and the structure of the thing represented. By contrast, a Fregean representation need have no such correspondence, since the structure of the representation reflects not the structure of the thing itself, but the structure of the procedure (thought process) by which that thing is identified. To understand a Fregean representation is to know how to interpret it so as to establish what it is referring to, basically by the method described by the logician Frege as applying *functions* to *arguments*. Analogical representations, however, are understood or interpreted by matching the two structures concerned (that is, of the representation itself and of the domain represented), and their associated inference-procedures, in a systematic way. Applying this distinction to our kingfisher cartoon, for example, the formula expressing Snell's law is a Fregean representation, whereas the diagram itself (with the lines representing the paths of light and constructing the relevant angles) is an analogical representation.

An example of the use -- and usefulness -- of analogical representation has been provided by B. Funt (1980), who has followed Sloman's suggestions by programming a system that can reason from visual diagrams. Funt utilizes the 2D space inherent in the hardware of the machine as an analogue of 2D paper, so that a diagram is embodied in the machine as a certain state of a 2D visual array, or "retina." The system's task, given a diagram like that of *Figure 4.2*, is to discover whether the arrangement of blocks depicted is stable, and -- if it is not -- to predict the movements (falling, sliding, motion ended by contact with another block) and the

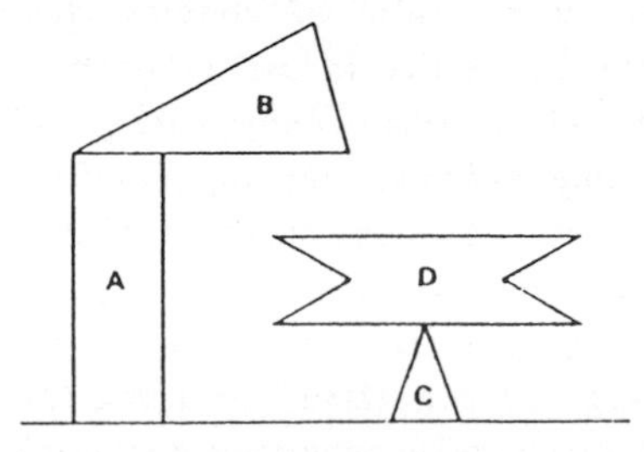

Figure 4.2

final state of the various blocks. The answers to these questions are discovered from the diagram (given certain simple diagrammatic transformations carried out by the system, which are structurally analogous to changes that would happen in the real world), rather than being computed in terms of abstract mathematical equations and specific numerical values.

Previous programs that could recognize stability or instability of putative block-structures did so by computing sophisticated equations of physics and quantitative parameters, and one such program (one of the planning programs previously mentioned) needed over 80% of its computational resources for these calculations alone (Fahlmann, 1974). What is more, those programs were unable to predict the specific structural changes that would follow on an instability. But, much as it is "obvious" to us from the diagram (though not from a verbal or mathematical description of the same state-of-affairs) that B will hit D, that D will then tilt with its left half moving downwards, and that B will end up touching both A and D but not the ground, so it is easily discoverable by Funt's program that this is what will happen.

Briefly, the program "imagines" gradual changes in the position of the blocks by exploiting the 2D nature of the retina in which the diagram is embodied. So for instance it imagines gradually moving an unstable block (such as B) downwards, pivoting on the relevant point of support. It studies "snapshots" of the successive positions, and so discovers specific points of contact with coincidentally present blocks (such as D) which will interrupt the fall that would have been predicted by a theoretical physicist from equations and measurements describing A and B. As in this case, many detailed relations between blocks are implicit in the diagrammatic representation which could be explicitly stated only with the greatest difficulty. To take another example of this advantage of the diagrammatic representation, consider the recognition of empty space. What space is initially empty, and what would remain empty after stabilization of the blocks, can be directly discovered from the diagram and the imagined snapshots. But previous "blocks-world" programs have had to rely on highly counterintuitive assumptions about empty space, and/or have had to make complex mathematical or logical calculations to deduce the empty space in the scene.

Funt's work is relevant to the topic of naive physics discussed earlier. He points out that the physical knowledge exploited by his system is comparable to that of the lay person rather than the physicist. Thus the system has simple computational procedures, or "perceptual primitives," which address the visual array in parallel so as to identify area, centre, point of contact, symmetry, and so on. These spatial notions are likely to be useful in many different problem domains. Also, the program has knowledge of *qualitative* physical principles relevant to its actual tasks, such as that if an object sticks out too far it will fall, and that it will pivot around the support point nearest to the centre of gravity. Moreover, since it is able to discover the empty space, and also those spaces that would remain empty throughout stabilization changes, it possesses a type of knowledge that would be crucial to an animal looking for a pathway or for a safe space through which to move. Leaping animals, at least those whose weight might cause significant changes in the terrain leapt upon, presumably have some understanding of support and of potentially dangerous or unstable structures. For instance, chameleons clambering in trees seem to be capable of making a number of these judgments, preferring thick branches to thin ones and avoiding blind ends or gaps. Experimental study might help show what types of instability various animals are able to recognize, and perhaps whether they are able to distinguish any class of scenes as the likely outcome of a specific sort of instability. Are they able, for example, to distinguish between unstable structures differentially likely to collapse onto a baby animal underneath?

Funt has shown that Sloman's distinction between types of representation can be exemplified in computational terms, but much as Sloman's work is suggestive rather than definitive, so Funt's work is exploratory only. Among its specific limitations, Funt mentions its total ignorance of velocities, acceleration, and momentum. He remarks that were these matters to be included in a future version of the program (which of course would enable types of prediction currently impossible to it), they would have to be represented in terms of equations. But it is not obvious that some useful qualitative distinction between *fast* and *slow* might not be available to some creatures incapable of formulating equations. Ethological evidence is in principle relevant to this question, but so also would be an abstract understanding of the computational power of Fregean and analogical representations. If principled results were to be arrived at within computational logic, expressing the advantages and disadvantages of these representational modes, we might be better able to understand the cognitive potential available at different points in the phylogenetic scale.

V. Objections: Validation, Parallelism, and Phenomenology.

The *Punch* cartoonist obviously recognized that, from the fact that *if* the kingfisher were consciously applying Snell's law its dives would be (as they are) appropriately placed, it does not follow that this is indeed the explanation of its diving ability. Analogously, if one had produced a computational model whose performance mapped onto the kingfisher's behaviour, one could not thereby be certain of having captured the bird's psychology. For, as is often pointed out by critics of cognitive psychology (Heil, 1981), there is always in principle more than one model capable of matching observed behaviour. However, this caveat is a special case of the general truth that *any* scientific theory is necessarily underdetermined by the evidence. That this underdetermination causes methodological problems is well-known to every practising scientist. Someone offering computational theories of animal behaviour would be no worse off on this account than any other psychologist faced with the task of testing theory against data.

The special difficulty is not how to choose between several alternative computational accounts, once we have got them, but how to arrive even at *one* in the first place. Psychologists and philosophers unfamiliar with AI typically underestimate the procedural-representational complexity of human and animal minds, and may not even realize that there are unsolved computational problems related to everyday psychological descriptions. That is, descriptions of perception and action are assumed to be unproblematic which in fact are deeply puzzling. Thus most ethologists take the existence of various interpretative and representational capacities for granted, and concentrate on asking which of these capacities are shared by which species. Theorists sympathetic to AI, by contrast, are primarily interested in how such capacities are computationally possible.

For instance, the experimental psychologists Premack (1978) and Savage-Rumbaugh (1978) have asked *whether* chimps can perceive the world as humans can, and do things which we can do. Can a chimp perceive a movie as representing a second individual trying to solve a problem, like reaching bananas or switching on a heater? Can a chimp plan ahead of time, either on its own behalf or on behalf of its fellow? Can two chimps cooperate in the solution of a task, perhaps using artificial symbols as publicly observable indicators of the tool that is required at a given stage of the problem?

Computationally-inclined psychologists or epistemologists, however, are more likely to be interested in *how* these things can be done, irrespective of which species manage to do them. How is it possible for a creature to perceive intentions, or to ascribe specific beliefs and difficulties to another individual? How is it possible for a creature to form means-end plans for reaching a desired object, plans within which other objects are represented as instruments to the overall end (Charniak and McDermott, 1984)? How is it possible for an external symbol, as well as one in the internal representational medium of the creature's mind, to be employed by one animal and perceived by another as a request for a specific tool? How is it possible for a creature to perceive apparent movement, or to distinguish visually between replacement, motion, and change?

It is this difference in theoretical focus which has led Sloman (whose work in analogical representation was mentioned above) to acknowledge the fascination of these recent studies of chimps, and yet to complain that such studies are premature:

> In the long run we shall all learn more if we spend a little less time collecting new curiosities and a little more time pondering the deeper questions. The best method I know of is to explore attempts to design working systems [i.e., programs] that display the abilities we are trying to understand. Later, when we have a better idea of what the important theoretical problems are, we'll need to supplement this kind of research with more empirical studies (Sloman 1978).

Ethologists may reply that they wish to discover *which* achievements are within the grasp of chimps, beavers, and bees. This is indeed a legitimate question: natural history should include comparative psychology, an account of what different animal species can do. Many such questions have remained unasked by professional ethologists, because of the inhibitory influence of behaviourism -- and even of the founding fathers of ethology, who were anxious to avoid sentimental anthropomorphism. But it remains true that a deeper understanding of animals' minds will require careful attention to the computational processes underlying their observed abilities.

However, perhaps there is a particular reason (over and above the a priori possibility of alternative theoretical models) for denying the relevance of AI ideas to animal psychology? Even in the human case, it may be said (Dreyfus, 1972), it is doubtful whether computations like those used in AI go on (except possibly during conscious logical or mathematical calculation). And animal minds, a fortiori, do not engage in this sort of computation (which is why the *Punch* cartoon would have been less funny if it had shown a human fisherman). Introspection does not reveal complex sequences of step-by-step formal reasoning. If anything, it suggests that many unformulated ideas influence our experience simultaneously. Perception in particular seems relatively immediate, and the notion that a bird's perceptions are laboriously constructed by strings of formal computations is absurd. What is more, the vertebrate brain appears to be capable of parallel processing, so programs written for digital computers are of questionable relevance to human or animal psychology. In short, the complexity of thought may be less than is assumed by AI-workers -- and, moreover, may be of a very different type. This objection is, in part, an appeal to ignorance. From the fact that a mental process does not appear in introspection one cannot infer that it does not go on at non-conscious levels of the mind. What is more, there is a great deal of empirical evidence (amassed over many years) suggesting that human perception is the result of a non-introspectible process of construction, a process that takes a measurable amount of time and that can be interfered with in specific ways. One might, of course, argue that all talk of non-conscious mental processes is so philosophically problematic as to outlaw cognitive psychology in general (irrespective of whether it uses AI-ideas) (Malcolm, 1971). This extreme viewpoint would deny to theoretical psychologists rights of extended language-use that other scientists enjoy (Martin, 1973). Short of this position, one must admit the possibility of non-introspectible mental processes.

But what of the objection that, while such processes probably do exist, they may be very different from the processes posited by AI? The first thing to be said in reply is a familiar Popperian theme: a clearly articulated hypothesis, which fails to match the facts in certain specifiable ways, can be a crucial stage in the development of a more satisfactory scientific understanding. So even if AI were incapable of modelling actual thought-processes, it would not follow that nothing of theoretical interest could have been learned from it. The second point is related to the first: the scientific research-programme that is AI includes a number of significantly different approaches. As the reference above to "parallel-processing" suggests, the logical-sequential approach is not the only possible form of a computational model.

Some very early work in AI attempted to model parallel processing, but the machines available were so primitive that little was learnt from this exercise. Most research in the field has concentrated on modelling logical-sequential computation, which is well suited to general-purpose digital computers. As a corollary, AI-workers have tended to play down the importance of neurophysiological knowledge about the brain. This is partly due to the fact that many significant computational questions can be pursued quite independently of hardware-considerations, but is also due to the fact that a "general-purpose" machine is precisely one whose hardware is capable of carrying out indefinitely many different types of computation. In recent years, however, neurophysiological and psychophysiological evidence has been taken more seriously by some AI researchers. Recent work takes a radically new approach to computation, wherein parallel processing by dedicated (as opposed to general-purpose) hardware is used -- or simulated -- in computing properties previously assumed to require highly abstract sequential processing.

One such property is shape, and it is pertinent to our topic to note that the perception of shape is being modelled in this way (Hinton, 1981). On this approach, a shape can be recognized as a whole without the constituent parts being recognized as such (a part is represented in a radically different way if it is seen as a Gestalt in its own right). We saw above that shape-perception is not needed for the recognition of object-identity, so that if we were to attribute identity-perception to a creature (perhaps because of its ability to follow a moving target) we would not thereby be justified in attributing shape-perception also. But, if we were to attribute shape-perception to a perceiver, what computational powers would we be crediting the subject with? If shape-perception required the application of high-level concepts, it would be implausible to say, for example, that a kingfisher can see the shape of a stickleback (which I described earlier as "a sort of narrow pointed ellipse with sharp projections on its upper surface", and which we could describe in many other ways). But this very recent AI work suggests that the bird might be able to perceive the shape of a stickleback despite being unable to represent it in terms of high-level concepts. Further, it implies that the kingfisher need not be able to articulate the image of the stickleback into independently recognized component parts: it could perceive a fish without being able to perceive a fin.

A computational model of this type that was initially developed for shape-perception is now being applied to the control of bodily movement (Hinton, 1984). The bodily skill of smoothly moving one's arm requires delicate compensatory movements in the various joints, as well as subtle control of velocities at different stages of movement. Earlier attempts to compute motor-control met with very limited success. For example, roboticists relying on sequential processing were unable to write programs capable of computing the subtly balanced flexions of shoulder, elbow, and wrist that would be necessary for smooth movement of a robot-arm. But these aspects of motor-control are now being modelled by these new AI techniques, with encouraging results. Many problems remain; for example, it is not yet known how to compute a path that will avoid an obstacle placed between the arm and the end-point of movement. But these new developments in AI should give pause to those philosophers (including phenomenologists) who complain that AI can have nothing to say about bodily skills, and so is questionably relevant to human and animal forms of life.

But if simultaneous perceptual processing and bodily skills are not wholly intractable to a computational analysis, what of experience itself? Can AI have anything useful to say about consciousness?

I remarked above that von Uexkull's pictures fail to express the phenomenological quality of what it is like to be a sea-urchin, fly, or dog -- or, one might add, a bat (Nagel,1974). Nor would different pictures have succeeded where these failed. As T. Nagel points out, the problem in understanding what it is like to be a bat rests on the difficulty of matching different subjectivities. Assuming (as we all do) that our experience is somehow intrinsically different from the bat's, how could we even conceive of what the bat's experience is really like -- that is to say, what it is like for the bat? No mere subtraction or addition of conceptualizable features could transform our own experience into the bat's.

Related points were made with respect to spatial perception, in the discussion of naive physics above. Since it is implausible to suppose that a creature's understanding of "inside" is independent of its manipulative abilities, a kingfisher's perceptual experience of containment would differ from a chimp's. Similarly, no dog could perceive a bone to be *inside* a box in the way in which human adults can, because a baby's sensorimotor understanding of spatial concepts is radically transformed (not merely added to) by the learning of language.

This difficulty in understanding different subjectivities casts doubt on the possibility of a theoretical phenomenology, and a fortiori seems to dash any hopes of a systematic comparative psychology concerned with the experience of animals. But at the end of his paper Nagel hints at the possibility of an objective phenomenology. Its goal would be to describe (at least in part) the subjective character of experiences in a form comprehensible to beings incapable of having those particular experiences. Structural features of perception, he suggests, might be accessible to objective description even though qualitative aspects are not. Unfortunately, Nagel gives no examples of what might be meant by "structural" features of phenomenology. Could AI ideas help to clarify these suggestive remarks?

Ullman's work is premised on the phenomenological fact that human beings can experience apparent movement in several different ways. From a subjective viewpoint, these differences do not seem to depend on linguistically represented concepts, and moreover are of such a general character that it is implausible to ascribe them uniquely to human perceivers. That is, if a creature's phenomenology has any dimension comparable to visual experience as we know it, we can intelligibly ask whether (and when) distinctions such as these are perceived by it. We can ask, for instance, whether it enjoys any or all of the experiences, "seeing the same thing moving", "seeing one thing being replaced by another", "seeing a thing of a particular shape" and "seeing a thing being transformed into another". Indeed, that such questions are intelligible is largely what is meant by saying that its phenomenology "has a dimension comparable to visual experience as we know it."

The relevance of Ullman's study is that he provides a theoretical account of differential (human) phenomenology that can be empirically investigated, and which if correct would explain how and why these distinct experiences arise when they do. This account is couched in computational terms, which are objective rather than subjective but which relate to what one might call the "structure" of phenomenology. That is, phenomenological distinctions such as those just listed can be intelligibly related to hypothetical underlying computations (whether Ullman's hypotheses are correct is not of central interest here). Moreover, "structural" relations between them can be clarified, by showing for instance that *this* computation is or is not a necessary prerequisite or accompaniment of *that* one. For example, Ullman's demonstration that computation of shape is not necessary for computation of identity gives theoretical support to the view that a creature might be able to experience identity without being able to recognize shape. So evidence (whether behavioural or biological) of a creature's ability or inability to perform certain computations could count as theoretical grounds for ascribing or denying experiences of certain types to it.

This is not to identify computation with consciousness. We know from the example of "blindsight" that "visual" computations can occur without any conscious phenomenology (Weiskrantz, 1977). Nor can one escape the difficulty that Nagel ascribes to all "reductive" theories of the mental, that their truth is logically compatible with the absence of any subjective aspect whatever. But Nagel himself -- like biologists and ethologists in general -- is content to take for granted that other creatures do have experiences. He does not require of an objective phenomenology that it provide a philosophical proof of this presupposition. Rather, he asks for what one might term a systematic study of the structural constraints on "seeing-as", a study which would illuminate our own subjective life as well as enabling us to say something about the experiences of alien creatures.

Again, Nagel gives no examples to show what sort of study this might be. Possibly relevant are computationally-influenced investigations of visual imagery that have sought to explain introspectively obvious but intuitively mysterious facts about our visual experience. For example, it has been found that some striking perceptual differences in viewing a wire-frame cube (including, for example, the ease with which certain mental images can be formed of it) depend on which alternative structural description of the object is assigned to it by the perceiver (Hinton, 1979). That is, an object seen as one sort of structure can be experienced (and imagined) in ways different from those made possible by seeing it as another sort of structure. This approach to visual perception illuminates the nature and generation of our own experiences, and could in principle provide theoretical grounds for saying that a bat, or a Martian, who applied a specific (objectively definable) structural description to an object would be more likely to experience a specific type of imagery accordingly.

In sum, computational ideas are in principle relevant to the psychology of animals, and to their phenomenology too. Counterintuitive though this may seem, AI might help us understand the intelligence and perceptual experience of kingfishers as well as kings.

References

Ackley DH, Hinton GE and Sejnowski TJ (1985) A learning algorithm for Boltzmann machines. Cognitive Science 9:147-169

Bobrow DG, ed. (1985) Qualitative reasoning about physical systems. Cambridge, Mass.: MIT Press. See also Artificial Intelligence 13:1980

Boden MA (1972) Purposive explanation in psychology. Cambridge, Mass: Harvard University Press

Boden M A (1977) Artificial intelligence and natural man. New York: Basic Books ch. 12 (2nd ed MIT Press 1986)

Charniak E and McDermott D (1984) Introduction to artificial intelligence. Reading, Mass: Addison-Wesley ch. 9

Collett TS and Land MF (1978) How hoverflies compute interception courses. J Comp Physiol 125:191-204

Davis R and King J (1976) An overview of production systems. In: EW Elcock and D M Michie eds., Machine Intelligence 8 New York: Wiley

Dreyfus H L (1972) What computers can't do: a critique of artificial reason New York: Harper and Row

Fahlmann SE (1974) A planning system for robot construction tasks Artificial Intelligence 5:1-50

Fikes RE, Hart PE and N Nillson. (1972) Learning and executing generalized robot plans. Artificial Intelligence 3:251-288

Funt BV (1980) Problem-solving with diagrammatic representations. Artificial Intelligence 13:201-230

Gibson JJ (1979) The ecological approach to visual perception. New York:Houghton-Mifflin

Griffin DR (1978) Prospects for a cognitive ethology. Behavioral and Brain Sciences 1:No. 4

Hardy SM (1984) Robot Control Systems. In T O'Shea and M Eisenstadt eds., Artificial Intelligence: Tools, Techniques, Applications. London: Harper and Row: 178-191

Hayes PJ (1979) The naive physics manifesto. In: D Michie ed., Expert systems in the micro-electronic age. Edinburgh: Edinburgh Univ Press. 242-270

Heil J (1981) Does cognitive psychology rest on a mistake? Mind 90:321-342

Held R and Hein A (1963) Movement-produced stimulation in the development of visually guided behaviour. J Comp Physiol Psychol 56:872

Hess EH (1956) Space perception in the chick. Scientific American 195:71-80

Hinton GE (1979) Some demonstrations of the effects of structural descriptions in mental imagery. Cognitive Science 3:231-250

Hinton GE (1981) Shape representation in parallel system. Proc Seventh Int Joint Conf AI (Vancouver):1088-1096

Hinton GE (1984) Parallel computations for controlling an arm. J Motor Behaviour 16:171-194

Hinton GE and JA Anderson eds. (1981) Parallel models of associative memory. Hillsdale, NJ: Erlbaum

Hobbs J and Moore RC, eds (1984) Formal theories of the commonsense world. Norwood, NJ: Ablex

Holst E von (1954) Relations between the central nervous system and peripheral organs. Brit J Animal Behaviour 2:89-94

Kohler I (1962) Experiments with Goggles. Scientific American 206:62

Lehrman DS (1955) The physiological basis of parental feeding behavior in the ring dove (*Streptopilia risoria*) Behaviour 7:241-286

Lehrman DS (1958) Effect of female sex hormones on incubation behavior in the ring dove (*Streptopilia risoria*). J Comp Physiol Psychol 51:142-145

Lehrman DS (1958) Induction of broodiness by participation in courtship and nest-building in the ring dove (*Streptopilia risoria*). J Comp Physiol Psych 51:32-36

Lettvin JY, Maturana HR, Pitts, WH and McCullough WS (1959) What the frog's eye tells the frog's brain. Proc Inst Radio Engineers 47

Lorenz K (1957) [Original 1937]. The nature of instinct: the conception of instinctive behaviour. In: C H Schiller, ed., Instinctive behavior. 129-175

Lorenz K (1977) Behind the mirror: a search for a natural history of human knowledge. London: Methuen

Malcolm N (1971) The myth of cognitive processes and structures. In: T Mischel ed. Cognitive development and epistemology New York: Academic Press. 385-3924

Marr DA (1980) Vision. San Francisco: Freeman

Martin M (1973) Are cognitive processes and structure a myth? Analysis 33:83-88

Mayhew J, Frisby J (1984) Computer vision. In T O'Shea, M Eisenstadt eds, Artificial Intelligence: tools, techniques, applications. New York: Harper and Row 301-357

McCarthy J, Hayes PJ (1969) Some philosophical problems from the standpoint of artificial intelligence. In B Meltzer, D Michie eds., Machine Intelligence 4. Edinburgh: Edinburgh Univ Press:463-502

Miller A, Johnson-Laird P N (1976) Language and perception. Cambridge, Mass: Belknap Press

Minsky ML, Papert S (1969) Perceptrons: an introduction to computational geometry. Cambridge, Mass: MIT Press

Nagel T (1974) What is it like to be a bat? Philosophical Review 435-451

Power RJ (1976) A model of conversation. Univ Sussex experimental psych lab: Working Paper

Premack D, Woodruff G (1978) Does the chimpanzee have a theory of mind? Behavioral and Brain Sciences 1:#4

Rock I, Tauber ES and Heller DP (1964 and 1966) Perception of stroboscopic movement: evidence for its innate basis. Science NY 147:1050-1052; and Science NY 153:382

Rosenblatt F (1958) The perceptron: a probabilistic model for information storage and organization in the brain. Psych Rev 65:386-408

Rumelhart DE and McClelland JL, eds. (1986) Parallel distributed processing: explorations in the microstructure of cognition. Cambridge, Mass.: MIT Press

Sacerdoti ED (1974) Planning in a hierarchy of abstraction spaces. Artificial Intelligence 5:115-136

Savage-Rumbaugh ES et al. (1978) Linguistically mediated tool use and exchange by chimpanzee. Behavioral and Brain Sciences 1:#4

Selfridge OG (1959) Pandemonium: a paradigm for learning. In: DV Blake and AM Uttley, eds., Proceedings of the Symposium on Mechanization of Thought Processes London H.M. Stationery Office 511-529

Sloman A (1978) Intuition and analogical reasoning. In: The computer revolution in philosophy: philosophy, science and models of mind. Hassocks, Sussex: Harvester Press. 144-176

Sloman A (1978) What about their internal languages? Behavioral and Brain Sciences 1:602-603

Sloman A (1979) The primacy of non-communicative language. In: M McCafferty, K Gray eds. The analysis of meaning. ASLIB and Brit Comp Soc

Sloman A (1980) What kind of indirect process is visual perception? Behavioral and Brain Sciences 3:401-404

Smith UK (1941) The effect of partial and complete decortication upon the extinction of optical nystagmus. J Gen Psych 25:3-18

Stratton GM (1896) Some preliminary experiments on vision. Psychol Rev 3:611

Stratton GM (1897) Vision without inversion of the retinal image. Psychol Rev 4:341

Sussman GJ (1975) A computer model of skill acquisition New York: Elsevier

Uexkull J von (1957) A stroll through the world of animals and men. In: CH Schiller, ed. Instinctive behavior: the development of a modern concept New York: International Univ Press 5-82 [Original 1934]

Ullman S (1979) The interpretation of visual motion. Cambridge, Mass.: MIT Press

Ullman S (1980) Against direct perception. Behavioral and Brain Sciences 3:373-415

Weiskrantz L (1977) Trying to bridge some neurological gaps between monkey and man. Brit J Psychol 68:431-435

Young R M (1976) Seriation by children: an artificial intelligence analysis of a Piagetian task. Basel: Birkhauser

Young RM (1977) Mixtures of strategies in structurally adaptive production systems: examples from seriation and subtraction. DAI Research Report 33 Dept AI Univ Edinburgh

AN EVOLUTIONARY EPISTEMOLOGICAL APPROACH TO THE EVOLUTION OF INTELLIGENCE

H.C. Plotkin
University College London

At the end of a lengthy survey of the comparative mental powers of man and other animals, Darwin (1871) concluded that "the difference in mind between man and the higher animals, great as it is, certainly is one of degree and not of kind." Some may wish to argue that differences of degree will lead to differences of kind, but almost no biologist would now question Darwin's main point, which was that man's mind, including those aspects that we call intelligence, has evolved. Such universal agreement, testimony to the power of the notion of evolution and the generality of the intuition as to what intelligence is, contrasts sharply with our lack of understanding as to the way in which this has occurred. In part this is a result of difficulties in interpretation of one of the disciplines that lies at the heart of the issue, viz. comparative psychology. In part it is because of the restrictive way in which the whole problem of the evolution of intelligence has been viewed. An evolutionary epistemological approach to intelligence sheds some light on both matters.

It shows, first, that adaptive behaviour, including intelligent behaviour, is shaped and controlled by a hierarchy of knowledge gaining and storing processes. This framework provides for the conceptual ordering of what previously had seemed to be refractory and counterintuitive comparative data. Second, the case will be made that intelligence itself is to be seen as a process of evolution. Intelligence is not merely a passively evolved trait of man and certain other animals, but is itself a dynamic focus of adaptive activity -- a causally significant part of the evolution of intelligent creatures. The aims of this paper, therefore, are both to recast the comparative approach, and to argue for intelligence as being both cause and consequence of evolution.

In the first part of the paper, some of the basic theoretical assumptions of evolutionary epistemology, together with certain aspects of the more specific model that I have been attempting to develop, will be outlined. The second part applies these principles to the issues of intelligence and its evolution. The third part is a brief appraisal of the implications of this approach to our understanding of certain aspects of human evolution, specifically the impact of the evolution of human culture on human intelligence.

NATO ASI Series, Vol. G17
Intelligence and Evolutionary Biology
Edited by H.J. Jerison and I. Jerison

Evolutionary Epistemology

Campbell's (1974) authoritative review has led to a burgeoning literature in evolutionary epistemology (see for examples Bradie, 1986; Callebaut and Pinxten, 1986; Riedl and Wuketits, 1987). There is a strong, and growing, philosophical contingent working in this area. What will be described here, however, is restricted to evolutionary epistemology as an approach to biology, and specifically to issues that bear on the problem of intelligence. The following is a necessarily restricted outline of the position that is detailed in Plotkin (1982; 1986) and Plotkin and Odling-Smee (1979; 1981; 1982).

In very general terms, evolutionary epistemology is the application of evolutionary theory to problems of knowledge. From such an extensive enterprise I will here cull a specific line of argument. This is that the capacity for knowing has not merely evolved, which it undoubtedly has, but that individual knowledge is the product of a process identical to that which is held by conventional theory to be the cause of evolution. But before enlarging upon this thesis, it is important briefly to place evolutionary epistemology in the context of other approaches to evolutionary theory.

The theory of evolution has changed significantly a number of times over the last two centuries. There is no reason to believe that the currently most widely accepted version, neodarwinian orthodoxy which is based on the notion that evolution occurs through changes in gene frequencies in breeding populations wrought by natural selection, is going to prove any more resistant to change than previous accounts. There is, though, an interesting difference in the nature of the pressure that is now being exerted on neodarwinism when compared to criticisms aimed at previous versions of the theory. The difference lies in the success of the theory -- it has become a victim of that success. Virtually all biologists now accept evolution as one of the most important theories of biology; and to many, it is the central theory of biology. As such, many lay claim to it as an explanatory device. When, therefore, some biologists, be they developmentalists, ecologists, molecular biologists, paleontologists etc., feel that they are not being adequately served by the theory, they seek to alter it in such a way that they are being so served. The result is that there is no one "correct" account of evolutionary theory. The nature of the theory that wins out at any time in the history of any science, or within a restricted group of scientists at a particular moment, is determined by its scope and the extent to which it encompasses phenomena that are held to be crucial at that time or by that group, including newly discovered or reinterpreted empirical findings.

This is how evolutionary epistemology is to be understood. It is an approach to evolution that has been fostered by biologists with a particular interest in cognition, intelligence and culture, and who feel that these are important phenomena for which no proper place has yet been found in evolutionary theory. It is cognitive science's way of placing individual mentation within a wider evolutionary framework. The best possible outcome that might be expected is an enrichment both of evolutionary theory and our understanding of cognition and culture.

For the purposes of this essay, I will argue the position through two main points. The first, already mentioned, is that the process by which knowledge or information is gained is universal and is a process of evolution. The second is that living things are complex, hierarchically organized systems, each level of which is defined by the presence of a knowledge gaining or evolutionary process.

The Evolutionary Analogy

In 1880 William James wrote an essay on "the remarkable parallel..(that)..obtains between the facts of social evolution and the mental growth of the race, on the one hand, and of zoological evolution, as expounded by Mr. Darwin, on the other" (Pp 441). James provided a lengthy argument on the similarities between the Darwinian conception of evolution as the outcome of natural selection acting on variant forms of a species, and the creative thinking of "great men", a manifestation, surely by any account, of intelligence. He concluded that "...new conceptions, emotions and active tendencies which evolve are originally produced in the shape of random images, fancies, accidental outbursts of spontaneous variation in the functional activity of the excessively unstable human brain, which the outer environment...selects...just as it selects morphological and social variations due to molecular accidents of an analogous sort." (p 456). The specific form of the analogy that James used, and which can be strongly criticized, is not at issue here. The point to be made is the general nature of the argument, which was that whatever the processes are that account for 'biological' evolution (adaptation and speciation), these same processes may help in understanding complex cognitive phenomena, such as creative thinking in man. This is what is meant by the phrase 'the evolutionary analogy'. James was not the first to use the analogy. Spencer had adopted a predarwinian version of it before the publication of "The Origin of Species". The use of Darwinian concepts in the analogy by T.H. Huxley, who drew the parallel between the struggle for existence between organisms and the struggle for existence within an organism of its developing parts, appeared over a decade before the James essay. But it was James who first used the analogy in its Darwinian form as the basis for an extensive analysis of thought and problem solving. Subsequently, scores of writers have employed it in this same context, including such diverse figures as Simmel, Baldwin, Piaget, Popper, Lorenz, Skinner, Simon and Campbell. It is an error to consider the analogy a recondite intellectual thread running barely detectably through the writings of a restricted and obscure group of theorists. It is also an error to think that it has been used in the same way by all who have written of it.

In general, it can either be used literally as an analogy. It is then purely a heuristic device by which what is known of the process of 'biological' evolution is used to illuminate and further our understanding of cognitive processes and cultural evolution by pointing to similarities and conformities, and suggesting further lines of analysis and experimentation. Or, what starts out as an analogy is hardened into an assumption of identity of process. For example, instead of saying that trial-and-error learning is like evolution in this and that respect, one says that trial-and-error learning and evolution are the product of identical processes. This, too, acts as a heuristic device by which both analogies and disanalogies between the various applications of the process can be established. However, it is also an assertion that all knowledge gain, which is the establishment of matching relationships between features of environmental order and aspects of organismic organization, contains at its core certain nontrivial, identical features of process, no matter how different the mechanisms embodying these might be, and no matter how different the forms of knowledge might seem. It is this strong form of the analogy, the assumption of identity of process, that is followed in this paper.

What then is this process? More correctly, it is a set of processes or subprocesses that can be described by a simple routine. Variants are generated (genes and genotypes in the case of 'biological' evolution), they are tested (by natural selection to stay with the case of 'biological' evolution), and the successful variants that have passed the test phase are regenerated (the further propagation of genes and genotypes). In the cognitive domain the variants that are generated are responses, associations, map references, inferences, solutions, hypotheses etc., depending upon the precise nature of the subset of cognitive processes being described. The test will be contiguity, excellence of fit to some deductive template, or some other form of match to an expectation or a forecast, depending again upon the nature of the cognition. And regeneration will take the form of the generation of further variants, some identical with those selected in the past and some entirely novel (cognitive mutations, in effect). The exact mix of old and new variants at the regenerate phase will vary again with the form of cognitive process being considered. In simple associative learning, for example, it will be very conservative with a low number of novel variants. Complex problem solving will have a much higher proportion of novel variants at the regenerate phase, which allows for what we commonly describe as "creativity" of cognition.

Not only can the analogy be used in this way in the cognitive domain of humans and other animals, but it can also be extended to artificial intelligence, and hence proferred as a truly general characteristic of intelligence. Dennett makes this point so well that he is worth quoting at length:

> "Some AI researchers have taken their task to be the simulation of particular cognitive capacities 'found in nature' .... but others take their task to be, not simulation, but the construction of intelligent programs by any means whatever. The only constraint on design principles in AI thus viewed is that they should work, and hence any boundaries the AI programmer keeps running into are arguably boundaries that restrict all possible modes of intelligence and learning. Thus if AI is truly the study of all possible modes of intelligence, and if generate-and-test is truly a necessary feature of AI learning programs, then generate-and-test is a necessary features of all modes of learning, and hence a necessary principle in any adequate psychological theory." (Dennett, 1981. p 80.)

Dennett's "generate-and-test", of course, is his shorthand description of a form of the evolutionary analogy. And it must be clear from his use of the analogy, as well as from the previous paragraphs, that the evolutionary analogy is not just applicable to "simple" forms of learning such as trial and error. The subprocesses of generate, test and regenerate are, it is assumed, essential to all forms of knowledge gain.

There is one further point that must be made about the analogy in its strong form. It follows from the assumption of identity of process that the analogy can be run in either direction. We can not only understand more about cognition by applying what we know about evolution, but the reverse holds as well. That is, we should be able to use an increasing understanding of the complexities of cognition to further knowledge about 'biological' evolution. Piaget was a powerful exponent of the use of the analogy "in reverse".

Some critics of the evolutionary analogy have argued that evolution is too simple a process on which to model cognition, especially human cognition. That criticism, however, makes the questionable assumption that evolution is a simple process. There is nothing known, or assumed, by evolutionists about 'biological' evolution that supports that assumption. Lewontin (1982) criticizes the analogy on other grounds. He argues that it perpetuates the error of contemporary evolutionary thinking that the environment "poses" problems for the phenotype which are then "solved" by the evolution of appropriate adaptations, whereas the relationship between organism and environment is more properly described as a "dialectic" in which organism and environment are locked into a close interrelationship with the phenotype being an active agent in the determination of the selection pressures that act on it. I think that Lewontin is correct in his criticism of evolutionary theory but incorrect in his criticism of evolutionary epistemology. Most cognitive scientists who use the evolutionary analogy start from the assumption that cognition is an active process and that knowledge is the outcome of both what is "out there" and what is inside the knower's head. Piaget's concept of equilibration typifies such a position, and as already pointed out, Piaget was a strong proponent of "running the analogy backwards". For Piaget, the dialectic of accommodation and assimilation, as he described it for human intelligence, also finds expression at more fundamental levels of evolution. It is precisely on such an issue that cognitivists have something of value to say to evolutionists.

A Hierarchy of Processes for Gaining Knowledge

As already noted, a criticism levelled from a number of different sources in recent years against the orthodoxy of neodarwinism is that it is limited in scope to genetic bookkeeping in breeding populations (Williams, 1986), in which shifting gene frequencies in the gene pools of breeding populations are accounted for by the effects of natural selection on the adaptive attributes of phenotypes. Apart from obvious claims as to the adaptive significance of attributes such as those in the psychological or sociocultural realm, the theory gives no substantial guidance when it comes to giving a causal explanation of events at the psychological or sociocultural level. It must be acknowledged that evolutionists like Williams would argue that that is not the role of a theory of evolution. But the obvious rejoinder is to ask why the population geneticist is in a privileged position when it comes to using biology's most powerful theory as a causal explanation, and the rest of us must do without such direct assistance and resort instead either to ad hoc postulates which allow us precariously to attach other biological phenomena to evolutionary theory, or simply to do without the theory.

An obvious alternative is to resist the limited claims of the current orthodoxy and somehow expand the theory. The most common way of doing this in recent years has been to use the concept of hierarchical organization (Eldridge and Salthe, 1985). This use of hierarchical notions by ecologists, systematists and developmentalists as a means of making evolutionary theory directly available to them accords well with evolutionary epistemology which, in its most recent form, has been cast in a hierarchical framework. Campbell's (1974) scheme, for example, envisages at least ten levels extending from the genetical to learning, thought and science. Here I will consider a four-level hierarchy. Both schemes are multiple-level models of evolution. The central notion of the model presented is that there is more than one evolutionary process; that each process defines a separate level in a hierarchical system; and that each is embodied in mechanisms that are sensitive to different frequencies of environmental change.

The four levels of the model are the genetic, the developmental, the individual learner and the sociocultural level. The model is based on two fundamental assumptions. The first is that the hierarchy gains knowledge, and that that knowledge is an important causal component of the form that adaptations take. The second is that the original, and hence most fundamental, level of knowledge gain and storage is the genetic. However, the rate at which this level can furnish adaptations to the phenotype is restricted temporally by the rate of generation turnover. That is, it is sensitive to events of only low frequency relative to the rate of generating turnover time. Changes of higher frequency cannot be detected by this level; and if such changes are significant in detrimentally affecting the fitness of an organism, then they become the selection pressures for the evolution of the subsidiary levels of the hierarchy, each of which is sensitive to higher frequencies of change. Each level of the hierarchy is thus a tracking device that is tuned to ever higher frequencies of change. For example, level three is able to gain knowledge of events that occur rapidly. Level four operates even more quickly. It takes longer to learn through one's own direct interaction with the environment (a level three process) than it takes to acquire the same information from a conspecific (a level four process).

The reason for these levels having the capacity for operating at increasing rates requires some amplification. The rate at which knowledge is gained in terms of the selection and propagation of adaptive phenotypic attributes that are wholly genetically determined is fixed absolutely by the parameter of generation time. Other variables may, and do, alter evolutionary rates, but none of these can make evolution at level one go faster than some upper limit that is set by that parameter. Levels two and three, by definition, comprise knowledge gaining and storing processes that, because they work by way of within-organism and not between-organism evolutionary processes, are freed from the limitation of generation time and they can furnish adaptations at higher rates. But can the argument be made for level four (socioculture) being a more rapid evolutionary process than level three (individual learning)? At first sight this seems unlikely for two reasons. First because level four, like level one, is partly dependent on the between-organism exchange of variants (there is a "mating system"); and second, because level four is dependent upon a particular form of individual (level three) learning, i.e. social learning. The first point, though, is nullified by the necessary existence of non-genetic information transmission between individuals which frees level four from the genetic generation time limitations of level one. The second is incorrect because although individual and social learning may occur at similar rates (which is not known, but it does seem to be a reasonable assumption), if learning processes are evolutionary processes, the central tenet of the model, then the rate at which a process of evolution can establish adaptations is not just determined by the frequency of variant generation, but also by the variables that Wilson (1985) refers to as the "basic equation of evolution". This is that "the rate of evolution within a population equals the number of mutations arising per unit of time multiplied by the fraction of those mutations destined to be fixed" (pp 155). Wilson's assertion can be applied not only to level one evolution, but to that of any level, including level four. At that level, knowledge is being gained and stored as a result of a dynamic interaction between more than one learner. Thus, in unit time, level four gives rise to more variant generation (and hence more mutations) than level three -- an animal that learns by observing the learned behaviour of another animal, is adding its own level three variants to that of another. Furthermore, the learning of the "other" animal has already been subjected to selection, and this selection of already selected variants will increase fixation rates as well. Thus, despite both levels requiring level three processes which have roughly the same variant-generation frequency, the overall rates of evolution at level four are higher than those of level three because of increased variant generation and mutation fixation rates that arise when two individual learners are sharing information that has itself been acquired by level three learning.

Given these four basic levels, within each level there are many possible forms of variants, and hence many possible ways for gaining and storing knowledge. This is a commonplace understanding. At the genetical level of a species, there are large numbers of genes and many genes occur in different allelic forms. At the developmental level, many possible developmental trajectories are potentially available to an organism, the ones being selected being a function of that organism's genotype and the nature of the environment in which development occurs. At level three, the cognitive level, there are many forms of learning and within each form, say simple associative learning, there are many kinds of separate associations that can be stored. And at level four, the socio-cultural level, the common assumption is of a very rich network of knowledge when this level is present in a species like man. Across these levels there exist many possible combinations of the variants of each level. I will refer to the vertical organization of the model as "modules". Each module is a hierarchically organized set of knowledge gaining processes. These modules may partially overlap, i.e. share variants, at some levels, and be independent at others. Thus no assumption is made about the independence of modules. All that this means, for example, is that two level 3 learning mechanisms (motor skills and spatial mapping, say) may share certain genetic and developmental bases.

Not all modules extend over all levels, but as is implicit in the hierarchical nature of the model, all must cover at least two levels. Some will extend across three levels, and in rare instances all four levels are covered. Whether a hierarchically organized module of knowledge gaining processes can exist independently of the genetical level is an important point of difference between various approaches to cognition and intelligence. The model adopted here assumes that the genetical level is always implicated.

The conception, then, is of many possible modules operating across various numbers of levels. The constant feature is that all variant-generating mechanisms at each level gain knowledge by the same generate-test-regenerate routine, albeit with differing mixes of old and new variants at the regenerate phase. Thus, for example, the development of orientation receptors in the visual system of a cat is the product of knowledge gained at both first (genetic) and second (developmental) levels. That same cat, which learns the spatial position of a resource, will be using knowledge processes at three levels, the third level taking the form of a spatial learning device. If and when that cat enters into social alliances with other cats such that it is able to gain knowledge of the world through the prior experience of those conspecifics, four levels are in action. The crucial difference between the third and fourth levels is that the former is a within-animal knowledge gaining device, and the precise knowledge gained cannot be transmitted to others. All that can be transmitted to others, genetically, is the adequacy of the device. The fourth level, on the other hand, is defined by the extragenetic transmission of knowledge between phenotypes -- they learn from each other and the knowledge gained may be quite specific.

A hierarchy of processes that gain and store knowledge about the organism and its world, which is a hierarchy of evolutionary processes, has some interesting characteristics. Limitations of space allow these only to be stated, their development being available in the references cited above. For one thing, it directly brings into the purview of evolutionary theory classes of phenomena previously left out. Development is the classic example of a central area of biology that, until recently, has been conceptually separated from evolution by most evolutionary biologists. The model presented here is just one of a number of attempts that "allows us to analyse a very simple array of separate but interacting causes constituting an hypothesis of a comprehensive evolutionary mechanism" that allows for exploration of "the possibility that asymmetries in the introduction of variation at the focal level of individual phenotypes, arising from the inherent properties of developing systems, constitutes a powerful source of causation in evolutionary change" (Thomson, 1986 p 221 and 222). The model presented here goes further in including level three (individual cognition) and level four (socioculture) as additional and significant sources of variation in evolving systems.

A second feature of this approach is that it provides an explanation for behavioural and psychological predispositions which, whilst well recognized, have been difficult to understand in the light of other theories of learning and cognition. What a hierarchical model does is provide a framework for understanding how knowledge, which appears as a priori at one level, has been gained a posteriori at another level. Third, and closely related to the issue of psychological predispositions, it begins to resolve the age-old nature-nurture issue, arguably the central problem for all the social sciences. It does this not by having recourse to the pseudo-explanatory device of "interactionism", but by an entirely different kind of conceptualization of nested processes, hierarchically organized. This resolves the seeming paradox of the innate capacity to learn only certain things; and it eliminates the self-defeating dichotomizing into nature and nurture that is inherent in the notion of "interactionism". Fourth, because the scheme allows adaptive load to be shared across levels, we need no longer feel impelled by our own narrow areas of expertness to place the burden of explanation for any and all adaptations at just one level.

Finally, and closely related to all of the previous points, evolutionary theory becomes more complicated, if more elegant as well, as an explanation suited to the massive complexity of living systems. The notion of cause is no longer restricted to just one (genetic) level with natural selection as the only filter. Each level is defined by the existence of an evolutionary process; each level has a selection device; and each level has a degree of causal autonomy relative to other levels.

Intelligence and Evolution

In order to consider the implications of the scheme presented in the first part of this paper to our understanding of the biology of intelligence, one must make an additional, linking, assumption and then spell out some of the rules by which the scheme might be thought to work. The assumption is that "intelligence is an adaptation" (Piaget, 1936), which allows us to invoke the multiple level model as an explanation of the evolution of intelligence. However, intelligence is a complex adaptation formed by knowledge gained from at least three levels of the hierarchy, and in many cases, spread horizontally across many modules, particularly at the third (cognitive) level. Making sense of this complexity requires some statement of the rules by which the system works. Such rules, in effect, are hypotheses about how the system operates. They must be translatable into minimal empirically testable claims.

Some rules by which the model works

The first rule is an optimizing assumption or trade-off rule common to many approaches to adaptation, viz. the maximizing of benefits and the minimizing of costs. We seek a mix of fast, reliable and cheap knowledge gain and utilization. For the hierarchy of knowledge gaining and storing processes just described, this takes the form of a trade-off between (1) levels tracking at the highest frequency necessary for a particular interaction between the organism and its environment, (2) those levels least prone to sampling errors and (3) those levels least subject to spontaneous changes of state of storage over the time span for which storage is required, and (4) those levels which are metabolically least expensive. The levels that provide the maximum trade-off of these four factors will be the levels that gain the knowledge for the formation of the behavioural adaptation that the particular organism-environment interaction requires.

The second rule, already partially stated, is specific to the multiple level model. It is that the dimension of increasing frequency of change to which the levels are tuned corresponds to a dimension of increasing proneness to errors of sampling and to spontaneous errors of store, and also to increasing metabolic cost. This rule means that although the genetic level is the slowest level of knowledge gain, it is the least subject to errors, the nature of the store is reliable relative to storage-time requirements, and it is metabolically cheap. The higher levels in the hierarchy operate faster, are more prone to errors of sampling and storage, and they involve the metabolically expensive central nervous system. Also, the greater the amount of knowledge that must be gained at these high frequencies, the larger that organ system must be.

These first two rules relate to the vertical distribution of knowledge gain across the levels of the hierarchy. The third rule is concerned with the horizontal or within-level, organization of the devices that generate the variants making up the modules of the hierarchy. It is a rule of inertia which asserts that the modules that have already evolved, and the variant generators of which they are composed, will be used for as many forms of knowledge gain as possible. New forms of variant generators and selectors, or new combinations of these comprising new modules, will evolve only if the costs of additional variance generators and selectors are consistently outweighed by benefits in the form of more efficient behavioural adaptations.

B. Some Implications of the Rules for Intelligence

Sharing Knowledge Across Levels

According to the model, intelligence manifests itself as knowledge gain and utilization at the third level of the hierarchy. However, the interpretation of the model given here implicates the first and second levels in every instance of intelligent behaviour. More than that, rules one and two mean that the levels that will gain and store the knowledge for the construction of adaptations, behavioural adaptations in this specific case, will always be the most fundamental possible. In other words, if an adaptation can be constructed by just the first and second levels, then it will be. This is, in principle, a firm empirical prediction, even if it is not original with this approach. G.C. Williams, working on "the principle of the economy of information", urged in 1966 that "all elements (of behaviour) that can be instinctive...will be instinctive". The prediction is firm only in principle because, for the most part, we do not know how to partition the world, and hence changes in that world, relative to an organism that is interacting with that world. For the moment, therefore, the calculations must be made at an intuitive level, but they are not difficult to understand.

Climatic changes, for example, if occurring at a very low frequency (such as ice ages) will lead to adaptations resulting from knowledge gained at just the first, genetic, level with possible minor tuning by the second, developmental, level. This is because all organisms have a generational turnover time that is very much less than the periodicity of such major climatic events. The periodicity of lesser seasonal changes, however, raises the ratio of the frequency of environmental changes to generational turnover time, and hence spreads more evenly the burden of the load of knowledge gain across first and second levels. Many insect species, for example, show facultative polymorphisms as a function of environmental conditions during development (Barrington, 1979). Such developmental effects occur because the conditions eliciting them are environmental fluctuations with a frequency higher than can be detected by the first level acting alone.

All of this may seem remote from intelligence, but it leads to the empirically testable assertion that intelligence, even once evolved, will not gain knowledge necessary for the adaptive solution to a problem posed by a changing world if those changes occur at a frequency that is detectable by the first and second levels. But why, if the third, cognitive, level already exists, should it not operate to detect and acquire knowledge about changes occurring at any frequencies lower than its upper operating limit? The answer lies partly in the rules given above as they operate in systems over long periods of time. More fundamental levels are cheaper and more reliable. However, there is also an answer in the immediate operation of a system that has evolved level three. This is that the third level cannot detect changes occurring at very low frequencies. There is the difficult borderline area of changes occurring with a frequency near the upper limits of the second level, but at the risk of labouring the point, it is an empirically testable matter. (The inability of a high frequency operating level to detect events occurring at frequencies lower than the operating characteristics of that level is probably a sufficiently general condition for it to be stated as a fourth rule by which the model works -- in this case, a rule concerned specifically with the gaining of knowledge once level three has already evolved. Level four, however, may not be governed by this rule because its "exosomatic extension" in man provides the ability to detect changes of very low frequency. The successive nature of ice-ages is known to science alone.)

Climatic effects seem reasonably easy to handle at a common sense level within the framework of the rules by which the model is held to operate. There are, however, many other forms of change that present us with real difficulties as to how to partition organism-environment interactions in terms of rates of change and the levels that operate to gain knowledge about such change. Prominent here are events that are "organism induced". Simple locomotion is a good example of such difficulty. An animal that alters it position in space in a non-uniform environment is generating change that may be significant to its survival. In some cases, level one tracking devices manifested as reflexes, kineses etc. will supply the appropriate adaptive behaviours. In others, a memory for spatial location operating at level three seems to be required. Furthermore, interactions with other forms of change, e.g. changing climatic events induced by migration, may pose very difficult problems in terms of the frequencies at which the organism is experiencing change. Although these are technical problems for the model, they certainly do place formidable obstacles between the model and its empirical testing.

Is Intelligence a General or a Species-typical Trait?

A feature of the model is that it goes some way to solving what has come to be called the "constraints on learning" problem. Over the last two decades a small storm has raged amongst interested biologists and psychologists as to the extent to which learning is to be viewed as a species-typical trait. For much of this century, learning theorists have been concerned to establish general principles of learning and have been either thoughtlessly or deliberately unmindful of possible ecological constraints on learning. Classical ethologists, on the other hand, assumed that learning must always be to some degree constrained by species-typical factors that have their roots in the ecology of that species, a view that they felt was vindicated by laboratory studies of learning the results of which began to be published from the late 1960s onwards, and which showed powerful constraints on whatever general processes might be shared by learners. The general process theorists responded by "stretching the parameters" of their theories in order to take account of such species-typical learning phenomena as were being reported. The proponents of the constraints were labelled biological boundaries theorists and were accused by the general process theorists of "loose speculation about adaptive significance" (Bitterman, 1975) and of being "neo-evolutionary learning theorists (who) aggrandize minor science, the study of the particular, at the expense of the extremely important, the study of the general" (Revusky, 1977).

The important points about the argument are, first, that there are now few who dispute the data as demonstrating that learning may display species-typical features; and second, it highlights the theoretical weakness of both general process theory, which was never constructed with possible species-differences in mind, and the biological boundaries theorists who merely waved their hands at the notion of adaptation as an explanation of their findings. Simply calling a phenotypic trait an adaptation is not an explanation, however, and that is what gave so much sting to Revusky's comments about minor as opposed to major science. Listing adaptations is an instance of the former, which is what the biological boundary theorists were doing. Constructing a theory of adaptations that would account for the learning constraints would be the latter, and that the biological boundary theorists did not do.

The model of a hierarchy of knowledge gaining and storing devices operating on the basis of selectional processes, as stated in the first part of this paper and according to the rules stated above, is a theory of adaptation that both predicts and explains the constraints on learning at both a general (i.e. constraints will always exist) and specific (i.e. the nature of the constraints) levels. Consider the well known example of nausea conditioning in the laboratory rat, a form of learning that shows certain characteristics that are different from other forms of conditioning in the same animal. One of the curious features of this learning is that it appears to be constrained by the a priori knowledge that it is the taste and smell of food that is to be associated with the illness, and not where the food is eaten. How can this be explained? In terms of our model, the relationship [food--- poisoning--- nausea and other symptoms of illness] is one that changes only very slowly. The knowledge of those relationships would be supplied by the first, genetic, level. The first level cannot, however, detect the relationships [which specific foods--- what specific tastes --- illness]. Here the changes governing these relationships occur at a higher frequency than can be detected by level one, or even level two. If food poisoning contributes significantly to the fitness of certain kinds of animals, then knowledge of that specific, relatively high frequency relationship of taste, food and illness can only be acquired by a knowledge gaining device that can operate at the appropriate frequency, which in this instance is level three. We therefore expect the total behavioural adaptation to be partitioned across levels: [food may lead to illness] is knowledge gained and stored at the first level in the form of instructions for "wiring" the central nervous system such that only the tastes of food can be easily associated with nausea; the wiring, however, is in the form of an "open programme" which means that it can acquire the more specific information [the salient characteristics of the particular foods that lead to illness] by the operation of level three.

The multiple level model not only postdictively explains the constraints on learning within the framework of a general theory of adaptation, but it predicts that all level three knowledge gain is nested under, and hence constrained by, levels one and two. According to the model, tabula rasa learning is impossible because the third level evolved out of a failure of the first and second levels to deal with changes occurring above a certain frequency; and having evolved out of this failure of levels one and two, it is rooted within and constrained by those more fundamental levels of the hierarchy. Hindsight, of course, is a weakness in science. Nonetheless, if the constraints on learning were not already an empirical certainty, the model would have predicted their existence.

There is an obvious counter scenario on the evolution of intelligence, and when intelligence operates, to the one presented above. This is the argument that even if intelligence has its origins within the kind of explanatory framework presented here -- or however it might have evolved, once in existence it may become detached not only from its original function but from any specific function at all: it is an all-purpose device for gaining knowledge of any events that occur with a frequency less than some upper, and more than some lower, limit. Furthermore, the argument goes, certain "contextual variables" such as differences in perception, motor capacity, motivation, and so on then determine the differences between different species within major taxonomic groups, all of whose constituent species share this free-floating intelligence. This view is presented in extreme form by Macphail (1985), who argues that there are no differences in intelligence between vertebrate species, with the exception of man.

Like the constraints on learning controversy, this is not an issue that is likely to be settled empirically because of the way the case has been framed. For this reason no attention will be paid to the literature that claims to present empirically validated differences in intelligence between different species of vertebrate. We need to consider the essence of the case itself. Take the extreme instance of a chimpanzee and goldfish comparison. In order to prove a difference in intelligence between these two species, we have to devise test procedures that do not advantage the chimpanzee and disadvantage the goldfish because "many of the most impressive achievements of the chimpanzee have involved the use of its limbs, and this poses for comparative psychologists the challenge of devising for vertebrates without comparable limbs tasks whose formal intellectual demands parallel those made in the tasks mastered by chimpanzees" (Macphail, 1985. p 46). This passage highlights the difference with the position adopted in this paper. The multiple level model of knowledge gaining processes considers the manual dexterity of chimpanzees not as a contextual variable but as one of the primary sources of chimpanzee intelligence. The argument goes as follows: chimpanzees have a high degree of manual dexterity. This allows them to manipulate and change their immediate environment at a high rate. The knowledge gaining processes of levels one and two, which gave rise to the manual dexterity in the first place, cannot track the rates of change which that selfsame trait generates. The chimpanzee, therefore, has evolved modules of knowledge gain that extend to the third level and hence can operate at the required frequency -- but the changes that are being assimilated and accommodated to are specific to hand-use. Because of such dexterity, some of the intelligence of the chimpanzee is concentrated around that skill. In the language of this paper, the chimpanzee has a nervous system with a high concentration of variance generation in the sphere of manual dexterity. Thus, if one removes the possibility of manual-original-dextrous-related variance generation from the chimpanzee, it may indeed appear to be less intelligent. This will be because it is less intelligent. But under normal, "ecologically" proper, circumstances manual dexterity (more correctly manual-visual dexterity) is an important part of that animal's interaction with its world. It is not appropriate to label it a contextual variable.

This line of reasoning applies also to man, whose intellectual differences from other vertebrates Macphail does allow. A considerable amount of human intelligence is related to language. This is presumably because the human brain is a prodigious generator of variants in the linguistic sphere. If this is a correct argument for humans, then it must also be correct for chimpanzees. Indeed, it must generalize to the specialized adaptations of all animals. The foci of intelligence in any intelligent species are the specialized forms of interaction with, and adaptation to, its environment that that species has evolved. This position, which is a direct implication of the model, is the polar opposite of the notion that intelligence is some kind of unattached adaptive potential. This line of argument does not, however, confine intelligence only and directly to its original source of adaptation. A translation from the original source of variants to a different, perhaps higher order, knowledge form is not ruled out by the model. Thus the visual skills of the monkey might form the basis of volume conservation (e.g. Pasnak, 1979), or the manipulatory skills of the chimpanzee might lead to analogical reasoning (Gillan, Premack and Woodruff, 1981). Nonetheless, the "original" variant sources will lie elsewhere, and this implication is again empirically accessible. Developmental studies are an obvious first choice for empirical verification.

The question as to which came first -- manual dexterity or intelligence in the chimpanzee, language and/or manual dexterity or intelligence in man -- may not be within empirical reach. It is not clear, though, that it makes any difference to the general argument being presented here. Indeed, as will be returned to below, it is likely that synergistic relationships hold between, say, manual dexterity -- intelligence -- greater manual dexterity -- increased intelligence etc. Whichever came first, level three is an evolutionary process requiring the generation of variants. Those variants cannot be generated in a vacuum -- they must be variants of something; nor can they be so multifaceted as to be adaptable to any adaptive demand made on an animal. To the contrary, variants are specific to pre-existing level one and level two adaptations. The model admits of no tabula rasa intelligence.

Some Problems of Comparative Analyses

There are two complicating factors in assessing comparative studies of intelligence.

The one is that intensive field studies of animal behaviour reveal ever more startling complexities, often from unexpected quarters. Recent evidence of maternal care in crocodiles (Pooley and Gans, 1976), tool use in Egyptian vultures (Lawick-Goodall, 1970), and prodigious feats of memory in several species of food-hoarding birds (Shettleworth, 1983) suggests that we are only just beginning to understand how little we know about complex behaviour in animals and its phylogenetic distribution. However, behavioural complexity is not in itself a measure of intelligence. That is the second complication. Impressive feats, such as migratory behaviour in various species of teleost fish, reptiles and birds can be largely accounted for by the operation of levels one and two alone. Indeed certain bizarre features of the migrations of eels and turtles can only be understood in terms of changes (in the form of continental drift) occurring so slowly that the behaviour is controlled by the low frequency detecting level one, to the point where, from a level three or four vantage point, it possesses features verging on the maladaptive.

Given these difficulties, how then should comparative studies of intelligence proceed in the light of a multiple level model of evolution? The shift away from arbitrary comparisons between convenient species using laboratory learning paradigms as tests of intelligence towards a comparative science based on substantive and explicit conceptual frameworks is one of the major achievements of behavioural biology in recent decades. One of these, the "ecological" study of learning (Johnston and Pietrewicz, 1985), places the problem firmly into the context of adaptational theory. Learning is seen as a process by which a better adaptive fit is gained between specific behaviours in relation to particular ecological demands. This is an approach that is obviously congruent with a multiple level model of evolution. But the latter goes beyond the view that intelligence is an isolated source of adaptive behaviours and hence any comparative study based upon it will be wider than establishing intelligence within just an ecological context.

Intelligence, it must be repeated, is defined in this paper as the operation of a subsidiary knowledge gaining level in a hierarchy of knowledge gaining processes, that level being capable of detecting changes in the environment, or of animal-environment interactions, that occur above a certain frequency. This frequency is not absolute but relative to fundamental species attributes such as longevity, developmental characteristics, mobility and so on. Thus intelligence, as a trait to be compared across species, is determined both by general life history strategies and very specific attributes such as olfactory acuity or manual dexterity.

Not only does a multiple level model of evolution provide a wider context for comparing intelligence across species, but it also raises the possibility of making quite specific quantifiable predictions about which levels will be involved in the evolution of adaptive responses to which features of the environment. The essence of the approach is to partition the environment and animal-environment interactions into sets of features "banded" in terms of their frequency of change. These "bands" must be matched by the generation of variants at a similar frequency from either level one (genetic), level two (developmental) or level three (intelligence).

In brief, what is required is (a) a detailed description of general life history strategies of an intelligent species (for example, longevity, period of ontogeny as a percentage of life expectancy, and reproductive strategies); (b) information as to specific phenotypic attributes (motor abilities, sensory acuities, dietary requirements); and (c) an account of environmental features (distribution of food resources, predators, conspecifics, and weather patterns, for example), especially their rates of change. Given all this, there is a potential for quantitatively derived predictions as to which behaviour is adaptively shaped by genetic mechanisms, which by development, and which by the operation of cognitive processes. The range of detailed knowledge that is required is very considerable, and it must cover numbers of different species if the model is to have any comparative use. The payoff is a comparative approach by which intelligence is seen within the more general biological economy of intelligent creatures.

Intelligence and Human Evolution

Intelligence, as depicted here, is a form of evolution. It has, of course, also evolved. The relationship of level three to levels one and two is one of nested subordinacy -- level three (intelligence) must act in support of level one (genetical) evolution in the long term. It need not in the short term. But over long periods of time, maladaptive intelligence will be selected against. Because level three operates to gain knowledge of events occurring with a relatively high frequency and hence is the source of behavioural adaptations to rapid environmental fluctuations, the overall adaptive competence of intelligent creatures is increased, and this may extend the range of niches which they can occupy. Extended niches must mean that novel selection pressures are confronted, and such novel selection may exploit previously unexploited and unselected genetical variants. The net result of the nesting of these evolutionary processes is that the overall rates of evolution, both at levels one and three, of an intelligent species must be increased. This is a version of Waddington's (1959) exploitive system, i.e. the possession by certain species of animals of cognitive mechanisms whose causal status in evolution is coequal with that of genetical, natural selective and developmental systems. It is also an empirically testable assertion, work on which is reported by Wyles et al (1983). They demonstrated a correlation between brain size, which they took to be a measure of the capacity for behavioural innovation together with the social propagation of acquired information, and rates of anatomical evolution in certain land vertebrates.

The same reasoning should apply to man as to other intelligent species. This is especially so given that man is not merely an intelligent animal but one that has evolved to a unique degree the fourth (sociocultural) level of evolution. As in the case of the other levels, the fourth is subordinate to more fundamental levels. But unlike level three, the rapidity with which knowledge is gained and utilized at level four is a consequence of a much greater degree of decoupling between level four and level one -- level four by-passes level one transmission routes, and at least in that limited sense is, as frequently noted, Lamarckian in character. Just as other levels mutually enrich the potential for each to evolve further, so too does level four (sociocultural evolution). Its special characteristics, however, makes a difference in two ways. The one is that whilst a synergistic relationship may well operate between levels one and four with level four "revealing" to new selection pressures previously unselected genetical variants, it may well evolve compensatory shifts at level four before level one is implicated. The other concerns the source of variants of level four, and which gives level four the high rate of evolution relative to level three ascribed to it in the first part of this paper.

Consider again the notion that the environment can be banded in terms of frequency of change into sets of features, E^a, E^b and E^c, where E^a are features of environmental order that change very slowly, E^b are features that change more rapidly, and E^c are events with high rates of change (remember that these frequencies are relative to certain important aspects of the life histories of individual phenotypes that characterize any species; notably longevity, generational turnover time and duration of ontogeny). Variant sets V^a, V^b, and V^c are generated at matching rates by level one (genetic), level two (developmental) and level three (intelligence) evolutionary processes such that some limited number of variants are selected from each level and enter into adaptive relationships E^a - V^a, E^b - V^b, and E^c - V^c.

In each case, the variants that are generated do not become features of the environment for other levels in the hierarchy. Thus no V^a (genetical variants) enter into E^b; and no V^b (developmental pathways) enter into E^c. This partitioning of variants at one level from environmental events at another does not, however, hold for levels three and four. Some V^c variants become the environmental features, EV_c^c, that enter into an adaptive match with variants V_c^c, to form EV_c^c - V_c^c. That is, sociocultural evolution involves the products of intelligence of one individual as part of the environment that has to be adapted to of another intelligent individual. It is this relationship of two, or more, intelligent individuals to one another -- where one intelligence becomes the "environment" of the other -- that provides levels three and four with such dynamic characteristics when compared to level three acting alone.

It is not difficult to maintain the distinction between levels three and four when one is comparing two species, one of which has only evolved level three but not level four, and the other has evolved both. But that distinction is less easy to maintain, and is very poorly understood within a species, like man, that has evolved both. There are no criteria available by which it is possible to know which V^c can become EV_c^c, and those V^c that cannot become EV_c^c. For example, it is conceivable that the individuals of one species may acquire via social learning information as to the position of a food resource, but cannot, except by individual learning, acquire knowledge as to whether that resource is toxic or not. In such a case, the domain of $V^c \dashrightarrow EV_c^c$ is rather small. In man there is an ability to share a much wider range of knowledge, i.e. the domain of $V^c \dashrightarrow EV_c^c$ is very large. This is a significant species characteristic of humans.

As the domain of shared variants at levels three and four increases, so must the synergism between them. Increases in one will be reflected in increases in the other, a relationship that must express itself as an increase in neural information processing capacity. Here, perhaps, lies the key to understanding the rapid evolution of brain size and intelligence in the genus Homo.

The mutual driving relationship between intelligence and culture in man is a view frequently expressed by social philosophers and social engineers. It is usually held to be a "soft" sentiment with little empirical support and no foundation in "hard" natural science. A brief attempt has been made to show here how that relationship can indeed be derived from fundamental principles and theory. The extent to which the evolutionary model of intelligence described in the preceding pages successfully bridges that gap between "hard" and "soft" science is a measure of the potential that evolutionary epistemology has for synthesizing data and theory across the divide of the social and biological sciences.

Acknowledgements

I am grateful to Celia Heyes, Harry Jerison, John Odling-Smee and Michael Ruse for reading and commenting on earlier drafts of this paper. My thanks also to the conferees of the NATO ASI at Poppi, Italy, some of whom presented me with interesting and demanding comment on the model presented above.

References

Barrington EJW (1979) Invertebrate Structure and Function. Nelson London

Bitterman ME (1975) The comparative analysis of learning. Science 188: 699-709

Bradie M (1986) Assessing evolutionary epistemology. Biology and Philosophy 1: 401-459

Callebaut W, Pinxten R (eds) (1986) Evolutionary Epistemology: a Multiparadigm Program. Reidl Dordrecht

Campbell DT (1974) Evolutionary epistemology. In PA Schilpp (ed) The Philosophy of Karl Popper. Open Court Publishing Chicago

Darwin C (1871) The Descent of Man and Selection in Relation to Sex. Random House New York

Dennett DC (1981) Why the law of effect will not go away. In Brainstorms, MIT Press Cambridge Mass. pp 71-89

Eldridge N & Salthe SN (1985) Hierarchy and evolution. Oxford Surveys in Evolutionary Biology 1:184-208

Gillan DJ, Premack D & Woodruff G (1981) Reasoning in the Chimpanzee: 1. Analogical reasoning. Journal of Experimental Psychology: Animal Behavior Processes 7:1-17

James W (1880) Great men great thoughts and the environment Atlantic Monthly 46: 441-449

Johnston TD & Pietrewicz AT (eds)(1985) Issues in the Ecological Study of Learning. Erlbaum Hillsdale New Jersey

Lawick-Goodall J (1970) Tool-using in primates and other vertebrates. Advances in the Study of Behaviour 3:195-249

Lewontin RC (1982) Organism and environment. In H C Plotkin (ed) Learning Development and Culture: Essays in Evolutionary Epistemology. Wiley Chichester

Macphail EM (1985) Vertebrate intelligence: the null hypothesis. Philosophical Transactions of the Royal Society London series B 308:37-51

Pasnak R (1979) Acquisition of prerequisites to conservation by Macaques. Journal of Experimental Psychology: Animal Behavior Processes 5:194-210

Piaget J (1936) The Origin of Intelligence in the Child Reprinted by Penguin Harmondsworth

Plotkin HC (1982) Evolutionary Epistemology and Evolutionary Theory. In HC Plotkin (ed) Learning, Development and Culture: Essays in Evolutionary Epistemology. Wiley Chichester

Plotkin HC (1986) Evolutionary Epistemology and the Synthesis of Biological and Social Science. In Callebaut W, Pinxten R (eds) Evolutionary Epistemology: A Multiparadigm Program. Reidl Dordrecht.

Plotkin HC, Odling-Smee FJ (1979) Learning, change and evolution. Advances in the Study of Behaviour 10:1-41

Plotkin HC, Odling-Smee FJ (1981) A multiple level model of evolution and its implications for sociobiology. The Behavioral and Brain Sciences 4:225-268

Plotkin HC, Odling-Smee FJ (1982) Learning in the context of a hierarchy of knowledge processes. In HC Plotkin (ed) Learning Development and Culture:Essays in Evolutionary Epistemology. Wiley Chichester

Pooley C, Gans C (1976) The Nile Crocodile. Scientific American 234:114-124

Revusky S (1977) Learning as a general process with an emphasis on data from feeding experiments. In NW Milgram L Krames & TM Alloway (eds) Food Aversion Learning. Plenum New York

Riedl R, Wuketits FM (eds)(1987) Die Evolutionare Enkenntnistheorie. Verlag Paul Parey Berlin

Shettleworth SJ (1983) Memory in food-hoarding birds. Scientific American 248:86-94

Thomson KS (1986) The relationship between development and evolution. Oxford Surveys in Evolutionary Biology 2:220-233

Waddington CH (1959) Evolutionary systems: animal and human. Nature 183: 1634-1638

Williams GC (1966) Adaptation and Natural Selection. Princeton University Press Princeton New Jersey

Williams GC (1986) A defence of reductionism in evolutionary biology. Oxford Surveys in Evolutionary Biology 2:1-27

Wilson AC (1985) The molecular basis of evolution. Scientific American 3:148-157

Wyles JS, Kunkel JG, Wilson AC (1983) Birds, behavior, and anatomical evolution. Proceedings of the National Academy of Sciences 80:4394-4397

COMPARATIVE NEUROANATOMY AND THE EVOLUTION OF INTELLIGENCE

William Hodos
Department of Psychology, University of Maryland
College Park, Maryland 20815 USA

The search for the biological determiners of animal intelligence often has focused on the size of the brain as a fundamental feature from which intelligence may be deduced. This search has been based on certain assumptions about the nature of brain organization and the nature of intelligence. I believe that these assumptions require re-evaluation in the light of data from comparative neuroanatomy and theories of intelligence. My hope is that such a re-evaluation may lead us in the direction of a better understanding of the relationship between brain and intelligence even though that road may be less direct and more rocky than we would like.

The post-Darwinian era has seen a continued interest in the evolution of the brain and the evolution of intelligence. The basic model that has guided research in this area has been the idea that the brain is the organ of intelligence and therefore as the one evolved in the direction of greater size and complexity so should the other. This model, however, rests upon a number of assumptions that appear rarely to have been challenged. These assumptions include:

1. Growth in the overall dimensions of the brain represents a proportional increase in the internal constituents of the brain.
2. There is general agreement about what intelligence is.
3. Intelligence is a quantifiable, unitary biological entity like body weight or height.
4. Intelligence is the same in different taxonomic groups.

Unless all of these assumptions are correct, many of the conclusions about the evolution of intelligence drawn from allometric studies of the brain will have to be modified quite considerably. In this paper, I shall try to make the case that all of them either are incorrect or are at least open to question.

Brain Size and Brain Organization

For purposes of our discussion of the brain and intelligence, let me distinguish between the "visceral brain," the "somatic" brain and the "intellectual brain." The visceral brain includes those brain components that are directly involved in the regulation of vegetative body functions, reflexes, instinctive behavior and fixed action patterns. The somatic brain consists of those portions of the brain that are involved in the sensory input from and motor outflow to the somatic musculoskeletal

NATO ASI Series, Vol. G17
Intelligence and Evolutionary Biology
Edited by H. J. Jerison and I. Jerison

system. The intellectual brain consists of those neuronal populations that are involved in the gathering, storage and retrieval of information about the environment as well as the application of this acquired information to those environmental challenges that can be responded to behaviorally. For example, a large proportion of the mass of the medulla and pons is occupied by cranial nerve nuclei that are basically visceral and somatic in function. Similarly, the size and weight of the diencephalon is affected by the development of the hypothalamus, which has many important visceral functions.

Jerison (1973) has proposed the "principle of proper mass," which states, in effect, that the mass of a given neuronal population that performs a particular function will be in proportion to the extent to which that function is performed. Thus, animals such as birds or primates, which make extensive use of vision, have well developed visual systems whereas animals that make relatively little use of vision, such as moles, have visual regions of the brain that are greatly reduced in size. Jerison's principle leads to the expectation that animals with large, well-developed body parts, such as limbs and tails or with sophisticated somatosensory systems would have a well-developed somatic brain. Likewise, animals that possess elaborate visceral systems or make extensive use of neuroendocrine mechanisms or instinctive behavior would be expected to have proportionately larger visceral brains. Finally, animals that made greater application to new circumstances of acquired information and skills, used reasoning and concept formation, constructed and used tools, estimated quantities, etc, would have a larger intellectual brain than those that had a lesser capacity to use these abilities.

A further deduction from Jerison's proper-mass principle would be that increases in the size of the brain beyond those that were required for somatic and visceral functions would reflect increases in the relative size of the intellectual brain. Thus, if two closely-related species with similar body sizes and which would be expected to have similar visceral and somatic functions were to differ in brain size, we would have to conclude that this difference was an indication of a difference in their use of the intellectual brain; i.e., that the larger-brained of the two species was the more intelligent.

The proper-mass principle leads to a reasonable working hypothesis about the relationship between brain size and intelligence if we can assume that evolution has produced no changes in the relationship between the visceral and somatic brains and the mass of the body; i.e. that body size is a reasonable indicator of the relative size of the sum of the somatic and visceral brains. While this assumption is probably valid for closely-related species, its validity becomes questionable for the more remotely-related species. For example, one such instance may be seen in those fishes that have an extremely well-developed gustatory system (Morita and Finger, 1985; Ariens Kappers, Huber and Crosby, 1936) that serves not only the visceral brain as the afferent branch of various feeding reflexes but the intellectual brain as well by providing information about the chemical attributes of the environment. As a consequence of this massive gustatory system, the visceral afferent and visceral motor columns of these fishes are quite well developed and in certain fishes (carps and catfishes) these regions attain massive proportions. In contrast, birds and some mammals have considerably less gustatory capacity and the comparable regions of the medulla are reduced accordingly. These relations are in no way reflected in body size. Similar considerations would apply to changes in respiration or digestion.

Encephalization and Intelligence

Many of our contemporary ideas about brain size and intelligence have their roots in the generalizations or "laws" that Marsh (1886) proposed to describe the evolution of the brain in reptiles, birds and especially mammals. The most influential of the generalizations proposed by Marsh were (1) relative brain size gradually increased through geological time; (2) the increase was largely confined to the cerebral hemispheres; (3) the increase was accompanied by an increase in the complexity of cerebral convolutions.

A term that has been frequently associated with the evolution of the brain and intelligence is "encephalization." Northcutt (1984) states that the earliest use of the term was to describe an increase in relative brain size through geological time. This is essentially the concept embodied in the first of Marsh's laws, which I shall refer to as type I encephalization. Contemporary discussions of encephalization, however, employ the term to encompass at least two additional concepts. Type II encephalization is probably derived as an inference from the second of Marsh's laws and "...may be defined as an evolutionary process in which the forebrain progressively takes over functions which, in more primitive forms, are organized at levels below that of the cerebrum." (Weiskrantz, 1961, p. 30); i.e., a progressive shift of function from lower brain regions (tectum and striatum) to isocortex (neocortex). Jerison uses encephalization in a third sense: "...an amount of brain size beyond that required by body size...." (1982, p. 767); "...an increase in processing capacity beyond that required for the control of basic body functions." (1985, p. 15). All three uses of the term "encephalization" are legitimate as descriptive statements, irrespective of whether they are consistent with the weight of evidence; but one should be aware the same term is used by different writers to represent very different concepts.

That the brain as a whole, and the forebrain in particular, have increased progressively in a number of vertebrate lineages (type I encephalization) appears to be one of the principal conclusions of the various allometric studies (Dubois, 1897; Jerison, 1955, 1970, 1973; Bauchot and Stephan 1966, Northcutt, 1981). Type I encephalization, Marsh's laws and Jerison's principle of proper mass suggest that increased relative size means increased function. Since the brain is the organ of intellect, a natural assumption would be increased intellectual functioning concomitant with increased relative brain size.

What of encephalization as a shift of function to the isocortex (neocortex)? If type II encephalization were not already a concept in the literature (Marquis, 1934; Noback and Moskowitz, 1962), would the pressure of contemporary data oblige us to invent it? I suspect not, inasmuch as the recent literature offers little to support it (Weiskrantz, 1961; Jerison, 1973; Macphail, 1982).

Jerison's definition of encephalization (type III) is similar to type I encephalization but is a much stronger statement. When viewed in the context of his definition of "biological intelligence" (Jerison, 1982) it suggests that the bulk of the variance in the increase in relative brain-size is attributable to increases in the information-processing components of the brain and relatively little to advances in its vegetative components. But equating information processing with intelligence would seem to broaden the use of the term "intelligence" to include behavioral mechanisms not usually considered as "intelligent." For example, motivation, aggression, biological rhythms, photoperiodicity, etc would have to be be subsumed under the rubric of the total information-processing capacity of the brain. If that were to be the case, we would have to change our conception of intelligence away from that of a cognitive entity involving attention, learning, memory, reasoning, concept formation, etc. to an entirely different concept that included highly-stereotyped fixed-action patterns, emotional and motivational behavior and biological rhythms.

The visceral brain includes not only hindbrain structures such as cranial nerve nuclei and descending reticular formation, but many forebrain structures such as hypothalamus and portions of the thalamus as well. Moreover, important constituents of the visceral brain are components of the limbic system, which is mainly a forebrain system (Nauta and Feirtag, 1986). Although a portion of the visceral brain is involved in purely vegetative functions and, like the somatic brain, should vary in size in direct proportion to body size or the extent to which body parts are used, a substantial portion of the visceral brain is not involved directly in the regulation of vegetative functions; many of its components are important for such behavioral processes as courtship, reproductive behavior, nest building and parental behavior, territoriality, aggression, motivation, sleep and wakefulness, as well as biological rhythms and photoperiodic effects (Guthrie, 1980; Demski, 1984; Carlson, 1980). What proportion of the relative brain growth that was the result of type III encephalization accounts for the evolution of these behavioral processes? These behaviors surely have not been unaffected by the action of natural selection and this must have been reflected in the relative sizes of their neural substrates. The usefulness of type III encephalization either as a descriptive statement or an explanatory principle would seem to depend on the size of the visceral brain being tightly linked to body size across vertebrate classes. To my knowledge, this is an assumption that has not been tested in a systematic way. My speculation is that the intellectual components of the brain are not the only ones that are responsive to adaptive pressures; the visceral brain, especially the limbic system, probably also has responded to adaptive pressures and increased in size relative to body weight but not necessarily in synchrony with the information-processing brain. To the extent that this has occurred, one might expect the correlation between overall brain size or weight and intelligence across classes to be weakened.

My point here is not that the type III encephalization concept is invalid, but that it must be refined to take into account evolution and specialization in the visceral brain. One such method of refinement might be to separate the visceral brain from the total when brain weight/body weight comparisons are made. To be sure, such a separation would add a large measure of additional complexity and uncertainty to these studies since difficult dissections would be required. One might be able to avoid separating out the somatic brain because it would, most likely, vary more directly with body size and, in any case, would be even more difficult so separate. The result, however, might be a more accurate representation of the size of intellectual brain.

Overall Brain Size and Internal Structure

Overall size or weight have been convenient indicators for brain allometry studies. Such values are readily available in reference books or can easily be obtained by investigators without resorting to dissection of the brain. Such overall values were the basis of the pioneering studies of Dubois (1897), Bauchot and Stephan (1966) and Jerison (1955, 1970, 1973), which demonstrated systematic relationships between brain size and body size. Sacher's (1970) study of the correlation between overall brain size and individual brain components is relevant to this problem. He reported nearly-perfect correlations between the overall size of the brain and the size of the cerebellum, isocortex (neocortex) and diencephalon in insectivores and primates. But these animals are all closely related mammals. A considerably lower correlation correlation coefficient would probably be obtained if data from fishes, birds and a wider representation of mammals were to be included. Sacher also reported that the correlation between overall brain size and the size of the olfactory bulbs was only +0.35. Whether this lower correlation represents contributions of the olfactory system to both the visceral and the intellectual brains is a matter for future research.

Dramatic non-uniformities in the relationship between overall brain size and its internal constituents may be seen when comparisons are made across classes. The cerebellum may be taken as a case in point. This structure is large and well developed in mammals, birds and sharks, but is quite small in amphibians and reptiles. In bony fishes, the cerebellum is modest in size in many orders, but in electric fishes, e.g., mormyrids, it attains gigantic proportions (Nieuwenhuys and Nichols, 1969).

Similar observations could be about about the optic tectum, which is rather modest in primates, but better developed in tree squirrels and tree shrews and is a rather prominent feature of the mesencephalon in the majority of the non-mammalian classes that reaches a pinnacle of development in birds (Ariens Kappers, Huber and Crosby, 1936). Moreover Northcutt (1981, 1985) reports data on the allometry of the brains of cartilaginous fishes that defy conventional expectation about relative brain size across vertebrate classes. The data that he reports indicate that the distributions of relative brain sizes in cartilaginous fishes exceed all but the largest brains of bony fishes and reptiles and overlap the distributions of the brain of birds and mammals. Indeed, the forebrains of cartilaginous fishes can achieve weights that would be expected for birds or mammals of equivalent body weight. To what extent developments in regional brain evolution reflect the relative development of the visceral brain and the intellectual brain remains to be seen in the light of future research.

Other students of the allometric approach have related the size or weight of individual brain components to body weight or overall weight of the brain. Thus, Radinsky (1975) and Passingham (1975) have related volume of isocortex (neocortex) to total brain volume in insectivore, prosimians, simians and anthropoid apes.

Both studies indicated that although humans have the largest volume of isocortex of any of the mammals studied, it was no greater than would be expected for a primate with such a large brain, which would seem to suggest that major changes in cortical organization have not occurred in the primate lineage. Passingham and Etlinger (1974) showed similar findings for the hippocampus. In contrast, the amount of motor cortex in raccoons and other procyonids varies with the extent to which the forepaws are used to manipulate food (Welker and Seidenstein, 1959; Welker and Campos, 1963). Similar correlations have been observed between manual dexterity and both somatosensory cortex and motor cortex in different species of otters (Radinsky, 1968). The approach used by Welker, et al. and Radinsky suggests an alternative that may lead us closer to the goal of relating brain structure to intelligence; namely to attempt to correlate specific behavioral abilities with the morphology of specific, circumscribed brain loci rather than attempting to relate global intelligence to global brain measures.

The difficulty of the global approach to the brain-and-intelligence problem is further compounded when one tries to make comparisons between mammals, which have a tightly laminated isocortex, to nonmammals. Although nonmammals lack this unique telencephalic organization, their telencephalon contains many of the same cell populations that are present in mammals, but organized in a looser laminar arrangement. Contemporary comparative neuroanatomists largely have abandoned the view that the telencephalon of non-mammals is a massive corpus striatum (Karten, 1969; Nauta and Karten, 1970; Northcutt, 1981). The corpus striatum has been identified in non-mammals by anatomical and histochemical techniques and has been found to occupy a rather modest proportion of the telencephalic mass. The remaining telencephalic regions, which earlier anatomists labelled with the unfortunate suffix "-striatum", contain many of the same cell populations as are found in mammalian sensory and motor isocortex and have the appropriate relations with the thalamus and brain stem. At present, one can only speculate on how these similarities and differences could impact upon global brain measures and their relationship to global intelligence.

Can studies of the gross brain reveal anything about the internal constituents of the brain? Some reasonable deductions probably could be made from the comparison of the brains of closely-related species so long as one of them has not evolved a unique specialization not found in the others of which it makes extensive use. In the latter instance, the principle of proper mass would predict an expansion of that specialization's neural substrate with a corresponding increase in gross brain size.

How could one falsify the assertion that brain size is related to intelligence? The answer at first seems simple enough: Plot intelligence and brain size in a coordinate space and look at the regression of the one upon the other. Yet in spite of all of the interest in brain size and intelligence, rarely have the two actually been subjected to a careful correlational study (Van Valen, 1974). One such study was conducted by Passingham (1979), in which he compared the cranial capacity of living humans (based on measurements of their skulls) with their scores on standardized tests of intelligence. He reported that cranial capacity accounted for less than 3% of the variance in intelligence and even less when the influence of the correlation between height and intelligence was removed from the data. Although the absence of a close correspondence between brain size and external cranial measurements could mask a small effect of brain size on intelligence (Passingham, 1982), certainly no robust relationship is evident.

Van Valen (1974) surveyed the literature on brain size and intelligence and concluded that the correlations generally were rather low between these two variables. Jerison (1982) argues that one should expect a low correlation between intelligence and brain size when the data are taken from a single species. Perhaps the low correlations between brain size and intelligence that have been reported are an indication that overall brain size or weight are not the relevant variables to correlate with intelligence test performance. Perhaps only the size or weight of the intellectual brain should be considered. Indeed, even the size or weight of the intellectual brain may be less important for a high correlation with intelligence scores than the details of its micro-organization and local circuitry. The presence of such micro-organizational features might have little or no correlation with overall brain size.

The Definition of Intelligence

Apart from human beings, we are unable to offer a quantitative evaluation of intelligence to match against our quantitative data on brain size. This lack of quantification is due, at least in part, to comparative psychologists' reluctance to define animal intelligence. Contemporary textbooks of comparative psychology or animal behavior generally have no chapters on intelligence and the word rarely appears in the index. A notable exception is a recent textbook of animal behavior by McFarland (1985), which devotes a chapter to a discussion of animal intelligence. The term "intelligence" does sometimes appear, in passing, in symposium volumes on animal cognition (Griffin, 1982, Roitblat, Bever and Terrace, 1984), in which a writer may imply that a particular behavioral paradigm is related to intelligence, but rarely is the assertion made outright. As a result, discussions of brain size and intelligence often omit a definition of intelligence altogether and instead refer to it in the layman's sense of the term (Steele Russell, 1979; Macphail, 1982). Jerison defined it in very general cognitive terms, "...the way one knows the world and uses that knowledge when adapting to changing situations...." (1973, pp.16-17). Similarly, Macphail, although sidestepping a formal definition stated, "Our interest, then, is to be the behaviour of animals in situations which make demands upon the subject that allow it to demonstrate the behavioural flexibility that is implied by intelligence" (1982, p.6).

Since specialists in animal behavior are reluctant to deal with intelligence, we must turn to experts in human behavior to see what they have to say on the subject. A recent reference work, Handbook of Intelligence (Wolman, 1985), devotes more than one third of its nearly one thousand pages to a discussion of the definition of human intelligence. This discussion, written by eight experts, reveals that the theorists still have not agreed on the resolution of a fundamental conflict, which dates back to the origins of intelligence testing; namely, is intelligence a single entity (i.e., general intelligence) that can account for an individual's performance in a wide variety of situations or is it a multidimensional phenomenon in which each independent dimension represents a unique intellectual ability? A briefer review of contemporary theories of human intelligence has recently been provided by Sternberg (1985a).

Many of us probably feel intuitively that intelligence must be a unitary entity. Indeed, our linguistic use of the term forces us in that direction. The very word "intelligences" does not lie easily on the tongue. But those who might intuitively lean towards the general-intelligence position, should consider that there is no way to measure general intelligence directly; its existence is inferred statistically from the pattern of intercorrelations between a large number tests of specific abilities. In other words, there is no single behavioral test for general intelligence; it can only be determined by searching for what is common in the results of a number of individual tests of diverse abilities. To complicate matters further, human intelligence theorists, such as Sternberg (1985b) have argued that there may be three or more types of general intelligence.

In my opinion, the search for the neural basis of animal intelligence and our understanding of the evolution of both the brain and intelligence will proceed more rapidly if we abandon attempts to correlate neural characteristics with a statistical construct (general intelligence) and concentrate our efforts on the neural basis of specific abilities. At some point in the future, when we would have sufficient data on the relationship between specific brain structures and specific abilities, we could then apply the techniques of statistical analysis to attempt to identify a general brain factor -- a statistical entity that would be the neural counterpart of general intelligence. Allometry between the general brain factor and a general intelligence factor based on intercorrelations between tests of specific abilities then would provide a meaningful way to examine the relationship between brain and intelligence within and between taxonomic groups. But until such time as we have a sufficient data base to carry out such analyses, we will only be able to speculate on the relationship between brain size and general intelligence without any supporting data. What are the specific abilities that frequently are included in models of human intelligence and which might be applicable to animal intelligence? Nearly always included in such lists are language skills. I have eliminated language from my list although I have included "communication in symbolic form," which would allow the consideration of the various ape "language" experiments without becoming embroiled in the issue of whether or not the findings truly represent language.

The following is a list of behavioral abilities that usually are considered as fundamental dimensions of human intelligence (Sternberg, 1985b; Humphreys, 1985; and Horn, 1985) and that seem appropriate as dimensions of animal intelligence. I have added "tool use" because it frequently is cited as an indicator of animal intelligence.

- speed of learning
- retrieval of information from long-term memory
- decision making
- problem solving
- communication in symbolic form
- counting
- spatial-relations ability

- concept formation
- rule learning
- tool use

Until not very long ago, the tendency was to consider that most of the above list were abilities that were attributes only of humans or, at the most, primates. The recent interest in animal cognition (Hulse, Fowler and Honig, 1978; Griffin, 1982; Mellgren, 1983; Roitblat, Bever and Terrace, 1984), however, has revealed that various of these abilities are widely distributed throughout the animal kingdom.

The task of collecting data from behavioral experiments that represent these abilities in vertebrates, sorting them out by orders or classes and producing tables of intercorrelations that could yield a general intelligence factor would be quite enormous. The bulk of the data that exist today surely are from mammals (especially rats and monkeys) and birds (mostly pigeons). Although some data exist for fishes and reptiles, studies on the cognitive capacity of amphibians are in very short supply. Any conclusions that we could come to from the present data base would be hopelessly biased by our results from this handful of species.

A more practical way to proceed would seem to be to confine our immediate attention to the specific abilities themselves. That would free us from the burden of having to have masses of data collected before we could begin to try to determine functional relationships. Let us take spatial relations as an example. The evidence is quite good that the optic tectum (superior colliculus) is involved in processing information about the organization of space around the individual (Goldberg and Robinson, 1978). Maps of auditory and visual space exist in register with each other in the tectum (Knudsen, 1982), and animals with other distance senses, such as boid and crotalid snakes, which possess infrared-detection ability, also have tectal sensory maps that are in register with the visual and auditory maps (Hartline, 1985). The tectum would seem to be a good candidate for a brain region that might correlate well with spatial-relations ability. Although the tectum is rather modest in size in most mammals, it generally is well developed in the other vertebrate classes and reaches a peak of size and differentiation in birds. How does spatial-relations ability vary with relative tectal size and differentiation? We have very little evidence at present, but one recent report offers some suggestive findings that may point the way to future studies of a more systematic kind. Hollard and Delius (1982) compared the ability of pigeons and humans to recognize previously unfamiliar, irregular, geometrical shapes in different orientations. They found that humans required progressively greater time to recognize these shapes as the orientation shift increased; whereas pigeons had the same short latencies irrespective of the orientation of the stimulus. In other words, the pigeons had no difficulty in recognizing the stimulus in any orientation. If the optic tectum can be demonstrated to play a crucial role in the ability to recognize shapes in different orientations, perhaps by means of lesion experiments or electrophysiological studies, this rather simple test could be applied to a broader sampling of vertebrates to provide information of the neural basis of the spatial-relations ability and its evolution.

The retrieval of information from long-term storage is another specific ability that lends itself well to investigation. Numerous learning and memory paradigms have been used successfully in every vertebrate class. The hippocampus, amygdala and related structures and pathways have been implicated in various aspects of memory (Milner, Corkin and Teuber, 1968; Olton, 1984; Mishkin, 1982). The role of these neural structures in memory phenomena has been investigated extensively in primates and rodents, but little interest has been shown in performing such studies in the rest of the vertebrate subphylum in which comparable structures have been identified.

Although speed of learning often is regarded by lay persons as a sign of intelligence in humans, it may not be especially useful in studies of animal intelligence. Bitterman (1965a,b) has pointed out that the speed of learning in animal experiments too easily can be affected by motivational and contextual variables to be useful as an indicator of learning ability. But even in studies of human intelligence, the speed of learning of simple tasks does not appear to be a good predictor of intelligence-test scores because of such factors as differences in attention, ability to follow instructions and the previous experience of the subjects (Sternberg, 1985b). Sternberg concludes that while learning surely is central to intelligence, it is not the speed of learning that is important; rather such factors as the extent to which what has been learned is applied to other situations and the extent to which what is learned is "ecologically valid." Applying the concept of ecological validity to animal intelligence would mean that we should avoid the simple, arbitrary stimuli and responses that have become traditional in laboratory studies of animal learning and instead make use of stimuli and responses that are important to the animal in the environment to which evolution has adapted the animal to survive. Rats, for example, perform poorly in comparison to primates in the visual learning-set task (a type of rule learning) in terms of their speed of acquisition and the final level of performance that they attain. But rats perform on a par with primates on the same task when the stimuli are olfactory (Slotnick and Katz, 1974). Nocturnal rodents rely very little on visual pattern discrimination and depend heavily on olfactory learning to survive in their environment.

Jerison (1982) distinguishes between "psychometric intelligence" and "biological intelligence." Psychometric intelligence would include the processes described above, whether they are independent dimensions or whether they are interrelated by a common, general factor. Biological intelligence is "...the behavioral consequence of the total neural information-processing capacity in representative adults of a species, adjusted for the capacity to control routine bodily functions." (Jerison, 1982, p.723). This definition is reasonable as a general definition of intelligence. Its weakness as a tool in the search for the biological basis of intelligence is that it is very similar to Jerison's definition of encephalization, which he uses as the evidence for biological intelligence. A definition of intelligence that is based on relative brain size cannot be used without bias to determine the relationship between relative brain size and intelligence.

The Quality and Quantity of Animal Intelligence

When we speak of the evolution of intelligence, the notion seems implicit that a single process or a single set of processes has evolved; i.e., that intellectual differences among animals are only quantitative, not qualitative. This idea appears to have been generally accepted without controversy. A notable exception has been Bitterman (1965a,b; Bitterman and Woodward, 1976), who has argued that we have very little evidence for a universal intelligence. His own comparative studies of learning have cast doubt on the existence of a single set of laws that can account for learning and performance in different vertebrate classes. Some of the differences in learning and performance may reflect differences in attention or differences in rule-learning or differences in response strategies (Brookshire, 1976). Attention and rule learning are factors usually considered in psychometric intelligence. While one might argue that these differences are quantitative and not qualitative, the right combination of quantitative differences in a mosaic entity such as intelligence easily could result in a qualitative difference.

A rather different position is taken by Macphail (1982) who offers evidence, based largely on a comparative survey of learning experiments in animals and humans, that there are no quantitative differences in intelligence among species, including humans, if allowances are made for the enormous advantages provided to humans by their ability to use language in a variety of intellectual activities. Macphail's position may be criticized on the grounds that it rests heavily on data from learning experiments, which constitute an assessment of only one segment of the intelligence mosaic. Nevertheless, his position must be given consideration by investigators of the brain-intelligence issue.

Another issue in the assessment of animal intelligence is the question of context effects (Macphail, 1982; Herrnstein, 1985). Elsewhere (Hodos, 1982) I have pointed out that the problems of comparing the intelligence of different classes of vertebrates and the problems of comparing the intelligence of humans from different cultures are very similar. Cross-cultural comparisons, like cross-species comparisons, often fail because the concept of intelligence and its measurement were intended as devices for comparing an individual to other members of the same population and not as a means of comparing different populations. A basic question might be how does a particular child compare in tests of intellectual performance with other children of his or her age? The advantage of the intelligence quotient (IQ) was that by dividing mental age by chronological age one could compare children of different ages (Chronbach, 1960). But once these methods were applied to children from different school populations or different cultures with different educational experiences and standards, the results were a dismal failure (Gould, 1981). Bitterman's (1965a,b) approach of a systematic variation of motivational conditions to minimize context effects has considerable merit, although it does require heroic efforts to achieve. Macphail (1982) discusses other approaches to dealing with the context problem, but cautions that none of them have offered a foolproof solution.

Animal Cognition

Macphail's (1982) recent survey of animal cognition and the other recent volumes on this subject (Hulse, et al., 1978; Mellgren, 1983; Roitblat, et al., 1984) reveal that nearly all of the research in animal cognition has been carried out in a very few species of mammals and birds. We should be very careful about generalizing from this limited sample lest we be guilty of overgeneralizing. Without more data on a much wider range of species, especially the anamniotes, we will have no way of knowing whether or not the cognitive similarities and differences that exist in rats, monkeys and pigeons represent the general case for mammals and birds as classes, much less how representative they are of amphibians, reptiles, sharks and bony fishes. We shall have to expand the comparative psychology data base if we wish to be able to make broad statements about the continuity of intellectual life (or the lack of it) in various lineages and how this might be related to the evolution of neural structure.

Conclusions

The low correlations that have been reported between overall brain size and intelligence in humans suggest that the development and organization of individual brain regions and the degree of elaboration of local circuitry within them may be more important determiners of intelligence than the gross size of the brain. Similarly, the notion of a unitary general intelligence that influences a great diversity of behavioral activities may not be a useful model for comparison with morphological data. General intelligence is a statistical concept that can only be measured by finding what is common to performance on tests of a wide variety of specific intellectual abilities. Our search for the biological bases of animal intelligence may progress more rapidly if we abandon the general-intelligence model and general brain indices such as total weight or total volume. My recommendation is that we concentrate our efforts on determining the relationships between specific morphological components of the brain and specific intellectual abilities. By accepting the multidimensional natures of both the brain and intelligence we will have better opportunities to uncover the relationships between the two.

Acknowledgment

The preparation of this article was supported in part by grant EY00735 from the National Eye Institute. The author is grateful to S. E. Brauth, H. J. Jerison, R. G. Northcutt, A. S. Powers, and R. E. Passingham for their valuable comments.

References

Ariens Kappers CU, Huber, GC, Crosby E (1936) The Comparative Anatomy of the Nervous System of Vertebrates, Including Man. Macmillan New York

Bauchot R, Stephan H (1966) Donnees nouvelles sur l'encephalisation des Insectivores et des Prosimians. Mammalia 30:160-196

Bitterman ME (1965a) The evolution of intelligence. Scientific American 212:92-100

Bitterman ME (1965b) Phyletic differences in learning. American Psychologist 20:394-410

Bitterman ME, Woodward WT (1976) Vertebrate learning: common processes. In: Masterton RB, Bitterman ME Campbell CBG Hotton N (eds.) Evolution of Brain and Behavior in Vertebrates. Erlbaum Hillsdale

Brookshire KH (1976) Divergences in learning. In: Masterton RB, Bitterman ME, Campbell CBG Hotton N (eds.) Evolution of Brain and Behavior in Vertebrates. Erlbaum Hillsdale

Carlson NR (1980) Physiology of Behavior. Allyn and Bacon Boston

Cronbach LJ (1960) Essentials of Psychological Testing. Harper New York

Demski LS (1984) The evolution of the neuroanatomical substrates of reproductive behavior: Sex steroid and LHRH-specific pathways including the terminal nerve. American Zoologist 24:809-830

Dubois E (1897) Sur le rapport du poids de l'encephale avec le grandeur du corps chez mammiferes. Bulletins de Societe d'Anthropologie de Paris. 8:337-376

Goldberg ME, Robinson DM (1978) Visual system: superior colliculus. In: Masterton RB (ed.) Handbook of Behavioral Neurobiology. Plenum New York

Gould SJ (1981) The Mismeasure of Man. Norton New York

Griffin DR (ed.) (1982) Animal Mind - Human Mind. Springer Berlin

Guthrie, DM (1980.) Neuroethology. Wiley New York

Hartline PH (1985) Multimodal integration in the brain: combining dissimilar views of the world. In: Cohen MJ, Strumwasser F (eds.) Comparative Neurobiology. Wiley New York

Herrnstein RJ (1985) Riddles of natural categorization. Philosophical Transactions of the Royal Society (London) B308:129-144

Hodos W (1982) Some perspectives on the evolution of intelligence and the brain. In: D. R. Griffin (ed.) Animal Mind - Human Mind. Springer Berlin

Hollard VD, Delius JD (1982) Rotational invariance in visual pattern recognition by pigeons and humans. Science 218:804-806

Horn JL (1985) Remodeling old models of intelligence. In: Wolman B (ed.) Handbook of Intelligence. Wiley New York

Hulse SH, Fowler H, Honig WK (1978) Cognitive Processes in Animal Behavior. Erlbaum Hillsdale

Humphreys LG (1985) General intelligence. In: Wolman B. (ed.) Handbook of Intelligence. Wiley New York

Jerison HJ (1955) Brain to body ratios and the evolution of intelligence. Science 121:447-449

Jerison HJ (1970) Brain evolution: New light on old principles. Science 170:1224-1225

Jerison HJ (1973) Evolution of the Brain and Intelligence. Academic Press New York

Jerison HJ (1982) The evolution of biological intelligence. In: Sternberg RJ (ed.) Handbook of Human Intelligence. Cambridge University Press Cambridge

Jerison HJ (1985) Animal intelligence as encephalization. Philosophical Transaction of the Royal Society (London) B308:21-35

Karten HJ (1969) The organization of the avian telencephalon and some speculations on the organization of the amniote telencephalon. Annals of the New York Academy of Sciences 167:164-179

Knudsen EI (1982) Auditory and visual maps of space in the optic tectum of the owl. Journal of Neuroscience 2:1177-1194

Macphail EM (1982) Brain and Intelligence in Vertebrates. Clarendon Press Oxford

Marquis DG (1935) Phylogenetic interpretation of the functions of the visual cortex. Archives of Neurology and Psychiatry 33:807-815

Marsh, OC (1886) Dinocerata. U.S. Geological Survey Monograph 10:1-243

McFarland D (1985) Animal Behavior. Benjamin/Cummings Menlo Park

Mellgren RL (1983) Animal Cognition and Behavior. North-Holland Amsterdam

Milner B, Corkin S, Teuber H-L (1968) Further analysis of hippocampal amnesic syndrome: 14 year follow-up study of H. M. Neuropsychologia, 6:215-234

Mishkin M (1982) A memory system in the monkey. Philosophical Transactions of the Royal Society (London), B298:85-95

Morita Y, Finger T (1985) Topographic and laminar organization of the vagal gustatory system in the goldfish, Carassius auratus. Journal of Comparative Neurology, 238:187-201

Nauta WJH, Feirtag M (1986) Fundamental Neuroanatomy. Freeman New York

Nauta WHJ, Karten HJ (1970) A general profile of the vertebrate brain, with sidelights on the ancestry of the cerebral cortex. In: Schmitt FO (ed.) The Neurosciences: Second Study Program. Rockefeller University Press New York

Nieuwenhuys R, Nichols C (1969) A survey of the general morphology, the fiber connections, and the possible functional significance of the gigantocerebellum of mormyrid fishes. In: Llinas R (ed.), Neurobiology of Cerebellar Evolution and Development. American Medical Association Chicago

Noback CR, Moskowitz N (1962) Structural and functional correlates of "encephalization" in the primate brain. Annals of the New York Academy of Sciences, 102:210-218

Northcutt RG (1981) Evolution of the telencephalon in nonmammals. Annual Review of Neuroscience, 4:301-350

Northcutt RG (1984) Evolution of vertebrate central nervous system: Patterns and Processes. American Zoologist, 24:701-716

Northcutt RG (1985) Brain phylogeny: Speculations on Pattern and Cause. In: Cohen MJ, Strumwasser F (eds.), Comparative Neurobiology. Wiley New York

Olton DS (1984) Working memory and serial patterns. In: Roitblat HL, Bever TE, Terrace HS (eds.) Animal Cognition. Erlbaum Hillsdale

Passingham RE (1975) Changes in the size and organization of the brain in man and his ancestors. Brain, Behavior and Evolution, 11:73-90

Passingham RE (1979) Brain size and intelligence in man. Brain, Behavior and Evolution, 16:253-270

Passingham RE (1982.) The Human Primate. Freeman Oxford

Passingham RE, Etlinger G (1974) A comparison of cortical functions in man and other primates. International Review of Neurobiology, 16:233-299

Radinsky L (1968) Evolution of somatic sensory specialization in otter brains. Journal of Comparative Neurology, 134:495-505

Radinsky L (1975) Primate brain evolution. American Scientist, 63:656-663

Roitblat HL, Bever TG, Terrace HS (eds.) (1984) Animal Cognition. Erlbaum Hillsdale

Sacher GA (1970) Allometric and factorial analyses of brain structure in insectivores and primates. In: Noback CR, Montagna W (eds.) The Primate Brain. Appleton New York

Slotnick BM, Katz HM (1974) Olfactory learning-set formation in rats. Science, 185:796-798

Steele Russell I (1979.) Brain size and intelligence: a comparative perspective. In: Oakley DA, Plotkin HC (eds.) Brain, Behaviour and Evolution. Methuen London

Sternberg, RJ (1985a) Human intelligence: The model is the message. Science, 230:1111-1118.

Sternberg RJ (1985b) Cognitive approaches to intelligence. In: Wolman B (ed.), Handbook of Intelligence. Wiley New York

VanValen L (1974) Brain size and intelligence in man. American Journal of Physical Anthropology, 40:417-424

Weiskrantz L (1961) Encephalisation and the scotoma. In: Thorpe WH, Zangwill OL (eds.) Current Problems in Animal Behaviour. Cambridge University Press Cambridge

Welker WI, Campos GB (1963) Physiological significance of sulci in somatic sensory cortex in mammals of the family Procyonidae. Journal of Comparative Neurology, 120:19-36

Welker WI, Seidenstein SS (1959) Somatic sensory representation in the cerebral cortex of the raccoon (Procyon lotor). Journal of Comparative Neurology, 111:469-501

Wolman B (ed.)(1985) Handbook of Intelligence. Wiley New York

THE FOREBRAIN AS A PLAYGROUND OF MAMMALIAN EVOLUTION

H.-P. Lipp
Institute of Anatomy, University of Zurich
Winterthurerstrasse 190
CH-8057 Zurich, Switzerland

Theories concerning the process of evolution commonly assume that the Neo-Darwinian principles of random mutation and subsequent selection apply to the evolution of brain and behavior as well as to other biological systems. Specifically, the evolution of behavior is thought to precede changes in body morphology, which are seen as adaptations to newly acquired behavioral habits (Piaget 1976, Mayr 1977). The best known example is probably the evolution of the Darwin finches. Yet, evolutionary theories barely touch the problem of how random mutations can result in new behavior patterns or capacities. Given that even small mammals and birds have millions of neurons and billions of synapses, it would seem difficult to imagine how random mutations can produce the ordered complexity typical for the mammalian brain, and even more difficult to see why the most complex biological system evolves faster than body morphology.

This paper attempts to show that the problem can be explained rather parsimoniously, using two generally accepted concepts from behavioral and developmental brain research, and two fairly trivial notions from developmental and population genetics. The conclusions are restricted to animals with some degree of encephalization, that is, to mammals and birds. The aim is to provide a simplified concept of how point mutations can result in heritable but ordered (i.e., non-deleterious) changes in behavior, and to explain that there must be a differential rate of selection for such point mutations, which results in canalized selection leading to a smooth reorganization of brain and behavior. This may set a neuronal framework according to which body morphology evolves phylogenetically.

The hypothesis is based on 5 key notions. The first is that there is extensive genetic variation of behavior, the second that mammalian brains are organized as hierarchies of "miniature nervous systems" (Jerison 1976), the third that this system hierarchy is built up by differential timing of maturation of specific brain parts ("Flechsig's rule"), the fourth that the epigenetic processes of brain development ("plasticity") can mask the consequences of point mutations affecting the development of brain systems mediating behavior, and the fifth that selective forces are most likely to act on genetic variation of top-ranking brain systems, since these are comparatively less buffered ("developmental last word effects").

NATO ASI Series, Vol. G17
Intelligence and Evolutionary Biology
Edited by H. J. Jerison and I. Jerison

Genetic Variation of Behavior

The set of problems may be discussed best by examining the potential effects of a point mutation on behavior produced by the complex brain of a mammal. Such single gene effects have been repeatedly postulated in studies of the behavioral genetics of mice, although they have never been demonstrated as clearly as, for instance, in *Drosophila* (Tully 1984). Regardless of the mode of inheritance, however, rodents show abundant genetic variation of behavioral traits (Fuller and Thompson 1979). Perhaps the most impressive phenomenon is the rapid success of breeding selection for behavioral traits as diverse as the propensity for ethanol dependence and sensitivity (McClearn and Kakihana 1973, Wilson et al 1984), nest building (Lynch 1982), aggression (Hyde and Ebert 1976), two-way avoidance (Driscoll and Baettig 1982), long and short sleep (Dudek and Abbott 1984), or winning and losing in a competitive situation (Masur and Benedito 1974), to mention only a few. Similarly, the efficacy of behavioral selection is familiar to any observer of dogs.

The relatively easy selection for behavioral traits implies that a significant proportion of their genetic variation is determined by only a few genes which apparently lack deleterious side-effects. It also follows that the modulation of a specific behavior could be based on rather simple mechanisms, for the message of a given gene is not complex. The effects of the message may vary according to the processes the gene can influence. Yet there is no need to assume that heritable behavioral traits depend upon multiple interactions of complex processes which, somehow magically, produce a predictable outcome, or even to scotomize significant behavioral effects of single genes on brain and behavior - 'because the workings of the brain are not understood yet'. However, genes are not more knowledgeable than scientists, but they obviously succeed in influencing behavior specifically. Thus, the chief problem here is reduced to the question of how a single allele (or, equivalently, any simple factor) can affect a specific behavior, without undesirable corollaries, how it achieves penetrance and reliable expression, and how such an allele, once in the gene pool of a species, is affected by selection.

Some Aspects of Cerebral System Hierarchies

The design of the mammalian brain is most likely that of a large system hierarchy, in which smaller components are organized as subsystems that are coordinated by superimposed systems (for an example with ethological background, see Figure 7.1). For the sake of simplicity, the smallest units may correspond to neurons, but, by the same token, a single synapse could be designated as the smallest unit as well. The term hierarchy does not necessarily imply a linear command hierarchy. The function of a superimposed neuronal system is primarily to *coordinate* the activity of a set of subordinate (sub)systems, regardless of the level in the hierarchy. In this aspect, the organization of a neuronal population resembles more a democracy than an army.

The simplest and least invasive way to change a brain of the complexity seen in mammals would be to leave the basic components intact, and to reorganize the functional organization of superimposed systems (Popper 1973). This bears an analogy

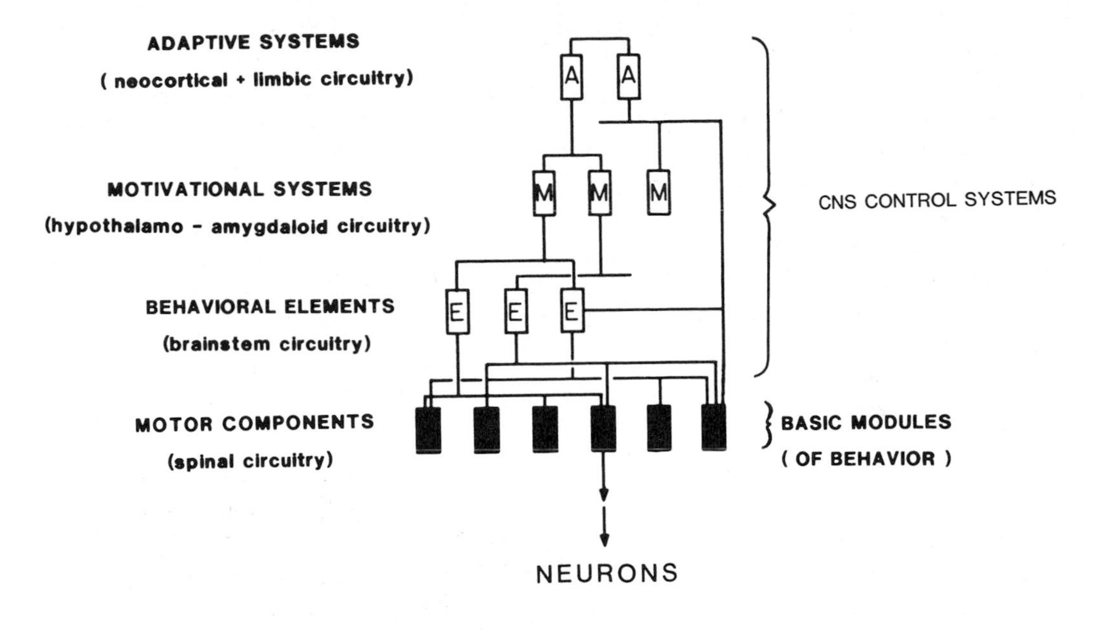

Figure 7.1. A possible hierarchy of neuronal systems. Ethological perspective.

to the task of adapting a complex computer program written in a high level language by an author failing to provide documentation. No programmer would start by rewriting the BIOS (basic input/output system), nor would he touch the assembler-written subroutines and/or alter the command words of the language. The strategy is to find the top menu (hopefully there is one) and to identify the addresses of the various subroutines, working from top to bottom. By recombination of subroutine addresses, the program might then be altered substantially and work without crashes.

The equivalents of subroutine addresses in the mammalian brain are the set-points and matching sites of cerebral systems, which may be perceived as the functional nodes of cerebral organization. As with a subroutine or subprogram, the actual functioning of the various subsystems need not be transparent to the addressing control system. Thus, reorganization of cerebral systems can be obtained by a differential connection of nodes (for instance, by a rerouting of axons), and variations of system interactions by altering the proportions of inputs converging on a given set-point (for example, by regulating the number of specific receptors). Yet, such changes are not complex and could well be produced by a single allele. If a point mutation is to have behavioral relevance and be non-deleterious for the organism, it should influence a set point.

Flechsig's Rule and Developmental Last-word Effects

Even if one assumes that a single allele could affect selectively a behaviorally relevant set-point, this does not necessarily imply behavioral changes. The mammalian brain has several mechanisms which buffer internal and external interferences. The main factors antagonizing the expressivity and penetrance (and, thereby, the observability) of a point mutation are (a) the inherent homeostasis of systems, (b) the mechanisms of structural reorganization (developmental and adult "plasticity", see also Katz and Lasek 1978), and (c) behavioral adjustment.

The presence of such powerful buffer mechanisms means, first of all, that a single gene can hardly be an absolute determinant of behavior. Its effects must remain probabilistic, that is, it can *bias* the occurrence of certain behavioral patterns through influencing the interaction of neuronal systems, but it cannot encode behavior per se. Its likelihood of producing a bias which is specific (expressivity in genetical terms) and highly predictable (penetrance) depends on the buffer capacity which must be overcome. It will be argued now that optimal expression and maximal penetrance must be seen in point mutations which influence the latest stages of brain differentiation.

This notion is derived from the chronobiological ideas developed in the last century by Flechsig and Hughlings Jackson. Flechsig's rule holds that the hierarchical status of brain systems is reflected by their maturational sequence during ontogeny (e.g., Flechsig defined the association cortices as the cortex regions with the most delayed onset of myelination), and corresponds to the familiar phenomenon that "simple" behavioral patterns and capacities appear earlier in ontogeny than the more "complex" ones. In other words, the building up of a system hierarchy is a temporally ordered process. Hence, through *late timing*, a point mutation can avoid neutralization by two of the developmental buffers above, and place itself in the *top coordinating systems.*

The temporal development of CNS buffer capacity and the likelihood of penetrance of a point mutation affecting behavior is shown schematically in Figures 7.2 and 7.3. For simplicity of description, the time course of brain differentiation will be divided into initial, early, middle and late phases (see also Lipp and Schwegler 1982). During the initial phase, the buffer capacity is probably small and penetrance high, because the systems most likely affected are those forming the modular components of higher order processes. Such mutations may become apparent as neurological mutants.

The capacity of buffering point mutations increases rapidly, however, during the early and middle phases. First, system homeostasis becomes effective, for more and more coordinating systems become superimposed, and this should result in an increased capacity for compensation of mutation effects. Second, developmental reorganization begins to take place. The differentiation of the mammalian brain is characterized by a set of processes known as "developmental rules" with the (probable) purpose of correcting gross deviations from a developmental pathway (Jacobson 1978). For instance, neurons compete for synaptic sites in a target region, with cells that fail to establish a minimal number of contacts dying prematurely. This strategy of establishing a surplus number of neurons normally guarantees a suffi-

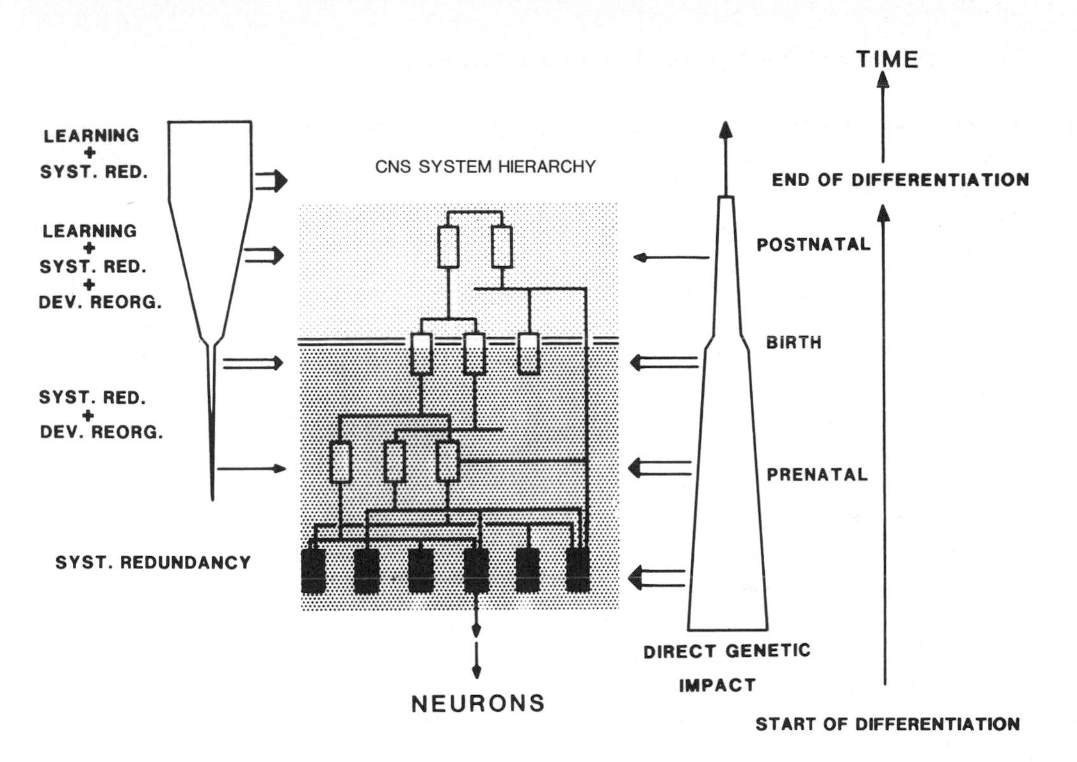

Figure 7.2. The developmental building-up of the system hierarchy seen in Fig. 7.1. Time arrow from bottom to top. Left: buffer mechanism in the developing CNS that may compensate mutation effects. Right: magnitude of genetic impact at different stages of development. Note that the neuronal systems at the top are the last to differentiate. Abbreviations: SYST. = system; RED. = redundancy; DEV. = development.

cient number of contacts, but may lead to an alteration of brain connectivity if the usual pattern of axonal ingrowth from different sources into a common target region is disturbed for genetic or experimental reasons. If such disturbances occur in early or middle phases of development, one may expect a "cascadic" reorganization of brain circuitry (a domino effect). Whether this results in compensation of mutation effects is not certain, but it is likely that their final consequences become less predictable the more intervening steps of reorganization are possible. Thus effects of mutations affecting an intermediate period of brain development are likely to be masked, or to be expressed as differential phenotypes of variable penetrance only.

Conversely, the effects of a point mutation become increasingly predictable, and are considerably less buffered, if they act in the latest phases of brain differentiation. At that time, the remaining buffers are the homeostasis of the top coordinating systems and behavioral adjustment, while the capacity of structural reorganization has been largely lost. A genetic message that is active during these latest stages of development thus has the property of a "*developmental last word effect*" which is more difficult to mask than mutation effects acting earlier.

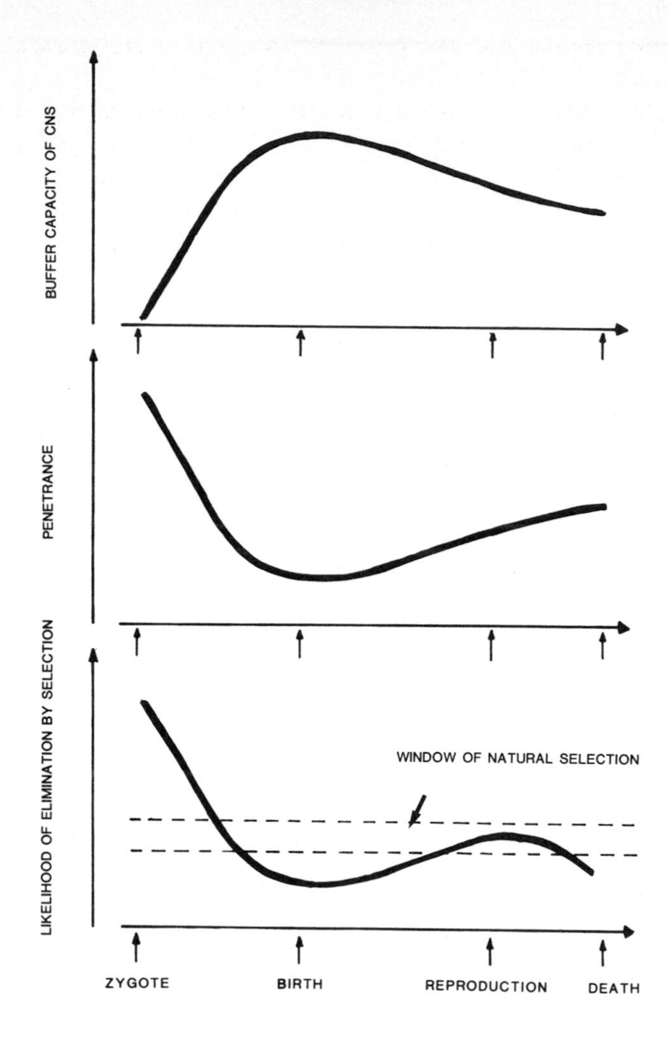

Figure 7.3. The buffer capacity of the CNS (top) changes during development. This determines the penetrance of a point mutation with behavioral consequences (middle diagram), depending on when it becomes active. The likelihood of elimination through selection is also a function of the relation above. A lowered position of the "window of selection" corresponds to increased selective pressure.

Somewhat paradoxically, it follows that the functionally top-ranking systems in the brain are the easiest and most sensitive targets for predictable effects by point mutations. In order to affect them specifically, a genetic message must simply wait (or last long enough) until these systems begin to form. In fact, genetic messages active in the very late stages of ontogeny have few other targets than brain systems coordinating the final capacities of the adult brain, together with the reproductive system, and the processes terminating the growth of various body parts.

This also implies that the late-differentiating (and top-ranking) systems must be equally sensitive to environmental interference: maximal heritability and maximal environmental sensitivity are two faces of the same token. An example from human psychopathology might be schizophrenia, which combines high heritability with high sensitivity to environmental stressors (McGue et al 1983), and which classically emerges during or after puberty.

A further corollary conclusion is that one of the most criticized assumptions of sociobiology - the occurrence of single genes determining the probability of traits such as altruistic behavior - is logically plausible.

Lastly and parenthetically, this notion of developmental last word effects may offer also a a simple solution for the loose yet pertinent association between facial and behavioral traits in humans. Perhaps the phenomenon of physiognomy simply reflects pleiotropy of late acting genes.

Differential Rates of Selection

Given a steady rate of point mutations with potential behavioral consequences, what is their fate in the gene pool of a species?

Their likelihood of survival presumably depends on their detectable influence on brain functions, and on their effects on biological fitness. Assuming equal biological fitness, the prerequisite for selection is visibility. Hence, the selection rate - a function of visibility - must co-vary with the buffer capacities and, accordingly, with the hierarchical levels of the CNS organization. (See Figure 7.3).

It follows that point mutations having early consequences are likely to be eliminated rapidly by natural selection. Consequently, their target systems in the brain will show little genetic variation, and appear invariant across species. To give an example, the extensor musculature of mammals shows much stronger tonic activity than the flexor system (because of gravitation). The tonus balance is regulated by brain stem structures receiving input from rostral brain parts inhibiting the extensor tonus. A mutation which reduces this inhibition (e.g., by local receptor deficiencies) may not cause overt changes in behavior, but it would require constant compensation by other brain systems. On critical occasions, this may interfere with adaptive responses such as rapid flight from a predator, the responsible allele being eaten. Nevertheless, such mutations must have occurred and, infrequently, survived. Otherwise, the reversed tonus balance in sloths is difficult to explain.

Point mutations affecting the "middle" layers of the system hierarchy are probably the most difficult to detect by observation of behavior, for their consequences are buffered by all three mechanisms (homeostasis, reorganization, behavioral adaptation), but it is difficult to judge their effects on biological fitness or ecological adaptation. It would seem possible, however, that many of these mutations will escape selection and contribute to the genetic variation of a species. For instance, guinea pigs (Valenstein and Nauta 1959) and some rabbit strains (Edinger and Wallenberg 1902) appear to have fornix projections that partially bypass the mammillary bodies, while this has not been observed in rats and mice (author's own studies). Given the vast possibilities of telencephalic regulation, such minute genetic variation would seem to have little influence on overt behavior. On the other hand, the neuronal populations in the hypothalamus are effective modulators of any form of motivational and emotional behavior. Thus mutation effects acting during the growth of hypothalamic connections have the *potential* to induce new anatomical links between differential motivational systems, or to bias their functional balance (see also Lipp 1979). However, their behavioral effects may be expressed poorly or unpredictably. Consequently, such alleles are difficult targets for selection, and may persist for long periods in the gene pool of a species. This situation also permits accumulation of genetic variation.

Point mutations which influence the top-ranking systems, however, are more vulnerable to selection, since they are easier to detect. Yet, it is likely that such alleles interfere minimally with biological fitness or reproductive success unless a highly specific selective pressure emerges. To give a hypothetical example in humans: Mathematical giftedness may depend on some specific alleles influencing the process of cerebral lateralization (Annett 1985) but does not necessarily provide for reproductive advantages. Should, however, permission for mating depend upon passing an exam in topological algebra, disabling alleles would be rapidly eliminated. In this context, one may notice that mice are easily bred for strong or weak lateralization, and that this trait is associated with strong cortical asymmetries, or lack thereof (Lipp et al 1984).

So, with time and generations, these differential rates of selection could result in *differential genetic variation* of brain and behavior, less in the lower levels of the CNS hierarchy (which tend to remain genetically conservative) and more in the higher levels, the top levels having, in addition, the best observability of gene effects.

Canalized Selection

The relatively high visibility of genetic variation in the top-ranking CNS systems implies that, after a change of environmental pressure, these alleles become the preferred *initial* targets for natural selection. These primary selection effects may then serve as a matrix, or Anlage, for subsequent canalized selection (Waddington 1957), like the growth of crystals around a germ configuration. Because of the (postulated) high genetic variability of the CNS systems forming the intermediate levels of the system hierarchy, selection may proceed very rapidly according to the selected system configuration at the top. So, new genetic combinations of traits and capacities can arise in a short evolutionary span - ahead of subsequent adaptation of body morphology. Moreover, it would seem that such a mechanism - an orderly reorganization from top to bottom - is the only efficient way to change a complex system hierarchy in a minimal time.

The Mammalian Forebrain as a Playground of Evolution

If the main arguments given above are not fundamentally wrong, one may postulate that encephalization is an evolutionary process which provides a species not only with intelligence but also with an increasing buffer capacity for accumulating genetic variability of brain and behavior - one of the main conditions of beginning speciation. The consequence would be that the more complex the brain of a species, the greater the potential of the species for *genetic* adaptation to new ecological situations, and, because of the non-randomness of the canalized selection, the faster its evolutionary transformation. So, mammals may maintain a sufficient rate of evolution, despite their low reproductive rates, and large and complex brains are no drag for evolution but rather a promoter. Perhaps the hypertrophy of the human frontal lobes forecasts a burst of further hominid speciation....

Acknowledgments

I wish to thank Bob Martin and Wilfried Neuhuber for critical reading. This work was supported by the Swiss National Foundation for Scientific Research (3.041-0.84).

References

Annett M (1985), Left, right, hand and brain: the right shift theory, Erlbaum, Hillsdale New York

Driscoll P, Baettig K (1982), Behavioral, emotional and neurochemical profiles of rats selected for extreme differences in active, two-way avoidance performance. In: Lieblich I (ed.) Genetics of the brain, Elsevier, Amsterdam, p 95

Dudek BC, Abbott ME (1984), A biometrical genetic analysis of ethanol response in selectively bred long-sleep and short-sleep mice. Behav Genet 14:1-20

Edinger L, Wallenberg A (1902), Untersuchungen über den Fornix und das Corpus mammillare. Arch f Psychiatr 35:1-21

Fuller JL, Thompson WR (1979), Foundations of behavior genetics, Mosby, St. Louis

Hyde JS, Ebert PD (1976), Correlated response in selection for aggressiveness in female mice. I. male aggression. Behav Genet 6:421-427

Jacobson M (1978), Developmental neurobiology, ed 2, Plenum Press, New York

Jerison HJ (1976), Evolution of brain and intelligence, Academic Press, New York London.

Katz MJ, Lasek RJ (1978), Evolution of the nervous system: role of ontogenetic mechanisms in the evolution of matching populations. Proc Natl Acad Sci USA 75:1349-1352

Lipp H-P (1979), Brain complexity enhances speed of behavioral evolution. Behav Br Sci 2:42

Lipp H-P, Schwegler H (1982), Hippocampal mossy fibers and avoidance learning. In: Lieblich I (ed) Genetics of the brain. Elsevier, Amsterdam, p 326

Lipp H-P, Collins RL, Nauta WJH (1984), Structural asymmetries in brains of mice selected for lateralization. Brain Res 310:393-396

Lynch CB (1980), Response to divergent selection for nesting behavior in *Mus musculus*. Genetics 96:757-767

McClearn GE, Kakihana R (1973), Selective breeding for ethanol sensitivity in mice. Behav Genet 3:409-410

McGue M, Gottesman II, Rao DC (1983), The transmission of schizophrenia under a multifactorial threshold model. Am J Hum Genet 35:1161-1178

Masur J, Benedito MAC (1974), Genetic selection for winner and loser rats in a competitive situation. Nature 249:284

Mayr E (1977), Populations, Species and Evolution, Belknap Harvard University Press Cambridge, Massachusetts England

Piaget J (1976), Le comportement moteur de l'evolution, Gallimard, Paris

Popper KR (1973), Objective knowledge: an evolutionary approach, Oxford University Press, London

Tully T (1984), *Drosophila* learning: behavior and biochemistry. Behav Genet 14:527-555

Valenstein ES, Nauta WJH (1959), A comparison of the distribution of the fornix system in the rat, guinea pig, cat, and monkey. J comp Neurol 113: 337-363

Waddington CH (1957), The strategy of genes, Allen & Unwin, London

Wilson JR, Erwin VG, DeFries JC, Petersen DR, Cole-Harding S (1984), Ethanol dependence in mice: direct and correlated responses to ten generations of selective breeding. Behav Genet 14:235-256

COMPARING THE STRUCTURE OF BRAINS: IMPLICATIONS FOR BEHAVIORAL HOMOLOGIES *

A. Fasolo and G. Malacarne
Dipartimento di Biologia Animale
Laboratorio di Anatomia Comparata
Via G. Giolitti, 34
TORINO

Intelligence has often been considered a global attribute, and brain evolution theories such as encephalization inevitably consider it as being global, because their data are mainly quantitative (Jerison, 1985).

Facts and theory in brain/mind/Intelligence interactions are the first difficulty one comes across.

Analytically, intelligence appears to be a metaphor for heterogeneous capacities, abilities and performances (i.e., problem-solving, language and learning ability, mental states or others) rather than something per se. These qualities are not exclusively human nor can they be made to fit into all-or-nothing patterns. They are found in several taxa of the animal kingdom, although they may be developed to a greater or lesser degree (see Bullock, 1982; Griffin, 1982). This suggests that brain and intelligent behaviours have evolved in the animal kingdom.

In the evolution of the brain and intelligence several widely accepted "naive" ideas may generate a lot of confusion. First, it is implicitly suggested that evolutionary trends follow linear progressions starting from the simplest and working their way up to the most complex. Secondly, there is a tendency to draw conclusions too easily when studying existing and different beings synchronically by creating possible evolutionary sequences in a diachronic perspective, while forgetting that each and every species is an end-branch of the evolutionary tree. Comparative psychology suggests functionally different learning patterns in animals and demonstrates the existence of marked evolutionary divergences (see the paper by Poli in this book) even as it attempts to show that general (common) laws still remain a promising approach (Bitterman, in this book).

Anyway, it seems obvious that some processes are related to environmental constraints, others emerge in response to the need for more flexible answers, and others still are part of a less specific and foreseeable ecological niche (see Brookshire, 1976; Hodos, 1982). Likewise, brain structures have developed along several lines, and one usually finds a "mosaic-like" pattern even within a particular line. In other words, an animal may have a high degree of specialization or effi-

*Dedicated to Professor Valdo Mazzi who inspired our comparative reflections.

NATO ASI Series, Vol. G17
Intelligence and Evolutionary Biology
Edited by H.J. Jerison and I. Jerison

ciency in some brain areas or behavioural patterns without its necessarily being the case for other parts (e.g. Armstrong, 1982).

References to "simple" and "complex" organisms without a context for evaluating behaviors or structures may be erroneous. This is why a comparative analysis may effectively contribute to the study of biological intelligence and artificial intelligence, provided it follows Bullock's three R's: roots, rules and relevance (Bullock, 1983). In fact, a comparative analysis:

(i) suggests alternative intelligence-models;
(ii) sets methodological constraints and sets limits on comparisons between animals and human beings;
(iii) increases our knowledge by suggesting the evolutionary mechanisms underlying I-processes in biological systems and how they are established. In the following review we shall be looking at qualitative changes in the CNS and their behavioural implications.

We shall try to demonstrate the following points:

(a) the need for a rationale in the comparative approach to appreciate "what" and "why" we are comparing but also causes of correspondence;

(b) the heuristic limits of homology and other related concepts;

(c) the importance of multi-level homological analysis, comprising behaviour, anatomy, morphofunctional organization, molecular biology and embryology. Comparisons may be at one level, while validation and explanation may be found at another one.

(d) the effects of fresh data from molecular and cell biology that can affect the understanding of brain evolution, and the impact of such changes on hypotheses concerning ontogeny/phylogeny relationships and the parsimony concept;

(e) the limitations of existing brain evolution theories. So far they have not explained the evolutionary mechanisms underlying changes in nervous and behavioural structures (and their heritability);

(f) the consequences of complexity theory and the self-organization concept in the epigenesis of cell networks.

Character Comparison. Concepts and Their Application

As Northcutt recently pointed out (1984), "the ability to describe phylogenetic changes (i.e., any definable attribute of an organism in any character) is based on the pattern of variation observed among different taxa. Equally important, elucidation of evolutionary mechanisms or processes is based on the kinds of character pattern that can be recognized. In both of these analyses it is critical to distinguish a character and its subsequent phylogenetic transformation (homologous characters) from other characters that may appear similar, but have different evolutionary histories (homoplasous characters) if errors in interpretation are to be minimized" (p. 701).

In the same review Northcutt discussed the concepts of character comparison depth, focussing on character similarity and common ancestry. He finally adopted Wiley's definitions (1981), as follows:

Homology: "A character of two or more taxa is homologous if this character is found in the common ancestor of these taxa, or, two characters (or a linear sequence of characters) are homologous if one is directly (or sequentially) derived from the other(s)."

Homoplasy: "A character found in two or more species is homoplasous (non-homologous) if the common ancestor of these species did not have the character in question, or if one character was not the precursor of the other."

"Convergence is the development of similar characters from different pre-existing characters".

Notwithstanding these apparently clear cut definitions, since Darwin homology has been defined one way and tested in another way, i.e., definitions of homology have been based on common ancestry, but criteria for homologue recognition have generally rested on phenetic similarity (Remane, 1956; Simpson, 1965; Bock, 1977).

In cladistic analysis, homologous features are considered to be those which characterized monophyletic groups (Patterson, 1982), and in order to recognize shared derived characters, one must determine the direction of change or polarity (i.e., primitive versus derived condition) of the characters that are suspected of being homologous on the basis of phenetic similarity. Three criteria are usually given: (1) out-group rule, (2) ontogenetic character precedence, (3) geological character precedence.

The first two criteria are useful in comparative neurology. The out-group rule, initially proposed by Hennig (1966), and here defined by Northcutt, states that "given two characters that are homologues and found within a monophyletic group, the character that is also found in the sister group is the primitive (plesiomorphic) character, whereas the character found only within the monophyletic group is the derived (apomorphic) character."

The theory of ontogenetic character precedence (von Baer's theorem) establishes character polarity based on comparison of developmental patterns rather than the distribution of characters among adults in closely related taxa. Von Baer's theorem states that members of two or more closely related taxa will follow the same course of development to the stage of their divergence. Thus characters observed to be more general are assumed to be primitive, whereas those that are less general are assumed to be derived.

Examples given in Northcutt's review (1984) include the out-group rule for corpus callosum in mammals and ontogenetic precedence for telencephalon development. It is often difficult to apply these criteria to the analysis of variations in CNS characters, as we will see in the following pages.

Moreover, several crucial points remain unclear:

(1) We need a definition and efficient use of the terms, "character" and "resemblance."

(2) Phenetic similarity might be an elegant way to by-pass some theoretical difficulties on the morphology/function duality, as clearly shown by Hailman (1976).

(3) The rank of the taxa under investigation may vary greatly (closely related, as between species in the same family, or distantly related as in different orders or classes). Accordingly, the use of homology changes varies substantially, and

many homologies in fact reflect grades of organisation rather than phylogenies (Gould, 1976 a).

(4) The homology for a single character can be used for a multivariate analysis either to build new taxonomical and phylogenetic trees of the species compared, or to prompt for phylogenies and infer evolutionary mechanisms for the single character itself through different taxa.

(5) The sampling used in the comparison might be a limitation, as the quantitative approach to the evolution of CNS clearly shows (e.g. in Martin, 1982). Moreover, how can we account for individual variability? Important intraspecific differences have been described, for instance, in the rodent CNS (Wahlsten, 1977; Oliverio, 1983, 1985).

(6) In functionally equivalent but independently derived systems "many of the CNS areas involved might be homologous even when the peripheral adaptations are clearly homoplasous" (Wilczynski, 1984).

(7) Since homological recognition always embodies ancestry (and "heritability"), the so called phenotypic modulation (Smith-Gill, 1983) must be considered.

(8) Homology might be a more powerful tool when studying poorly known structures, as it may suggest functions (and relations).

(9) Finally, in many cases the similarities between characters at one level (e.g. behavioural) are based on evidence for a different level, as in Hodos's definition (1976).

"Behaviors are considered homologous to the extent that they can be related to specific structures that can, in principle, be traced back through a genealogical series to a stipulated ancestral precursor irrespective of morphological similarity" (p. 156).

Several of these difficulties will be discussed in detail later on in the analysis of homology recognition at different levels. For the sake of brevity we shall now proceed by examples.

Homology and the study of animal behaviour

Eibl-Eibesfeldt wrote (1970): [Homologous] "behaviour patterns are of great taxonomic value and can help to elucidate the natural relationships among animals." In fact, the anticipated contribution of classical ethology to systematics and evolution has been of little value when considered by itself, whereas, when integrated into the comprehensive biology of a taxonomic group (e.g. Kessel,1955), it has thrown light on adaptive radiation processes.

It is useful to note in considerations of the evolutionary approach to behaviour how quickly we are moving away from classical ethological models and theory in attempts to understand the unschematic and never totally predictable variability of behaviour. Instinct and the consequent ethogram concept (an exhaustive repertoire of behaviour patterns of a species) have been and are powerful operative concepts for practical ethological work (Schleidt et al., 1985). However, they can easily lead to the concept of animals' inheriting stereotyped movements in response to simple environmental stimuli.

"Descent in most cases implies a direct genetic relationship, where the information, which concerns the adaptiveness of the behaviour pattern in question, is passed through the genome" (Eibl-Eibesfeldt, 1970). Support for this model came from studies of simple Mendelian segregation (see the example of the hygienic bee, Rothenbuler, 1964) or from hybrid studies (see the example of *Agapornis personata x roseicollis* Dilger, 1962). If we judge the relevance of a discipline by its fruits we must say that "Mendelian" studies on behaviour and the hybrid technique approach have grown weaker.

With sociobiology and behavioural ecology a more correct approach to evolutionary biology has developed: the concept of natural selection at the individual level as the major force in evolution. Behavioural ecology is not particularly interested in observing behavioural homologies but rather in convergent adaptations (i.e., analogies). The principal aim of this comparative approach is to describe the associations between morphological, physiological and behavioural traits. In contrast to classical ethology, however, only gross behavioural and ecological traits are compared in order to formulate laws predicting causal or relational links between variables (Jarman, 1981).

The adaptationist paradigm is the guideline for a neodarwinist researcher; the central problem at the heart of this discipline remains the relationship between genes and particular behavioural patterns. Selection will not lead to evolutionary change unless there are genetic differences that in some way induce physiological structural or behavioural differences among individuals that affect their fitness. The concept of heritability derived from quantitative genetics is a powerful tool for the understanding of natural evolutionary changes even if this concept is not the solution to the nature-nurture problem (Arnold, 1981).

But according to some authors most workers find the concept of heritability confusing. "Heritability is the portion of phenotypic variance in a population that is attributable to genetic variance but does not measure the importance of genes in determining an individual phenotype" (Paul, 1985). A trait may have a high degree of heritability and be extremely sensitive to environmental changes. Conversely, even in traits with low but statistically significant heritabilities artificial selection usually alters phenotypic means (Cade, 1984). In behaviour genetics recent contributions are focusing on the interaction of genes at different loci during behavioural development. According to Bateson and D'Udine (1986) "genetic analysis can be reconciled with the modern notion of behavioural development once the prevalence of inter-locus interaction has been accepted."

In considering the functional approach to behaviour we must ask whether behavioural ecology can be a firm reference-point in the gradual progress of behavioural sciences in an evolutionary perspective.

The comparative approach, rarely used in the past in a completely scientific manner now provides a reliable method of verifying testable predictions, even though some problems persist. But there are other problems (Felsenstein, 1985): (i) the dissimilarity of bibliographical data and the selection of variables to consider; (ii) the choice of a taxonomical level in order to isolate independent phylogenetic convergences; (iii) interpretation of relationships where cause-effects links are not always clear; (iv) assumption of independence in regression analysis, whereas species connected in an evolutionary tree are not fully independent. Solutions can be found if our understanding of the topology of branching in the phylogenetic tree improves. This is the only way the divergent and the convergent comparative approach can come together to explain organic diversity.

Another point concerns the approach of behavioural ecology to proximate causes of behaviour. By choice, this discipline is not concerned with genetic, ontogenetic and physiological causes of behaviour, acknowledging its limitations in these respects. One may foresee, however, that its black-box approach will give way to models dealing with morphophysiological mechanisms so that phylogenetic inertia and developmental constraints to adaptation can be considered in more detail and not merely acknowledged. Causal and functional approaches to biology can remain conceptually autonomous and with equal rank, but the one needs the other in a multi-disciplinary approach. In this perspective neurobiological issues (e.g. on CNS plasticity and development) can contribute to the comprehension of behaviour in an evolutionary perspective. Thus research in the next few years should reach a higher level of maturity, keeping in mind that evolutionary biology includes many theoretical models based on vague ideas about natural selection. James and McCulloch (1985) remind us that congruence of observations with predictions is not sufficient justification for accepting a theory, and theoretical models such as natural selection can be studied only experimentally.

Anatomical level

Classical comparisons (and recognition) of homologous structures in comparative neurology relied on anatomy. Morphological criteria, such as sulci, embryonic grooves and cell masses (Campbell, 1982) were employed whatever the goal of comparative investigation, whether evolutionary or any other. These studies had been very successful at one time, and have recently been revived in Rudolf Nieuwenhuys's topological study of vertebrate brain stems (Nieuwenhuys, 1972, 1974).

The goals of such morphological analyses vary with the individuals. In some cases they are directed toward understanding changes in a portion of the nervous system by sampling representatives of different classes on the basis of existing taxonomy (e.g. Nieuwenhuys, 1974). Or they can be constructions of tentative phylogenetic trees built with multiple brain traits (often biologically heterogeneous as in the case of Johnson et al. [1982, a, b, 1984]).

Such pattern analysis is however tainted by a certain logical bias tied both to a reasoning "polarity" (primitive versus complex) and some circularity between homological recognition and taxonomical allocation of the species under investigation. For instance, two contrasting ways of reasoning were used for the evolution of the amphibian forebrain. Hodos (1982) states that "the relatively poorer differentiation of the forebrain structures of these animals (modern amphibians) led [early investigators] to believe that this represented the primitive or primordial state. However, an examination of nontetrapod vertebrates indicates that their forebrain structures are quite well differentiated. The seemingly simpler organization of the amphibian forebrain thus appears to be a derived state rather than an ancestral condition" (p. 46).

On the other hand, Northcutt and Kicliter (1980), on the basis of comparisons with living anamniotic forms, brain-body ratio and so on, concluded that "all these data strongly suggest that the telencephalon of amphibians is not degenerate (i.e., secondarily simple) but that the amphibian pattern of telencephalic organization is characteristic of a primitive anamniotic level of telencephalic organization that repeatedly produced advanced forms in which the telencephalon underwent pallial hypertropy and differentiation" (p. 247).

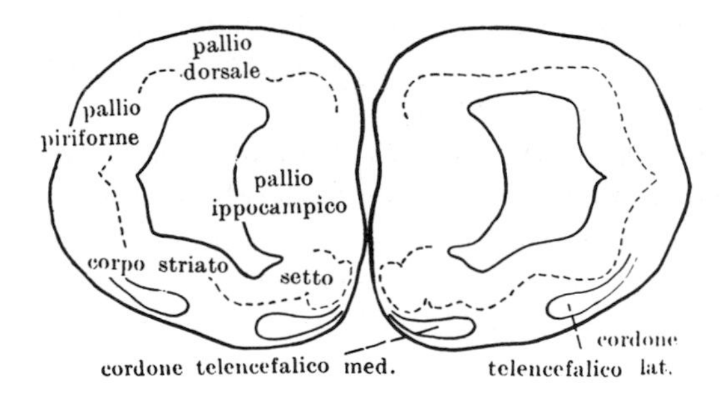

Figure 8.1. The "Bauplan" concept. Highly diagrammatic drawing of a crosssection from the telencephalon in *Necturus* showing the basic pattern of telencephalic organization in vertebrates (from Herrick, redrawn by N. Beccari, 1943. *Neurologia comparata*, Sansoni, Firenze).

The use of morphological criteria and the reference to recognizable patterns of nervous organization address the concept of "morphological pattern (Bauplan)" (Figure 8.1). The "morphological pattern" follows Cuvier's argument and represents a central point of the idealistic school. The Gegenbaur School of Comparative Anatomy, on the other hand, stressed the evolutionary significance of morphology in strong contrast to idealistic approaches.

More recently, the "neo-classical" school of idealistic morphology originating in Jacobshagen's (1925) work stressed the concept of "Bauplan" as an abstraction or conceptualization of all significant spatial configurational relationships shared by a group of organisms (Kuhlenbeck, 1967, p. 161). In this approach the Bauplan "must be properly understood as a justifiable and very useful fiction with a wide range of operational validity (Kuhlenbeck, p. 163), and it might be applied more strictly to macrotaxonomy.

In this instance, the Bauplan theory comprises only those configurations that are actually shared by groups of organisms. The concept can be translated into a topological interpretation of morphology. Homology is very important and can be considered bound to two (or more) morphological configurations (patterns) that are "essentially" (i.e., in the aspect under consideration) "the same", because there is a one-to-one correspondence between them that obtains in topological structures embodied in the system of interlocking (open) sets. We are dealing with the general problem of the relationship between parts (a concept comprised in the homology concept) but which the Bauplan placed within a topological space.

The great comparative neuroanatomist J.C. Herrick repeatedly attacked the "mystical realm" of the classical idealistic approach (in Kuhlenbeck, 1967), but he used the concept of a basic pattern, though with aims different from those of the formal-analytic school. He wrote:

"But when the vertebrate phylum is viewed as a whole, the nervous apparatus shows a wider range of adaptive structural modifications of this common plan than is exhibited by any other system of organs of the body. In order to understand the significance of this remarkable plasticity and the processes by which these diverse patterns of nervous organization have been elaborated during the evolutionary history of the vertebrates, it is necessary to find out what were the outstanding features of the nervous system of the primitive ancestral form from which all higher species have been derived." (Herrick, 1948, p. 13)

"It is probable that none of the existing Amphibia are primitive in the sense of survival of the original transitional forms and that the urodeles are not only aberrant but in some cases retrograde (Noble, 1931; Evans, 1944); yet the organization of their nervous systems is generalized along very primitive lines, and these brains seem to me to be more instructive as types ancestral to mammals than any others that might be chosen ....

"In brief, the brains of urodele amphibians have advanced to a grade of organization typical for all gnathostome vertebrates, *Ambystoma* being intermediate between the lowest and the highest species of Amphibia. This brain may be used as a pattern or template, that is, a standard of reference in the study of all other vertebrate brains, both lower and higher in the scale." (Herrick, 1948, pp 16-17)

In conclusion, the "basic pattern" is very useful as a frame of reference for comparative analysis in the evolutionary approach as well, but it is transformed into an operational tool employed in a strategy where primitive (or degenerated forms) are chosen as a template for understanding ancestral brain organizations, and it refers more to analogy than to true homological recognition. Thus it appears to be a less powerful approach for derived characters and for the development of hypotheses about evolutionary processes.

Morphofunctional Level

Criteria such as neural connections or histochemical properties of cell groups (i.e., morphofunctional grounds) have been used recently in attempting to identify homologies.

Examples of new approaches are research by degeneration (e.g. in sharks, Ebbesson, 1980), transport degeneration techniques (e.g. in amphibians, Northcutt and Kicliter, 1980, and in fishes Northcutt and Bradford, 1980, Echteler and Saidel, 1981). However, these valuable trends in comparative neurology introduce a functional criterion rather than being strictly anatomical or morphological. The identification of ascending spinal pathways in vertebrates (Northcutt, 1984) is based on the final targets of these pathways.

Histochemical comparison was also employed, e.g. in the analysis of acetylcholine esterase (AChE) distribution in the telencephalon of amniotic vertebrates (Parent and Olivier, 1970) and in the proposed homology of the dorsomedial telencephalon of teleost with the corpus striatum of land vertebrates, both on a histochemical basis (high concentration of acetyl cholinesterase and heavy catecholaminergic innervation) and on topographical data (Northcutt and Bradford 1980). Such histochemical analysis is approximate, however, since it recognizes only large topographical brain areas, and it is usually unable to deal with the many functionally different neuron types present within such areas.

Recently developed cytofunctional methods seem more powerful. Cytochemical procedure (Falck and glyoxylic acid) for amines and immunocytochemistry (and allied immunochemical procedures), in particular, can identify neuronal systems and their comparative and anatomical-functional relationships (see for instance, Northcutt and Reiner, 1985). (Figure 8.2). Moreover, cytofunctional methods can support homologies based on histochemistry (as for evolution of the amniote basal ganglia, Reiner et al., 1984; Fasolo et al., in preparation).

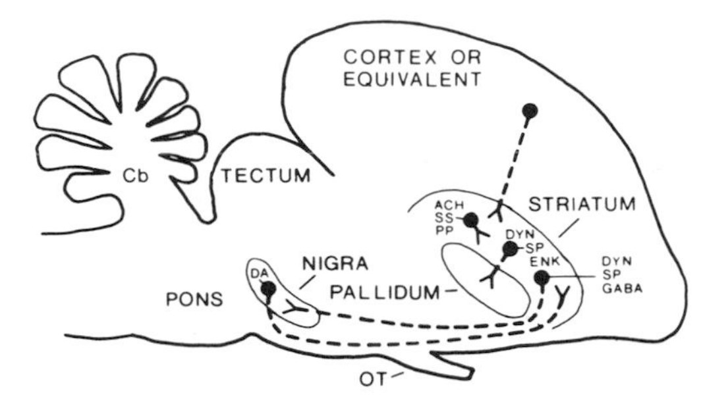

Figure 8. 2. Sagittal view of a schematized amniote brain showing the major cell types and connections of the basal ganglia that are common to all amniotes studied. The "striatal" subdivision of the basal ganglia receives an input from neocortical neurons (in mammals) or from neurons found in the field homologue of neocortex (in birds and reptiles). Three types of neurons are found in the striatum, namely, intrinsic neurons, striatopallial projection neurons and striatonigral projection neurons. These neuron types may not always be mutually exclusive, e.g. striatopallidal neurons may also project to the nigra. Three transmitter specific-intrinsic neuron types have been identified in the amniote striatum, namely, acetylcholine-containing (ACh) neurons, somatostatin-containing (SS) neurons and pancreatic polypeptide-like immunoreactivity-containing (PP) neurons. Among striatopallidal neurons, separate populations of substance-P-containing (SP), enkephalin-containing (ENK) and dynorphin-containing (DYN) neurons appear to project from the striatum to the pallidum in amniotes. Some substance-P-containing and dynorphin-containing "striata" neurons project to the nigral cell group of the midbrain. It is presently unknown if the same individual peptidergic striatal neurons project to both the pallidum and the nigra. In birds and mammals GABAergic striatonigral projection neurons have also been demonstrated. Neurons of the nigra contain dopamine (DA) and project back to the striatum. Abbreviations: Cb-cerebellum (from Reiner et al., 1984).

In general, the results obtained so far lead to three possible conclusions: (a) constancy; (b) trends toward complexity; (c) heterogeneity (Fasolo et al., 1985 a, b).

(a) Good examples of *constancy* in tetrapod Vertebrates are, for instance, the Substance P system in the spinal dorsal horns and serotonin cells in raphe and periventricular regions.

(b) A good example of a *trend toward complexity* is hypothalamic organization. In a study of representatives of Tetrapod classes, where we see different grades, both in classic neuro-secretory and in parvicellular nuclei, from the simple periventricular gray of amphibians to the nuclei (i.e., segregated cell masses) of mammals and birds.

(c) *Heterogeneity* is even more frequent. For instance, Substance P, which is involved in many neuroendocrine functions, is seemingly present in the median eminence of the newt, *Xenopus*, frog, quail, man and primates, but not in the bullfrog, the lizard, the pigeon, the rat (see Fasolo et al., 1985; Stoeckel and Porte, 1985)). Many of these discrepancies may be due to the different antibodies employed or to other technical problems. Nevertheless, such heterogeneity is puzzling, to say the least.

As an example let us consider the distribution of neuropeptide Y (NPY) in the neurohypophysial complex. In anurans (Danger et al., 1986) it innervates the pars intermedia, where it seemingly serves as an MSH (melanocyte stimulating hormone) inhibiting factor, while in urodeles (Fasolo et al., in preparation) the pars intermedia lacks such NPY innervation. We can explore two hypotheses: a) NPY exerts different neuroendocrine and neurochemical actions in different amphibia, and b) in different amphibia it exerts the same functions by means of different neuroanatomical structures. Our most recent data confirm the first hypothesis, since NPY does not inhibit the release of alpha-MSH *in vitro* from newt hypophysis.

Developmental Level

The theory of ontogenetic character precedence is a fundamental phylogenetic criterion. An example of its use is the study of neuronal development of the telencephalon in order to understand primitive versus derived conditions in amniotic vertebrates (Northcutt, 1981, 1984). In general, however, embryological criteria other than sulci or cell masses have been rarely used in inferring homology. Moreover, little attention has been paid to the role of *developmental constraints* (defined as biases on the production of variant phenotypes or a limitation on phenotypic variability caused by the structure, character, composition and dynamics of the developmental system [Maynard-Smith et al., 1985]) as possible sources for similarities different from homology or convergent evolution. Another growing area of interest is *developmental plasticity* (Smith-Gill, 1983). This problem is important not only for the elucidation of the relationship between ontogeny and phylogeny but raises another interesting question.

According to Smith-Gill (1983), it can be assumed that:

> "Developmental plasticity may be of two forms: developmental conversion or phenotypic modulation. In developmental conversion, organisms use spe-

cific environmental cues to activate alternative genetic programs controlling development. These alternative programs may either lead to alternative morphs, or may lead to the decision to activate a developmental arrest. In phenotypic modulation, nonspecific phenotypic variation results from environmental influences on rates or degrees of expression of the developmental program, but the genetic programs controlling development are not altered. Modulation, which is not necessarily adaptive, is probably the common form of environmentally induced phenotypic variation in higher organisms, and adaptiveness of phenotypic plasticity, therefore, cannot be assumed unless specific genetic mechanisms can be demonstrated."

In the case of the central nervous system recent data support the hypothesis of phenotypic modulation. For instance, surgical adrenalectomy or drug treatment induces the coexpression of CRF (corticotropin releasing factor) and vasopressin in the paraventricular nucleus of adult brains, whereas in other (normal) conditions the two neuropeptides are produced by separate cells (Wolfson et al., 1985). It remains to be seen where such mechanisms represent a nonspecific answer and where they are significant and, in the second instance, how they can be selected.

Molecular Level

Molecular comparisons have been widely used in order to understand either the evolution of a molecule within living beings or the so-called molecular phylogeny (the use of the biochemical approach to elucidate relationships). The two approaches are complementary in some ways. The evolution of a protein molecule can be guessed by comparing the primary (i.e., the amino acid sequence) or the secondary and tertiary structure (Niall, 1982) of such molecules in different species. Molecular homologies can be drawn on the basis of these comparisons. Moreover mechanisms of genetic diversification or requisites for different functional roles in various species can also be inferred. Having thus provided the tempo and mode for the evolution of a molecule (e.g. a peptide hormone or a globulin, or an enzyme), phylogenetic trees can be constructed for related species.

Another, more recent, approach, based on genetic engineering, which uses restriction mapping of DNA or direct DNA sequencing, has proved very valuable in creating evolutionary trees (e.g. for the apes see discussion in Cherfas, 1982).

Several enzyme molecules which are important in information transfer have been studied and often show a high homology (e.g. K, Na-ATPase, Cantley, 1986; tyrosine-hydroxilase, Joh et al., 1983) in different taxonomical groups.

The potency of new DNA technologies currently makes it easier to learn sequences (and genes) than to understand their coded products. This seems true for the highly conserved homeoboxes and for the *per* locus of *Drosophila* (Shin et al., 1985). The locus has a fundamental involvement in the expression of biological rhythms in this insect, and mutations at this locus can shorten, lengthen or eliminate a variety of rhythmic activities that range from circadian behaviour (i.e., eclosion, locomotor activities) to short-period behavior, such as the 55-s-rhythm of the courtship song. Apparently the *per* sequence is highly conserved in several vertebrate genomes (i.e., man, mouse, toad, chicken), and it is thought to contribute to the organization of biological rhythms in many species. More information is available for neuropeptides and neurotransmitters in the nervous system.

TACHYKININ PEPTIDES

Peptide	Sequence	Source
	1 5 10	
Physalaemin	pGlu-Ala-Asp-Pro-Asn-Lys-Phe-Tyr-Gly-Leu-Met-NH_2	Amphibian
[Lys^5,Thr^6]-Physalaemin	——— Lys-Thr ———	Amphibian
Uperolein	—— Pro ——— Ala ———	Amphibian
Phyllomedusin	pGlu-Asn ——— Arg —— Ile ———	Amphibian
Kassinin	Asp-Val-Pro-Lys-Ser- Asp-Glu —— Val ———	Amphibian
Enterokassinin	Asp-Glu-Pro-Asn-Ser- Asp-Gln —— Ile ———	Amphibian
[Glu^2,Pro^5]-Kassinin	Asn-Glu-Pro-Lys —— Asp-Glu —— Val ———	Amphibian
Hylambatin	Asp-Pro-Pro ——— Asp-Arg ——— Met ———	Amphibian
Enterohylambatin	Asp-Pro-Pro-Asn —— Asp-Arg ——— Met ———	Amphibian
Eledoisin	—— Pro-Ser-Lys- Asp-Ala —— Ile ———	Molluscan
Substance P	Arg-Pro-Lys —— Gln-Gln —— Phe ———	Mammalian
Substance P-like Peptide	Asp-Ile- Pro-Lys —— Asp-Gln —— Phe ———	Amphibian
Neurokinin A	His-Lys-Thr-Asp-Phe —— Val ———	Mammalian
Neurokinin B	Asp-Met-His- Asp-Ser —— Val ———	Mammalian

MAMMALIAN LHRH:

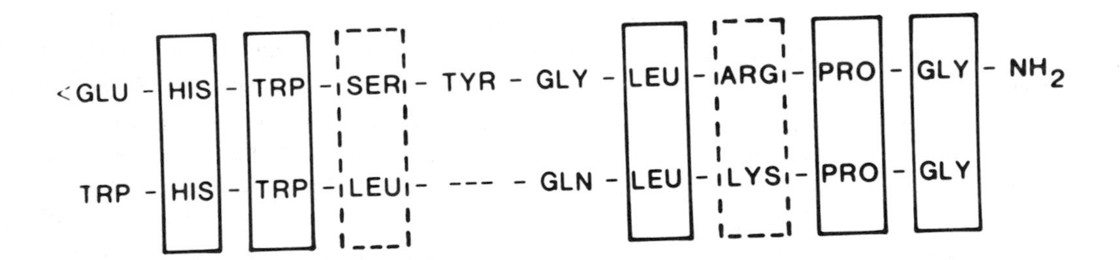

YEAST α FACTOR:

Figure 8.3. Molecular similarities in active peptides. (a) The tachykinin family (from Lazarus et al., 1985). (b) Similarity of vertebrate and microbial sex factor. Amino acid sequences of mammalian luteinizing hormone- releasing hormone (LHRH) and the N-terminus of alpha-mating factor of the common yeast, *Saccaromyces cerevisiae* (from Le Roith and Roth, 1984).

The results of the past decade show that intercellular communication in vertebrates is not unique to the endocrine and nervous systems. The messenger molecules that are associated with these systems are produced by other cells in vertebrates, multicellular invertebrates, unicellular organisms and higher plants. The biochemical elements of intercellular communication seem to have originated evolutionarily in unicellular organisms, possibly in microbes. These fundamental biochemical mechanisms seem to have been highly conserved throughout evolution (the paleocentric theory, Le Roith and Roth, 1984, Roth et al., 1985). Several examples can be cited, e.g. the distribution of Substance-P-like peptides (the tackychinin family) in vertebrates and invertebrates (Fasolo et al., 1985; Gaudino et al., 1985) (Figure 8.3).

Another relevant example is the striking similarity of vertebrate and microbial sex factors. The Alpha factor, a mating ectohormone of the yeast (*Saccharomyces cerevisiae*) not only has extensive sequence homology with the decapeptide LHRH (luteinizing hormone releasing hormone), but also can bind to LHRH receptors of rat pituitary gonadotropes inducing LH release (Loumaye, 1982). This may be the effect of strong evolutionary pressure on these molecular mechanisms (not only the messenger molecules themselves, but also processing enzymes and receptors).

The role of neuropeptides in evolution is open to speculation. For instance, Joosse (1986) emphasizes the differences between classic neurotransmitters, which are processed locally in neurons, and the neuropeptides, which are directly gene-related. The apparent conservative pattern for basic neuropeptide families (and their genes) may suggest further hypotheses. Gene coding for peptides is less likely to have been repeatedly formed anew during evolution. If such genes were retained for long periods, there are two possibilities: (a) the genes have remained non-expressed and represent the "archives" of peptide memory (Acher, 1984), in which case these non-expressed regions would escape the pressure of selection; (b) they have always been expressed, i.e., the coded peptides have always been functional (see also Lazarus et al., 1985). Joosse stresses the role of neuropeptides in evolutionary plasticity of neuron systems. He believes that peptides are much more appropriate for experiments with environmental changes. In addition, the proven coexistence of several neuropeptides and neurotransmitters in the same neuron (Hokfelt et al., 1984) as well as the plasticity of neurotransmitters at the molecular level (Black et al., 1984) represent further steps in the development of different signalling systems.

To sum up, subtle changes in neuropeptides (and their receptors) may have been favoured by evolutionary pressures, while the basic pattern of messenger molecules is quite ancient.

The conservative pattern exhibited by informational molecules shifts our interest from molecular to higher organization levels. Le Roith and Roth (1984) "suggest that the biochemical elements and the functional elements are highly conserved and that it is the anatomical elements that permit more sophisticated uses of these systems which have undergone evolution." According to Medawar (in Schmitt, 1984), "Endocrine evolution is not an evolution of hormones but of the uses to which they are put". In other words, when we discuss the evolution of CNS (or anything else) we have to turn to supracellular organization and to cell differentiation within the framework of socio-cytological interactions.

Also in this context we encounter difficulties in envisaging molecular homologies. The development of new assays and modern techniques has made possible a new biochemical assault on the classic problem of cell adhesion and how it relates to both morphogenesis and primary processes of development. Three cell adhesion molecule (CAM) types have been detected and isolated thus far (Edelman, 1985), and evidence is building of strong evolutionary conservation of CAM structure and function in all vertebrates and perhaps also Metazoa (Rutishauser and Goridis, 1986). "It is important to note that these observations do not indicate that CAM's from different species are non specific but rather that the mechanism by which they contribute to form and histogenesis must be dynamic and hierarchical, involving regulatory mechanisms as well as protein binding specificity" (Edelman, 1985, p. 152). In summary, these data reflect the twin "conservatism and opportunism of evolution" (Simpson, 1959).

Neurobiology and Theories of CNS Evolution

"Critical evaluation of current hypotheses concerning CNS evolution reveals that these hypotheses generally describe patterns of character variation and rarely address processes" (Northcutt, 1984, p. 712). After a cogent criticism of the most popular theories on CNS evolution in vertebrates (encephalization, parcelling, invasion, equivalent cell group hypotheses), Glenn Northcutt concludes his review with a syncretic touch: "Migration, invasion, and loss have all undoubtedly occurred in CNS evolution, but none of these phenomena is sufficient, by itself, to explain CNS evolution and each can be discounted in specific cases....These conclusions, however, do not address the equally important questions of what selective pressure resulted in these phenomena and what advantages were conferred by these changes." (p. 713).

In our opinion, this would be possible only after a cross-level analysis of heritability focusing on a specific problem and a more subtle understanding of mechanisms leading to neural specificity. At present the developmental analysis of identified neuronal systems both *in vivo* and *in vitro* may be the most promising field.

The increasing possibility of obtaining *in vitro* organotropic cultures of neurons has led us to envisage the phenotypic plasticity of neurons (e.g. in neural crest lineages, Le Douarin, 1986, Le Douarin et al., 1985) as well as some definite peculiarities in their function and morphology. An elementary behavioral circuit (Montarolo et al., 1986) *in vitro* was reproduced in cultures of certain neurons of *Aplysia*. The neuritic properties shown by some neurons in culture (intrinsic neurons of the olfactory bulb or dopaminergic neurons of the nigro striatal pathway, Denis-Donini and Estenoz, 1985) and the specific pattern of fibre outgrowth from transplants (as in mouse hippocampi, Zhou et al., 1985) imply that a given neuronal population shares a certain general character as well as molecular cues for establishing neuronal connectivity. *In vivo*, the concepts of morphogenetic field (Goodwin, 1985) and the "equivalence group" (Sulston and White, 1980) could be very useful to understanding how neuron groups sharing some common properties are modulated.

"Equivalence groups," or ontogenetic buffering mechanisms, refer to a control model .system. An equivalence group as suggested for *Drosophila* or *Caenorbalditis elegans* would consist of a set of neighbouring precursor cells which, according to the simplest hypothesis, are all initially at the same state of determination. The final differentiation might depend on several different cues, both genetic and environmental, including sociocytological interaction and cell cooperativity.

We may gain some deeper insights from research on the now classic models, such as worms, slugs and even *Drosophila.* There we find an apparent paradox: In such simple animals, which produce only simple behaviours, a small number of neurons and cells generally serve a large number of functions. While the *C. elegans* (a 1,000-cell organism) has at least 100 cell types, a mammal (a many billion-cell organism) has less than an order of magnitude more (several hundred) cell types. According to White (1985), the 302 neurons of the *C. elegans* NS may be placed in 118 classes according to morphology and synaptic connectivity. Although this is undoubtedly an overestimate of cell types, clearly, for all its paucity of neurons, *C. elegans* NS has a rich variety of neuron types. In contrast, a typical 1 cerebellum (e.g., of the cat) has billions of neurons in 5 neuron classes. In a certain sense, therefore, in multicellular structures a thousand cells are the equivalent of a single cell in the worm.

It is tempting to conjecture about the two different models, i.e., vertebrates *versus* invertebrates, and the differences in their data handling capacities (and Intelligence?). It would seem that the limited number of neurons of many different types is based on a rather rigid genetic determination, whereas in vertebrates, possibly due to limits on the expansion of genetic control (i.e., a cellular constraint), epigenetic control is overwhelming (see Changeux, 1983. At the molecular level there may be much more diversity in the vertebrate brain. Thus, in the rat the number of 1 RNA's unique to the brain has been estimated at nearly 30,000 [Sutcliff et al., 1984; Nauta and Feirtag, 1986]).

The existence of ontogenetic buffering mechanisms has been suggested as a way of maintaining concordance in the developing organism and of integrating discordant mutations with viable organism that can then be perpetuated during evolution (Katz et al., 1981, Katz, 1982). If these molecular buffering mechanisms could be demonstrated, we could exploit a new trend toward understanding how ontogeny channels phylogeny.

Certain elegant and in some ways complementary solutions have recently been proposed (Lipp; Deacon, this Volume). Lipp, in particular, suggested a high penetration of genetic variation of top-ranking CNS coordinating systems, implying that after a change of environmental pressure these alleles become the preferred initial targets for natural selection. He postulated that encephalization is an evolutionary process that provides a species not only with intelligence but also with increased buffer capacity for accumulating genetic variability of brain and behaviour, which is an important precondition for speciation. Consequently, the more complex the brain of a species, the greater the potential of the species for genetic adaptation to new ecological situations, and, because of the non-randomness of the canalized selection, the faster its evolutionary transformation. Deacon, on the other hand, posited that internal human brain organization is determined by brain and body size disproportion, possibly due to some early embryological events. Such changes

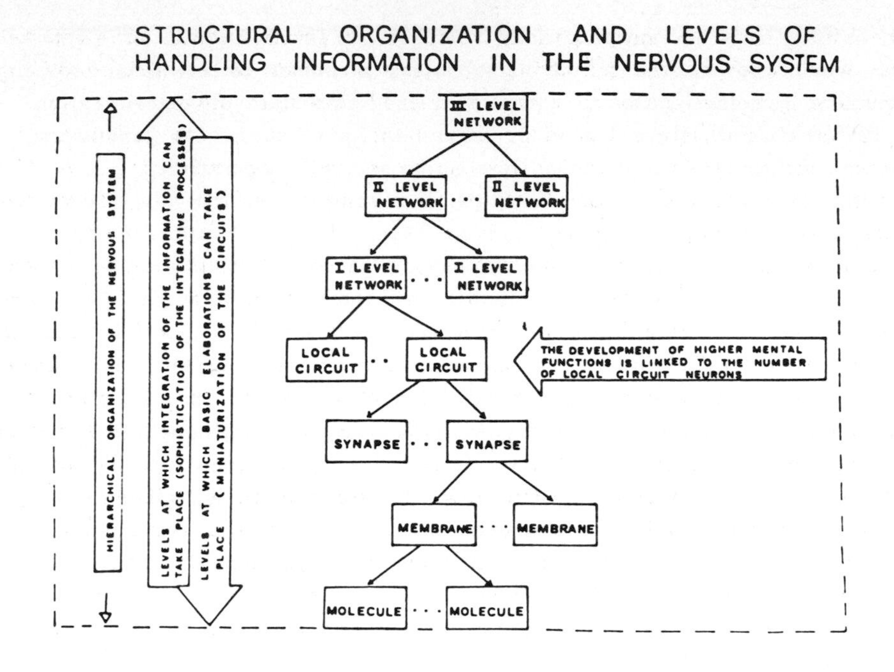

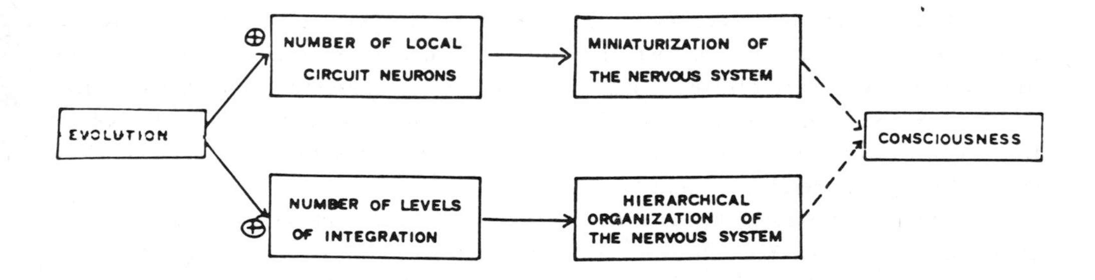

Figure 8.4. (a) Schematic illustration of hypothesis on the structural organization and levels of handling information in the nervous system. (b) Schematic illustration of hypothesis on the evolution of the nervous system (from Agnati and Fuxe, 1984).

in the early stages of morphogenesis result in relatively higher neuronal proliferation, and biases in the parcelling of disproportionate brain structures lead to enhanced neurological and behavioral properties of language centers.

These two contributions do not seem mutually exclusive. On the contrary, they strengthen the idea that low-rank early mutations can affect the buffer size of CNS, favouring the "last word effects" of genetic variation acting on late top-ranking systems.

The emphasis on epigenetic mechanisms fits in well with some new concepts on the structure of neuronal networks, such as the local circuit model, the concept of miniaturization and hierarchical organization of CNS, receptor-receptor interactions (Agnati and Fuxe,1985; Fuxe et al., 1985) (Figure 8.4), and parasynaptic modulation (Schmitt, 1984). It compels us, however, to discard the concept of "genetic program" (Goodwin, 1986), which is fundamentally deterministic, implies a programmer, and is not capable of explaining evolutionary novelty during its production (Atlan, 1986). Evolutionary novelty is produced by a nondeterministic, self-organizing complex system, where its meaning is organized "ex post" within a given context of observation (Atlan, 1985).) The "genetic program" metaphor does not explain how functional concordance (i.e., the appropriate matching) among all the interacting products of different genes can be maintained after genomic changes.

Models and Laws in Brain Organization

New data on the cellular and molecular bases of neuronal functions have revived interest in the causes of similarity. This means that for a single character a multilevel analysis is necessary to understand the relationships between phenetic similarity and the generative program of such similarity. In fact, if homology is resemblance caused by continuity of information (Van Valen, 1982), such continuity should be tested independently of other parameters so as to validate the homological recognition by an independent criterion. This approach, in turn, creates the problem of the extent of similarity due to developmental, cellular or genetic constraints which produce the structure in space and time.

A tempting solution would be the discovery of some general laws for cellular, embryological and genetic controls. The results are not promising. After a careful analysis of behavior and homologous neurons in the insect brain, Franz Huber concluded, "The implications of insect neuroethology for the study of vertebrates are, therefore, restricted" (1983, p. 127).

Even in connection with an extremely simple differentiated organism, such as the worm *Caenorhabditis elegans* (about 1,000 cells), we can agree with Sydney Brenner, who "...lost faith in the perfection of biological systems a long time ago....Anything that is produced by evolution is bound to be a bit of a mess!" No simple general statement can be made in terms of symmetrical features, global trends, gross morpho-genetic invariants or the like, for 1,000 cell organisms. Brenner continues, "I'm not sure that there necessarily is anything more to understand than what it is.....That is a possibility for this level of complexity" (in Lewin 1984).

Conclusions

In comparative studies on the CNS homology has usually been employed in a loose way. At the moment, the formal-analytical approach offers a more plausible explanation of morphological findings and of homology, but it is unable to deal with subtler similarities (and the molecular "causes" of homology). In contrast, the evolutionary approach can explain the causes, but it is unable to deal rigorously with similarities. A multilevel approach to similarity and a multivariate analysis of changes and invariants might resolve contradictions in results, but a new definition for homology is needed anyhow.

As in the case of other useful concepts, "the heuristic value of homology is not without its limits and has become nearly exhausted," as Ernst Scharrer pointed out as long ago as 1946. Considering the current expansion of functional approaches and the choice of representatives in different taxa, the use of analogy with respect to grades in neural organization (Gould, 1976b) seems more suitable. In some ways von Cranach's position (1976) could be acceptable, i.e., that homology represents a limited and peculiar case of the concept of analogy (see also Dessi'-Fulgheri, 1979).

As Stephen Jay Gould (1976b, p.176) said, an analog, by which he means an organ or organism similar to a particular fossil in what it does rather than where it has come from, offers some useful cues when it is considered as "most like", a replicate, or a reinforcer of the functional theme. And it also offers us a chance to compare different generative paths in similar structures.

Varius multiplex multiformis -- Marguerite Yourcenar

Acknowledgment With financial support from M.P.I. grant (40% Fasolo)

References

Acher R (1984) Evolution of proteins: from gene determination to cellular integration. In: B Guerra (ed) Biomembranes. 357-403, Plenum, New York London

Agnati L F, Fuxe K (1984) New concepts on the structure of the neuronal networks: the miniaturization and hierarchical organization of the central nervous system. Bioscience Reports 4:93-98

Armstrong E (1982) Mosaic evolution in the primate brain: differences and similarities in Hominid thalamus. In: Armstrong E, Falk D (eds) Primate brain evolution. 131-161. Plenum, New York London

Arnold S J (1981) Behavioral variation in natural populations. I. Phenotypic, genetic and environmental correlations between chemoreceptive responses to prey in the garter snake, *Thamnophis elegans*. Evolution 35: 489-509

Atlan H (1985) Complessita', disordine e autocreazione del significato. In Bocchi G, Ceruti M (eds) La sfida della complessita'. 158-178. Feltrinelli, Milan

Bateson P, D'Udine B (1986) Exploration in two inbred strains of mice and their hybrids: additive and interactive models of gene expression. Animal Behav. 34: 1026-1032

Black IB, Adler JE, Dreyfus CF, Jonakait GM, Katz DF, La Gamma EF, Markey KM (1984)

Neurotransmitter plasticity at the molecular level. Science 225: 1266-1270

Bock WJ (1977) Foundations and methods of evolutionary classification. In Hecht MK, Goody PC, Hecht BM (eds) Major patterns in vertebrate evolution. 851-895. Plenum, New York London

Brookshire KH (1976) Vertebrate learning: evolutionary divergences. In: Masterton R B, Bitterman ME, Campbell CBG, Hotton N (eds) Evolution of brain and behavior in vertebrates. 191-216. Erlbaum, Hillsdale, New Jersey

Bullock TH (1982) Afterthoughts on animal minds. In Griffin DR (ed) Animal mind-human mind. 407-414. Springer-Verlag, Berlin Heidelberg New York

Bullock TH (1983) Implications for neuroethology from comparative neurophysiology. In Ewert J P, Capranica R R, Ingle D J (eds) Advances in vertebrate neuroethology. 53-75. Plenum, New York London

Cade WH (1984) Genetic variation underlying sexual behavior and reproduction. Amer. Zool. 24: 355-366

Campbell CBG (1982) Some questions and problems related to homology. In Armstrong E, Falk D (eds) Primate brain evolution. 1-11. Plenum, New York London

Cantley L (1986) Ion transport systems sequenced. Trends in Neurosci 9:1-3

Changeux JP (1983) L'homme neuronal. Fayard, Paris

Cherfas J (1982) Proving the pattern of life. In: Cherfas J (ed) Darwin up to date. IPC Magazines Ltd. London, pp. 39-42

Cranach M Von (1976) Methods of inference from animal to human behaviour. Aldine, Chicago

Danger JM, Leboulenger P, Guy J, Benyamina M, Pelletier G, Tonon MC, Vaudry H (1986) Le NPY localisé dans le système hypothalamo-hypophysaire des amphibiens agit comme un facteur MIF. In: Colloque National des Neurosciences, Bordeaux 22-26 Avril 1986

Davis R E, Northcutt RG (eds.)(1980) Higher brain areas and functions. Fish neurobiology II. Univ of Michigan Press

Denis-Donini S, Estenoz M (1985) Studio dei fattori che controllano la crescita neuritica durante lo sviluppo del sistema nervoso centrale. Atti Associazione Biologia Cellulare e del differenziamento. 4:56

Dessi-Fulgheri F (1982) Il problema del confronto tra il comportamento animale e quello umano. In Parisi V, Robustelli F (eds.) Il Dibattito sulla sociologia. Pubbl. Istituto Psicologia CNR 39-52

Dilger WC (1962) The behavior of lovebirds. Scientific American 206:88-98

Ebbesson SOE (1980) On the organization of the telencephalon in Elasmobranchs. In: Ebbesson SOE (ed) Comparative neurology of the telencephalon. 11-14. Plenum, New York London

Echteler SM, Saidel WM (1981) Forebrain connections in the goldfish support telencephalic homologies with land vertebrates. Science 212: 683-684

Edelman GE (1985) Cell adhesion and the molecular processes of morphogenesis. Ann. Rev. Biochem. 54: 135-169

Eibl-Eibesfeldt I (1970) Ethology. The biology of Behavior. Holt, Rinehart, Winston. New York

Evans FG (1944) The morphological status of the modern Amphibia among the Tetrapoda. J. Morphol. 74: 43-100

Fasolo A, Gaudino G, Franzoni MF (1985 a) Cytofunctional methods in comparative neurology: results, perspectives, difficulties. Atti Accademia Scienze di Torino 119:123-126

Fasolo A, Gaudino G, Franzoni MF, Andreone C (1985 b) Immunocytochemical distribution of neuropeptides in the hypothalamus of submammalian vertebrates: comparative and evolutionary considerations. X Int. Symposium of Comparative Endocrinology. Copper Mountain

Felsenstein J (1985) Phylogenies and the comparative method. American Naturalist 125: 1-25

Fuxe K, LF Agnati, K Andersson, M Martire, SO Ogren, L Giardino, N Battistini, R Grimaldi, C Farabegoli, A Harfstrand and G Toffano (1985, in press) Receptor-receptor interactions in the central nervous system. Evidence for the existence of heterostatic synaptic mechanisms.

Gaudino G, A Fasolo, G Merlo, LH Lazarus, T Renda, L D'Este and F Vandesande (1985) Active peptides from amphibian skin are also amphibian neuropeptides. Peptides 6 (Suppl.): 209-213

Goodwin BC (1985) La traduzione della complessità biologica in una sottile semplicità. In: G Bocchi, Ceruti M (eds.) La sfida della complessità. 246-258. Feltrinelli, Milano

Gould SJ (1976 a) Grades and clades revisited. In RB Masterton, W Hodos and HJ Jerison (eds.) Evolution, Brain, and Behavior, Persistent Problems. 115-122. Erlbaum, Hillsdale, New Jersey

Gould SJ (1976 b) In defense of the analog: a commentary to N Hotton. In RB Masterton, W Hodos and HJ Jerison (eds) Evolution, Brain, and Behavior, Persistent Problems. 175-179. Erlbaum, Hillsdale, New Jersey

Griffin D R (ed.) (1982) Animal mind-Human mind. Springer-Verlag, Berlin Heidelberg New York

Hailman JP (1976) Homology: logic information and efficiency. In RB Masterton, W Hodos, HJ Jerison (eds.) Evolution, brain and behavior, Persistent problems. 181-198. Erlbaum, Hillsdale, New Jersey

Hennig W (1966) Phylogenetic systematics. University of Illinois Press, Urbana Illinois

Herrick CJ (1948) The brain of the Tiger salamander (*Ambystoma tigrinum*). University of Chicago Press, Illinois

Hodos W (1976) The concept of homology and the evolution of behavior. In RB Masterton, W Hodos, HJ Jerison (eds.) Evolution, brain and behavior. Persistent problems. 153-167. Erlbaum, Hillsdale, New Jersey

Hodos W (1982) Some perspectives on the evolution of intelligence and the brain. In: DR Griffin (ed.) Animal mind-human mind. 33-56. Springer-Verlag, Berlin Heidelberg New York

Hokfelt T, Johansson O, Goldstein M (1984) Chemical anatomy of the brain. Science 225: 1326-1334

Huber F (1983) Implications of insect neuroethology for studies on vertebrates. In: Ewert JP, Capranica RR, Ingle DJ (eds.) Advances in vertebrate neuroethology. 91-138. Plenum, New York London

Jacobshagen E (1925) Allgemeine vergleichende formenlehre der tiere. Klinkhardt, Leipzig

James FC, McCulloch CE (1985) Data analysis and the design of experiments in ornithology. In F Johnston (ed.) Current Ornithology. 2:1-63. Plenum, New York London

Jarman PJ (1981) Prospects for interspecific comparison in sociobiology. In: King's College Sociobiology Group, Cambridge (ed.) Current problems in sociobiology. 323-342. Cambridge University Press, Cambridge UK

Jerison HJ (1985) Animal intelligence as encephalization. Phil. Trans. Roy. Soc. Lond. 308: 21-35

Joh TH, EE Baetge, ME Ross (1983) Similar gene coding regions for catecholamine biosynthetic enzymes: possible existence of ancestral precursor gene. 5th Int. CA Symposium, Gothenburg, Sweden

Johnson JI, JAW Kirsch, RC Switzer (1982 a) Phylogeny through brain traits: fifteen characters which adumbrate mammalian genealogy. Brain Behav. Evol. 20: 72-83

Johnson JI, RC Switzer, JAW Kirsch (1982b) Phylogeny trough brain traits: the distribution of categorizing characters in contemporary mammals. Brain Behav. Evol. 20: 97-117

Johnson JI, JAW Kirsch, RC Switzer (1984) Brain traits through phylogeny: evolution of neural characters. Brain Behav. Evol. 24: 169-176

Joosse J (1986) Neuropeptides: peripheral and central messengers of the brain. In CL Ralph (ed.) Comparative endocrinology. Developments and directions. 13-32. Liss, New York

Katz M, RJ Lasek, R Kaiserman-Abramof (1981) Ontophyletics of the nervous system: Eyeless mutants illustrate how ontogenetic buffer mechanisms channel evolution. Proc. Natl. Acad. Sci. (USA) 78: 397-401

Katz M (1982) Ontogenetic mechanisms: The middle ground of evolution. In JT Bonner (ed.) Evolution and development. 207-212. Springer-Verlag, Berlin

Kessel EL (1955) Mating activities of balloon flies. Systematic Zoology 4: 97-104

Kuhlenbeck H (1967) The central nervous system of vertebrates. Vol. 1: Propaedeutics to comparative neurology. S. Karger, Basel, New York

Lazarus LH, WE Wilson, G Gaudino, BJ Irons, A Guglietta (1985) Evolutionary relationship between nonmammalian and mammalian peptides. Peptides 6 (Suppl 3):295-307

Le Douarin NM (1980) The ontogeny of the neural crest in avian embryo chimaeras. Science 286: 663-669

Le Douarin NM (1986) Cell line segregation during peripheral nervous system ontogeny. Science 231: 1515-2522.

Le Douarin NM, ZG Xue, J Smith (1985) *In vivo* and *in vitro* studies on the segregation of autonomic and sensory cell lineages. J. Physiol. (Paris) 80: 255-261

LeRoith D, J Roth (1984) Vertebrate hormones and neuropeptides in microbes: evolutionary origin of intercellular communication. In L Martini, WF Ganong (ed.) Frontiers in Neuroendocrinology 8:1-25. Raven, New York

Lewin R (1984) Why development is illogical? Science 224: 1327-1329

Loumaye E, J Thoner, KJ Calt (1982) Yeast mating phenomene activates mammalian gonadotrophs: evolutionary conservation of a reproductive hormone. Science 218: 1323-1325

McGinnis W, RI Garber, J Wirz, A Kuroiwa, WJ Gehring (1984) A homologous protein-coding sequence in Drosophila homeotic genes and its conservation in other metazoans. Cell 2: 37

Martin RD (1982) Allometric approaches to the evolution of the primate nervous system. In E Armstrong and D Falk (eds.) Primate Brain Evolution. Plenum Press, New York, London, 39-56

Maynard-Smith J, R Burian, S Kauffman, P Alberch, J Campbell, B Goodwin, R Lande, D Raup, L Wolpert (1985) Developmental constraints and evolution. The Quarterly Review of Biology 60: 265-287

Montarolo PG, P Goelet, VF Castellucci, J Morgan, ER Kandel and Schacher (1986, in press) Protein and RNA synthesis inhibitors block long-term heterosynaptic facilitation of *Aplysia* gill-withdrawal reflex

Nauta WJH, Feirtag, M (eds) (1986) Fundamental neuroanatomy. Freeman and Co., New York

Niall HD (1982) The evolution of peptide hormones. Ann. Rev. Physiol. 44: 615-624

Nieuwenhuys R (1972) Topological analysis of the brain stem of the Lamprey (*Lampetra fluviatilis*). J. Comp. Neur. 145: 165-178

Nieuwenhuys R (1974) Topological analysis of the brain stem: a general introduction. J. Comp. Neurology. 156: 255-276

Noble GK (1931) The biology of the Amphibia. Mc Graw Hill, New York

Northcutt RG, Bradford MR Jr (1980) New observations on the organization and evolution of the telencephalon of actinopterygian. In Ebbesson SOE (ed) Comparative neurology of the telencephalon. 41-95. Plenum, New York London

Northcutt RG (1981) Evolution of the telencephalon in nonmammals. Ann. Rev. Neurosci. 4: 301-350

Northcutt RG (1984) Evolution of the vertebrate central nervous system: patterns and processes. Amer. Zool. 24: 701-716

Northcutt RG, Kicliter EE (1980) Organization of the amphibian telencephalon. In: Ebbesson SOE (ed) Comparative neurology of the telencephalon. 203-255. Plenum, New York

Northcutt RG, A Reiner (1985) An immunohistochemical study of the telencephalon of the african lungfish. Abstracts USA Society for Neuroscience. Vol. 11: 1310

Oliverio A (1983) Genes and behavior: an evolutionary perspective. Advances in the study of behavior 13: 191-217

Oliverio A (1985) Geni e comportamenti: un approccio evoluzionistico. In L Bullini, M Ferraguti, F Mondella and A Oliverio (eds.) La vita e la sua storia. Stato e prospettiva degli studi di genetica. Scientia, PP. 253-264

Parent A, Olivier A (1970) Comparative histochemical study of the corpus striatum. J. Hirnforsch. 12: 75-81

Patterson C (1982) Morphological characters and homology. In: KA Joysey and AE Friday (eds) Problems of phylogenetic reconstruction (Systematics Association Special Volume 21). Academic Press, New York, pp. 21-74

Paul DB (1985) Textbook treatments of the genetics of intelligence. The Quarterly Review of Biology 60: 317-326

Reiner A, SE Brauth and HJ Karten (1984) Evolution of the amniote basal ganglia. Trends in Neurosci 7: 320-325

Remane A (1956) Die grundlagen des naturlichen systems der vergleichenden anatomie und phylogenetik. Geest und Portig KG Leipzig

Roth J, D Le Roith, MA Lesniak, E Collier. (1985) Molecules of intercellular communication in vertebrates, metazoa, and microbes: do they share common origins? Abstracts of Int. Symposium on Nonmammalian Peptides, Roma May 11-15. pp.13

Rothenbuhler WC (1964) Behavior genetics of nest cleaning in honey bees. IV. Responses of F, and backcross generations to disease-killed brood. American Zoologist 4: 11-123

Rutishauser U, C Goridis (1986) NCAM: the molecule and its genetics. TIG March, 72-76

Scharrer E (1946) Anatomy and the concept of analogy. Science 103: 578-579

Schleidt WM, G Yakalis, M Donnelly and J McGarry (1984) A proposal for a standard ethogram, exemplified by an ethogram of the bluebreasted quail (*Coturnix chinensis*). Z. Tierpsyethol 64: 193-220

Schmitt FO (1984) Molecular regulators of brain function: A new view. Neuroscience 13: 991-1001

Shin HS, TA Bargiello, BT Clark, FR Jackson and MW Young (1985) An unusual coding sequence from a Drosophila clock gene is conserved in vertebrates. Nature 317: 445-446

Simpson GG (1959) Anatomy and morphology: classification and evolution: 1859 and 1959. Proceedings of the American Philosophical Society, 103: 286-306

Smith-Gill SJ (1983) Developmental plasticity: developmental conversion versus phenotypic modulation. Amer. Zool. 23: 47-55

Stoeckel ME, A Porte (1985) Immunocytochemical localization of substance P in neurohemal zones. Gunma Symposium on Endocrinology 22: 157-169

Sulston JE and White JG (1980) Regulation and cell autonomy during postembryonic development of *Caenorhabditis elegans*. Dev. Biol. 78: 577-597

Sutcliff JG, JR Milner, JM Gottesfeld and W Reynolds (1984) Control of neuronal gene expression. Science 225: 1308-315

Van Valen LM (1982) Homology and causes. J. Morphol. 173: 305-312

Wahlsten D (1977) Heredity and brain structure. In: Oliverio A (ed) Genetics, environment and intelligence. 93-115. Elsevier, North Holland

White JG (1985) Neuronal connectivity in *Caenorhabditis elegans*. Trends in Neurosci 8:277-283

Wilczynski W (1984) Central neural systems subserving a homoplasous periphery. Amer. Zool. 24: 755-763

Wiley E O (1981) Phylogenetics. Wiley, New York

Wolfson B, R W Manning, L G Davis, R Arentzen and F Baldino (1985) Co-localization of corticotropin releasing factor and vasopressin mRNA in neurones after adrenalectomy. Nature 315: 59-61

Zhou C F, G Raisman and R J Morris (1985) Specific patterns of fibre outgrowth from transplants to host mice hippocampi, shown immunohistochemically by the use of allelic forms of HY-1. Neuroscience 16: 819-833

LANGUAGE, INTELLIGENCE, AND RULE-GOVERNED BEHAVIOR

Philip Lieberman
Brown University, Providence RI 02912, USA

One of the odder phenomena of the past thirty years is that Chomsky's transformational, i. e., "generative" linguistic theory denies the major premises of modern biological thought but nonetheless claims to have a biological basis. Chomsky and his adherents claim to hold a nativist biological position. Generative grammar supposedly reflects the species-specific neural constraints of the human brain. Chomsky (1980a), for example, claims that human beings all have a "universal grammar" that derives from a neural mechanism, a "fixed nucleus," that he places in the left, dominant hemisphere of the brain. Given this strong biological claim, one would expect that linguists were working in a biological framework, where the properties of the human brain would constrain linguistic theories concerning the ontogenetic development language, the relationship between language and cognition, and the evolution of language. We might expect linguists to take account of the findings of neuroanatomy, developmental and comparative psychology, and evolutionary biology. However, it is clear these biological data are irrelevant to linguists because present linguistic theories are, at best, disjoint from any actual characteristics of the human brain, human biology or human evolution.

The Present Linguistic Model

The standard transformational model as developed by Chomsky (1980a, 1980b) and his associates (e. g. Fodor, 1980) makes the following claims:

1. The brain consists of a number of independent "modules." The neural processing performed in a given module determines a particular mode or component of cognitive behavior, e. g., syntax, the linguistic lexicon, speech perception, vision. The modules are each disjoint from other modules and other aspects of human cognitive behavior. Thus the "language organ" is disjoint from any other aspect of human cognition.

Chomsky and Fodor would probably deny that their theory derives from mechanistic rather than biological models. It is supposedly based on "logical" design principles. Principles that would lead to the simplest, most logical and "economical" theory for the neural mechanisms that underlie human linguistic ability. However, their neural theory appears to be based on the design logic and details of present-day electronic computers rather than that of biological brains. Modular design is logical when one builds a computer. It is advantageous to assemble computer systems

NATO ASI Series, Vol. G17
Intelligence and Evolutionary Biology
Edited by H. J. Jerison and I. Jerison

that are made up of a set of independent "modules." This simplifies their design and facilitates their repair. The serviceman can run a diagnostic test, pull out the defective module, and replace it. However, there is no particular reason why the human brain should be designed this way. We can't change defective parts. Moreover, the design of the human brain is not "logical." Like any other aspect of any other living organism, it reflects its evolutionary history (Changeux, 1980). Its design logic is the historical logic of evolution which is coded in our genetic endowment. There is no reason to believe that a modular system is biologically "simpler" than an integrated system in which neural mechanisms that are adapted for cognitive behavior function for both language and thought. Behavioral data, moreover, point to a connection between linguistic ability and intelligence. The data derived from literally hundreds of studies of human intelligence, for example, demonstrate that the scores obtained on standard tests of intelligence like the Stanford-Binet are closely correlated with the size of the vocabulary of an individual (Sternberg, 1985). In the limiting condition, language is absent in severely retarded people (Wills, 1973).

2. Since the standard generative grammatical theory claims that the neural determinants of cognitive behaviors like language or the "capacity to deal with the number system" are modular, they "logically" could not have evolved by means of Darwinian natural selection. Chomsky (1980a), for example, argues that since there would be no selective advantage for a partially evolved faculty, the faculty could not have evolved. It presumably had to spring forth, fully developed, at the "current stage" of human evolution. The only model of human evolution that would be consistent with the present standard generative theory is a sudden coordinated set of saltations that simultaneously furnished human beings with the neural bases for various aspects of cognitive behavior and language. There would have to be a set of saltations because the neural modules are independent. The genetic changes would have to be extreme, abrupt and coordinated. Since human beings like all other living organisms are put together in genetic "bits and pieces" there would have to be a coordinated series of saltations - a most unlikely possibility (Mayr, 1982).

3. The neural modules that underlie language and cognition supposedly do not mature in the course of ontogenetic development. Fodor (1980), for example, claims that "higher" cognitive structures cannot develop from "simpler" ones. Human infants supposedly are born with a fully developed linguistic "competence" in place (Pinker, 1984).

4. According to current linguistic theory, all human beings supposedly have the same linguistic competence. Since linguistic competence hypothetically derives from an innate neural "language organ," linguists must claim that linguistic ability is transmitted by means of genes that never vary throughout the total human population. This would be unlike any other aspect of the morphology of any living organism. Every observation since Darwin proposed the theory of evolution by means of natural selection, demonstrates that genetic variation always occurs for every aspect of biology in every living population (Mayr, 1982). The linguistic model is clearly an example of essentialistic thinking, which as Mayr demonstrates, is simply inappropriate for characterizing the biologic endowment of any living organism.

Linguists when challenged on this issue often respond with facile arguments like, "all people have noses." They miss the obvious point that is as plain as your nose or my nose - all human noses are different. Noses, hearts, lungs are all different for different people. The study of variation is an inherent and central aspect of biology and physiology. If the biological bases of language were indeed uniform across the human population, it would place them outside the explanatory range of modern science. However, it is clear that this is not the case. Recent data, e. g., Lieberman et al., (1985) show variations in the perception of speech that probably reflect genetic variation. Some individuals are unable to identify sounds that other individuals readily identify; the situation is similar to that which occurs in the visual domain for color-blindness. Syntactic ability also appears to vary for different people. In the limiting condition it is almost absent in severely mentally retarded individuals (Wills, 1973).

5. Linguists and a great many psychologists accept the assumption that data relevant to linguistic "competence" can somehow be differentiated from other data that relate to "performance." The distinction between competence and performance is based on the individual linguist's intuition. The only data that formal linguists really use involve intuitive judgments of a sentence's "grammaticality." The discussions of linguists have become remarkably similar to those of medieval schoolmen (or rather a parody staged by Monty Python). Two competing "theories" are subjected to a crucial test - which theory predicts that a peculiar sentence is "grammatical." The joke is that the "crucial" sentence is usually incomprehensible and is "grammatical" to one linguist, ungrammatical to the other.

The competence-performance distinction has become a convenient way of dismissing data that are not in accord with the particular transformational theory of the moment. The perceptual data of Miller and Isard (1963) were once regarded as crucial. They demonstrated the psychological reality of the details of the transformational theory conveyed in *Syntactic structures* (Chomsky, 1957). However, these data became trivial and uninteresting manifestations of "performance" when transformational theory was reformulated in *Aspects of the theory of syntax* (Chomsky, 1965). As Bunge (1985) notes, linguistic theories unlike scientific theories, are not tested against actual data. Linguistic theories are instead tested against theoretical "competence" data. The ultimate determinant of competence is the linguist's intuition. The linguistic enterprise violates the principles that scientists have attempted to follow since the time of Francis Bacon. Intuitive judgements, in themselves, are worthless as data.

A Biological Model

I would like to advance a model for the biological bases and evolution of human language and cognition. The model (Lieberman, 1984, 1985) attempts to take account of as great a range of biological phenomena as possible. For example, though the physiology of the brain is not understood, the human brain like all other organs undergoes a process of maturation as an infant develops into an adult. We therefore should expect linguistic ability to increase as a child grows precisely *because* there may be an innate biological component to human linguistic ability. Though the present form of the human brain is obviously species-specific, it has an evolutionary history and much is known concerning this history. We therefore should expect to find homologues between human beings and other animals. We also should *not* expect to find extreme modularization of the brain. Some aspects of linguistic and cognitive ability may reflect common underlying neural mechanisms. Like all parts of living organisms, the physical characteristics of the human brain are genetically transmitted and people's brains therefore must vary. We therefore would expect to find variation in human linguistic ability.

The biological model proposes that human linguistic ability derives from a number of biological "mechanisms." Some of these biological mechanisms like the lexicon are present in reduced form in other living species. We probably do not differ qualitatively from closely related animals like chimpanzees in this regard. The linguistic lexicon is the mental "dictionary" in which the meanings and sounds or gestures that signal words are stored. The various chimpanzee language studies (Gardner and Gardner, 1984; Savage Rumbaugh et al., 1985) demonstrate, beyond a reasonable doubt, that apes use, acquire, and transmit concepts by means of words. The neural base of the lexicon probably is a Hebbian "distributed" neural network. A distributed neural network inherently has "associative" properties (Anderson, 1972). The "intuitive" aspects of cognition also may be linked to this same distributed neural network, i. e., a neural system that is not located in one small part of the brain.

In contrast, other aspects of human language and cognition may involve localized, species-specific neural mechanisms. Human speech, human logical thinking and the syntactic "rules" of human language appear to involve specialized parts of the brain that are presently found only in human beings. The "unique" species-specific aspects of human language appear to be speech and syntax. These elements give human language its speed and complexity. Human speech (and probably human syntax) appear to have evolved by means of Darwinian natural selection over the last 500,000 to 1.5 million years.

Human Speech

Muller in 1848 demonstrated that the production of human speech involves the modulation of acoustic energy by the supralaryngeal airway. This acoustic energy is generated by the larynx or by turbulent airflow at a constriction in the airway. Until the 1960's, however, it wasn't realized that human speech is an important component of human linguistic ability. Speech allows us vocally to transmit phonetic "segments" (the letters of the alphabet approximate phonetic segments) at a rate of up to 25 per second. In contrast, it is impossible to identify other non-speech data at rates that exceed 7-9 items per second. A short, 2 second-long sentence can contain about 50 sounds. The previous sentence, for example, which consists of approximately 50 phonetic segments can be uttered in two seconds. (The 50 orthographic symbols and numbers would yield about 50 phonetic segments.) If this sentence had to be transmitted at the non-speech rate, it would take so long that a human listener might well forget the beginning of the sentence before hearing its end. The high data transmission rate of human speech is thus an integral part of human linguistic ability, allowing complex thoughts to be vocally transmitted within the constraints of short-term memory.

Although sign language can also achieve a high data transmission rate, the signer's hands can't be used for other tasks. Nor can viewers see the signer's hands except under restricted conditions. Though visual hand signs still function as part of the linguistic code (McNeill, 1985), the primary linguistic channel is vocal. Vocal language represents the continuation of the evolutionary trend towards freeing the hands for carrying and tool use that started 6 million years ago with upright bipedal hominid locomotion.

The key biological mechanisms that are necessary for human speech are:

(1) the supralaryngeal vocal tract and "matching" neural mechanisms that

(2) govern the complex articulatory maneuvers that underlie speech, and

(3) decode the acoustic cues for linguistic information in the speech signal.

The process by which the high transmission rate of human speech is achieved involves the generation of "formant-frequency" patterns and rapid temporal and spectral cues by the species-specific human supralaryngeal airway and its associated neural control mechanisms. Formant frequencies are simply the frequencies at which maximum acoustic energy will get through the supralaryngeal airway, which acts in a manner similar to the way a pipe organ lets maximum acoustic energy through at certain frequencies. Both the pipe organ and the supralaryngeal airway act as "filters," letting relatively more acoustic energy through at the formant frequencies. During the production of speech we continually change the shape and make small adjustments in the length of the supralaryngeal airway, thereby generating a changing formant frequency pattern. The "sounds" of human speech differ with respect to their formant frequency patterns, as well as with respect to temporal factors and the acoustic source filtered by the supralaryngeal airway. The fundamental frequency of phonation, which reflects the rate at which the vocal cords of the larynx open and close, determines the "pitch" of a speaker's voice. The fundamental frequency pattern can convey linguistic information like the pitch "tones" that differentiate words in languages like Chinese, independent of the formant frequency pattern.

Figure 9.1 shows the typical non-human airway in which the tongue is positioned entirely within the oral cavity where it forms the lower margin. The midsagittal view shows the airway as it would appear if the animal were sectioned on its midline from front to back. The position of the long, relatively thin tongue reflects the high position of the larynx. The larynx moves up into the nasopharynx during respiration, providing a pathway for air from the nose to the lungs that is isolated from any liquid that may be in the animal's mouth. Non-human mammals and human infants, who have this same morphology until age 3 months, can simultaneously breathe and drink. The ingested fluid moves to either side of the raised larynx which resembles a raised periscope protruding through the oral cavity, connecting the lungs with the nose. Figure 9.2 shows the human configuration. The larynx has lowered into the neck. The tongue's contour in the midsagittal plane is round, and it forms the anterior (forward) margin of the pharynx as well as the lower margin of the oral cavity. Air, liquids and solid food make use of the common pharyngeal pathway. Humans thus are more liable than other terrestrial animals to choke when they eat because food can fall into the larynx, obstructing the pathway into the lungs.

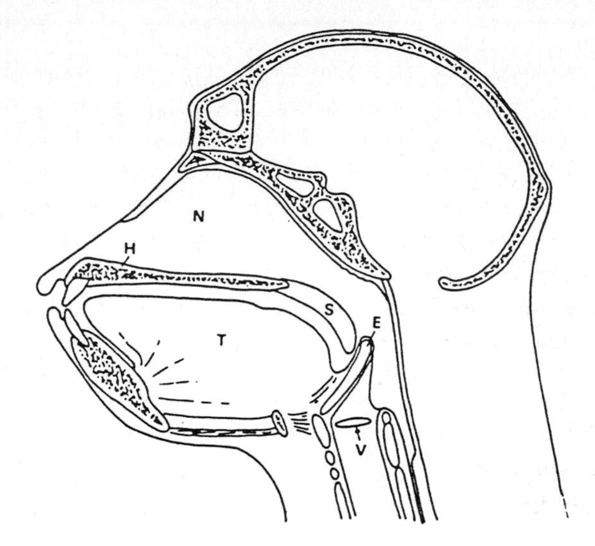

Figure 9.1. Midsagittal view of the head and supralaryngeal vocal tract of an adult chimpanzee. N; nasal cavity; H: hard palate; S: soft palate; E: epiglottis; T: tongue; V: vocal cords of larynx. Note that the tongue is long and thin and lies within the mouth of the chimpanzee. The larynx can lock into the nasal cavity providing a sealed airway from the nose to the lungs. (Adapted from Laitman and Heimbuch, 1982.)

The ontogenetic development of the human vocal tract.

During ontogenetic development the human palate moves back with respect to the bottom, i. e., the base of the skull. The base of the human adult skull is restructured in a manner unlike that of all other mammals to achieve this supralaryngeal airway (Laitman and Crelin, 1976). The adult human configuration is less efficient for chewing because the length of the palate (the roof of the mouth) and of the mandible (lower jaw) has been reduced compared to non-human primates and archaic hominids (Laitman, et al. 1979). The reduced length of the palate and mandible also crowds our teeth, presenting the possibility of infection due to impaction - a potentially fatal condition until the advent of modern medicine. These vegetative deficiencies are offset, however, by the increased phonetic range of the human supralaryngeal airway. The round human tongue moving in the right-angle space defined by the palate and spinal column can generate formant frequency patterns like those that define vowels like [i], [u], and [a] (the vowels of the words *meet, boo,* and *mama*) and consonants like [k] and [g]. These sounds have formant frequency patterns that make them and sounds like the consonants [b], [p], [d], [t] better suited for vocal communication than other sounds. They occur more often in different human languages and are acquired earlier by children. The error rate for misidentification of the vowel [i] is particularly low and it can serve as an optimum cue in the perceptual process of vocal tract "normalization."

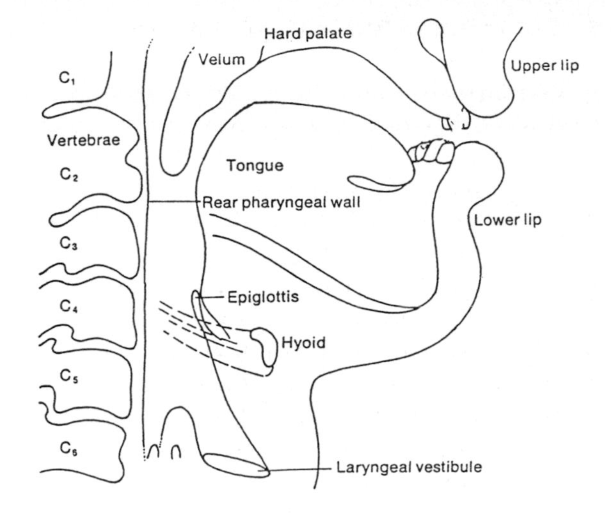

Figure 9.2. Midsagittal view of the adult human supralaryngeal vocal tract. The shape of the tongue has changed from that of the nonhuman primate. Note that the tongue forms the lower and front boundaries of the mouth and pharynx. It is round in cross-section. The position of the larynx is low and it can not form a sealed airway from the nose to the lungs. The round shape of the human tongue makes it possible to produce sounds like the vowels [i] and [u] which facilitate the perceptual "decoding" of human speech.

The human supralaryngeal vocal tract thus is maladapted for respiration, swallowing and chewing. It would be useless for speech unless matching neural mechanisms were also present. No other living species has these neural mechanisms or the human supralaryngeal vocal tract. The presence of the human supralaryngeal in various extinct hominids thereby serves as an "index" for the presence of these neural mechanisms (Lieberman, 1984, 1985).

Speech perception

The perception of human speech is a complex process that appears to involve general auditory mechanisms as well as neural mechanisms that have been "matched" to the constraints of the human supralaryngeal airway by means of natural selection. Specialized neural mechanisms operate at different stages in a "speech mode" of perception in which human listeners appear to apply different strategies or mechanisms to a speech signal than they would if it was a non-speech signal. First, human listeners are able to "extract" the formant and fundamental frequencies of speech signals, even when these signals have been degraded by telephone circuits or noise. The process by which human beings extract the formant frequencies appears to involve the listener's "knowing" the filtering characteristics of the supralaryngeal airway at some internal neural level of representation.

Second, human listeners must also form an estimate of the probable length of the speaker's supralaryngeal airway in order to assign a particular formant frequency pattern to an intended phonetic target. Human beings have supralaryngeal airways of different lengths. Those of young children are, for example, half the length of those of adults. There is overlap in the mapping of formant frequency patterns with intended phonetic targets owing to this dispersion in anatomical airway length. The word *bit* spoken by a large adult male speaker can have the same formant frequency pattern as the word *bet* produced by a smaller individual. The longer supralaryngeal airway will produce lower frequency formant frequencies for the "same" phonetic element. Human listeners must take this factor into account. Human beings when listening to speech also integrate an ensemble of acoustic cues and contextual constraints, e. g., rate, phonetic environment, etc., that are related by the physiology of speech production. They assign patterns of formant frequencies and short-term spectral cues into discrete phonetic categories in a manner that is consistent with the presence of neural "detector" mechanisms. These neural detectors appear to be "matched" or tuned to respond to the particular acoustic signals that the human speech-producing anatomy can produce.

The evolution of the human supralaryngeal airway and matching perceptual mechanisms is probably similar in kind to the match between anatomy and auditory perception that has been demonstrated in other species such as crickets, frogs, and monkeys. Human speech first makes use of anatomical structures and neural perceptual mechanisms that also occur in other species, e. g., the larynx is similar for all hominoids, rodents possess the auditory mechanisms for certain linguistic cues (voice onset time), etc.

The anatomy and neural control mechanisms that are necessary for the production of the complex formant frequency patterns that typify human speech, however, appear to have evolved comparatively recently, in the last 1.4 million years. Comparative and fossil studies indicate that the evolution of the human supralaryngeal vocal tract may have started in hominid populations like that of the *Homo habilis* KNM-ER 3733 fossil. More recent hominid lineages differ with respect to the presence of a modern supralaryngeal airway. Hominids like classic Neanderthal retained the nonhuman supralaryngeal vocal tract; other hominids contemporary with Neanderthal had human supralaryngeal airways.

Is speech the selective force for the human vocal tract?

The effects of malformations of the supralaryngeal vocal tract have long been noted. The effects of cleft palate have been studied in great detail (Folkins, 1985). The studies of anomalies of the vocal tract can constitute "experiments in nature" which allow us to assess the selective value of different aspects of vocal tract morphology. Apert's and Cruzon's syndromes result in anomalous supralaryngeal vocal tracts. The palate is positioned in a posterior position relative to its normal orientation with respect to the face and the pharynx is constricted. Landahl and Gould (1986) use acoustic analysis, psychoacoustic tests, and computer modeling based on radiographic data to show that the phonetic output of subjects having these syndromes is limited by their supralaryngeal vocal tracts. The Apert's and Cruzon's subjects attempt to produce normal vowels but are unable to produce the normal range of formant frequency values. Psychoacoustic tests of their productions yield a 30 percent error rate for vowels. Acoustic analysis shows that they are unable to produce the quantal vowels [i] and [u].

Figure 9.3 shows the results of a computer modeling of an Apert's subject. The loops of the Peterson and Barney (1952) study of normal subjects are shown on the F1 versus F2 formant frequency plot. The vowel symbols clustered towards the center of these loops represent the formant frequency patterns that the Apert's supralaryngeal vocal tract can produce when the experimenters attempted to perturb it towards the best approximation of a normal vocal tract's configuration for each vowel. Note the absence of [i]'s or [u]'s. The computer modeled plot in which the Apert's vocal tract was pushed to its phonetic limits is virtually identical to the formant frequencies that the Apert's subject actually produced.

These data indicate that the normal human supralaryngeal vocal tract is optimized to achieve a maximum phonetic output. They are consistent with the hypothesis that the selective force on the restructuring of the human supralaryngeal vocal tract in the course of human evolution was its phonetic output. The palate in Apert's syndrome continues to move back on the base of the skull, past its "normal" position during ontogenetic development. The "normal" configuration of the human supralaryngeal vocal tract yields the maximum formant frequency range. Configurations in which the palate is anterior (Lieberman et al., 1972) or posterior (Landahl and Gould, 1986) yield a reduced range of formant frequency patterns. The anomalous posterior positions do not appear to be detrimental to upright bipedal posture and locomotion. The anomalous posterior and anterior positions clearly make it impossible to produce sounds like [i] or [u]. The restructuring of the human supralaryngeal vocal tract thus appears to be driven by its phonetic function.

Speech Production

The production of human speech likewise involves species-specific neural mechanisms. Broca first identified the area of the brain in which the motor programs that control the production of human speech are stored. The articulatory maneuvers that underlie the production of human speech are perhaps the most complex that human beings attain. Until age 10 years, normal children are not up to adult criteria for even basic maneuvers like the lip positions that produce different vowels (Watkins and Fromm, 1984). Human speakers can execute complex coordinated articulatory maneuvers involving tongue, lips, velum, larynx and lungs, directed towards linguistic goals, e. g., producing a particular formant frequency pattern. Lesions in Broca's area result in deficits in speech production. Though able to move individual articulators, or use his tongue and lips to swallow food, the victim cannot produce particular speech sounds. Lacking Broca's area, pongids likewise cannot be taught to control their supralaryngeal airways to produce any human speech sounds. Although the non-human pongid airway is inherently unable to produce all of the sounds of human speech, it could produce a subset of human speech sounds. Non-human primates are unable to produce even these sounds in an intentioned manner.

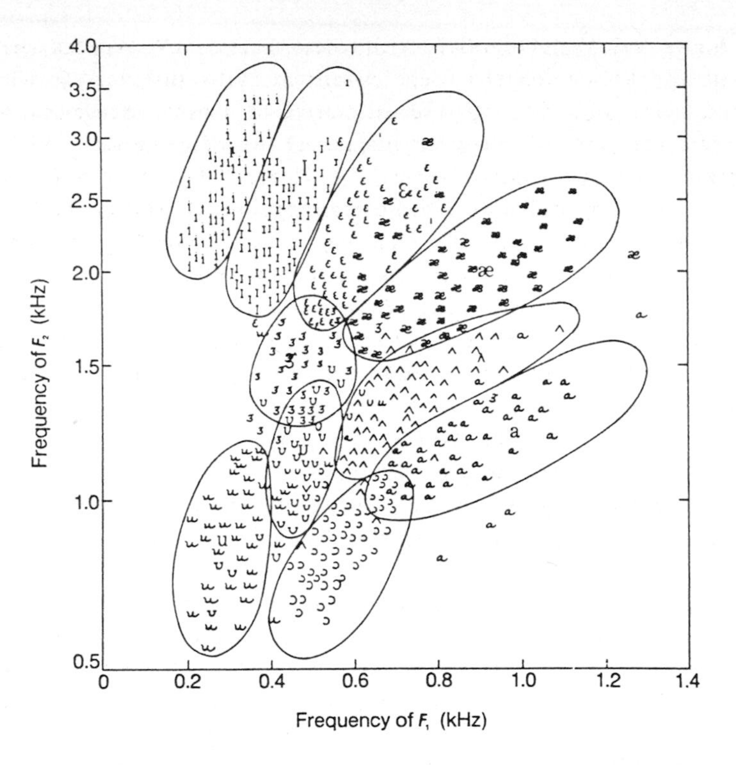

Figure 9.3. Formant frequency space derived by computer modeling of an Apert's subject. The loops of the Peterson and Barney (1952) study of normal subjects are shown on the F1 versus F2 formant frequency plot. The vowel symbols clustered towards the center of these loops represent the formant frequency patterns that the Apert's supralaryngeal vocal tract can produce when the experimenters attempted to perturb the anomalous Apert's supralaryngeal vocal tract towards the best approximation of a normal vocal tract's configuration for each vowel. Note the absence of [i]'s or [u]'s. These data indicate that the normal human supralaryngeal vocal tract is optimized to achieve a maximum phonetic output. (after Landahl and Gould, 1986.)

The Evolution of Syntax and Rule-governed Behavior

The evolution of the neural mechanisms that govern the rule-governed syntax of human language may involve the Darwinian process of preadaptation. The lateral frontal cortex is adapted to the regulation of orofacial gestures. The gestural patterns that human beings use to transmit certain aspects of affect appear to be innate (Meltzoff and Moore, 1977). This area of the brain is elaborated in the course of hominid evolution (Deacon, 1985) and ultimately yields Broca's area which facilitates the production of speech. Broca's area is also implicated in the comprehension of syntax. Thus it is possible that the speech production and syntactic deficits of chimpanzees, and the agrammatic victims of Broca's aphasia, are functionally linked (Lieberman, 1984,1985).

Recent data relating the acquisition and deterioration of complex speech-motor-control and syntactic comprehension are consistent with this theory which claims that the neural substrate that evolved to facilitate speech motor control is the preadaptive basis for human rule-governed syntactic and cognitive ability (Lieberman, 1984, 1985). In a recent study, Emery (1982) showed that aged people who are cognitively "normal" show deficits in the comprehension of written and spoken sentences that feature certain syntactic constructions. Subjects are not able, for example, to correctly identify the subject in passive sentences like,

Susan was kissed by Tom.

Nor are they able to comprehend sentences that have embedded clauses. Subjects generally are able to use semantic and pragmatic information and have no difficulty with sentences like,

Bill ate the apple.

The deficits noted by Emery appeared to mirror ones that have been noted in young children. Emery's data also showed deficits in Piagetian cognitive tasks that also mirror the cognitive development of young children. These phenomena occurred in "normal" aged people, who were *not* demented.

We have relocated some aspects of Emery's study and have furthermore correlated syntactic deficits with certain aspects of speech production. Our pilot experiment involves 20 residents of the Jewish Home for the Aged of Rhode Island. We assessed their syntactic comprehension and anticipatory coarticulation in the production of speech. Each person was tested using the Rhode Island Test of Syntactic Ability (Engen and Engen, 1982). This test was originally devised for hearing impaired persons. It consists of 100 sentences which are each read to the subject. The person being tested is asked to point to one of three sketches. One sketch correctly illustrates the meaning of the sentence. The sentences differ in syntactic complexity. However they all make use of similar vocabulary. The test has been designed to avoid any meaning distinctions that are conveyed by morpheme differences; these distinctions are hard to perceive and would introduce errors for hearing impaired subjects. The scoring of the Rhode Island Test furthermore reveals "clustered" error patterns which can reveal difficulty with a particular syntactic distinction. The test has been administered to over 4000 individuals and normative data are available. In general, normal children over the age of 6 years make no errors. Our data replicate Emery (1982) and Kynette and Kemper (1986) in that we find errors in syntactic comprehension. However, we find individual differences using the Rhode Island Test of Syntactic Ability and an error pattern that differs from that of young children. Our aged subjects do not appear to have any vocabulary problems.

We also had each aged person repeat the syllables [si] and [su], [di] and [du], [ti] and [tu] as well as a series of sustained vowels and sentences. We obtained 5 tokens of each syllable. The speech samples were analyzed using our computer system and the consonants that proceeded each vowel were excised. Tapes were prepared for a psychoacoustic test in which listeners were asked to identify the vowels that followed each consonant when they heard only the consonant. Normal adults, when they produce these syllables, anticipate the vowel that follows each consonant and modify their articulation producing acoustic cues that enable a listener to identify the vowel that's coming when she/he hears the isolated consonant.

Children do not effect this "anticipatory coarticulation" under age 3 years. They gradually acquire these articulatory maneuvers between ages 3.5 and 6 years (Sereno et al, in press).

The aged people that we tested showed strong individual differences on these two tests. Five persons, ages 65 - 80, had very low or no syntactic errors. Thirteen, ages 65 - 90, had errors ranging from 20 to 35 percent. (Test data were incomplete for two people.) We have completed the coarticulation test for 9 subjects, 3 who had no syntactic errors and 6 who had 30 percent error rates. The persons who had high syntactic error rates showed deficits in anticipatory coarticulation similar to that of a 3.5 year old child. The persons who had no syntactic errors had near normal coarticulation. The correlation coefficient was 0.80. These data indicate that many aged people who have excellent linguistic competence have extremely high receptive syntactic errors. The correlation between deficits in coarticulation and syntax are consistent with the theory that we have been discussing concerning the neural substrate that underlies rule-governed behavior in human beings.

The Acquisition of Syntax by Children

As I noted earlier, many linguists believe that the possible form of the syntax of any human language is constrained by an innate neural syntax "organ." This theory is supported by regularities in the manner in which children learning different languages acquire control over syntax, and also by similarities that can be seen in the formal representation of the syntactic structures of many different languages. Children also appear to be predisposed to learn rapidly the syntax of their native language without any formal instruction; some linguists have used these data to support the existence of a modular neural "language acquisition device" that is specifically designed to facilitate the acquisition of language. However, though children acquire language rapidly they also rapidly acquire all other aspects of human culture. The neural substrate that underlies the acquisition of the syntax of human language may also be involved in the acquisition of other aspects of language and cognitive behavior. Piagetian theory, for example, links the general "sensorimotoric" intelligence in the first year of life with the later development of language in children. Social interaction clearly appears to be a necessary component of linguistic and cognitive development (Bates, 1976). Other recent data show that semantic development and cognitive development are linked (Gopnick and Meltzoff, 1985, 1986). The relationship between cognitive development and linguistic development, and the general relationship between language and cognition are still open questions. I think that the available data point to a close relationship rather than to disjoint morphology, development and evolution.

Conclusion

In conclusion, human linguistic ability appears to involve a number of genetically independent biological components of which some appear to be species-specific. At this point there is no reason to believe that the neural bases of human linguistic and cognitive ability are completely independent. Present generative linguistic theory is not consistent with the basic premises and data of modern biology. Moreover, the evolution of human linguistic and cognitive ability is within the range of explanation of the Darwinian, synthetic theory of evolution.

References

Anderson, J. A. (1972) A simple neural network generating an interactive memory. Mathematical Biosciences 14:197-220.

Bates, E. Language and context: the acquisition of pragmatics. New York: Academic Press.

Bunge, M. (1984) Philosophical problems in linguistics. Erkenntnis 21:107-173.

Changeux, J.-P. (1980) In Language and Learning: The debate between Jean Piaget and Noam Chomsky. ed. M. Piattelli-Palmarini, 185-202, Cambridge Mass: Harvard University Press.

Chomsky, N. (1957) Syntactic structures. The Hague: Mouton.

Chomsky, N. (1965) Aspects of the theory of syntax. Cambridge Mass.: MIT Press.

Chomsky, N. (1980a) Initial states and steady states. In Language and Learning: The debate between Jean Piaget and Noam Chomsky. ed. M. Piattelli-Palmarini, 107-130, Cambridge Mass: Harvard University Press.

Chomsky, N. (1980b) Rules and Representations. Behavioral and Brain Sciences 3:1-61

Emery, O. B. 1982. Linguistic patterning in the second half of the life cycle. unpublished PhD dissertation, University of Chicago

Engen, E. and T. Engen (1983) Rhode Island test of language structure Baltimore: University Park Press.

Fodor, J. (1980) On the impossibility of acquiring "more powerful" structures, In Language and learning: the debate between Jean Piaget and Noam Chomsky, ed. M. Piatelli-Palmarini, 142-162, Cambridge, Mass.: Harvard University Press

Folkins, J. W. 1985. Issues in speech motor control and their relation to the speech of individuals with cleft palate. Cleft Palate Journal. 22, 106-122.

Gardner, R. A. and B. T. Gardner, 1984. A vocabulary test for chimpanzees (*Pan troglodytes*). Journal of Comparative Psychology, 4: 381-404.

Gopnick, A. and A. Meltzoff (1985) From people, to plans, to objects: Changes in the meaning of early words and their relation to cognitive development. Journal of Pragmatics 9:495-512.

Gopnick, A. and A. N. Meltzoff (1986) Words, plans, things, and locations: Interactions between semantic and cognitive development in the one-word stage. In The Development of Word Meaning Eds. S. A. Kuczaj and M. D. Barrett. New York: Springer-Verlag.

Kynette, D. and S. Kemper, (1986) Aging and the loss of grammatical forms: A cross-sectional study of language performance. Language and Communication 6: 65-72

Laitman, J. T. and E. S. Crelin (1976) Postnatal development of the basicranium and vocal tract region in man. In Symposium on development of the basicranium Ed. J. Bosma. 206-219. Washington D. C., : U.S. Government Printing Office.

Laitman, J. T., R. C. Heimbuch, and E. S. Crelin (1979) The basicranium of fossil hominids as an indicator of their upper respiratory systems. American Journal of physical Anthropology 51:15-34.

Landahl, K. and Gould, H. J. 1986. Congenital Malformation of the speech tract: Preliminary investigation and model development. Proceedings of the Association for Research in Otolaryngology, 1986 meeting.

Lieberman, P., K. S. Harris, P. Wolff, and L. H. Russell (1972) Newborn infant cry and nonhuman primate vocalizations. Journal of Speech and Hearing Research 14:718-727.

Lieberman, P. (1984) The Biology and Evolution of Language, Cambridge, Mass: Harvard University Press

Lieberman, P. (1985) On the evolution of human syntactic ability: Its pre-adaptive bases - motor control and speech. Journal of Human Evolution 14:657-668

Lieberman, P., R. H. Meskill, M. Chatilllon and H. Schupack. (1985) Phonetic speech deficits in dyslexia. Journal of Speech and Hearing Research. 28:480-486.

Mayr, E. (1982) The Growth of Biological Thought, Cambridge, Mass.: Harvard University Press

Meltzoff, A. N. and M. K. Moore (1977) Imitation of facial and manual gestures by human neonates. Science 198:75-78.

McNeill, D. (1985) So you think gestures are nonverbal? Psychological Review 92: 350-371.

Miller, G. A. and S. Isard (1963) Some perceptual consequences of linguistic rules. Journal of Verbal Learning and Verbal Behavior 2:217-228.

Pinker, S. (1984) Language learnability and language development. Cambridge Mass:Harvard University Press.

Savage-Rumbaugh, S., D. Rumbaugh and K. McDonald (1985) Language learning in two species of apes. Neuroscience and Biobehavioral Reviews. 9:653-665.

Sereno, J., S. R. Baum, G. C. Marean and P. Lieberman (in press) Acoustic analyses and perceptual data on anticipatory labial coarticulation in adults and children. Journal of the Acoustical Society of America.

Sternberg, Robert J. (1985) Beyond IQ:A triarchic theory of human intelligence. New York: Cambridge University Press.

Wills, R. H. (1973) The institutionalized severely retarded. Springfield Ill: Charles C. Thomas.

THE EVOLUTION OF HUMAN CEREBRAL ASYMMETRY

Jerre Levy
Committee on Biopsychology
University of Chicago IL 60637

Although considerably less extensive and obvious than functional cerebral asymmetries in the human brain (see Bradshaw & Nettleton, 1983), recent research has demonstrated beyond question that the two cerebral hemispheres are laterally specialized in species as widely separated as passerine birds, rodents, and monkeys (see Glick, 1985, for reviews). The presence of cerebral asymmetry in such diverse groups can be explained by a parallel evolution in response to common selective pressures for an efficient use of neural space, in which different neural programs that serve different functions are organized in opposite sides of the brain. A neural organization that is optimally designed to serve one set of functions cannot normally be optimally designed to serve a different set of functions, and lateral differentiation allows the development of two different operational programs that serve different and equally important processes.

There is evidence that the *nature* of functions served by each side of the brain has a certain similarity across very distantly related groups. Thus, specialization for sequencing complex vocalizations is present in the left hemisphere of both passerine birds and people, and emotional functions are specialized to the right hemisphere in both rats and people. Quite possibly, certain functional organizations can develop more readily in the left hemisphere and others in the right hemisphere, due to effects of molecular asymmetries that are common to all living organisms.

The foregoing considerations suggest that although human cerebral asymmetry shares certain remarkable similarities with cerebral asymmetries in other animals, it is reasonable to believe that, except for our primate relatives, these are analogies, rather than homologies, that are due to common selective pressures for the efficient use of neural space, constraints on neural programs, and biases in lateralized development that derive from the asymmetry of biological molecules. An understanding of the evolution of cerebral lateralization in the human brain depends, therefore, on a consideration of the properties of the human left and right hemispheres in relation to those of monkeys and apes.

Phyletic Versus Phylogenetic Comparisons

Ideally, researchers who seek to describe the evolution of the human brain would have a time machine that allowed them to travel back in time some 30 or 35 million

NATO ASI Series, Vol. G17
Intelligence and Evolutionary Biology
Edited by H. J. Jerison and I. Jerison

years to the common primate ancestor, whose path of evolution would be followed through the emergence of the common ape-human ancestor, and forward until the appearance of *Homo sapiens.* Phylogenetic comparisons of the modern human brain to the brains of human ancestors would then reveal the nature of changes that occurred in neurology and the behavior it controls. Unfortunately, no such time machine is available and no such phylogenetic comparisons are possible. Rather, human neuropsychological evolution can only be inferred through phyletic comparisons between people and the latest models of monkeys and apes.

Common characteristics that are present in both monkeys and people imply that such characteristics appeared prior to differentiation of the primate lines that led to modern monkeys and apes. Characteristics that we share with apes but not monkeys can be inferred to have arisen after branching of the common ape-human line, but prior to separation of human and nonhuman evolutionary pathways. However, the modern ape, like the modern human being, has undergone considerable evolutionary changes compared to the common ancestor, and in some instances, may be as specialized and different from that ancestor as our species. Indeed, it is quite possible that certain ancestral characteristics of primates could be present in monkeys and elaborated in people, but lost in some highly specialized ape.

As an analogy, consider that we are much more closely related to whales than we are to lizards, yet both we and lizards have the four-limbed external morphology of our common amphibian ancestor, whereas whales have developed an external morphology that resembles that of fish. We could therefore learn much more about the evolution of human locomotion by studying lizards than by studying whales, since the whale, due its own specialized evolution, has lost the functional homology with human walking. In brief, there is no basis for an a priori assumption that some modern ape species is comparable to the common ape-human ancestor. In some characteristics, it may be more similar to that ancestor than are we, but in other characteristics, it may be more dissimilar.

Consider that this common ancestor not only gave rise to certain highly muscularized, typically quadripedal, large-boned, and robust species of great apes, and not only to our species, but also to the lithe, gracile, and bipedal gibbons and siamangs. Why suppose that the common ancestor had the very heavy bones and muscles, the massive brow ridge, the strongly sloping brow, the musculature, the robustness, the barrel chest, the knuckle walking, and the relatively infrequent and undifferentiated vocalizations of a modern *Pan troglodytes*? It seems far more likely that these characteristics are specialized modifications of a common ancestor who was smaller, more gracile, and more vocal (like monkeys) and more often bipedal than a modern common chimpanzee. Such a creature would have characteristics that were similar in some respects to the lesser apes, in other respects, to the great apes, and in other respects, to human beings. The emergence of the various modern groups would then reflect elaborations of certain ancestral characteristics and loss of others, which would differ for the modern species.

Some characteristics of human beings can therefore be expected to be elaborations of those that were present more than 30 million years ago in the common primate ancestor and that are still present in some groups of monkeys, but which may have been lost in the evolution of certain ape species. Phyletic comparisons are therefore considerably more complex than phylogenetic comparisons, since the degree of genetic relation between groups does not necessarily determine the degree of functional or even anatomical similarity. This depends entirely on the extent to which two closely related species both preserved and elaborated the same ancestral characteristic or on the extent to which each became a highly specialized modification of the ancestral form. Quite possibly, the evolution of the human speech system was based in part on an elaboration of a very vocal monkey and ape ancestor, whose social life demanded frequent and highly differentiated conspecific vocalizations. If so, the relative silence of *Pan troglodytes* would represent a specific specialization of that species that entailed loss of an ancestral characteristic. In this case, more vocal primates might be able to tell us more about the evolution of human speech than would common chimpanzees. In summary of this section, it is important to keep in mind that phyletic comparisons are *not* phylogenetic comparisons. This means that there is no guarantee that we can identify the substrates of human neurobehavioral characteristics more readily in our closer than in our more distant primate relatives.

Human Cerebral Asymmetry

According to ideas that first emerged in the 19th century, and that are still held by at least some current researchers, any intellectual superiority of human over nonhuman primates derives entirely from processes of the human left hemisphere. For almost a century after Broca's (1861) and Wernicke's (1874) discoveries that language was dependent on the left hemisphere, there was a widespread view that *all* human psychological functions were specialized to the left side of the brain, including analysis of sensory information, planning of behavior, thinking, and even consciousness itself (see Bogen, 1969). The right hemisphere was thought to be a mere relay station that transmitted sensory information in the left half of space to the left hemisphere for processing and motor plans, derived from the left hemisphere, to the muscles on the left side of the body.

This lopsided perspective was comforting to a species that had been removed from the center of the universe, whose ancestors proved to be apes, and whose noblest thoughts were only the top of an iceberg of hidden and vulgar motives. After all, only human beings could speak, and if language could be claimed to be synonymous with thought and consciousness, all restricted to the human left hemisphere, then an absolute distinction between human and other animals could be maintained. If the mute right hemisphere were allowed its own thinking and consciousness, there would be little basis for denying these to non-speaking animals.

Gazzaniga (1983), in discussing split-brain patients, states that "Indeed, it could well be argued that the cognitive skills of a normal disconnected right hemisphere without language are vastly inferior to the cognitive skills of a chimpanzee" (p. 536) and that "...the price of lateral specialization for language on the left is a state of rudimentary cognition for the right hemisphere, which is revealed only if the latter has to serve alone following brain bisection or left-brain damage" (p. 536). This would say that when the early hominids first hit the evolutionary scene, there was the beginning of enormous growth of intelligence in the left hemisphere during the next several million years, while, at the same time, the right hemisphere not only failed to evolve in intelligence, but actually underwent a regression. Under this model, processes of the human left hemisphere, but not those of the right, are responsible for the cognitive superiority of people over apes. Gazzaniga's (1983) statements even imply that the human right hemisphere is not only not superior in intelligence to the ape brain, but is radically *inferior*. Under this model, human mental evolution was an evolution of the left hemisphere, which is the sole source of human intelligence, in association with a cognitive regression of the right hemisphere.

In the large majority of right-handers, the human left hemisphere is essential for speech, phonetic representation and analysis, and syntactical understanding. It is also superior to the right hemisphere in discriminating the temporal order of rapidly changing sensory events and in programming skilled, serially ordered movements, either of the hands or of the articulatory apparatus (see Bradshaw & Nettleton, 1983). During speech, the right hand manifests coordinated communicative gestures more than does the left hand, which almost certainly reflects the left-hemisphere superiority in regulation of both the oral apparatus and praxic manual skills, especially in social communication (Kimura, 1982). However, the left-hemisphere specialization for speech, phonetics, and grammar, and its coordinated specialization for regulating serially ordered oral and manual movements, does not mean either that the right hemisphere plays no role in language or that the right hemisphere is without its own highly developed cognitive specializations.

The Language of the Human Right Hemisphere

A great deal of evidence shows that the human right hemisphere is not only vastly superior to the left hemisphere in a variety of highly complex visuospatial skills, but even has certain types of language capacities. Human split-brain studies have revealed that the isolated right hemisphere has an extensive comprehension vocabulary, at the level of a 12-year-old child, in spite of its lack of speech (Zaidel, 1976) and its deficiencies in phonetic representation (Levy & Trevarthen, 1977; Zaidel, 1978) and in syntactical understanding (Gazzaniga & Hillyard, 1971; Zaidel, 1977). Through nonvocal means, it can also generate words by arranging plastic letters or even, in some instances, by printing (Levy, Nebes, & Sperry, 1971).

Studies of brain-damaged patients also show that right-hemisphere lesions impair the ability to discriminate and encode intonational qualities of the voice (Tucker, Watson, & Heilman, 1977; Heilman, Scholes, & Watson, 1974) and to generate appropriate emotional intonation in speech output (Ross & Mesulam, 1979). Right-hemisphere lesions are also associated with impairments in story recall (Wechsler, 1973; Wapner, Hamby, & Gardner, 1981), in judging the connotative meaning of words (Gardner & Denes, 1973), in the appreciation of humorous material (Gardner, Ling, & Flamm, 1975), and in the comprehension of metaphor (Winner & Gardner, 1977). The story-recall deficits are due both to selective omissions of emotional material (Wechsler, 1973) and to a failure to comprehend the story structure (Wapner et al., 1981). D. Zaidel & Sperry (1974) have also found that there are memory deficits for verbal material in split-brain patients as compared to control subjects matched for

IQ, which indicates that the left hemisphere, when isolated from the right, manifests verbal deficiencies that are similar to those in patients with right-hemisphere damage.

The various observations reveal that although the right hemisphere does not generate phonetic representations, cannot speak, and has serious deficiencies in understanding grammatical rules and the meaning of syntactical dimensions (e.g., tense, voice), it has an extensive representation of the meaning of words and plays a special role in certain aspects of language. It is superior to the left hemisphere in interpreting the emotional intonation of speech and in regulating appropriate intonation in speech output. Its specializations are also important for appreciating humor, verbal emotional descriptions, metaphor, and the general schemata and global structure of stories. The linguistic specializations of the left hemisphere pertain specifically to speech, phonetic representation, syntactical understanding, and the propensity to acquire and use this vocal and gestural language system in interaction with the social world.

The Visuospatial Capacities of the Human Right Hemisphere

Although the right hemisphere clearly has a rather high competence in representing words and is important for certain other verbal communicative skills, its most advanced and specialized abilities lie in the visuospatial domain. In a study by Levy-Agresti & Sperry (1968), split-brain patients were given a small wooden block to feel, with either the left hand (right hemisphere) or right hand (left hemisphere) on each trial, in which block stimuli varied in shape and surface textures of some sides. After feeling the object, patients were shown 3 drawings of the 2-dimensional "architectural plans" of objects, one of which, if mentally folded, would reconstruct the block. Except in one patient, the left hemisphere performed at chance, whereas the right hemisphere was well above chance in selecting the correct drawing. Even in the one patient whose left hemisphere was above chance, the right hemisphere was greatly superior. Further, in this patient, the two hemispheres used different strategies to solve the problems. For the left hemisphere, item difficulty was determined by how difficult it was to differentiate the choice drawings by verbal description. For the right hemisphere, item difficulty was determined by how difficult it was to differentiate the mentally-folded choice drawings from a visual perspective.

Franco and Sperry (1977) showed split-brain patients sets of 5 objects, in which all objects in the set shared some unspecified geometric or topological invariant, although all objects were physically distinct. Patients were told to look at the inspection set (seen by both cerebral hemispheres) and then to feel 3 objects, with either the left hand (right hemisphere) or right hand (left hemisphere) and select the one that "went with the set". The task required the identification of a geometric or topological invariant that characterized all physically different members of the inspection set and the identification of this invariant in one of the 3 choice objects. The right hemisphere was much more accurate and faster than the left hemisphere, and indeed, for sets that had a topological invariant, the left hemisphere performed at chance. Two hemispherectomy patients, one with only a right hemisphere and one with only a left hemisphere, performed like the isolated right and left hemispheres, respectively, of split-brain patients.

The differences between hemispheres on the Franco and Sperry (1977) task were qualitative in nature. There were 4 types of sets, in which relations were based on either Euclidean, affine, or projective geometries, or on topology. The 4 types of sets vary, as Franco and Sperry say, in "the number of defining spatial constraints to which each is subject, the greatest number being present in the Euclidean forms with progressively fewer in the affine, projective, and topological sets, respectively" (p. 109). Thus, any of several spatial features may define a Euclidean set, but in moving to affine, projective, and topological sets, specific feature characteristics become progressively less defining, and the extraction of global, abstract relations becomes progressively more demanded.

Franco and Sperry (1977) found that the right hemisphere performed equally accurately on all types of sets, which indicates that global relations were being extracted from all, without a reliance on specific feature characteristics. However, the left hemisphere performed as well as the right on Euclidean sets, but showed a monotonic decline in performance from Euclidean to affine to projective to topological sets, and performed at chance on the latter. The above-chance performance for the left hemisphere on the geometrical sets may, therefore, be attributed to reliance on matching of specific features, without an extraction of the defining relation of sets.

Nebes (1971, 1972) has shown that the right hemisphere greatly surpasses the left at visuotactile matching tasks in which an arc must be matched to a circle or an "exploded" to a unified shape. On control tasks, which involved arc-arc, circle-circle, or unified-unified shape matching (cross-modality physical-identity matches), the two hemispheres were equal. In brief, the right hemisphere is superior to the left specifically at those visuospatial tasks that require a mental transformation or translation of spatial relations or that entail the extraction of abstract invariants that relate one set of spatial relations to another, whether the transformation involves a mental folding, an extraction of pure abstract relations, a mental completion of an incomplete representation, or a mental unification of fragmented parts.

In contrast to Gazzaniga's (1983) contentions, and to views that it is the left hemisphere only that differentiates human mental capacity from that of nonhuman primates, the evidence demonstrates conclusively that the human right hemisphere has evolved complexities and levels of intelligence in the visuospatial domain that far exceed what has been demonstrated for apes. These right-hemisphere abilities underlie the human capacity to make and read maps, to design architectural plans, to envision how these plans will be realized as 3-dimensional shapes, to represent 3-dimensional space in 2-dimensional drawings, to invent and comprehend geometries and geometric relations, and to decipher how some apish ancestor looked from bits and pieces of fossil bone. Such abilities are as complex and uniquely human as is the language of the human left hemisphere.

It was not merely the human left hemisphere that evolved in intelligence and special capacities, along with its evolution in size, but also the human right hemisphere. Thus, both hemispheres of our primate relatives may be expected to manifest special abilities, superior and complementary to those on the other side, that represent the substrates for the further elaborations that created the human brain.

Cerebral Asymmetries in Primates

In a majority of human brains, the right frontal lobe protrudes more anteriorly than the left frontal lobe, and the frontal width is wider on the right than left, whereas the left occipital lobe protrudes more posteriorly than the right, and the occipital width is wider on the left than right (LeMay & Culebras, 1972). The various asymmetries are identifiable from skull markings and are present not only in fossil hominids, but also in modern apes (LeMay, 1976), and even in baboons (Cain & Wada, 1975). Evidently, these gross neuroanatomical asymmetries emerged in the primate line prior to the separation of monkeys and apes. Although their functional significance is not yet known, it is reasonable to infer that they are indicative of functional asymmetries that are shared by monkeys, apes, and people.

In the human brain, the posterior language regions of the left hemisphere are reflected in a larger left than right temporal plane (Geschwind & Levitsky, 1968), an asymmetry that is present at birth (Chi, Dooling, & Gilles, 1977; Wada, Clarke, & Hamm, 1975; Witelson & Pallie, 1973), and indicated not only by direct examination, but also by the conformation of the Sylvian fissure (LeMay & Culebras, 1972; Rubens, Mahowald, & Hutton, 1976). The left Sylvian fissure is longer, more horizontal, and terminates at a lower point on the lateral surface of the brain than the right Sylvian fissure. The Sylvian-fissure asymmetry is present in apes, but not in monkeys (LeMay & Geschwind, 1975; Yeni-Komishian & Benson, 1976). Thus, it appears that the evolution of the human posterior language regions was based on a preadapted organization that first appeared in the hominoid line, after separation from the monkeys.

Unfortunately, no studies of functional cerebral asymmetries have yet been done in the great apes, although monkeys have been examined. Hamilton and his colleagues (Ifune, Vermeire, & Hamilton, 1984; Vermeire, Ifune, & Hamilton, 1984; Hamilton & Vermeire, 1985; Hamilton, 1986) have shown that in split-brain pigtail macaques, the right hemisphere is superior to the left in discriminating faces and facial emotions and in controlling facial expressions, as is seen in people. The left hemisphere is superior to the right in learning to discriminate two sloped lines that differ in slope by 15 degrees. Further, in 12 monkeys who were tested on both the line-slope task and the face-discrimination task, 8 showed a strong reversal of hemispheric asymmetries on the two tasks, an additional 3 showed relative differences between the two hemispheres in the expected direction, and only 1 failed to show the group-typical pattern.

Hamilton (1983) states that there is a right-hemisphere superiority in people for line-slope discriminations, which is opposite in direction to the asymmetry observed in monkeys. In fact, however, the lateralization of line-slope discrimination in people is contingent on the complexity of the task. White (1971) found a *left-hemisphere* superiority for discriminating 4 line slopes (at 0, 45, 90, and 135 degrees) in people, a result that was replicated by Umilta, Rizzolatti, Marzi, Zamboni, Franzini, Camarda, & Berlucchi (1973). Umilta et al. (1973) also showed that there was *no* hemispheric asymmetry for discriminating 4 different slopes that excluded the vertical and horizontal. A right-hemisphere superiority only emerged when 8 different line slopes had be discriminated. The varying asymmetries in people as a function of task complexity were attributed to a covert labeling and categorization of stimuli, dependent on the left hemisphere, that was easier in the simpler than in the more complex tasks.

Although monkeys cannot verbally label stimuli, it is likely that in the 2-slope task with which they were presented, monkeys assigned the two slopes to differing semantic categories (closer or more distant from horizontal, or closer or more distant from vertical). Thus, the monkeys' left-hemisphere superiority on the 2-slope discrimination is likely to derive from an underlying semantic categorization for which the left hemisphere is superior, not only in monkeys, but also in people. In brief, the direction of asymmetries in monkeys on both the face task and the line-slope task is the same as in people.

Further evidence for hemispheric specialization comes from studies of Japanese macaques. Petersen, Beecher, Zoloth, Moody & Stebbins (1978) and Beecher, Petersen, Zoloth, Moody, & Stebbins (1979) found that when conspecific calls were presented monaurally to one ear or the other, they were better discriminated through the right ear (left hemisphere). Petersen, Beecher, Zoloth, Green, Marler, Moody, & Stebbins (1984) established that the right-ear superiority derived from the communicative significance of stimuli, not from their purely physical characteristics. Additionally, left-hemisphere lesions disrupt these discriminations more than do right-hemisphere lesions (Heffner and Heffner, 1984). Although simian vocalizations differ from human speech in major ways, they both serve social communication via auditory signals. It is possible, therefore, that the development of the human language system was built upon a preadapted left-hemisphere specialization for discriminating and producing socially significant vocalizations.

The left hemisphere of monkeys also appears to be specialized for a primitive form of propositional reasoning. Damage to the temporal cortex of the left hemisphere, but not comparable damage to the right hemisphere, produces deficits in the ability of monkeys to perform conditional discriminations in which the response to a visual stimulus is conditional on the nature of a preceding auditory stimulus (Dewson, 1977). Of central importance is the fact that this asymmetry in the effects of lesions evaporates for a task in which the delay between the auditory and visual stimulus is gradually introduced (Dewson, 1978). A gradual introduction of the delay interval could, in principle, give the animal an opportunity to develop a strategy in which the two successive cues are integrated into a single sensory representation. In this case, the conditional aspects of the task would become converted by the monkey into a mere associative-memory task, since two pairs of cues are associated with a reward and two are not.

If an animal can actually judge that a stimulus implies *X* only if a preceding stimulus has some specified characteristic, but implies non-*X* otherwise, a much higher-order representation and processing is needed than for mere associative learning. The predominance of the monkey's left hemisphere for true conditional learning and the symmetry of the hemispheres for simple associative learning may well represent a preadaptation of the simian brain for the evolution of human propositional reasoning in the left hemisphere.

The right-hemisphere specialization of the monkey brain for producing facial emotions, for differentiating facial emotions, and for recognizing faces is comparable to that of the human right hemisphere (see Bradshaw & Nettleton, 1983). Although these abilities are fundamental for social interactions, they involve cognitive abilities that, with greater elaboration, are crucial for nonsocial processes also. The ability to identify an emotional facial expression displayed by different faces, or the same face in different orientations or with different expressions, requires an extraction of relational invariants that remain after physical transformation. This is not to suggest that the ability to interpret facial expressions or the ability to recognize faces is necessarily correlated with general spatial ability, since specific programs for facial analysis may not be available for application to nonfacial stimuli. It is to suggest, however, that design characteristics of the right hemisphere that are apparent in the monkey's facial perception skills may have become extensively elaborated in the evolution of the human right hemisphere to permit the extraction and generation of highly complex and abstract geometric relations that retain invariance under purely mental transformations.

The monkey left hemisphere manifests specializations for three separate processes that may have been crucial for the subsequent evolution of human language. First, the left-hemisphere superiority for discriminating two line slopes, one more vertical than the other, indicates a categorizing ability that exceeds that of the right hemisphere. Second, the left-hemisphere specialization for discriminating meaningful conspecific calls on the basis of their communicative value indicates a special ability for interpreting vocal social signals. Finally, the left-hemisphere superiority for learning true conditional discriminations, but hemispheric symmetry for simple associative learning, suggests a specialization for pre-propositional reasoning. It is not difficult to perceive how these abilities could become elaborated, differentiated, and integrated together to form the basis for the evolution of human language, including speech, phonetic analysis, and syntactical organization.

Although studies on cerebral asymmetries of function in monkeys are few in number, they reveal cognitive differences between the two hemispheres that either directly or conceptually can be related to those of the human brain. Whether even greater similarities in lateral organization would be seen in the great apes depends on whether these animals have preserved and elaborated ancestral characteristics or whether, instead, their evolution has involved the emergence of new, specialized characteristics that distinguish them equally from monkeys and people. The Sylvian fissure asymmetry that characterizes the human brain is, as noted, present in apes and people, but not monkeys, which suggests that the posterior language regions began their evolution toward the human form in the common pongid-hominid ancestor. However, postcentral regions of the brain are involved predominantly in sensory analysis, not the programming of motor activity, which depends on precentral regions. The relative paucity of vocalization in common chimpanzees, as compared either to monkeys or people, especially given the left-hemisphere specialization in monkeys for conspecific calls, raises the question as to whether *Pan troglodytes* might have evolved a left hemisphere that is less developed in social-auditory-vocal communication than the left hemisphere of the ancestral form. This question is addressed in the final section, which considers studies of language-trained apes.

Language in Chimpanzees

The ape-language studies (Fouts, 1972; Gardner & Gardner, 1971; Premack, 1971; Rumbaugh, 1977; Savage-Rumbaugh, 1984, 1986; Terrace, 1979) have shown that chimpanzees can develop a rich representation of the meaning of the symbols (e.g., manual hand signs, plastic chips, printed lexigrams) of their language, including the creation of new names for objects or events by the combination of known symbols, although they cannot speak, show no evidence of phonetic representation and analysis, and fail to develop syntactical rules and grammatical understanding (Terrace, 1979). The absence of syntactical regulation in ape language has led some researchers (Terrace, 1979) to conclude that symbol usage in chimpanzees is not language at all. A more accurate description is that symbol usage in chimpanzees is qualitatively distinct from the language of the human *left* hemisphere. However, the rich symbol representation in association with the lack of speech, phonetics, and syntactics is qualitatively very similar to the language of the human *right* hemisphere.

Common chimpanzees *Pan troglodytes*, unlike human children, require formal instruction for language acquisition, rarely use their language system for spontaneously naming objects or events, simply for the fun of naming, and do not spontaneously learn the meaning of spoken words (Savage-Rumbaugh, McDonald, Sevcik, Hopkins, & Rubert, 1986). Further, the fact that common chimpanzees do not spontaneously acquire the meaning of spoken words when training is focused on visual symbols suggests that, unlike human children, there is a bias to attend to visual, rather than auditory, symbols. Quite possibly, the human right hemisphere gains an extensive auditory vocabulary through interactions with the left hemisphere, which acquires this vocabulary spontaneously during development, and without formal instruction. If symbol usage in apes is reflective of their highest cognitive capacities, these capacities, as great as they are, are less developed than those of the human right hemisphere. Symbol representation in common chimpanzees might either reflect a differential reliance on the right hemisphere or, alternatively, a differential reliance on a left hemisphere that lacks certain critical specializations that are present in the left hemisphere of human brains.

The evolution of human left-hemisphere language may have been an elaboration of the comprehension and expression of social vocalizations, in association with communicative gestures, that became progressively differentiated, progressively dependent on social learning, and progressively integrated with the rudiments of propositional thinking and categorical representation. Both human hemispheres have a rich and highly developed representation of meaning, and both have symbolic representations, but it is the left hemisphere only that can speak, can acquire the phonetic representations that can be translated to articulatory outputs, that eventually learns the grammatical rules that govern verbal social interactions, and that communicates orally in coordination with asymmetrically right-handed gestures.

When common chimpanzees are supplied with symbols to represent the objects and events of their lives, they learn the referential meanings, but what they learn may be more similar to the representations of the human right than of the human left hemisphere. Terrace's (1979) conclusion that ape language is not the language of human beings is analogous to the conclusion that the language of the human right hemisphere is not the language of the human left hemisphere. Nonetheless, the human right hemisphere, as well as common chimpanzees, has referential symbolic representations, which certainly form an important, though incomplete, part of a language system. The quite fundamental and qualitative distinction between human left-hemisphere language and the language of the common chimpanzee suggests that even the most detailed analysis of the language of these animals may tell us little regarding the evolution of the linguistic specializations of the human left hemisphere. Even if the language of common chimpanzees is asymmetrically represented in their left hemisphere, it is a left hemisphere whose linguistic abilities do not even equal those of the human right hemisphere.

However, recent studies indicate that symbol usage and acquisition in pygmy chimpanzees *Pan paniscus* is quite different from that in common chimpanzees and considerably more similar, in several domains, to that of the human left hemisphere (Savage-Rumbaugh, Rumbaugh, & McDonald, 1985; Savage-Rumbaugh, McDonald, Sevcik, Hopkins, & Rubert, 1986). As in children, there appears to be critical period for symbol acquisition, since Matata, a wild-caught adult pygmy chimpanzee, has manifested considerably more difficulty in lexigram acquisition than either Mulika or Kanzi, juvenile pygmy chimpanzees, who were born in captivity and have been exposed to language symbols since birth. Neither Mulika nor Kanzi has required explicit instruction for lexigram acquisition, in contrast to common chimpanzees (Sherman and Austin) reared under similar conditions.

Further, in the absence of formal instruction, both Mulika and Kanzi have acquired the meaning of spoken English words, simply from exposure. Under controlled and context-free conditions, they can readily point to a picture or a lexigram that is designated by a spoken word, whereas Sherman and Austin perform randomly in matching pictures to spoken words, in spite of their nearly perfect performance in matching pictures to lexigrams. The pygmy chimpanzees, much more often than common chimpanzees, engage in spontaneous naming of objects and activities. Kanzi often makes requests of the form *(agent-verb-recipient)* when he is neither the agent nor recipient of an action. This contrasts with the common chimpanzee, who assumes that he is always the recipient of an action and that the addressee is the agent, and thus neither designates the agent nor the recipient. Also, pygmy chimpanzees vocalize considerably more, and with highly differentiated vocalizations in association with communicative gestures, than do common chimpanzees. Amazingly, Kanzi even attempts to duplicate the tonal patterns of human speech and calls a peanut an "ih-uh".

The qualitative differences between the language acquisition and usage of common and pygmy chimpanzees suggests the possibility that the two species are relying on differently organized neural systems. In particular, the language behaviors of pygmy chimpanzees are more likely to reflect human-like left-hemisphere processes than are those of common chimpanzees. Possibly, the left hemisphere of pygmy chimpanzees, but not that of common chimpanzees, has an integration between symbolic representation and social vocalization and gesturing that forms a critical stage in the evolution of language in the human left hemisphere. Future studies on the lateralization of symbol systems and of other cognitive processes in pygmy and common chimpanzees will be of critical importance in gaining an understanding of the evolution of human left-hemisphere language.

Pygmy chimpanzees are more "human" in a number of ways than are common chimpanzees (see Susman, 1984). Anatomically, the brow ridge is less pronounced in pygmy than common chimpanzees, the skull is more vaulted, the face is flatter, and the chest is less barrel-like. The female genitalia are rotated anteriorly, and ventro-ventral copulation is common, in association with prolonged eye-to-eye contact. Additionally, copulation occurs throughout the month, and not merely during the ovulatory phase (Savage & Bakeman, 1978). The pygmy chimpanzee father shares in the rearing of young, and pygmy chimpanzees are more often bipedal than common chimpanzees, vocalize more, and have more differentiated vocalizations and associated gestures. The greater similarity of pygmy chimpanzees and people than of common chimpanzees and people, not only with respect to symbol usage and acquisition, but in other behaviors as well, and in anatomy, has two possible evolutionary explanations. First, it might be supposed that the evolutionary line of *Pan troglodytes* separated from the common pongid-hominid ancestral form *prior* to differentiation of separate evolutionary lines of pygmy chimpanzees and people. However, this would mean not only that people are more genetically similar to pygmy than to common chimpanzees, but that pygmy chimpanzees are more genetically similar to people than to common chimpanzees. In brief, the number of neutral base substitutions in DNA that distinguish people from pygmy chimpanzees would be fewer than the number that distinguish pygmy from common chimpanzees. Although possible, such an evolutionary scenario seems extremely unlikely.

The second, and more likely, possibility is that the common ancestor of *Pan troglodytes*, *Pan paniscus*, and *Homo sapiens* was morphologically and behaviorally more similar to the modern pygmy chimpanzee than to the modern common chimpanzee, that these ancestral characteristics were preserved in *Pan paniscus* and elaborated in the genus *Homo*, but underwent considerable evolutionary change and specialization in *Pan troglodytes*. Under this model, the pongid and hominid evolutionary lines separated prior to differentiation of pygmy and common chimpanzees, so that the number of neutral base substitutions that differentiate people from common chimpanzees equals the number that differentiate people from pygmy chimpanzees.

Our greater morphologic and behavioral similarities to pygmy than to common chimpanzees would then be due to preservation of certain ancestral characteristics in both pygmy chimpanzees and people and to loss of these characteristics in favor of new specializations in common chimpanzees. The greater vocalization of pygmy than common chimpanzees, and the more highly differentiated vocalizations, indicates that a social vocalization system is more important in their behavioral world than in the behavioral world of common chimpanzees. Possibly, the left hemisphere of common chimpanzees lost an auditory-vocal bias that had been present for millions of years in primates and that is still retained in monkeys, pygmy chimpanzees, and people. The great qualitative distinctions between symbol usage and acquisition by common chimpanzees and the human left hemisphere may reflect a left-hemisphere organization in *Pan troglodytes* that is highly specialized departure from the ancestral form, and one that is ill-designed for the acquisition of a language system that is similar to that of the human left hemisphere. We and the pygmy chimpanzees may have a more "primitive" left hemisphere than do common chimpanzees, in the sense that it is more similar, in certain functional characteristics, to the ancestral pongid. The retention of such "primitive" characteristics, however, in the human evolutionary line, was essential to the evolution of the human left-hemisphere language system.

Future studies of the functional specializations of the left and right hemispheres of common and pygmy chimpanzees, both for symbol representation and for other cognitive abilities, will be of major importance in understanding the evolution of human language and cognition. It might be that human left-hemisphere language was an elaboration and integration of specializations for discriminating and programming rapidly changing, serially-ordered events, a social auditory-vocalization system, and a proto-propositional neural program. The substrates of such elaboration and integration may be present in pygmy chimpanzees, but absent in common chimpanzees. If so, the learning of visual symbols in common chimpanzees may rest on inherently limited language capacities of either the right or left hemisphere, whereas symbolic representation may rest on more developed language capacities in the left hemisphere of pygmy chimpanzees.

In summary, language-trained apes manifest impressive capacities for referential symbol representation and usage, but these capacities, as examined to date, do not exceed and are even inferior to those of the human right hemisphere. However, as Kanzi and other pygmy chimpanzees reach adulthood, when mature abilities can be assessed, it may turn out that these animals surpass the human right hemisphere in the comprehension of grammatical meaning, even if grammatical rules do not govern their productive capacities. Zaidel (1977) has shown that the isolated right hemisphere of split-brain patients performs well above chance, but is inferior to 5-year-olds, and is only at the level of 4-year-olds, in following such instructions as, "After touching the red circle, pick up the green square." It is quite conceivable that adult pygmy chimpanzees, reared from birth in a human verbal environment, could outperform the adult human right hemisphere in following such grammatically complex instructions. If so, a reasonable conclusion would be that the common ancestor of apes and people possessed special programs in the left hemisphere for syntactical understanding that exceeded what would become possible for the right hemisphere of *Homo sapiens*.

Although the specialized language capacities of the human left hemisphere and the specialized visuospatial capacities of the human right hemisphere far outstrip the linguistic and visuospatial abilities of apes, these laterally specialized skills did not arise *de novo* in the hominids. Rather, they represent elaborations of more basic and restricted cerebral asymmetries that are even apparent in the monkey brain. The human elaborations depended on a great expansion of relative brain size, which permitted an extensive lateral differentiation of associative processes that were superimposed on demands for symmetry of basic sensorimotor mapping functions. Those fundamental symmetries must necessarily engage a rather large fraction of the monkey brain and a lesser, but still substantial, fraction of the ape brain, as compared to possibilities for the human brain. We, like our primate relatives, are compelled to have a veridical map of our bilaterally symmetric bodies and experiential worlds, but we possess far greater numbers of "excess neurons" (Jerison, 1973), beyond those needed for the symmetric model, that can be devoted to asymmetric higher associative processes.

Those "excess neurons" yield models of reality that can, in principle, encompass the universe from its birth to its imagined death, and that can represent the most fundamental quanta of reality, as well as the boundaries of the distant galaxies. The temporal, spatial, and causal domains that can be generated by the living human computer seem to be as extensive as time, space, and causality itself. Time, even for the most brilliant ape, is his personal past and perhaps tomorrow. Space is the world that he personally perceives. Causality consists of the causal relations he personally experiences. Human beings are indeed special. Yet, the human potential was there in some ancient primate and has left its marks in the calls of Japanese macaques, in the recognition of faces of conspecifics in Pigtail macaques, in the neuroanatomical asymmetries in brains of baboons and apes, and in the gestures, vocalizations, smiles, lexigraphic naming, oral comprehension, and imitations of human tonal speech patterns by the eerily human Kanzi.

Acknowledgments

The support of The Spencer Foundation is gratefully acknowledged. To Sue Savage-Rumbaugh and Duane Rumbaugh, I express my deep appreciation for their generosity in allowing me to observe and interact with Kanzi, Austin, and other apes at the Language Research Center in Atlanta. I was much enamored of Austin (*Pan troglodytes*), who held my fingers through the wires of his cage and shared his popcorn with me, yet Austin still remained an ape in my emotional world. In contrast, Kanzi (*Pan paniscus*) became transformed into a human child, and it was only with intellectual effort that I could retain awareness that he was a juvenile ape. Perhaps the similarity of other species to us is specified not only by their behavior, but also by our own emotional responses to them.

References

Beecher M, Petersen M, Zoloth S, Moody D and Stebbins W (1979) Perception of conspecific vocalizations by Japanese monkeys *Macaca fuscata.* Brain, Behavior, and Evolution 16:443-460

Bogen JE (1969) The other side of the brain. II: An appositional mind. Bulletin of the Los Angeles Neurological Society 34:135-162.

Bradshaw JL and Nettleton NC (1983) Human Cerebral Asymmetry. Englewood Cliffs, NJ: Prentice-Hall

Broca P (1861) Remarques sure le siege de la faculte du langage articule suives d'une observation d'aphemie. Bulletin Societe Anatomique 6:330-357

Cain DP and Wada JA (1979) An anatomical asymmetry in the baboon brain. Brain, Behavior, and Evolution 16:222-226

Chi Je G, Dooling EC and Gilles FH (1977) Left-right asymmetries of the temporal speech areas of the human fetus. Archives of Neurology 34:346-348

Dewson JH (1977) Preliminary evidence of hemispheric asymmetry of auditory function in monkeys. In: S Harnad, RW Doty, L Goldstein, J Jaynes and G Krauthamer eds., Lateralization in the Nervous System. New York: Academic Press

Dewson JH (1978) Some behavioral effects of removal of superior temporal cortex in monkey. In: D Chivers and J Herbert eds., Recent Advances in Primatology Vol. 1 Behaviour. London: Academic Press

Fouts RS (1972) The use of guidance in teaching sign language to a chimpanzee. Journal of Comparative and Physiological Psychology 80:515-522.

Franco L and Sperry RW (1977) Hemisphere lateralization for cognitive processing of geometry. Neuropsychologia 15: 107-114

Gardner BT and Gardner RA (1971) Two-way communication with an infant chimpanzee. In: A Schrier and F Stollnitz eds., Behavior of Nonhuman Primates Vol. 4. New York: Academic Press

Gardner H and Denes G (1973) Connotative judgments by aphasic patients on a pictorial adaptation of the semantic differential. Cortex 9:183-196

Gardner H, Ling PK, Flamm L and Silverman J (1975) Comprehension of humorous material following brain damage. Brain 98:399-412.

Gazzaniga MS (1983) Right hemisphere language following brain bisection. American Psychologist May 525-537

Gazzaniga MS and Hillyard SA (1971) Language and speech capacity of the right hemisphere. Neuropsychologia 9:273-280

Geschwind N and Levitsky W (1968) Human brain: Left-right asymmetries in temporal speech region. Science 461:186-187

Glick SD ed., (1985) Cerebral Lateralization in Nonhuman Species. New York: Academic Press

Hamilton CR (1983) Lateralization for orientation in split-brain monkeys. Behavioral and Brain Research 10:399-403

Hamilton CR (in press 1986) Hemispheric specialization in monkeys. In: C Trevarthen ed., Brain Circuits and the Mind: Festschrift for RW Sperry. Cambridge: Cambridge University Press

Hamilton CR and Vermeire BA (1985) Complementary hemispheric superiorities in monkeys. Society for Neuroscience Abstracts 11:689.

Heffner HE and Heffner RS (1984) Temporal lobe lesions and perception of species-specific vocalizations by macaques. Science 226:75-76.

Heilman KM, Scholes R and Watson RT (1974) Auditory affective agnosia. Journal of Neurology and Neurosurgery 38:69-72

Ifune CK, Vermeire BA and Hamilton CR (1984) Hemispheric differences in split-brain monkeys viewing and responding to videotape recordings. Behavioral and Neural Biology 141:231-235.

Jerison HJ (1973) Evolution of the Brain and Intelligence. New York: Academic Press

Kimura D (1982) Left-hemisphere control of oral and brachial movements and their relation to communication. Philosophical Transactions of the Royal Society London B 298:135-149

LeMay M (1976) Morphological cerebral asymmetries of modern man, fossil man, and nonhuman primate. Annals of the New York Academy of Sciences 280:349-366

LeMay M and Culebras A (1972) Human brain morphologic differences in the hemispheres demonstrable by carotid angiography. New England Journal of Medicine 287:349-366

LeMay M and Geschwind N (1975) Hemispheric differences in the brains of great apes. Brain, Behavior and Evolution 11:48-52

Levy J, Nebes, R, and Sperry RW (1971) Expressive language in the surgically separated minor hemisphere. Cortex 49-58

Levy J and Trevarthen C (1977) Perceptual, semantic, and phonetic aspects of elementary language processes in split-brain patients. Brain 100:105-118.

Levy-Agresti J and Sperry RW (1968) Differential perceptual capacities in major and minor hemispheres. Proceedings of the National Academy of Sciences 61:1151

Nebes RD (1971) Superiority of the minor hemisphere in commissurotomized man for the perception of part-whole relations. Cortex 7:333-349

Nebes RD (1975) Dominance of the minor hemisphere in commissurotomized man in a test of figural unification. Brain 95:633-638

Petersen M, Beecher M, Zoloth S, Moody D and Stebbins W (1978) Neural lateralization species-specific vocalizations by Japanese macaques (*Macaca fuscata*). Science 202: 324-327

Petersen MR, Beecher MD, Zoloth SR, Green S, Marler PR, Moody DB, and Stebbins WC (1984) Neural lateralization of vocalizations by Japanese macaques: Communicative significance is more important than acoustic structure. Behavioral Neuroscience 98:779-790.

Premack D (1971) On assessment of language competence in the chimpanzee. In: A M Schrier and F Stollnitz eds., Behavior of Nonhuman Primates Vol. 4. New York: Academic Press

Ross ED and Mesulam MM (1979) Dominant language functions of the right hemisphere? Archives of Neurology 36 144-148

Rubens AB, Mahowald MW and Hutton JT (1976) Asymmetry of the lateral sylvian fissures in man. Neurology 26:620-624

Rumbaugh DM ed., (1977) Language Learning by a Chimpanzee: The LANA Project. New York: Academic Press

Savage S and Bakeman R (1978) Sexual morphology and behavior in *Pan paniscus*. In: D J Chivers and J Herbert eds., Recent Advances in Primatology. New York: Academic Press

Savage-Rumbaugh ES (1984) *Pan paniscus* and *Pan troglodytes*: Contrasts in preverbal communicative competence. In: RL Susman ed., The Pygmy Chimpanzee. New York: Plenum

Savage-Rumbaugh ES (1986) Ape Language: From Conditioned Responses to Symbols. New York: Columbia University Press

Savage-Rumbaugh S, McDonald K, Sevcik RA, Hopkins WD and Rubert E (1986) Spontaneous symbol acquisition and communicative use by pygmy chimpanzees *Pan paniscus*. Journal of Experimental Psychology: General 1986 115:211-235

Savage-Rumbaugh ES, Rumbaugh DM, and McDonald K (1985) Language learning in two species of apes. Neuroscience and Biobehavioral Reviews 9:653-665

Susman RL ed., (1984) The Pygmy Chimpanzee: Evolutionary Biology and Behavior. New York: Plenum

Terrace HS (1979) Nim. New York: Alfred A. Knopf

Tucker DM, Watson RT and Heilman KM (1977) Discrimination and evocation of affectively intoned speech in patients with right parietal disease. Neurology 27:947-950

Umilta C, Rizzolatti G, Marzi CA, Zamboni G, Franzini C, Camarda R and Berlucchi G. (1973) Hemispheric differences in normal human subjects: Further evidence from study of reaction time to lateralized visual stimuli. Brain Research 49:499-500

Vermeire BA, Ifune CK and Hamilton CR (1984) Laterality in monkeys watching and reacting to television. Society for Neuroscience Abstracts 10:314

Wada JA, Clarke R, and Hamm A (1975) Cerebral hemispheric asymmetry in humans. Archives of Neurology 132 239-246

Wapner W, Hamby S and Gardner H (1981) The role of the right hemisphere in the apprehension of complex linguistic material. Brain and Language 14:15-33

Wechsler AF (1973) The effect of organic brain disease on recall of emotionally charged versus neutral narrative texts. Neurology 23: 130-135

Wernicke C (1874) Der aphasische Symptomenkomplex. Breslau: Cohn and Weigert

White MJ (1971) Visual hemifield differences in the perception of letter and contour orientation. Canadian Journal of Psychology 25:207-212

Winner E and Gardner H (1977) The comprehension of metaphor in brain-damaged patients. Brain 100:717-729

Witelson SF and Pallie W (1973) Left hemisphere specialization for language in the newborn: Neuroanatomical evidence of asymmetry. Brain 96:641-646

Yeni-Komshian GH and Benson DA (1976) Anatomical study of cerebral asymmetry in the temporal lobe of humans chimpanzees and rhesus monkeys. Science 192:387-389

Zaidel D and Sperry RW (1974) Memory impairment after commissurotomy in man. Brain 97:263-272

Zaidel E (1976) Auditory vocabulary of the right hemisphere following brain bisection or hemidecortication. Cortex 12:191-211.

Zaidel E (1977) Unilateral auditory language comprehension on the Token Test following cerebral commissurotomy and hemispherectomy. Neuropsychologia 15:1-18.

Zaidel E (1978) Lexical organization in the right hemisphere. In: PA Buser and A Rougeul-Buser eds., Cerebral Correlates of Conscious Experience. Amsterdam: Elsevier/North-Holland

THE EVOLUTION OF INTELLIGENCE: A PALAEONTOLOGICAL PERSPECTIVE

Martin Pickford
Institute of Anthropology
University of Florence
50122 Florence, Italy

The main contribution of the fossil record concerning the evolution of intelligence, as I perceive it, is the time perspective: the time depth.

There are two major approaches that permit us to delve into the fossil record to learn something about the development of intelligence. The first of these, about which there is a comprehensive literature, concerns the gross morphology of the neurocranium and endocranial casts of fossil and living species. The second approach concerns behavioural traces left by animals, including early human and prehuman precursors.

Both lines of inquiry when plotted against stratigraphic order yield information concerning sequences of events (Figure 11.1). In addition it is possible to obtain a notion of rates of change in some parameters if the fossil record is complete enough (Figure 11.2). A synthesis of sequence and rate data can yield a picture which reveals whether there occurred clumping of events, tandem evolution or major palaeontological events such as faunal turnovers (Figure 11.6, 11.7). Comparison of such events with non-fossil data, including palaeoenvironmental or palaeo-

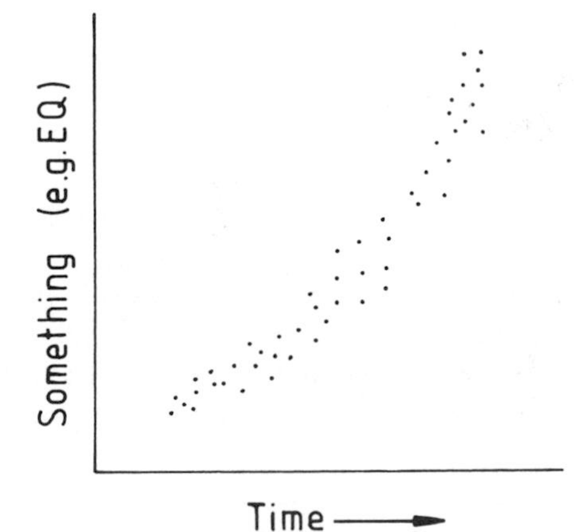

Figure 11.1. Bivariate plot to indicate changes in a measure over time. Changes in morphology, size and ranges of variation over time are easy to construct, easy to read, and easy to document in such graphs. In this hypothetical example, the covariate ("something, e.g., EQ") would have been measured in a large number of fossils (the dots) for which appropriate datings were also available.

NATO ASI Series, Vol. G17
Intelligence and Evolutionary Biology
Edited by H. J. Jerison and I. Jerison

climatic data can provide hints as to the external driving forces behind evolution (Figures 11.4, 11.5). For example, on the basis of correlations between fossil and non-fossil data we can examine suggestions that evolution might have occurred as a result of environmental forcing (see Vrba 1985 for a discussion of forcing models of evolution.)

Sequence of Events

In palaeontological circles, a sequence of events is reconstructed by integrating relative and absolute radiometric datings and stratigraphic order with other information about fossil samples (Figure 11.1). In the context of this paper, an obvious sequence to reconstruct is that of Brain/Body relationships (Jerison, 1973). Other morphological changes can also be measured, including changes in the external morphology and size of endocranial casts (Jerison, 1976, 1983; Holloway, 1974; Radinsky, 1974; Falk, 1982) and these might also be plotted against a time axis in order to appreciate the sequence of events and the rates of change. The external morphology of endocranial casts is more difficult to quantify and is controversial in other ways (Holloway, 1984; Falk, 1983). In view of these uncertainties my analysis is limited to brain size/body size relationships about which there is a reasonable consensus (Jerison,1983).

The Nature of the Data Base

The order Primates comprises one of eighteen extant orders of mammals. There are approximately 52 living genera and 182 living species of primates. Authorities differ on the exact number (Napier & Napier, 1985) and if the tree shrews (Tupaiidae) are included, the numbers increase by three genera and about 12 species respectively.

Over 104 genera of fossil primates have been described, and this number increases steadily as palaeontologists delve into the fossil record in poorly known time periods and relatively unexplored geographical areas. No doubt the total number of primate taxa that existed during the past 60-65 million years was vastly greater than this figure (Martin, 1986). Numerous taxa await discovery while many presumably never entered the fossil state. Given the fact that known fossil taxa are at least twice as diverse as extant primates (Szalay & Delson, 1979) anyone who ignores the fossil evidence does so at the risk of making grave errors and losing valuable data.

Brain and body sizes of fossil species are determined by various means: by estimating endocast volume either directly when specimens are complete and undistorted or by calculation in fragmentary or distorted specimens and estimating body weights from skeletal measures such as molar size, femur length, skull length, or total body length.

In living primates, brain size and body weight are usually determined by direct measurement (Harvey and Clutton-Brock, 1985), although the less direct methods used on fossil samples can also be undertaken. These differences in approach to determining brain size/body weight measures can give rise to problems in the data base as well as in interpretation. Errors in measurements obtained for fossil samples are usually greater that are error margins for extant species, partly because determinations in fossils are indirect and samples are usually small, while for extant species they tend to be by direct measurement on larger samples. However, body weight measures in extant species are by no means error-free (Pickford, 1986) especially for sexually dimorphic species, and in cases in which there are seasonal variations in body weight. Furthermore, individual body-weights can vary considerably depending on ontogenetic age, state of health and so on.

Over 100 fossil skulls belonging to more than 36 species of primates are complete enough to provide reasonably confident estimates of brain/body relationships. This sample of fossil primate endocasts is very limited. For the past 60 million years there is on average one endocast for every 600,000 years, and one species is represented by at least one endocast for every 1.8 million years during the past 60 million years. The problem is to get reliable evidence concerning changes in brain/body relationships from this kind of sample.

Rate of Change (Figure 11. 2)

Since the earliest days of palaeontology it has been noted that rates of evolutionary change were not the same in different lineages, and that even within a lineage there occurred periods of more rapid change interspersed with periods of less rapid change (Simpson, 1944).

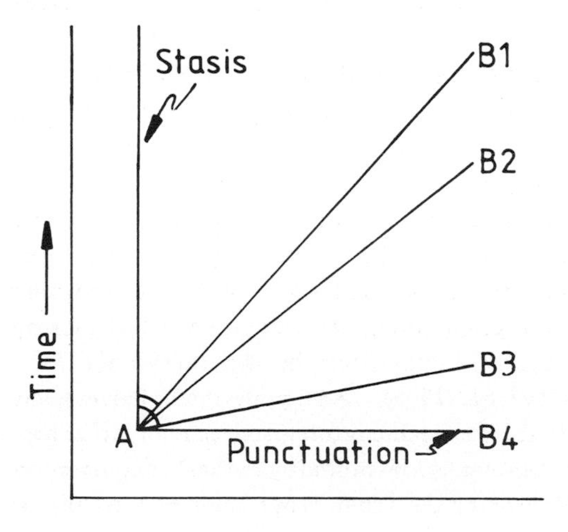

Figure 11.2. Bivariate plots to illustrate patterns of morphological change over time. Extremes are absence of change (stasis) and "infinitely" rapid change (A - B4, a "punctuation"). Stasis alternating with punctuation is evolution by saltations. Most changes in the fossil record can be described by intermediate rates (A - B2) as "gradualistic", though A - B3 may be "tachytelic" (rapid) and A - B1 "bradytelic" (slow) in Simpson's (1944) terms.

It has also been noted that periods during which rapid evolution occurred in a lineage were often the same or nearly the same as periods when rapid evolution occurred in other lineages. Although there is a certain amount of debate about the reality of such correlations, particularly regarding the degree of resolution possible with an imperfect fossil record, there appear to have been several periods when major events occurred contemporaneously in numerous lineages of organisms in diverse parts of the globe. Such 'clumping' of evolutionary changes has been interpreted in a variety of ways, usually being ascribed to some form of non-biological event. In the recent literature one finds reference to global changes induced by cometary impact, rearrangement of continental plates and oceanic circulation patterns, large-scale long-duration perturbations in atmospheric conditions brought about by orogenic (mountain-building) events and so on. In these models, a change in the physical environment is perceived to induce changes in the biosphere, with the result that some palaeontologists have introduced the notion of environmental 'forcing' of evolution (Vrba, 1985).

The history of encephalisation in Primates is amenable to 'rate-of-change' analysis, and such an analysis can be compared to non-biological data including palaeotemperature determinations, to see whether there might be some degree of correlation between the two (Figure 11.5).

The Primate Fossil Record

Figure 11.3 provides an overview of the fossil evidence regarding primates and their brains. All of the fossil material that is relevant for the record of encephalisation is represented in this illustration, which indicates where and when the fossils occurred. Each species (identified by number in the caption) for which encephalisation can be estimated is plotted as a numbered circle. Where the same group of primates is known from other parts of the stratigraphic record but without data regarding brain size, their range extensions are depicted by lines. The data are arranged regionally along the horizontal axis, the vertical axis representing time.

The earliest known primates of modern aspect are about 55 million years old. Many species of archaic primates are known from deposits older than this, but there is some debate about their status as primates. These more primitive species and genera are found as far back in time as 65 million years ago. Up to now five families of archaic primates have been recognised (Paromomyidae, Picrodontidae, Plesiadapidae, Carpolestidae and Saxonellidae) all currently classified in the infraorder Plesiadapiformes (Gingerich, 1976; Szalay and Delson, 1979). Although these Palaeogene families are considered to be primates by several palaeontologists, particularly on the basis of the dentitions, they appear to lack many if not all of the synapomorphies ["shared derived traits" that distinguish primates from other mammals as opposed to symplesiomorphies, i.e., more primitive traits shared with other primitive orders of mammals. Ed.] which characterise living primates (Martin, 1986).

Gingerich (1976) suggested that plesiadapids were rather similar to modern rodents in terms of their palaeobiology, in particular being comparable to ground squirrels and marmots. The encephalisation quotient of *Plesiadapis* is not very se-

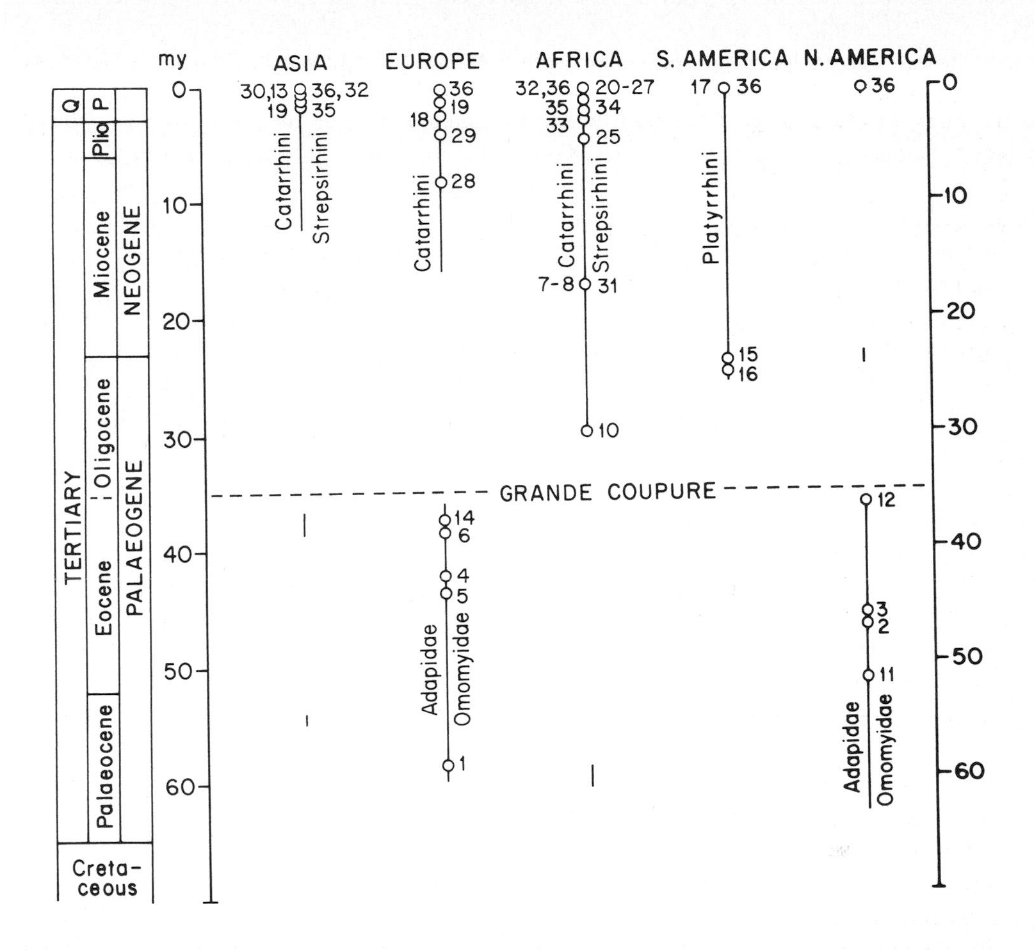

Figure 11.3. Fossil evidence on primates in space and time. Geological periods identified at left (Q = Quaternary; P = Pleistocene; Plio = Pliocene) on a scale of millions of years (my). Major taxa and continents on which fossils have been found indicated on chart. Numbered key to taxa as follows:

1. *Plesiadapis*
2. *Northarctus*
3. *Smilodectes*
4. *Pronycticebus*
5. *Leptadapis*
6. *Adapis*
7. *Komba*
8. *Mioeuoticus*
9. *Madagascan lemurs*
10. *Aegyptopithecus*
11. *Tetonius*
12. *Rooneyia*
13. *Tarsius*
14. *Necrolemur*
15. *Dolichocebus*
16. *Tremacebus*
17. *Extant platyrrhines*
18. *Paradolichopithecus*
19. *Macaca*
20. *Parapapio*
21. *Papio*
22. *Dinopithecus/Gorgopithecus*
23. *Cercocebus*
24. *Theropithecus*
25. *Libypithecus*
26. *Cercopithecoides*
27. *Paracolobus*
28. *Mesopithecus*
29. *Dolichopithecus*
30. *Hylobates*
31. *Proconsul*
32. *Pan/Gorilla/Pongo*
33. *Australopithecus*
34. *Homo habilis*
35. *Homo erectus*
36. *Homo sapiens*

curely based, complete fossil skulls or endocasts being unknown, but calculations by Jerison (1983) and Radinsky (1977) suggest that *Plesiadapis* was encephalised about as much as the average for extant rodents. Compared with its contemporary Palaeogene mammals, however, *Plesiadapis* was much more encephalised than archaic ungulates, and was perhaps similar to Eocene primate genera such as *Tetonius, Smilodectes*, and *Adapis*. Although *Plesiadapis* was about as encephalised as some of the Eocene Omomyidae and Adapidae, it had a lower encephalisation quotient than any living prosimians.

Plesiadapiformes are recorded from Palaeocene and Eocene deposits, the youngest records being about 38 m.y., prior to the grande coupure (for definitions see Table 11.1, below). Their heyday in terms of diversity however was in the period between 60 and 50 m.y. ago, during the Palaeocene/Eocene heating maximum (Figure 11.4 cf. Table 11.1), and only one lineage (the Paromomyidae) survived the Palaeocene/Eocene faunal turnover (Szalay and Delson, 1979) about 55-54 m.y. ago.

The demise of the Plesiadapiformes was followed closely by the rise of the Adapidae and Omomyidae, two families that fit comfortably into the order Primates as defined by Martin (1986). These two families appeared in North America and Europe rather abruptly at the beginning of the Eocene period and are usually assumed to have evolved in tropical regions of the globe during the Palaeogene and to have migrated northwards into higher latitudes during the Palaeocene/Eocene heating maximum. Recent discoveries in the Palaeocene of Africa may throw some light on the very early history of these two families (Sudre, 1875) but thus far no skulls or endocasts have been found.

The omomyids and adapids survived in Europe and North America for a considerable period (55-36 m.y. ago) only to disappear from northerly latitudes prior to or at the 'grande coupure'. A few fossils identified as adapids from the late Miocene of India and China represent the youngest known survivors of the family Adapidae (if the family identification is correct, Pickford, 1987). The omomyids, on the other hand, are unknown in Neogene deposits (Miocene to Recent); there is a late Oligocene *Ekgmowechashala* in North America, which is classified as omomyid *faute de mieux*. In any case its brain size is unknown. Although there is some debate about the origins of the extant haplorhines and strepsirhines vis-a-vis Eocene primates, there seems to be general consensus that the modern groups had their roots in Eocene forms. Unfortunately, the fossil record of primates from equatorial regions of the globe is particularly poor for the Palaeogene, and this is presumably where much of the evolutionary action took place. The conflicting hypotheses about anthropoid origins were discussed by Rosenberger et al. (1985) who summarised the various possibilities that have been proposed. These fall into three general groups: a) anthropoids are derived from adapids (e.g., Gingerich, 1980); b) anthropoids are derived from omomyids (e.g., Rosenberger et al. 1985) or c) anthropoids are paraphyletic with roots in both the adapids and omomyids (e.g., Simpson, 1963). For the moment there seems to be no clear-cut solution to the conflicting views.

After the demise of the Adapidae and the Omomyidae in northerly latitudes prior to or at the 'grande coupure' the primate fossil record is rather poor. North African forms such as *Qatrania, Propliopithecus, Apidium, Oligopithecus* and *Aegyptopithecus* may represent the earliest catarrhines, but if so they are exceedingly primitive (cf., Fleagle and Simons, 1982). In terms of its encephalisation, *Aegyptopithecus* had reached only the lemuroid grade (Jerison, 1983) being appreciably less encephalised than the least encephalised living hominoid (or cercopithecoid for that matter). Whether this alone would disqualify it from classification as a hominoid or catarrhine needs examination; postcranially and cranially it is not particularly hominoid-like. It is perhaps only in its dentition that *Aegyptopithecus* recalls living and Miocene hominoids.

The primate fossil record improves dramatically during the early Miocene, principally in East Africa, where hominoids, oreopithecoids, cercopithecoids and lorisoids have been recovered in significant quantities. Lower Miocene deposits in South America, in contrast, have yielded few fossil primates, which are usually assigned to extant families of platyrrhines, the Cebidae and the Atelidae (see, e.g., Szalay and Delson, 1979).

The Neogene history of primates in Europe begins with hominoids at about 16 m.y. (Pickford, 1986) which survive there until the late Miocene cooling period about 9-8 m.y. ago. Cercopithecoids were, in contrast, relative late-comers to Europe, the earliest being about 11 m.y. old, but they were able to survive in Europe well after the hominoids had become extinct there. During the Pliocene in particular, cercepithecids were widespread in southern Europe.

Hominoids were able to migrate to mid-latitude Asia during the middle Miocene, presumably as a consequence of the establishment there of warm conditions during the mid-Miocene heating event (Figures 11.3, 11.4), and in this respect the northwards shift in their zoogeographic range was similar to what occurred in Europe at about the same time. Monkeys, however, appear to have colonised Asia much later than they did Europe, the earliest records being late Miocene to Pliocene (Pickford, 1987).

During the late Miocene cooling events, hominoids became extinct in Europe, but were able to survive in the tropical parts of Asia, in particular in the Malayan/Indonesian archipelago. It was at this period (8-7 m.y. ago) or probably earlier that the Asian hominoid gene pool was definitively severed from the African gene pool, although there may well have been genetic isolation for some considerable period prior to zoogeographic isolation.

The Asian monkeys, which appear to have been more tolerant of colder climes than were the hominoids, as indeed some of them are today (Napier and Napier, 1985), probably had the possibility of some genetic connections with Africa and Europe until the Pleistocene period. Whereas the lower and middle Miocene can be considered to have been the heyday of the hominoid primates in terms of diversity, the cercopithecoids experienced their greatest diversity during the Pliocene and Pleistocene. A major radiation started about 6 million years ago, possibly accelerated by changes in global climate which led to the Messinian Crisis and subsequent cooling during the Plio-Pleistocene (4-2 m.y.). It was during this period too, that some African hominoids developed bipedal locomotion possibly in order to cover enormously enlarged territorial ranges. These experiments eventually culminated in the rise of the hominids, a history which is reasonably well documented from Pliocene and Pleistocene deposits of Africa and Eurasia.

Major Events in Primate Palaeontology (Table 11.1, Figure 11.4)

There are several marked coincidences between the palaeodistribution of primates and the $d^{18}O$ curve for the Tertiary Period. The number and positioning of these coincidences suggest that there was a causal relationship between the two histories, indicating that major changes in primate distribution patterns were 'forced' by global-scale climatic changes, continental configurations permitting. To some extent the appearance of new grades of primates in the fossil record occurs soon after significant or major events in the $d^{18}O$ curve, suggesting that there has been an element of climatic 'forcing' of evolution in the primates. The climatic events appear to precede the appearance of each new grade of primates by one or two million years.

Table 11.1 Climates and primate evolution. Letters and numbers as in Figure 11.4.

Climatic event	Palaeoprimatological event
1. Cretaceous-Tertiary boundary	A. Plesiadapiformes in North America and Europe. No data from southern continents.
2. Late Palaeocene/early Eocene warming, accompanied by northward spread of tropical and sub-tropical climatic regimes.	B. Northwards spread of Omomyids and Adapids from Africa (and possibly southern Asia) into Europe and North America followed by radiation of both groups.
3. End Eocene cooling event known in Europe as the 'grande coupure', results in southwards shift in tropical and sub-tropical climatic regimes. Temperate conditions come to all of Europe and much of Asia.	C. Extinction of Omomyids, Adapids and Plesiadapids in Europe and North America, except perhaps for a small pool of primates in southernmost North America. D. Origins of extant Primate groups such as Platyrrhines and Catarrhines follows this event. (Evidence from N.Africa and South America). Possible primate refuge in S-E Asia Development of enhanced, stereoscopic vision, and beginning of selection for colour vision.

(Table 11.1, Cont.)

4. Late Oligocene heating event, followed by a cooling event.	E. *Ekgmowechashala* to South Dakota, ?from the south. Cebidae spread southwards to southern South America. Both range increases rather short, ended by cooling event that followed heating episode. Northwards spread of primates into Europe, climatologically possible , did not occur because of Tethys barrier.
5. Mid-Miocene heating event (Langhian in Europe; Maboko event in Kenya)	F. Pliopithecid and hominoid ranges increase northwards from Africa into Europe and Asia, followed by range decrease during mid-Miocene cooling event.
6. Mid-Miocene cooling event.	G. Monkeys cross into Europe from Africa, initially just colobines. Decrease in ranges of hominoids.
7. Messinian Crisis in Mediterranean region.	H. Major cercopithecoid radiation. Cercopithecines to Europe and Asia. Colobines to Asia. Hominoids extinct in Europe and over much of Asia. Definitive geographic separation of African and Asian hominoid gene pools.
8. Initiation of Glacial periods. Major cooling of oceanic waters.	I. Reduction in latitudinal ranges of all Eurasian primate species toward the south. Primates extinct in Europe and over the bulk of Palaearctic Asia. Oriental Realm is refuge for lorisoids, tarsioids, hylobatids, hominoids, and cercopithecoids (colobines and cercopithecines). Africa is refuge for hominoids, cercopithecoids, lorisids, while Madagascar remains distinct with lemuroid fauna. Spread of humans worldwide.

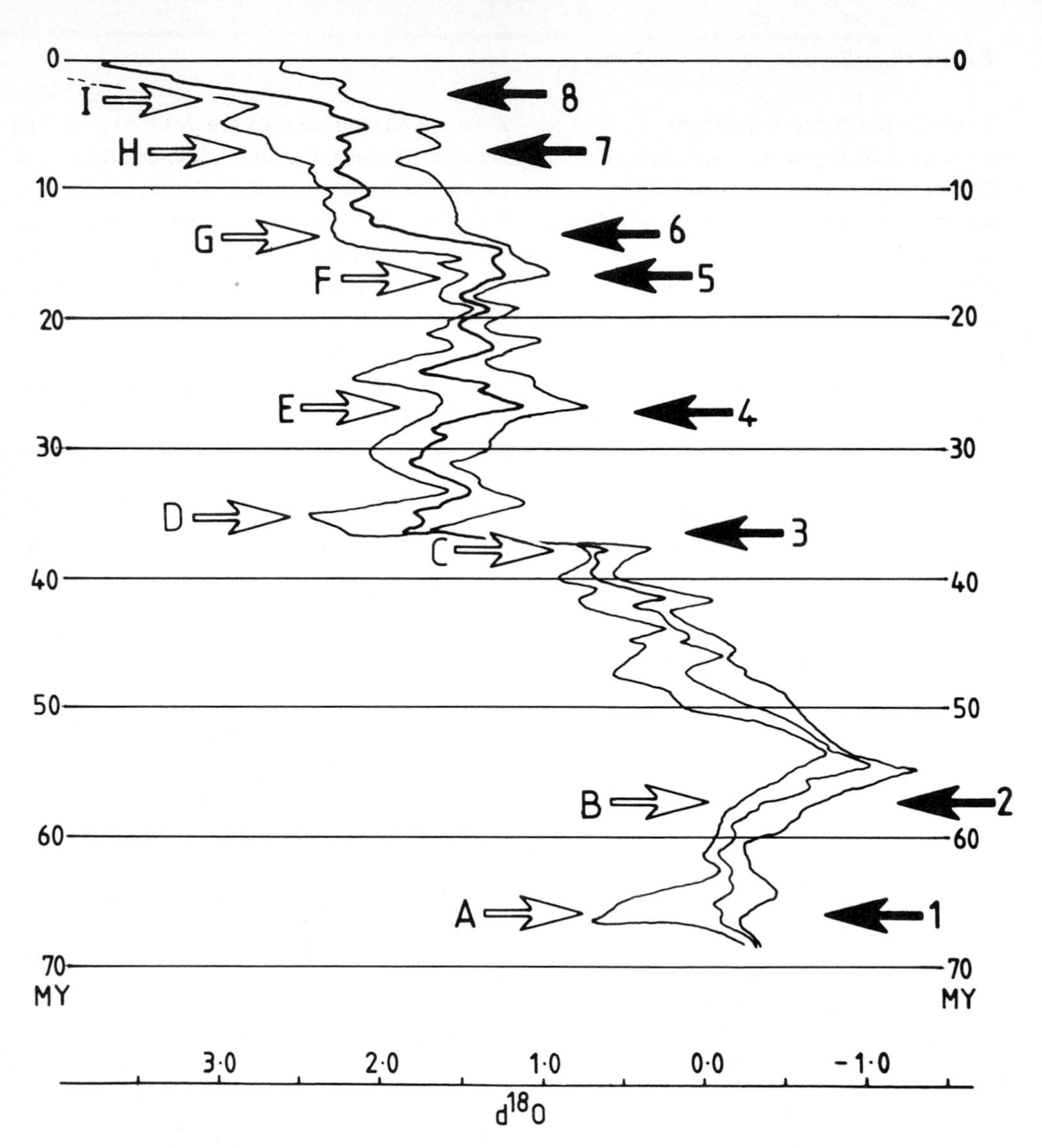

Figure 11.4. Major climatological and palaeoprimatological events of the Tertiary, and the Tertiary d^{18}O (oxygen concentration, inversely related to water temperature) curve for Atlantic and Pacific ocean benthic Foraminifera from Miller and Fairbanks (1985). Letters and numbers refer to descriptions in Table 11.1, above.

Encephalisation History of the Primates

The encephalisation scale in Figure 11.5 is an arbitrary division of the living and fossil primates into 12 grades that are related to zoogeography and encephalisation. Plesiadapiformes are included as a thirteenth grade, although their status as primates is doubtful (Martin, 1986; Gingerich, 1976, 1986). If this were a taxonomic exercise based on encephalisation it could be criticized for both excessive 'lumping' and 'splitting': some ceboids, for example, are about as encephalised as some cercopithecoids. I treat them separately because encephalisation increases in platyrrhines were achieved independently of increases in catarrhines.

On the other hand, there are various grades of encephalisation found within the ceboids and cercopithecoids; baboons, for example, being more encephalised than the guenons. The grades chosen for this diagram represent a compromise between these

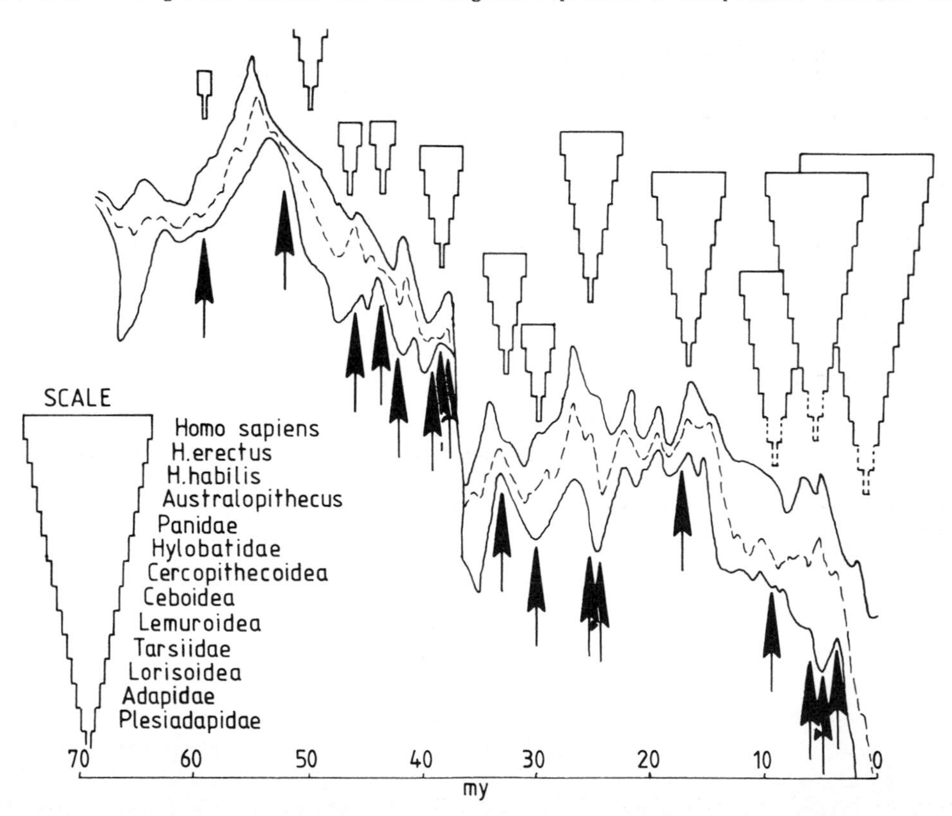

Figure 11.5. Primate encephalisation grades (ordinal units: width of inverted pyramids as shown on scale) that are represented in or suggested by the fossil record. Arrows depict ages of species of which the skull or endocranial cast is known or can reasonably be inferred. Increases in encephalisation are depicted by broadening of the inverted pyramids, the widths of which are proportional to the grade of encephalisation. Evidence is arranged either side of the $d^{18}O$ curve (Miller and Fairbanks, 1985). Encephalisation has increased with time and with drop in temperature of oceanic surface waters.

brain size criteria and evidence from zoogeographic occurrence and phylogenetic position. Despite the obvious danger of this approach, the resulting groupings are helpful for organizing the available data.

The primate fossil record falls into two major chronological categories; pre- and post- grande coupure. Palaeocene/Eocene fossil primates and their brains are best known from European and North American strata older than the grande coupure (Stehlin, 1909) a period of major faunal change in high latitude regions of the globe (see Figures 11.3, 11.4). In contrast Oligocene and Neogene fossil primates and their brains are known principally from tropical parts of the Old World.

The early Palaeogene primate record for tropical parts of the globe (Africa, Southern Asia) is exceedingly restricted, and nothing is known about the brains of these animals. Primates disappeared from the European and North American fossil records during the grande coupure, and did not reappear there for a considerable period. For Europe the earliest known Neogene primate is only 16 m.y. old, while for North America, except for a single occurrence in the late Oligocene, primates were absent until the late Pleistocene when humans colonized the continent.

In contrast, the primate fossil record of Africa, which is so poor prior to the Eocene/Oligocene boundary (37-36 m.y. ago) is comprehensive after it, being considerably more informative than those of all the other continents put together.

Primates were present in tropical Africa during the early Palaeogene (Sudre, 1975) but their poor fossil record mirrors the generally limited fossil record for all African mammalian groups during this period.

Primates in South America, Europe, and Asia during the late Palaeogene and Neogene are probably representatives of lineages that evolved in Africa and subsequently spread to neighbouring continents (Fleagle, 1986) but details of their origins remain obscure.

In Figure 11.4, encephalisation data are correlated with a curve representing $d^{18}O$ history during the Tertiary. Measures of $d^{18}O$ in benthic Foraminifera (Miller & Fairbanks, 1985) are considered to yield information regarding palaeotemperatures of the surface waters of the oceans where the Foraminifera lived. As such the $d^{18}O$ curve is considered to provide a partial history of oceanic climate for the Tertiary epoch. Since the world's oceans are a major driving force for global atmospheric weather systems, changes in oceanic conditions probably affected global atmospheric climates. These in turn would have affected the biosphere.

Throughout the Tertiary there was a trend in the primates for increase in maximum grades of encephalisation. In Figure 11.5 the inverted pyramids depict the grades of encephalisation known or inferred to have existed at various times during the past. This was generally established not by replacement of lower grades by higher grades; instead there was accretion of grades throughout the Tertiary, with the exception of the relatively small- brained adapids, which apparently became extinct at the end of the Eocene. Most grades of encephalisation established in earlier times persisted in some later species that lived to survive alongside other later established species that were at higher grades. Thus, for example, during the early Palaeogene, 58 m.y. ago, the most encephalised mammal was probably *Plesiadapis.* Somewhat later in time, the most encephalised species was probably the omomyid *Tetonius*, which occurs alongside plesiadapids. By 38 m.y. ago the most encephalised species was a later omomyid *Necrolemur* which occurs with plesiadapids and adapids. By lower Miocene times (18 m.y.) the ape-like *Proconsul* may have approached the great ape grade of encephalisation (Walker et al., 1983). *Proconsul* is found side by side with the prosimian *Komba*, in which the brain was comparable to that of living prosimians. We have ample evidence of several increases in grade of encephalisation in the hominids during the Plio-Pleistocene, each species at an advanced grade generally occurring contemporaneously with less encephalised species. Finally, in the spectrum of living primates we can find all but the two lowest and the second highest grades of encephalisation that have been suggested in Figure 11.5.

In summary, on the basis of over 100 fossil skulls belonging to more than 36 species of fossil Tertiary primates one may infer that there were accretionary increases in grades of encephalisation. In this model, grades of encephalisation already in existence are not replaced by subsequent increases in encephalisation, but are added to. Thus the number of grades of encephalisation increased throughout the Tertiary. Omitting the plesiadapids, the primate status of which is contentious, the fossil record indicates that during the Eocene there were three grades of encephalisation, during the Oligocene three additional grades appeared, during the Miocene there were two further additions and during the Plio-Pleistocene there was an increase of four grades.

Climate Forcing of Encephalisation (Figures 11.5, 11.6, 11.7)

Although the amount of data is limited, there are some patterns to be observed in the evolution of grades of encephalisation in the primates. There was apparently a gradual increase in grades from about 70 to 44 m.y., followed by a period of more rapid increase in encephalisation, after which there was a relatively stable period lasting until the grande coupure about 37 m.y. ago. There were some increases in grades of encephalisation until the Pliocene period, although details are too scat-

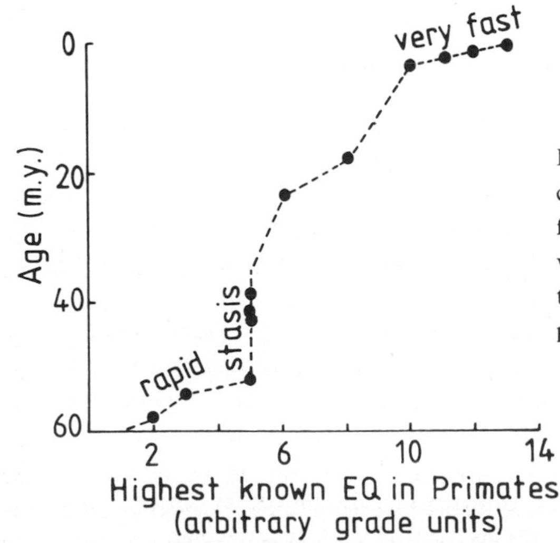

Figure 11.6. Rate of evolution of encephalisation (EQ) in primates; data from fossil endocasts. Increases in EQ were not steady or gradual but at rates that varied from slow stasis to fast punctuation (cf. Figure 11.2

tered for the purposes of determining whether there were any departures from gradualism during the Oligo-Miocene period. In contrast, the relatively extensive data on Plio-Pleistocene hominids indicate rapid increases in encephalisation grades from about 4 m.y. to about 0.1 m.y. ago.

The story is not clear for long periods of the Tertiary, but the two main periods of rapid increase in grades of encephalisation, which can be demonstrated to have occurred on the basis of reasonable evidence, coincide in timing with global cooling events which followed abruptly after heating trends. The mid-Eocene event at about 54 m.y. is one of these while the other is the Plio-Pleistocene cooling event following the Messinian heating event.

Although there appears to be a concordance in the data (Pickford, 1986), it has not yet been demonstrated that such global climatic changes caused selection for increased encephalisation in the primates. The idea could be attractive however, to scientists who would like to have evidence in support of climatic 'forcing' models for explaining the driving force behind major palaeontological events, for example high density faunal turnovers related to the grande coupure, the mid-Miocene event (Pickford, 1987), and other events during the Neogene (Vrba, 1985).

Even though the evidence is scanty there appears to be a possibility that primates were no less susceptible to these global scale environmental events than were other mammalian groups, and that selection for increased encephalisation may have been one of the responses by the primates at least twice, and perhaps more times during their history.

Primate Adaptations and Encephalisation

It is postulated that these global-scale climatic changes led to establishment of new niches into which primates ventured, in the process undergoing evolutionary changes. For example, desiccation of Africa during and after the Messinian crisis led not only to a major radiation of the cercopithecoids but also to the development of the earliest hominids. The hominids in particular occupied more open relatively dry country compared to the usual great ape habitat.

The usual assumption is that entry into these niches was accompanied by selection for larger brains, presumably because the information that had to be handled could not be handled by small efficient brains adapted to familiar environments. Primates with established grades of encephalisation presumably remained in niches that they were already adapted to or they possibly entered new but familiar niches where there was no strong premium on additional, unusually organized, processing capacity.

The origin of the anthropoids was related to and possibly caused by, a move from nocturnal to diurnal activity rhythms. Most prosimians are nocturnal, and even the diurnal species have poorly developed or undeveloped colour vision (Davis, 1976; De Valois, 1965). In addition, their stereoscopic vision is less well developed than that of anthropoids, the overlap of the visual fields of the two eyes being less than in anthropoids. All extant anthropoids, on the other hand have well developed colour vision (at least in all species tested) and the overlap between the visual fields in the visual cortex is marked.

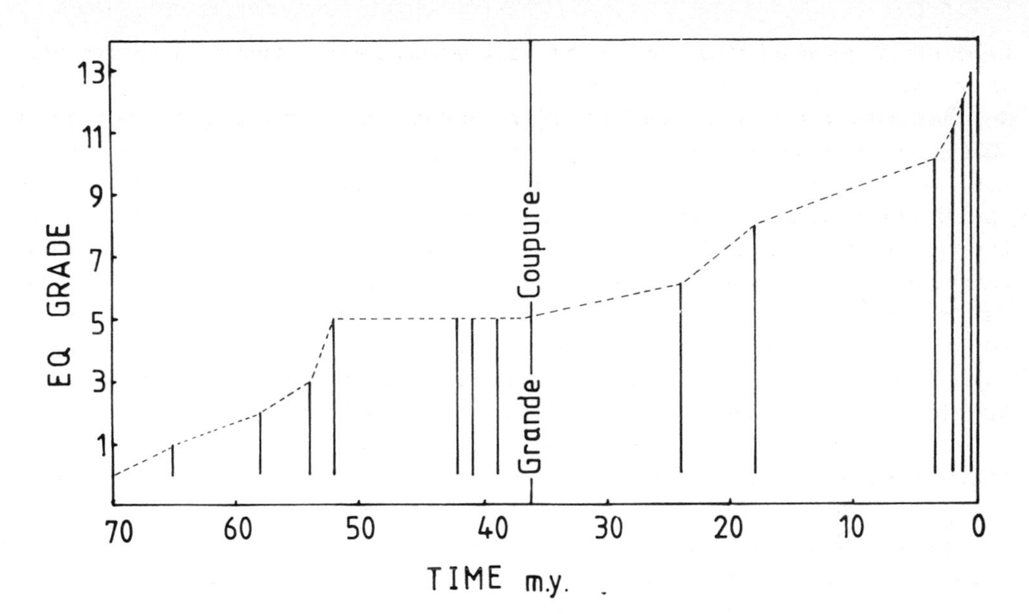

Figure 11.7. Summary of current information on ordinal grades of encephalization (cf. Figure 11.6) in primates during the Tertiary. Two periods of relatively rapid increase can be discerned, the earlier one at about 54-52 m.y. ago, and the second one between 4 m.y. ago and the present day. Vertical lines represent the estimated maximum grades of encephalization available from the fossil record.

The development of more complete stereoscopic colour vision appears to have resulted in two major changes in the skull of anthropoids compared to those of prosimians. The first is that a better developed binocular visual system requires a proportionally enlarged processing capacity in the brain. The same applies to colour vision, as opposed to non-colour visual systems. The result as far as the brain is concerned is probably that there was an increase in brain size related to the increased processing load, and this is recorded in the increased encephalization of anthropoids compared with prosimians.

Secondly, a stereoscopic visual system, which requires the accurate fusion of two undistorted images, would require that the eyeballs remain undistorted during visual activity. In prosimians, the eyeballs may be affected by the temporal muscle during chewing because the workings of the temporal muscle could alternately compress and relax pressure on the retinal part of the eyeball (Cartmill, 1980). In anthropoids, in contrast, the eyeball is completely isolated from the effects of the temporal muscle by the bony post-orbital septum and until distorted artificially, for example by pressing the eyeball with a finger, the eyeballs remain undistorted by most bodily activities. Thus the presence of a postorbital septum is probably related to the acquisition of enhanced stereoscopic colour vision in anthropoids.

The earliest evidence of the existence of post-orbital septa in primates is provided by *Aegyptopithecus* from early Oligocene deposits of Egypt. The Fayum strata, from which *Aegyptopithecus* came, postdate the grande coupure by about 3 to 4 m.y. Thus, it is inferred that stereoscopic vision of anthropoid type had developed by that time, although the evidence of the brain size in *Aegyptopithecus* indicates that it is little more encephalised than lemurs. The advance in stereopsis suggests that *Aegyptopithecus* may have been diurnal, but data on its encephalisation suggest that it may not have had significant colour vision.

If this scenario is correct, then improved stereoscopic vision followed the acquisition of diurnal activity rhythms by anthropoids early in the Oligocene, and the ability to see in colour was developed later. If *Aegyptopithecus* is representative of early anthropoids, then it would follow that the anthropoid type of stereoscopic visual system evolved more or less at the time of the grande coupure (36 to 37 m.y. ago) but that colour vision evolved later than 32 m.y. ago.

Distribution of Bone Mass in Hominid Skulls

It has been demonstrated that in Primates, as in other orders of mammals, skeletal mass scales isometrically with body weight (Potter, 1986). Within close limits, an increase in bone mass in one organ system of an evolving lineage will be accompanied by a corresponding decrease in bone mass in another part of the skeleton, in order to maintain the isometric skeleton weight/body weight scaling.

Rak (1987) has pointed out that the so-called hyper-robust australopithecine species *Australopithecus boisei* is in reality only hyper-robust in its splanchnocranium. Its neurocranium is hyper-gracile, the lineage having invested most of its cranial bone mass in the masticatory system at the expense of the neurocranial system. In so doing, *A. boisei* has had to adopt a series of apomorphic design strategies in order to maximise the structural effectiveness of the neurocranium utilising a minimum mass of bone. Such strategies are particularly evident in the region of the squamosal suture, the asterionic region, the shape of the braincase itself and the design of the supraorbital region.

In *Homo* species, in contrast, the bulk of the cranial bone mass is invested in the neurocranium, the splanchnocranium being reduced to a minimum consistent with the maintenance of an effective masticatory system.

During the early Pliocene some early hominids possessed crania in which the distribution of bone was not remarkably different from that seen in great apes (Ferguson, 1983, 1986; White, Johanson and Kimbel, 1983). Subsequent to the early Pliocene, two major trends occurred in hominoid evolution (McHenry 1984). Both of these entailed modification of the masticatory system, presumably following a shift in dietary strategies; one resulted in the development of hyper-robust splanchnocrania and hyper-gracile neurocrania, the other in hyper-gracile splanchnocrania and hyper-robust neurocrania. Lineages engaged in the former strategy were employing an evolutionary strategy which severely limited their potential for further augmentation of brain size.

The *A. boisei* lineage effectively reached the limits of encephalisation of about 500 cc brain capacity under the dual requirements of a highly developed masticatory system and its brain size requirements. By adopting a revised dietary strategy that emphasised a massive masticatory apparatus, the *A. boisei* lineage effectively blocked greatly increased encephalisation. As the brain in this lineage got bigger, the calotte became thinner, the resultant of these antagonistic pressures on bone mass being effected in *A. boisei* in which a cranial capacity of about 500 cc was attained. An additional design strategy partly employed by australopithecines was to pneumatize the neurocranial bones, not to lighten the skull as previously thought, but to increase the strength of the calotte without adding large quantities of bone. In an analogous way bird bones are strong but light, with maximum strength being achieved by adopting appropriate design of available bone mass rather than by increasing the amount of bone. Apart from small areas of the nuchal and squamosal bones however, such a strategy seems not to have been adopted as viable by robust australopithecines.

In contrast, the second major evolutionary pathway entered by hominids during the latter portion of the Pliocene (and perhaps earlier, if *Homo* occurs in the Hadar deposits (Ferguson, 1984; Senut, 1986)), concerned increases in encephalisation (Jerison, 1973), a correlate of which was increased investment of bone mass in the neurocranium, which not only became larger in surface area but thicker in section. As a result of the isometric scaling of skeletal weight to body weight, bone mass was redeployed from other parts of the skeleton, in particular the splanchnocranium, which became ultra-gracile.

The latter strategy has culminated in *Homo sapiens*, but there is a suggestion that the process of bone mass redeployment has not ended. For example, the trend in modern humans towards non-development of wisdom teeth and foreshortening the alveolar processes of the maxilla and mandible, (a ready source of income for dentists), may well be related to a continuation of the process of redeployment of bone mass from the splanchnocranium into the neurocranial region. The result is that in modern man, a hyper-gracile splanchnocranium is hafted onto a hyper-robust neurocranium. Such a great investment of bone in the neurocranium is an expected correlate of increased encephalisation, and there would probably have been concomitant selection for dietary behaviour that would maintain the species with a minimum of masticatory apparatus. Considering the available bone mass in the human cranium, the human lineage has probably reached its limit as far as further encephalisation is concerned, unless bone mass is to be redistributed from the post-cranial skeleton. That this might have happened in the past is suggested by the gracilization of the human post-cranial skeleton during the Palaeolithic period, but clearly, more evidence is required concerning this latter point.

In any event, it appears that hominids used two different evolutionary strategies during the Pliocene; one led to the development of a hyper-robust splanchnocranium emphasizing mastication, the other to a hyper-robust neurocranium emphasizing increase in brain size. In view of the isometric nature of skeletal/body scaling, each evolutionary strategy led to drastic rearrangements in the distribution of available bone mass of the cranium within the limits set by the requirement that all cranial functional systems should work efficiently and effectively.

In this perspective, the splanchnocranium and the neurocranium, which are often considered separately in morphometric and cladistic studies, are seen to be closely related as part of an overall bodily requirement to maintain the finely tuned balance between skeletal weight and body size. Increased encephalisation could only occur within the scope of an adequate masticatory system, even a reduced one such as occurs in *Homo.* The focus on increased bone mass in the masticatory system of robust australopithecines, on the other hand, limited their evolutionary potential for increased encephalisation to the point where their neurocrania became extremely thin, and in which architectural modifications were employed to their limit in order to maintain the mechanico-structural integrity of the skull as a whole. In *Homo*, in contrast, apart from the tendency to develop a more globular neurocranium, the architectural plasticity observable in modern human skulls (dolichocephaly, brachycephaly etc.) indicates that, within limits, the detailed architecture of the neurocranium is not of crucial evolutionary significance to the genus. Indeed, artificial modification of the shape of the neurocranium, such as for example that undertaken by head-binding societies, seems to have little effect, either on the intelligence or on the masticatory efficiency of the individuals so modified. However, in a large sub-spherical structure such as the human cranium, there is a major requirement for strength in the covering which protects the brain, and this has been achieved by the dual strategy of thickening the cranial vault and making it globular in overall shape.

Further back in geological time, it is suggested that the great ape grade of encephalisation had evolved by lower Miocene times (*Proconsul africanus* skull described by Walker et al. 1983). Is it coincidence that the great ape grade of encephalisation coincides in terms of definition and development with the redeployment of skeletal bone away from the tail, to the extent that hominoids are definable as the tailless catarrhines?

Taking a step further back into geological time, it is seen that the snout of *Aegyptopithecus* is large relative to the neurocranium in contrast with those of most monkeys and apes, in which the snout is relatively reduced. In the latter groups, the neurocranium is relatively enlarged due mainly to the fact that encephalisation increased from a lemuroid grade in *Aegyptopithecus* to cercopithecoid and pongid grades in the lower Miocene catarrhines. This hyper-development of the neurocranium was at least partly achieved by redeployment of bone mass from the snout, which became relatively reduced in Neogene catarrhines compared with those from the Palaeogene and possibly partly by redeployment of caudal bone mass.

Viewed within the framework of isometric scaling of skeleton weight to body size, the evolution of the various catarrhine lineages, including the hominids, becomes more understandable, at least as regards their heads and possibly their tails.

Behavioural Evidence Concerning the Evolution of Intelligence

Few traces of behavioural aspects of extinct primate species can be found to shed light on the evolution of intelligence. It isn't until the Plio-Pleistocene period that traces of the manufacture and use of stone tools are found from which we can infer certain levels of technical and behavioural skills in (we presume) our hominid precursors.

In most instances the evidence that we find is usually only of an indirect nature. For example, it has been shown that there is an unequal representation of left handed and right handed flakes in many stratigraphic levels yielding stone tools (e.g., Oakley, 1972) with most flakes being struck by right handed manufacturers. The scenario is that the left hand was generally the 'gestalt' hand, balancing and orienting the core, while the right hand and forearm undertook the striking action with the hammer stone (Wundrum, 1986). For left handed operators, the contralateral limbs would perform the differentiated tasks. It has been suggested that such simultaneous differential use of the hands provides evidence for cortical motor asymmetry and furthermore that such bimanual motor asymmetry follows the development of bipedalism but predates the earliest manufacture of stone tools in the 'hominisation' process (Wundrum, 1986), although it has also been suggested that tool making behaviour was the ultimate cause of motor asymmetry in hominids (Frost,1980).

In any case cerebral asymmetry is an ancient acquisition among vertebrates (Levy, 1977, this volume), and it is a big jump from demonstrating the possible existence of motor asymmetry in early hominids to understanding its implications for evolution of intelligence and language.

At other stages in the past other traces of behavioural and intellectual abilities become evident, but the timing of their development is difficult if not impossible to pin down. When, for example, did hominoids learn that they could produce fatal or crippling traumas on other species at a distance? Accurate stone throwing at stationary or moving targets with a view to killing or maiming prey presumably evolved early during hominid evolution, and would have required quite sophisticated visual and motor control in individuals who had developed the ability to anticipate the immediate future actions of mobile prey, but it isn't until we find evidence of bows and arrows, spears and so on, that we can actually demonstrate the existence of such abilities in hominids. In contrast, the stone and stick throwing undertaken by chimpanzees in experimental situations is inaccurate, not very powerful, and sufficiently slowly undertaken that most moving targets would have little difficulty escaping injury.

The archaeological record is replete with evidence concerning developmental stages in material culture. We have reasonably good evidence that hafting of tools followed (and possibly caused) the replacement of the Acheulian tradition (1.5 - 0.4 m.y.) of stone implement making with the 'middle stone age' (0.5 to 0.1 m.y.). It was a genius of a discovery, but the intelligence capable of making and appreciating the discovery was presumably already developed. The control of natural fire, and the subsequent discovery that fire could be manufactured, falls into a comparable category. And one can think of many other examples. The point is that by Acheulian times at the latest, hominid intelligence was probably of a comparable order to that of extant humans, at least as regards non-linguistic capacity. The apparent rapid spread of cultural innovations during the latter part of the Acheulian period would suggest however that language was already well developed by Acheulian times if not earlier.

It has been suggested that the use of fire by hominids dates back to the early Pleistocene, where claims have been made for the discovery of fire-baked clays in the same levels as Oldowan tool kits 1.4 m.y. ago (Gowlett et al., 1981). However, the evidence is not clear cut, and natural, non-human agencies have not definitively been ruled out, especially in view of the widespread evidence of clay-baking by a lava flow which overlies the locality (Chesowanja, Kenya). Nevertheless, it is quite conceivable that hominids had the ability to control fire by that remote period, even if they had not discovered how to manufacture it. By the middle Pleistocene, the evidence for fire used by hominids is more compelling (e.g., at Choukoutien, China (Binford & Ho 1985)), and its control may well have been influential in permitting early man to colonize the cooler latitudes of the globe. Goudsblom (1986) discusses the socio-cultural implications of domestication of fire by humans.

Although the traditional view that the apes were all forest-adapted mammals has recently been modified with the discovery that chimpanzees and even gorillas can inhabit quite open environments the basic adaptation of all the apes is for forest habitats (Napier & Napier, 1985). Hominids on the other hand are usually perceived as having been adapted to more open, even semi-arid environments. All ape species for which the territory sizes are known, have ranges which are less than 200 km^2 and usually considerably less, being about 20-30 km^2 in area (Yamagiwa, 1986). Within these ranges, food resources are scattered, but not particularly difficult to find, the bulk being easily accessible and requiring minimal preparation prior to ingestion. A few selected food items require special mention, including cryptic foodstuffs such as termites which are frequently fished for by chimpanzees. Even though nuts may not be cryptic, the edible contents are difficult to get at and may be considered so from the point of view of their exploitation (Boesch & Boesch; 1983, King 1986). In semi-arid environments, in contrast, food resources are usually much more scattered and in many cases, such as tubers and other geophytes, are not only cryptic, but also require a great expenditure of energy for their extraction.

It is estimated that the ranges of hunter-gatherers in semi-arid environments are several orders of magnitude greater than those of any of the apes. Territories of up to 1200 km^2 are not unusual, and foragers or hunters moving through their territory may cover 20-30 km per day. Chimpanzees, gorillas and gibbons in contrast frequently move less than five km per day, and often less than one kilometer.

Under such different conditions the details of strategies for exploiting natural resources are expected to be rather different. For the apes, food is generally abundant and easy to obtain. For hominids inhabiting semi-arid environments food is surely more scattered, often cryptic and difficult to obtain, and in many cases is seasonal in nature. Whilst chimpanzees and gorillas may move around in coherent groups which exploit food resources as a group, in situations where the food items are scattered, difficult to find and require long distance movements, such coherent full-time grouping may not represent a viable strategy. Under such circumstances a more satisfactory solution may be to subdivide the group into sections each of which forages, not only for itself but for the others as well. Such a strategy requires however that the sections reunite frequently (on a daily basis) so that resources gathered can be shared among them. Furthermore, apprenticeships required to become proficient in search and exploit methods for many different varieties of widely spread (especially fluctuating) resources would have required increased information processing capacity in early hominids in comparison with other hominoids. Thus although extractive foraging may not have played a significant role in the evolution of primate intelligence (King, 1986), the particular strategy adopted by hominids for exploiting the wide variety of resources scattered over large territories might well have required additional processing capacity.

Summary and Conclusions

For the purposes of this study the primates were subdivided into 12 groups corresponding to grades of encephalisation. Three of these grades originated soon after the Eocene/Oligocene boundary and four of them during the Plio-Pleistocene. Thus more than half the grades evolved in less than 10% of Tertiary time indicating a clumping of events. That these events coincide with or just postdate major changes in $d^{18}O$ values obtained from benthic Foraminifera, suggests that there may be an element of cause and effect between dramatic changes in oceanic climates (and hence atmospheric climates) and evolution in the primates. These coincidences, if valid, would tend to support evolutionary models in which there is a component of 'climatic forcing'. From the point of view of primate evolution, each major palaeoceanographic event possibly resulted in the opening up of new niches into which some lineages of primates moved, in the process undergoing modification of their morphology, behaviour and cerebral processing capacity. In the meantime, already established lineages remained in their respective or in new but familiar niches and the long term result was that there was an accretion of niches over time. From the point of view of encephalisation, the fossil evidence suggests the existence of three grades during the Eocene, six during the Oligocene, eight during the Miocene and twelve during the Pleistocene.

Although the fossil record of primate skulls is generally poor, there are enough specimens preserved to permit us to obtain glimpses of the development of brain and body in primates during the Tertiary. This can be analyzed as data on encephalisation, about which there is a fair consensus.

The timing of the clumps of encephalisation grade increases appears to coincide reasonably closely with episodes of major climatic change as deduced from the $d^{18}O$ curves for oceanic surface waters. It is postulated that these global-scale climatic changes led to the establishment of new niches into which primates ventured, in the process undergoing evolutionary changes.

The initiation of increases from great ape to hominid grades of encephalisation may have begun during the Messinian Crises (7-6 m.y. ago) when it is suggested that Africa suffered aridification during which vast savannah to semi-desert regions become established at the expense of forests and moist woodlands, which shrank in area. The African great apes apparently remained more or less within their established habitats (tropical forest to Guinean woodland) while the proto-hominids ventured out into new, ecologically more demanding habitats. In doing so, they would have increased their ranges by several orders of magnitude (up to 2500 km^2 for hominids compared with about 30 km^2 for chimpanzees, and about 20 km^2 for gorillas). Not only were their ranges larger, requiring an efficient, endurance-oriented locomotion system but food resources were more scattered within the ranges and were, in many cases, cryptic (e.g., tubers) or markedly seasonal (fruits and nuts). The mental map for such increased territorial ranges would presumably require more brain tissue than would mental environmental maps in, e.g., gorillas or chimps in which the home ranges are small and food resources more easily accessible.

The requirement to map the extended range, the memory required for locating scattered food resources, the necessity to detect cryptic foods and unusual predators or prey, and the likely requirement for new kinds of social interaction with conspecifics in the enlarged ranges, might very well be behavioural adaptations in the new niches that would be impossible without additional neural control systems housed in enlarged brains.

Acknowledgements

It is a pleasure to acknowledge the help of Prof. Norbert Schmidt-Kittler and the Deutsche Forschungsgemeinschaft who provided support while the research for this study was being undertaken.

References

Binford LR, Chaun Kun Ho (1985) Taphonomy at a distance: Zhoukoudian, The Cave Home of Beijing Man? Current Anthropology 26:413-442

Boesch C, Boesch H (1983) Optimisation of nut-cracking with natural hammers by wild Chimpanzees. Behaviour 83:265-285

Cartmill M (1980) Morphology, function, and evolution of the anthropoid postorbital septum. In: (RL Ciochon and AB Chiarelli eds) Evolutionary Biology of the New World Monkeys and continental drift. pp 243-274 New York, Plenum

Davis RT (1974) Primate behaviour vol 3: Monkeys as Perceivers. New York, Academic Press

De Valois RL 1965 Analysis and coding of colour vision in the primate visual system. In: Sensory Perception. Cold Spring Harbor Symp Quant Biol 30:567-579

Falk D (1982) Mapping fossil endocasts In: E Armstrong and D Falk (eds) Primate brain evolution: methods and concepts. pp 216-217 New York, Plenum

Falk D (1983) The Taung endocast: A reply to Holloway. American Journal of Physical Anthropology 60: 479-489

Ferguson W (1983) An alternative interpretation of *Australopithecus afarensis* fossil material. Primates 24:397-409

Ferguson W (1984) Revision of the fossil hominid jaws from the Plio/Pleistocene of Hadar, in Ethiopia including a new species of the genus *Homo* (Hominoidea: Homininae). Primates 25:519-529.

Ferguson W (1986) The taxonomic status of *Preanthropus africanus* (Primates: Pongidae) from the late Pliocene of Eastern Africa. Primates 27:485-492

Fleagle J (1986) Early anthropoid evolution in Africa and South America. In Else JG, Lee PC (eds) Primate Evolution 1:133-142 Cambridge, Cambridge University Press

Fleagle J, Simons EL (1982) The humerus of *Aegyptopithecus zeuxis*: A primitive anthropoid. American Journal of Physical Anthropology 59:175-193

Frost G J (1980) Tool behavior and the origins of laterality. Journal of Human Evolution 9:447-459

Gingerich PD (1976) Cranial anatomy and evolution of early Tertiary Plesiadapidae (Mammalia, Primates). Papers on Palaeontology 15:1-140 University of Michigan

Gingerich PD (1980) Eocene Adapidae, paleobiogeography, and the origin of the South American Platyrrhini. In R Ciochon, B Chiarelli (eds) Evolutionary Biology of the New World Monkeys and continental drift. pp 123-138 Plenum, New York

Gingerich PD (1986) *Plesiadapis* and the delineation of the Order Primates. In B Wood, L Martin,P Andrews (eds) Major topics in primate and human evolution. pp 32-46 Cambridge University Press

Goudsblom J (1986) The human monopoly on the use of fire: its origins and conditions. Human Evolution 1:517-523

Gowlett JAJ, Harris JWK, Walton D, Wood BA (1981) Early Archeological Sites, hominid remains and traces of fire from Chesowanja, Kenya. Nature 249:125-129

Harrison T (1986) New fossil anthropoids from the middle Miocene of East Africa and their bearing on the origins of the Oreopithecidae American Journal of Physical Anthropology 71:265-284

Harvey P, Clutton-Brock T (1985) Life history variation in Primates. Evolution 39:559-581

Holloway RL (1974) The casts of fossil hominid brains. Scientific American, 231:106-114

Holloway RL (1984) The Taung endocast and the lunate sulcus: a rejection of the hypothesis of its anterior position. American Journal of Physical Anthropology 64: 285-287

Jerison H (1973) Evolution of the brain and intelligence. Academic Press, New York

Jerison H (1976) Paleoneurology and the evolution of mind. Scientific American 234:90-101

Jerison H (1983) The evolution of the mammalian brain as an information- processing system. In: (J Eisenberg and D Kleiman eds) Advances in the study of mammalian behaviour: pp 113-146 Spec Publ 7, American Society of Mammalogists

King BJ (1986) Extractive foraging and the Evolution of Primate Intelligence. Human Evolution 1:361-372

Levy J (1977) The Mammalian brain and the adaptive advantage of cerebral asymmetry Annals of the New York Academy of Science 299:264-272

Martin R (1986) Primates: A definition. In: B Wood, L Martin and P Andrews (eds) Major topics in primate and human evolution pp 1-31 Cambridge University Press

McHenry H (1984) Relative cheek-tooth size in *Australopithecus* American Journal of Physical Anthropology 64:297-306

Miller KG, Fairbanks RG, (1985) Cenozoic $d^{18}O$ record of climate and sea level South African Journal of Science 81:248-249

Napier J, Napier PH (1985) The natural history of the Primates. British Museum of Natural History, London pp 1-200

Oakley KP (1972) Skill as a human possession. In: (S Washburn, P Dohlinow (eds) Perspectives on human evolution 2: 14-52 New York Holt Rinehart & Winston

Pickford M (1986a) Geochronology of the Hominoidea: Post Conference Summary. In: JG Else & PC Lee (eds) Primate Evolution. 1:123-128 Cambridge University Press

Pickford M (1986b) On the origins of Body size dimorphism in Primates In: M Pickford and B Chiarelli (eds) Sexual Dimorphism in living and fossil Primates: 77-91 Florence Il Sedicesimo

Pickford M (1986c) Major events in Primate Palaeontology: possible support for climatic forcing models of Evolution Antropologia Contemporanea 9:89-94

Pickford M (1987a) Concordance entre les evenements paleoceanographiques et paleontologie continentale de l'Afrique de l'est. Comptes rendues de l'Academie de Sciences (in press)

Pickford M (1987b) Affinities of *Sivaladapis, Indraloris* and *Sinoadapis*. Human Evolution 2:91-92

Pickford M (1987c) The diversity, zoogeography and geochronology of Monkeys Human Evolution 2:71-89

Potter B (1986) The allometry of primate skeletal weight International Journal of Primatology 7:457-466

Radinsky L (1974) The fossil evidence of Anthropoid brain evolution. American Journal of Physical Anthropology 41:15-28

Radinsky L (1977) Early Primate Brains: Fact and Fiction. Journal of Human Evolution 6:79-86

Rak Y (1987) What's so robust about the hyper-robust *Australopithecus boisei* ? American Journal of Physical Anthropology 72:244

Rosenberger A, Strasser E, Delson E (1985) Anterior dentition of *Notharctus* and the adapid-anthropoid hypothesis. Folia Primatologica 44:15-39

Senut B (1986) Long bones of the primate upper limb: Monomorphic or Dimorphic? In: M Pickford, AB Chiarelli (eds) Sexual dimorphism in living and fossil primates. pp 7-22 Florence, Il Sedicesimo

Simpson GG (1944) Tempo and mode in Evolution. New York, Columbia University Press pp 1-237

Simpson GG (1963) Principles of animal taxonomy. Columbia University Press, New York

Stehlin HG (1909) Remarques sur les faunules de mammiferes des couches eocenes et oligocenes du Bassin de Paris. Bulletin de la Societe Geologique de la France (Ser 4) 9:488-520

Sudre J (1975) Un prosimien du paleogene ancien du Sahara nord-occidentale: *Azibius trerki* . Comptes Rendus de l'Academie des Sciences (Paris) 280:1593-1542

Szalay F, Delson E (1979) The evolutionary history of the primates. New York, Academic Press

Vrba E (1985) Environment and evolution: alternative causes of the temporal distribution of evolutionary events. South African Journal of Science 81:229-236

Walker A, Falk D, Smith R and Pickford M (1983) The skull of *Proconsul africanus*: reconstruction and cranial capacity. Nature 305:525-527

White T, Johanson D, and Kimbel W (1983) *Australopithecus africanus*: its phyletic position reconsidered. In: RL Ciochon and RS Corruccini (eds) New interpretations of ape and human ancestry. pp 721-780 New York, Plenum

Wundrum IJ (1986) Cortical motor asymmetry and hominid feeding strategies. Human Evolution 1:183-188

Yamagiwa J (1986) Activity rhythm and the ranging of a solitary male mountain gorilla. Primates 27:73-282

ALLOMETRIC ANALYSIS AND BRAIN SIZE

Paul H. Harvey
Department of Zoology
University of Oxford
South Parks Road
Oxford OX1 3PS, U.K.

My aim in this paper is to demonstrate how quantitative allometric analyses have recently been used to generate and test theories that explain the reasons for differences in brain size among birds and mammals. The work I shall report is aimed at answering four major questions. First, how does adult brain size change with adult body size across species and higher level taxa? Second, what categories of animals have larger brain sizes for their body sizes than others? Third, how do differences in the ontogeny of the brain correlate with adult brain size? Fourth, how have different explanations for the evolution of brain size differences fared when confronted with the data?

Methods

Allometric Analysis

Brain size $[BR]$ increases with body size $[BO]$ according to the allometric relationship:

$$[BR] = (a)([BO])^b$$

where a is the allometric coefficient and b is the allometric exponent. The relationship can be made linear for analysis by taking the logarithm of both sides, so that:

$$\log [BR] = \log a + (b)(\log [BO])$$

A variety of procedures is available to fit straight lines to data sets. Their various merits are discussed at length elsewhere (e.g. Harvey and Mace, 1982; Seim and Saether, 1983). Model 1 regression, major axis, and reduced major axis analyses are the most commonly used. Model 1 regression assumes that statistical error is confined to measurements made on the dependent variable and, as a consequence (if the correlation coefficient is not 1.0), different regression lines will result depending on whether brain or body size is chosen as the independent variable. Since we have no a priori reason for deciding whether to regress log$[BR]$ on log$[BO]$ or

NATO ASI Series, Vol. G17
Intelligence and Evolutionary Biology
Edited by H.J. Jerison and I. Jerison

vice-versa, I have given low priority to the results of regression analyses in this paper. However, since the alternative regression lines make extreme assumptions about the distribution of error, it is always wise to compare major axis or reduced major axis lines with the extreme regression lines.

The reduced major axis is the geometric mean of the two regression lines and, like the major axis, lies between the extremes. For practical reasons, I have used major axis instead of reduced major axis results; the reduced major axis does not incorporate a covariance term and the sign of the slope b is not determined. Average slopes are calculated in the more complex analyses and no straightforward procedure has yet been devised to perform the equivalent of analysis of covariance using reduced major axis analysis (Harvey and Mace, 1982). However, such techniques are straightforward when using major axis analysis. In practice, although the results of regression analyses often differ markedly from those of reduced major axis or major axis analyses, the latter two give similar answers over a wider range of conditions.

Taxonomic Level of Analysis

There are two reasons why it is important to consider carefully the taxonomic level at which allometric analyses are performed.

First, for statistical analysis it is often important that we identify independent evolutionary events (Ridley, 1983). Large brain size might have evolved in a common ancestor of a particularly species genus and, coincidentally, so might monogamy. If neither character has been lost through evolutionary time, it would be incorrect to assume that each species within the genus provides independent data for analysis. A single generic average would be more appropriate. However, for phylogenetic reasons, it could also be true that all genera within a family are similar for particular characters. Since we do not have an accurate evolutionary tree of the mammals and since brain and body size are continuous variables, a conservative approach seems to be to identify the taxonomic level at which variance appears in the data (Harvey and Clutton-Brock, 1985). Species within genera, and genera within families may be very similar, whereas families within orders may differ markedly. A nested analysis of variance provides an appropriate procedure for identifying a suitable taxonomic level for correlational analysis (Sokal and Rohlf, 1981).

The second reason for evaluating the taxonomic level for analysis, is the problem of so-called grade shifts (Martin, 1980; Harvey and Mace, 1982). It is often argued that some taxa have different allometric coefficients than do others (Martin, 1980) and, equally problematically, that lower-level taxa have lower allometric slopes relating brain to body size than do others (Gould, 1967; Lande, 1979; Martin and Harvey, 1985). For example, primates have relatively larger brains for their body sizes than do carnivores (a difference in allometric coefficient), while species within a genus often have a lower allometric exponent linking brain to body size than do genera within a family.

Clearly, the taxonomic level at which we perform an allometric analysis may influence our estimates of both allometric exponents and allometric coefficients. Furthermore, if allometric lines are not correctly determined, deviations from the lines which are used to measure relative brain sizes (encephalisation quotients) will also be incorrect and lead to apparent ecological or behavioural correlates of relative brain size which are, in fact, statistical artifacts.

A Problem with Allometric Analyses

We must be wary of placing too much faith in the exact nature of allometric relationships. For example, if different parts of the brain scale allometrically with body size, then only if their allometric coefficients are identical can the whole brain scale allometrically with body size. In fact, when allometric equations are fitted to components of the whole brain, exponents differ. This means that either the whole brain does not scale allometrically or its components do not. In practice, this problem is not as worrying as it at first sight appears. Allometry is merely a tool for summarising data. Since correlations between logarithmically transformed variables are never perfect, we can never view allometric equations as anything more than useful approximations.

The Relationship between Brain and Body Size among Adults

For many years, the exponent linking adult brain to adult body size across species of mammal was thought to be 0.67 (e.g. Jerison, 1973). Recently, a number of authors using good data sets and sound statistical analyses have pointed out the exponent is, in fact, about 0.75 (see Harvey and Bennett, 1983; Figure 12.1). How-

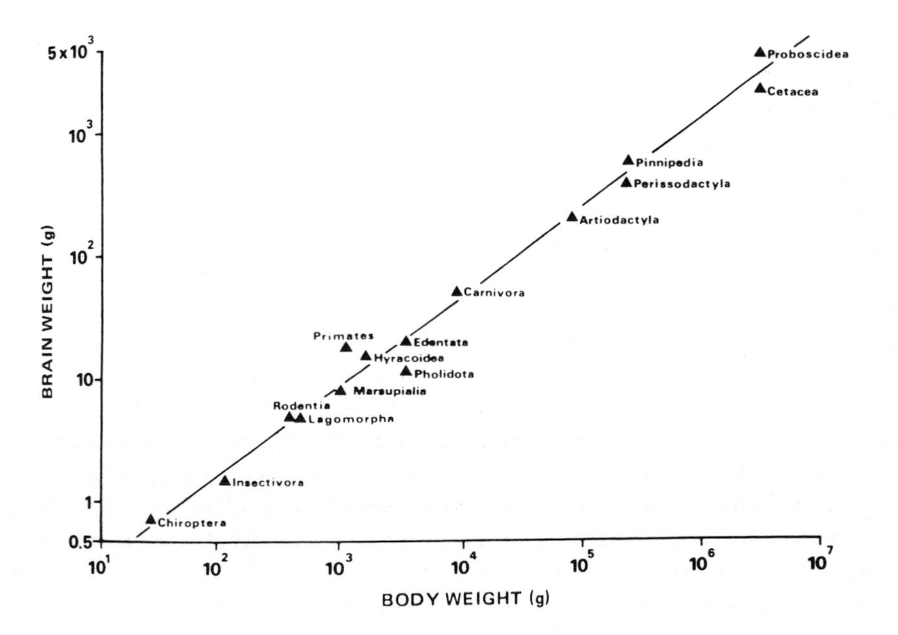

Figure 12.1. Allometric relationship between brain and body weight across 15 orders of mammals. Order points are derived from family averages that are calculated from generic values based on 883 species measurements. After Martin and Harvey (1985).

ever, it has also been claimed for many years that the exponent linking brain to body size among mammals changes with taxonomic level - the most common claim is that the exponent for congeneric species typically lies between 0.2 and 0.4 (Lande, 1979). Formal statistical analyses using nested or hierarchical slope-fitting procedures confirm earlier speculation (Martin and Harvey, 1985), although slopes for congeneric species are not as low as was previously supposed (see Figure 12.2).

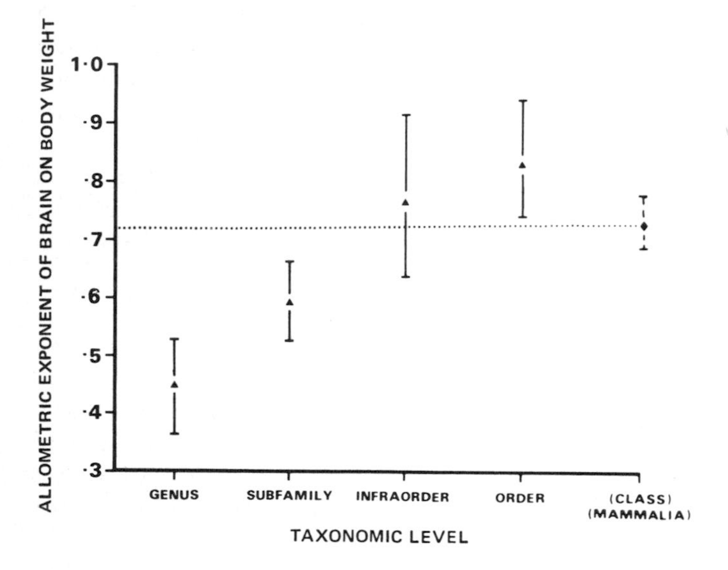

Figure 12.2 Common major axis slopes and 95% confidence limits at different taxonomic levels for primates. The dotted line and class estimate shown are from the inter-order comparison plotted in Figure 1. Generic level slopes for other orders are 0.53 (Artiodactyla), 0.56 (Carnivora), 0.57 (Chiroptera), 0.60 (Insectivora), and 0.37 (Rodentia). After Martin and Harvey (1985).

Nevertheless, at the lowest taxonomic level, among adults of a species, the data provide very little evidence for exponents greater than zero: brain size generally varies independently of body size (Martin and Harvey, 1985).

Martin (1981) also pointed out that the exponent linking brain to body size across bird species is about 0.56. Is there evidence for a change in exponent with taxonomic level in this group also? Using Portmann's (1946, 1947a, b) data, Bennett and Harvey (1985a) were able not only to look for changes in exponent of the whole brain but also of brain parts. Portmann provided tables from which the size of four brain parts (brain stem, optic lobes, cerebellum, hemispheres) could be calculated. The results of our analysis (see Figure 12.3) reveal (1) that there is evidence for change in allometric exponent with taxonomic level for the whole brain and for each of its components, (2) that the exponents do not merely increase with increasing taxonomic level but seem to decrease at the highest levels, and (3) that, irrespective of taxonomic level, exponents for some parts of the brain are consistently higher than for other parts.

Martin (1981) also demonstrated that interspecific scaling of brain on body size across reptiles resulted in a 0.56 exponent. I have looked for a change in exponent with taxonomic level among reptiles, and can find no evidence for any. Indeed, if any trend is apparent in the data it is for a decrease in exponent with increasing taxonomic level (see Figure 12.4).

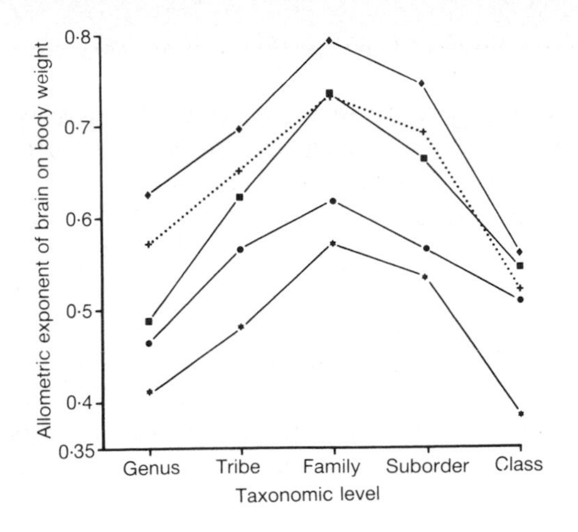

Figure 12.3. Common major axis slopes for the whole brain and the various brain parts at different taxonomic levels across birds. Symbols refer to whole brain (cross), hemispheres (diamond), cerebellum (square), brain stem (circle), optic lobes (star). After Bennett and Harvey (1985a).

Which Animals Have the Largest Brains?

Brain size increases with body size and, therefore, a more interesting phrasing of the above question is: which animals have the largest brains for their body sizes? There are a range of answers to this question. First, we know of taxonomic correlates of large relative brain size (I shall use the expression 'relative brain size' to refer to deviations from the brain to body size relationship. Animals with a large relative brain size are those that have a large brain size for their body size). Primates have relatively larger brain than carnivores, while mammals have relatively larger brains than birds or reptiles. The second type of answer to the question concerns the ecological, morphological, behavioural, or even ontogenetic correlates of relative brain size. If we know that some particular traits are always associated with large or small relative brain sizes, independent of phylogeny, then we may get some clues about to the selective forces acting on the size of the brain or its parts.

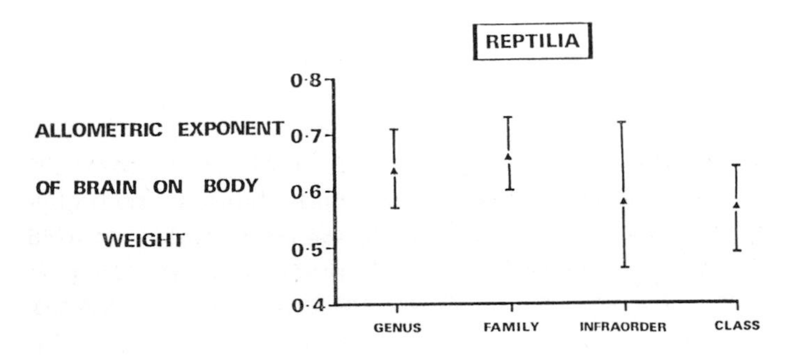

Figure 12.4. Common major axis slopes for whole brain at different taxonomic levels across reptiles.

Over the past decade many ecological correlates of relative brain size have been demonstrated within particular taxa. Some of the correlates are definitely not taxonomic. For example, leaf eating primates in different families have relatively small brain compared with fruit or insect eaters (Clutton-Brock and Harvey, 1980). Other dietary associations with relative brain size have been noted in rodents (Mace, Harvey and Clutton-Brock, 1981) and bats (Eisenberg and Wilson, 1978) although the latter is also a taxonomic association. However, attempts to generalize these findings to other taxa have been less successful. For example, within the mammals diet does not seem to be correlated with relative brain size across the order Carnivora, despite wide dietary differences among species (Gittleman, 1986). Nor, within birds, are there consistent dietary correlates of relative brain size (Bennett and Harvey, 1985b). Other relationships with relative brain size, such as arboreality versus terrestriality among the didelphid marsupials (Eisenberg and Wilson, 1981), also seem restricted to particular taxonomic groups and, therefore, to lack generality.

One non-consistent relationship does, however, pay further investigation. Among birds but not mammals, altricial species have relatively large brains as adults (Bennett and Harvey, 1985b). This correlation is very striking (Figure 12.5) and is independent of phylogeny across birds. That is, the relationship is not only present among species within the class, but it can also be detected among orders. Orders containing altricial species contain adults with relatively larger brains than those containing precocial species. Furthermore, although we can detect many ecological and behavioural correlates of relative brain size in birds at all taxonomic levels, these seem to result for the most part from confounding associations with mode of development (Bennett and Harvey, 1985b).

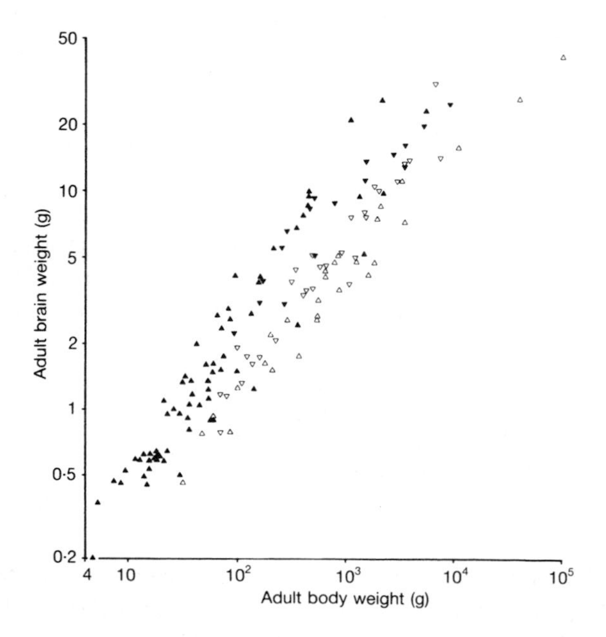

Figure 12.5 Adult brain weight vs. adult body weight plotted on logarithmically scaled axes for bird genera of the four developmental types. Symbols refer to altricial (closed triangle), semi-altricial (closed inverted triangle), precocial (open triangle), semi-precocial (open inverted triangle). After Bennett and Harvey (1985a).

Because of this developmental correlate of relative brain size among birds, we were led to examine the ontogeny of the brain/body relationship in both birds and mammals (Bennett and Harvey, 1985a). A further pattern emerged which suggests an evolutionary trade-off in brain development among both birds and mammals. I shall outline our findings in the next section.

The Ontogeny of the Brain and Adult Brain Size

We know that there is usually little or no neuronal division after birth. This means that neonatal brain size must, to some extent, set a limit to adult brain size. We also know that altricial birds have relatively large brains as adults. How much, then, does the brain develop before and after birds hatch? Fortunately, Portmann (1947b) has measured hatchling brain size for a number of bird species. A reasonable measure of prenatal brain development is to examine the size of the hatchling brain and compare it to the mother's body weight, since it is the mother who provides nourishment for the hatchling. Precocial birds give birth to relatively large brained hatchlings. We can then compare hatchling and adult brain sizes. We find that, for their hatchling brain size, altricial birds subsequently enlarge their brains disproportionately. Among birds, these two components of brain development are negatively correlated with each other (see Figure 12.6). Precocial birds go in for considerable prehatching brain development but little subsequent brain growth. Altricial birds, on the other hand, show the opposite pattern, with little prehatching compared with posthatching brain growth. The fact that adult altricial birds have relatively large brains implies that their extended posthatching brain development in some sense allows them to develop relatively large brains.

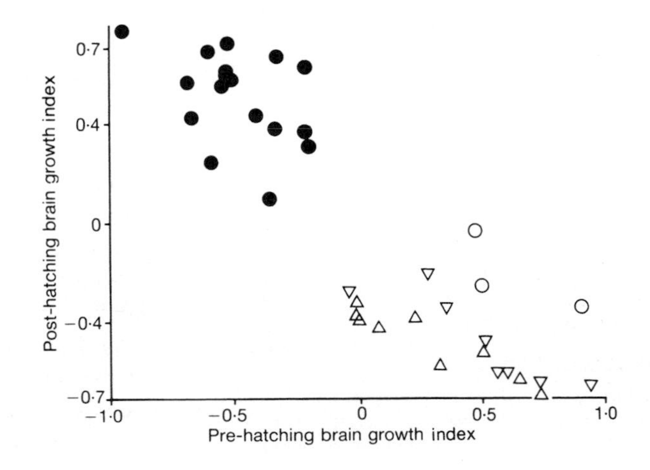

Figure 12.6. Pre-hatching vs. post-hatching brain growth for bird genera. Symbols refer to altricial (closed circle), semi-altricial (open circle), precocial (triangle), semi-precocial (inverted triangle). After Bennett and Harvey (1985a).

How does the mammalian brain develop? We can use equivalent indices of prenatal and postnatal brain development as we used for birds. When we do this, we find another sharp dichotomy between altricial and precocial forms, and a similar apparent trade-off between prenatal and postnatal brain development (Figure 12.7). However, in contrast to birds, extended postnatal brain development does mean that altricial mammals end up with larger brains than precocial mammals (Figure 12.8).

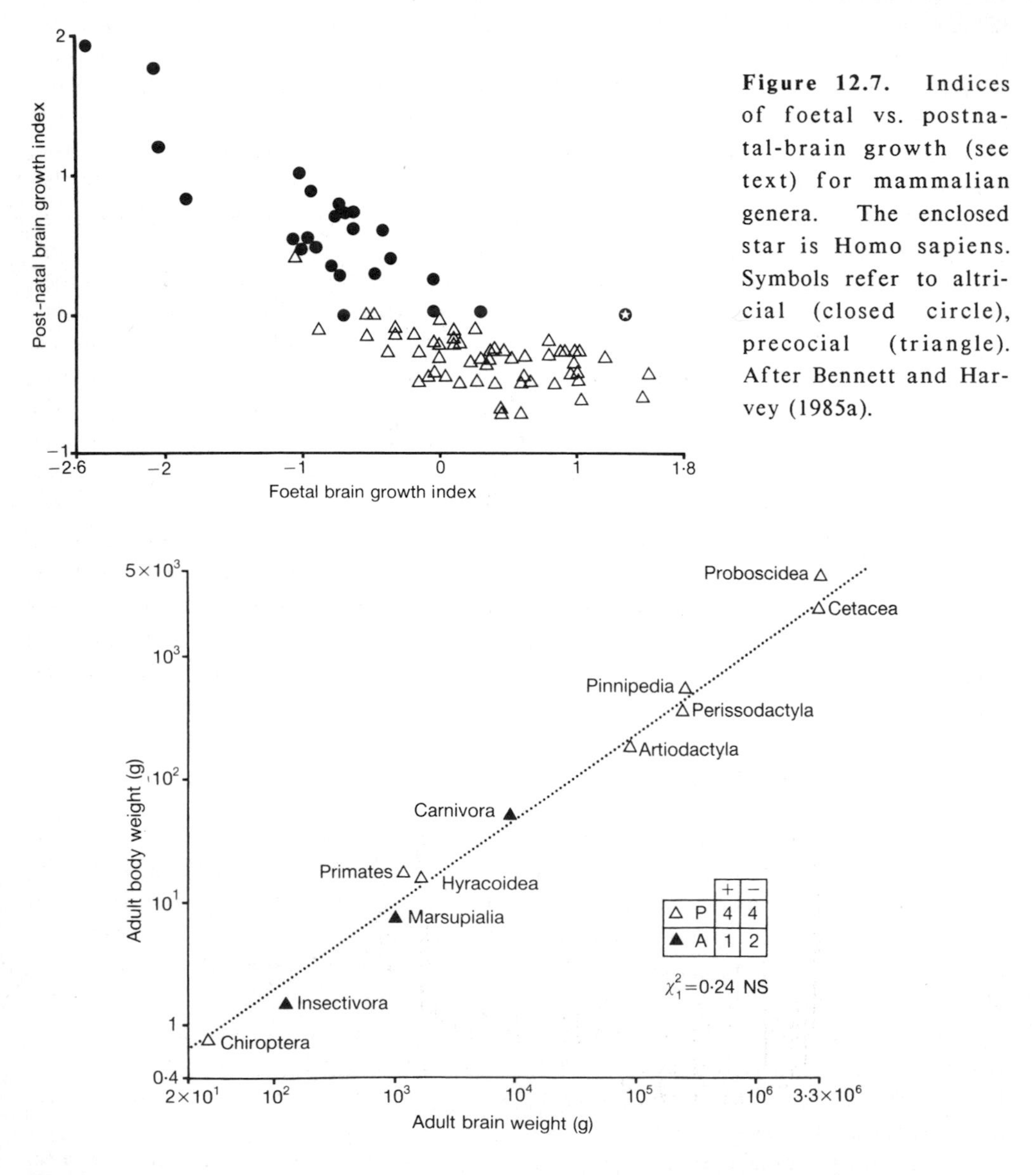

Figure 12.7. Indices of foetal vs. postnatal-brain growth (see text) for mammalian genera. The enclosed star is Homo sapiens. Symbols refer to altricial (closed circle), precocial (triangle). After Bennett and Harvey (1985a).

Figure 12.8. Adult brain weight vs. adult body weight for altricial and precocial orders of mammals. The line is the common major axis among the two developmental types. Symbols refer to altricial and precocial families (closed and open triangles). Cf. Bennett and Harvey (1985a) on altricial and precocial rodents.

For both of the above analyses, we have plotted one derived index against another. This can be a dangerous procedure, particularly when both indices are derived in part from the same measure (neonatal brain size). We have plotted the deviations from the linear plot of neonatal brain size on adult body weight. Statistically, this is equivalent to calculating the partial correlation between adult body size and adult brain size with the effects of neonatal brain size removed, and then reversing its sign. In other words, there is a strong positive correlation between adult brain and adult body size when the effects of neonatal brain size are removed. Our conclusion holds: those birds and mammals that develop their brains more before hatching or birth, develop them less postnatally.

The partial correlation interpretation of the trade-off between prenatal and postnatal brain development reveals one possible interpretation of it that is based on measurement error. Imagine that there is a perfect relationship between adult brain size and adult body size and that neonatal brain size is measured with some error. It follows that those species giving birth to (apparently) relatively large brained neonates will have (apparently) less postnatal brain development. This is because those species whose neonatal brains have been measured incorrectly as too large will then appear to have relatively large brains as neonates and to develop those brains less postnatally. This, however, is not the cause for the relationships given in Figures 12.6 and 12.7 because those Figures also distinguish between precocial and altricial birds and mammals by their positions on the graphs. This would not be so if their positions were due to error.

Why, then, should altricial birds end up with larger brains than precocial birds while the same is not true for mammals? Recent studies indicate that there may be neuronal division after hatching in birds but that there is no neuronal division after birth in mammals. Perhaps this taxonomic difference constrains altricial mammals to more limited postnatal brain development than is possible for altricial birds?

Explanations for the Evolution of Brain Size Differences

The exponent linking brain to body size among species of mammals is about 0.75. The incorrect 0.67 exponent which was accepted for many years led to the general assumption that a surface area explanation was required. Naturally, the discovery of a 0.75 exponent led to accounts that invoked some sort of metabolic interpretation (Martin, 1981; Armstrong, 1983), since basal metabolic needs scale to the 0.75 power of body size (Kleiber's law).

While the extremely tight relationship linking brain to body size begs functional interpretation, it is clear that no entirely satisfactory models have yet been produced. Both Armstrong's and Martin's explanations are contradicted by the data, though to different extents.

Armstrong argues that adult brain size is, in some sense, directly tied to adult metabolic needs. Differences in relative brain size between different taxonomic groups are either given vague post-hoc justifications, or are incorrectly explained away. An example of the former is Armstrong's explanation for why primates have relatively larger brains than other mammals: it is because primates divert more of

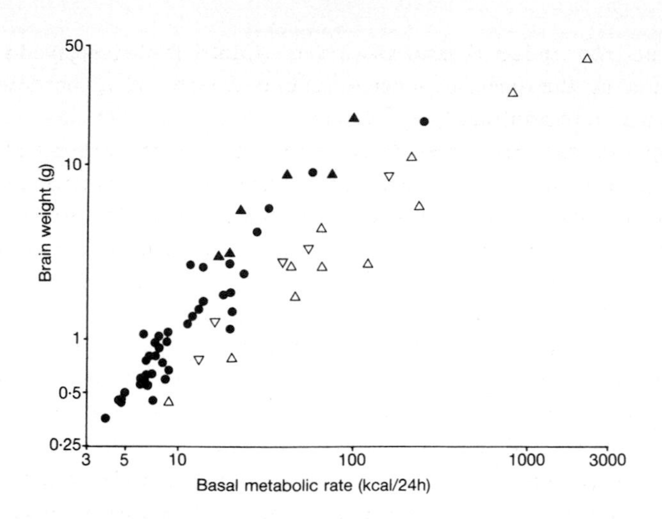

Figure 12.9. Adult brain weight vs. adult basal metabolic rate plotted on logarithmically scaled axes for bird genera of the four developmental types. Symbols refer to altricial (closed circles), semi-altricial (closed triangles), precocial (open triangles), semi-precocial (inverted open triangles).

their energetic resources to the brain (Armstrong, 1981). An example of the latter is provided by Armstrong and Bergeron's (1985) claim that, although passerines have relatively large brains for their body size when compared with other birds, this difference vanishes when brain size is plotted against metabolic needs. The point that Armstrong and Bergeron fail to appreciate is that those orders of birds (of which the passerines are just one) with relatively large brains for their body size are also altricial species. When all the available data are correctly analyzed altricial genera also have relatively large brains for their metabolic needs (Figure 12.9). Clearly mode of development is a factor that must be taken into account when explaining differences in relative brain size. Basal metabolic needs alone will not do the job.

Martin (1981), on the other hand, recognises that although the brain and metabolic needs scale to body size with a similar exponent, there are clear differences between altricial and precocial species in the development of the brain. These differences led to Martin's theory that adult metabolic turnover determines neonatal brain size over evolutionary time, and that neonatal brain size determines adult brain size. Thus, for mammals, there is a two stage process with neonatal brain size as a crucial intervening variable between adult body size and adult brain size. Neonatal brain size in a constant proportion of adult brain size within developmental categories.

Further weight was apparently lent to Martin's idea by the fact that birds and reptiles, which normally lay eggs, have an approximate 0.56 exponent linking brain to body size. Martin argued that egg size is determined by the mother's metabolic turnover (i.e. scaled to the 0.75 power of mother's body size), that hatchling brain size is constrained by the metabolic rate of the egg (i.e. scaled to the 0.75 power of egg size), and that adult brain size is a simple multiple of hatchling brain size. The product of the three exponents 0.75, 0.75 and 1.0 gives the exponent relating adult brain size to adult body size (=0.56).

Martin's theory was, to some extent, produced to accommodate the available data. How has it faced up to subsequent developments? For example, does neonatal brain size of birds within developmental categories scale in direct proportion to the metabolic rate of the egg? And is adult brain size a constant multiple of hatchling brain size in birds with the same mode of development? The answer to the second question is no, and the answer to the first may be yes. The exponent linking adult brain size to neonatal brain size is 1.45 across avian genera within developmental categories (95% confidence limits = 1.27-1.45: see Bennett and Harvey 1985a), which is significantly greater than Martin's expected value of 1.00. To test the idea that the metabolism of the egg constrains fledgling brain size is difficult because the egg provides energy for the embryo throughout development. However, across an available sample size of just 9 species, there is an estimated exponent of 1.0 relating hatchling brain weight to the metabolic rate of the egg just prior to hatching (see Bennett and Harvey 1985a).

In addition to the problem of the incorrect scaling of adult brain on hatchling brain size, Bennett and Harvey (1985a) mentioned two other difficulties that face Martin's hypothesis:

(1) As I have already mentioned, altricial birds hatch with small relative brain sizes but grow to have relatively larger brains as adults than do precocial species. Martin's ideas do not predict this phenomenon.

(2) At some taxonomic levels, the exponent linking brain size to body size is greater than 0.56. This finding is not surprising in the light of the exponent linking adult to hatchling brain size.

Clearly, Martin's ideas are attractive and may easily be modified in light of the above findings. However, their attraction lies in a simplicity that fails to specify the costs and benefits of having small or large brains at particular times in an animal's life, or the nature of the constraints acting on brain development. Furthermore, allometricians tend to view the brain as a whole but, for birds at least (see Figure 3), brain parts have different allometric exponents which also vary according to the taxonomic level of analysis (Bennett and Harvey 1985a). To paraphrase an earlier comment in this essay, allometric analyses may help us to understand the brain, but they have their limits.

References

Armstrong E (1983) Relative brain size and metabolism in mammals. Science 220:1302-1304

Armstrong E, Bergeron R (1986) Relative brain size and metabolism in birds. Brain Behav Evol 27:141-153

Bennett PM, Harvey PH (1985a) Brain size, development and metabolism in birds and mammals. J Zool 207:491-509

Bennett PM, Harvey PH (1985b) Relative brain size and ecology in birds. J Zool 207:151-169

Clutton-Brock TH, Harvey PH (1980) Primates, brains and ecology. J Zool 190:309-323

Eisenberg JF, Wilson DE (1978) Relative brain size and feeding strategies in the Chiroptera. Evolution 32:740-751

Eisenberg JF, Wilson DE (1981) Relative brain size and demographic strategies in didelphid marsupials. Am Nat 118:1-15

Gittleman JL (1986) Carnivore brain size, behavioral ecology and phylogeny. J Mamm 67:23-36

Gould SJ (1975) Allometry in primates, with emphasis on scaling and the evolution of the brain. Contrib Primat 5:244-292

Harvey PH, Bennett PM Brain size, energetics, ecology and life history patterns. Nature 306:314-315.

Harvey PH, Clutton-Brock TH (1985) Life history variation in primates. Evolution 39: 559-581

Harvey PH, Mace GM (1982) Comparisons between taxa and adaptive trends: problems of methodology. In: King's College Sociobiology Group (ed) Current problems in sociobiology. University Press, Cambridge, p 343

Jerison HJ (1973) Evolution of the brain and intelligence. Academic Press, New York

Lande R (1979) Quantitative genetic analysis of multivariate evolution, applied to brain:body size allometry. Evolution 33:402-416

Mace GM, Harvey PH, Clutton-Brock TH (1981) Brain size and ecology in small mammals. J Zool 193:333-354

Martin RD (1980) Adaptation and body size in primates. Z Morph Anthrop 71:115-124

Martin RD (1981) Relative brain size and basal metabolic rate in terrestrial vertebrates. Nature 293:57-60.

Martin RD, Harvey PH (1985) Brain size allometry: ontogeny and phylogeny. In: Jungers WL (ed) Size and scaling in primate biology. Plenum, New York, p 147

Portmann A (1946) Etudes sur la cerbralisation chez les oiseaux. I. Alauda 14:2-20.

Portmann A (1947a) Etudes sur la cerebralisation chez les oiseaux. II. Les indices intracerebraux. Alauda 15:1-15.

Portmann A (1947b) Etudes sur la cerebralisation chez les oiseaux. III. Cerebralisation et mode ontogenetique. Alauda 15:161-171.

Ridley M (1983) The explanation of organic diversity. University Press, Oxford

Seim E, Saether BE (1983) On rethinking allometry: which regression model to use? J Theoret Biol 104:161-168

Sokal RR, Rohlf FJ (1981) Biometry. Freeman, San Francisco

MAMMALIAN DOMESTICATION AND ITS EFFECT ON BRAIN STRUCTURE AND BEHAVIOR *

Dieter Kruska
Institut für Haustierkunde
Universität Kiel
Kiel FRG

The evolution of the vertebrate brain along with the function of this fascinating organ must be considered as a particular biological phenomenon many aspects of which have certain implications (Jerison 1973).

As a zoologist with a preference for comparative anatomical questions, I am especially interested in the changeability of the mammalian brain and in quantitative as well as qualitative changes in its proportions. In comparative neurology, the size of the brain and of some of its parts and the degree of differentiation of nervous tissue are the anatomical parameters for evaluating the functional importance and capacity of these regions. For this reason I am also interested in functional variations due to such changes. Differences in brain size have been ascertained in fossil as well as in recent mammals in many studies. According to such analyses mammals have reached different species-specific levels of cerebralization (encephalization).

Furthermore, studies, and particularly the numerous, very detailed, and accomplished studies of Dr. Stephan and co-workers (e.g. Stephan 1975; Stephan et al. 1982, 1986), have demonstrated interspecific differences in proportions of the brain within certain phylogenetic lines. These differences should be seen as evolutionary changes during phylogeny, but also as adaptive changes to certain ecological niches and specific modes of life.

Today I would like to draw your attention to a special field of zoology, "Haustierkunde" in German, which means "research on domestic animals." We have known at least since Darwin (1873) that mammals have changed over long periods of time as a consequence of domestication. When we compare the recent offspring of ancestors still living in the wild with domestic forms, e. g. the European boar versus the domestic pig, wild versus domestic sheep, the wild cat versus the domestic cat, or wolf versus dog, we see that domestic animals often look very different from their progenitors.

---

* [To maintain the flow of the text, the large number of figures in this report follow the appendices. We regret the omission of *Umlauten* and other diacritical marks in much of the text. Eds.]

NATO ASI Series, Vol. G17
Intelligence and Evolutionary Biology
Edited by H.J. Jerison and I. Jerison

It is important to note in this connection that wild animals and related domestic varieties can interbreed. This means that both forms belong to the same species, although they differ greatly in morphological features and in behavior. New species never developed during domestication. The various forms commonly known as races demonstrate only enormous intraspecific variability in a zoological sense. It is necessary to add that domestic mammals developed at different times, and the intensity of the domestication process during the initial phase, but also throughout domestication, may be of some importance when considering brain size (Kruska 1987). The oldest remains of some species from the Near East have been dated to approximately 10,000 or even 12,000 years ago (Zeuner 1963; Mason 1984). In other species domestication occurred later; still others are being domesticated now. But on the whole, compared with phylogenetic events, the change from wild to domestic mammals has taken place over relatively short periods of time.

Domestic forms originate in different mammalian orders (Appendix 13.A). Several species of the orders Lagomorpha, Rodentia, Carnivora, Artiodactyla and Perissodactyla have been domesticated. Since their wild progenitors have brains at different levels of cerebralization, domestication must have started with differently evolved and differently adapted forms. On the other hand, man keeps domestic forms in different ways and for different purposes. Breeding objectives are manifold and sometimes changeable.

Yet despite these differences all domestic mammals seem to share some conditions different from those of their counterparts in the wild. They all live in the special circumstances and care, however diverse, of domestication. They all originated in small isolated groups of wild animals. Once man chose breeding objectives, natural selection was replaced by artificial selection, and social organization, reproductive biology, environment and nutrition changed to satisfy these objectives. The question is what happened to the brain and to behavior as a result.

Comparison of Brain Size

Some ancestral forms of domestic animals are extinct, or their numbers have been so greatly reduced in the wild that it is impossible to study their brains. The majority of species, however, are available for comparative studies. It is possible to get an overview comparison of brain sizes of wild and related domestic forms by applying the methods of allometry to the relationships of brain to body weight. This allometry has been somewhat misunderstood or misinterpreted in the literature. For that reason, some general remarks on this method are in order in the context of domestication research.

In general, the brain weight of mammals is dependent on body size in a certain way. Direct comparisons of brain weights give information about real differences only when we compare equally sized mammals. In differently sized mammals, however, the relationship of brain to body weight follows a rule that can be described by the allometric formula. The logarithmic form of this formula, $\log y$ (brain weight) $= a \log x$ (body weight) $+ \log b$, represents a line in a double logarithmic scale with a certain slope (a-value). In analyses of brain and body weight, data for several mammals are described by a distribution ellipse in a double logarithmic scale. We can calculate allometric lines as a regression line, a reduced major axis, or a major axis. This last best fits the position of the ellipse (Rempe 1962). Disagreement in studies of mammals mostly concerns the a-value, and particularly its constancy for all mammals.

When comparing brain sizes of mammals, we must decide which of the several allometries with different a-values -- ontogenetic, intraspecific or interspecific -- best applies to the problem at hand. To answer questions of phylogeny and evolution, for example, we must make interspecific allometric comparisons of mammalian brain sizes. In domestication research, on the other hand, we must look at intraspecific allometries, since, as already mentioned, here we are making comparisons within species.

In calculating intraspecific allometric relations of brain to body size, the broadest possible range of adult body sizes is of the utmost importance. But the range is relatively limited within most wild mammalian species, which is why intraspecific allometries of brain-to-body weight relationships have long remained unclear. In domestication, however, the range in body sizes of adult individuals has become broader within many species. In the domestication of dogs, for example, we see extremes of "dwarfs" and "giants," such as the Chihuahua and the St. Bernard, a phenomenon observable to a lesser degree also in other species.

It is for this reason that intraspecific allometries were first confirmed for domestic mammals (Dubois 1914; Lapique 1908), but they occur also in wild species. In order to get insight into such allometries, I have analysed data from animals of known ages with respect to brain to body weight relationships during postnatal ontogenesis in the altricial species *Procyon cancrivorus* (Kruska 1975a) and *Mustela vison* f. dom. (Kruska 1977). I found a relationship between brain and body growth of individuals along two allometric lines. The first shows a rather steep slope of about $a = 0.80$, which suggests rather rapid brain development in relation to body growth during the first phase of early postnatal ontogenesis. The brain, however, reaches its final size earlier than the body, at a different species-specific rate. In the *Procyon cancrivorus*, for example, the brain becomes fully grown relatively early, at about 70 days, while the body continues to grow. The resulting allometric line of brain to body weight relationships has no slope at all ($a = 0$) for the second phase of ontogenesis.

In addition, single individuals from the same population show gradual differences within certain limits. Those whose genetics and/or nutrition have determined greater full size reach heavier final brain weights at about the same stage of the first ontogenetic phase. Consequently, their second phase allometric lines lie parallel to and above those of smaller specimens. Figure 13.1 describes the results for three individuals of different sizes.

With data on adults of different weights, the values of brain and body weights are arranged along several transposed lines of the second ontogenetic phase. Data for a very large sample of individuals are fitted to a distribution ellipse, where the major axis represents the intraspecific line. Let me stress again that this intraspecific line does not reflect growth allometry but rather describes the relationship between brain and body size of adult individuals within a given species.

Although this is assumed to be a normal case in mammalian species, certain exceptions may occur, especially in early development. In the *Mustela* species mink and ferret, for example, we found an unusual "overshooting" of brain weight, brain case size and caudal skull height during the juvenile life span (Kruska 1977, 1979; Apfelbach and Kruska 1979). In this case, brain and body growth follows the typical first ontogenetic line with a steep slope until the brain reaches a certain size. But, for as yet unknown reasons, in six to nine month old minks, brain weight decreases by approximately 15 % in both sexes. The skull, although already massively built up without sutures, shrinks accordingly. A comparable effect, known as Dehnel's phenomenon, has been observed in certain soricid species. Nevertheless, in minks from nine months on, the values for brain and body size follow second-phase allometry with increasing age. This may produce confusion concerning intraspecific allometries in certain species, especially when the ages of individuals in the sample are not known. However, even in these species, data for differently sized adult individuals are associated with an intraspecific allometric line.

In order to determine the effects of domestication, intraspecific allometric calculations of wild ancestral and related domestic forms have been performed (Ebinger 1972, 1974, 1975a; Ebinger et al. 1984; Espenkötter 1982; Frick and Nord 1963; Herre and Thiede 1965; Kruska 1970, 1972, 1973 a,b, 1975 b,c, 1980, 1982; Kruska and Röhrs 1974; Kruska and Schott 1977; Kruska and Stephan 1973; Röhrs 1955; Röhrs and Ebinger 1978; Röhrs and Kruska 1969; Schleifenbaum 1973; Schumacher 1963). These studies have produced similar results in several species. The most important (cf. Figures 13.2-4) are:

1. The slopes of intraspecific allometric lines in this relationship are less steep than those for interspecific comparisons, and nearly identical, with values between a = 0.20 and a = 0.30, and occasionally, a = 0.40.

2. Data for wild forms are associated with one allometric line, and those for related domestic forms are associated with another line. The two lines have identical slopes, that is, are parallel.

3. Domestic mammals often show a greater variability in brain size than corresponding wild forms. Their data on brain and body weights are reflected in wider distribution ellipses.

4. On the average, domestic mammals have smaller brains than their wild counterparts.

We see, therefore, that mammalian brains show a reduction in size as a consequence of domestication. This finding leads us to the conclusion that decrease in brain size may also indicate a reduction in the total functional capacity of the central nervous system.

The extent of average decrease in brain size due to domestication within several species can be obtained from the log b values of intraspecific allometric lines with brain sizes of the wild forms = 100 %. The analysis shows that the degree of decrease in brain sizes in domestic forms is species-specific and different relative to their wild counterparts. The dimensions of quantitative reduction vary considerably. To date we have obtained decrease values from 0 % to more than 30 % (Table 13.1).

In this connection it is important to note that brains of the generally less evolved rodent and lagomorph species show size decrease of only 0 % to 13.4 %, whereas the more evolved artiodactylean and carnivore species show reduction values between 17.6 % and 33.6 %. This suggests that the evolutionary level, which in the case of the brain means the cerebralization level of a given species, may influence the extent of reduction in size caused by domestication. This conjecture can be tested in interspecific allometric studies of brain to body weight relationships.

The problem of how to do this, and particularly what kind of slope would apply to an interspecific relationship, has long puzzled investigators. Dubois (1914) and numerous authors since have found a relatively constant interspecific a-value of 0.56 - 0.67 for several mammalian groups. But in contrast to these findings some recent studies (Martin 1981; Harvey and Bennett 1983; Martin and Harvey 1985; Gittleman 1986) have either chosen a higher value of about 0.75 for overall analysis of species from all mammalian orders, or else found different values specific to groups. This approach was used already by Wirz (Wirz 1950). Such sort of analysis removes any doubt about the results; they are to be expected from the method. Furthermore, Martin (1981) assumes that in scaling with an exponent of 0.75 there may be a connection between brain to body size and metabolic rate to body size.

We take issue, however, with Harvey and Bennett (1983, p.314) when they write that "until recently the interpretation of variation in relative brain size among mammals was dominated by a false fact - an incorrect value of the exponent scaling brain weight to body weight." Their statement totally ignores basic knowledge about the anatomy of the brain.

We know from comparative neuroanatomical research that, generally speaking, mammalian brains are differently constructed both quantitatively and qualitatively (Kappers et al. 1967; Starck 1962, 1982). Many species have heavier brains than others of the same body size, with certain parts bigger and greatly expanded with more neurons and higher levels of differentiation.

This is known from numerous comparisons. For example, a common rodent species, the Norwegian rat, has body weight = 214 g and brain weight of only 2.14 g (Kruska 1975 b), whereas a carnivore species, the stoat, with the slightly lower body weight of 185 g, has brain weight of 4.5 g (Schumacher 1963). Furthermore, it has been recognized that the telencephalon and the neocortex within it have evolved to predominance as parts of the brain in highly evolved mammals. Neocorticalization is the principal phenomenon of mammalian brain evolution. But neoencephalization also occurs in other brain regions; enlargements of neocortex always imply the expansion of correlated tissue. This becomes clear when we compare the relative values in our example.

In the rat the telencephalon accounts for only 56 % and the neocortex for 30 % of total brain size, whereas the corresponding values in the stoat are 68 % and 45 % respectively. Figure 13.5 shows comparable sections through the telencephalon of the rat and the stoat as an illustration of the differences in the size and differentiation of the neocortex.

We know, however, that brain size in mammals is mainly determined by two factors, namely, body size and the level of cerebralization. In a certain sense, this seems like a disastrous situation. Interspecific allometric calculations of several brain and body weight data will always lead to different a-values, depending on sample size and accidents of sample selection. Even with many data, calculating an overall allometric line for mammals does not make biological sense when it comes to brain size comparisons, no matter how reliable statistically. The influence of body size and cerebralization always affects the results.

In attempting to circumvent this problem, brain allometrists first took into account the influence of body weight on brain weight. This can be done only by a priori exclusion of the influence of cerebralization and only by using data from differently heavy species that are closely related phylogenetically and nearly identical biologically. While this is not possible for all mammalian groups, it has been tested on many families (see Kruska 1980). The analysis has shown that fairly constant slopes of interspecific allometric lines of 0.56 or 0.63 are valid for all mammalian groups studied, as well as for those within which domestic forms have developed.

To obtain information on cerebralization levels of ancestral wild forms it seems best, therefore, to calculate the distances of species-specific brain and body weight data-points from medium allometric lines that have the above mentioned slopes within the orders dealt with. In this way we can obtain cerebralization indices independent of the influence of body size. This has been done with a large sample of data on brain and body weights of adult individuals from institutional samples and literature (Kruska, unpubl.).

In this connection it seems worth mentioning that some recent studies have used data on brain case sizes from museum materials and average body weights of species from the literature (Gittleman 1986; Martin and Harvey 1985) for allometric calculations. This is permissible with fossil material (Jerison 1973) but should be avoided with recent material, especially since brain case size does not reflect brain weight of small and large mammals equally (Röhrs and Ebinger 1978). Just the same, I have graphed average interspecific allometric lines and plotted species data for the orders Rodentia, Artiodactyla and Carnivora as given in Figures 13.6 - 13.8.

From these interspecific allometric calculations I arrived at cerebralization indices (CI), in order to quantify the level of brain evolution in ancestral forms within their systematic groups (Kruska 1980). The indices provided the means for the further study of a possible relationship between the gradual differences in brain size decreases in domestic as compared with ancestral forms and differently highly evolved ancestral forms. Appendix 13.B shows average brain and body weights as well as cerebralization indices for several species of fissiped carnivores. For comparison I have scaled cerebralization indices in Figure 13.9. Here it becomes clear that, seen as a whole, species of Ailuridae, Procyonidae, Canidae and Felidae have reached nearly the same levels of cerebralization. This also holds true for Ursidae except for the Sun bear with its unusually large brain. Species of Viverridae, Herpestidae and Hyaenidae, on the other hand, show rather low indices. So do most Mustelidae, although their arboreal and water-adapted forms have reached higher cerebralization levels.

Species from other orders can also be compared in this way. For ancestral forms of domestic mammals we can contrast the range of the cerebralization index with the range of values representing the decrease in brain size due to domestication (Table 13.1). The results are as follows:

Rodentia: *Rattus norvegicus* (CI=62); *Mus musculus* (CI=66); *Cavia aperea*

(CI=90). No relationship shows up with respect to brain size decrease in domestic forms (Table 13.1). While the less cerebralized rat and the more highly cerebralized wild cavy in domestication show a certain degree of brain size reduction, the *Mus*, with intermediate cerebralization, does not.

Artiodactyla: *Sus scrofa* (CI=97); *Ovis ammon* (CI=117); *Lama guanacoe* (CI=127). Within this order a certain relationship obtains (Table 13.1), namely, that the reduction in brain size due to domestication decreases, the higher the cerebralization level of the ancestral forms.

Carnivora: *Mustela putorius* (CI=55); *Felis silvestris* (CI=104); *Canis lupus* (CI=119). In this order again no relationship seems to be present (Table 13.1). Brain size decreases from the wild ancestral to the domestic forms are nearly identical among these species, although ancestral forms are highly diverse in cerebralization. Thus no conformity within the orders seems to obtain with respect to this relationship.

We can also compare the orders themselves. Fig 13.10 shows their interspecific allometric lines compared with the line for Basal Insectivora according to Bauchot and Stephan (1966). They represent medium and order-specific cerebralization levels.

There are two main gaps between these transposed lines: Rodentia and Lagomorpha have reached an identical medium cerebralization level, whereas Artiodactyla and Carnivora show a common higher level. Independently of body size, rodents and lagomorphs generally have brain sizes about 2 1/2 times larger than the most primitive living eutherians, and carnivores and artiodactyles have brain sizes about 2 1/2 times larger than rodents and lagomorphs. Nonetheless, here a relationship seems to exist between the degree of cerebralization and brain size reduction due to domestication. Mammals with larger brains generally show a greater decrease in size during

Table 13.1. Average change (reduction) in brain size in domestic forms compared to wild living "ancestral" species at comparable body weight.

Rabbit	-13.1%	Fischer 1973
Laboratory rat		
Wistar albino	-8.1%	Kruska 1975 b
DA- pigmented	-12.0%	Kruska unpubl.
Laboratory mouse	0.0%	Frick and Nord 1963
Guinea pig	-13.4%	Ebinger et al. 1984
Pig	-33.6%	Kruska 1970
Sheep	-23.9%	Ebinger 1974
Llama/Alpaca	-17.6%	Kruska 1980
Ferret	-29.4%	Espenkötter 1982
Domestic cat	-27 .6%	(after data of Röhrs and Ebinger 1978; Bronson 1979)
Dog	-29 .0%	Röhrs and Ebinger 1978

domestication than do those with small brains. It is clear from within-order comparisons, however, that the level of brain evolution may not be the only factor influencing size reduction caused by domestication. We can also conclude that species-specific differences in the degrees of brain size reduction may mean a gradually different diminution in total function.

Comparison of the Size of Parts of the Brain: Evolutionary and Functional Implications

Do certain brain parts change differently from wild ancestral to domestic forms? This has been investigated in some species, using the method of total serial sections. This method has been described in detail elsewhere (Stephan 1960; Stephan et al. 1970; Kruska 1970; Kruska and Stephan 1973). I will briefly recapitulate it here.

Several brains of ancestral and domestic forms are cut into 10 or 20 micron thick sections in the frontal plane. Out of the total, about 250 equidistant sections are stained with cresylviolet for Nissl bodies of neurons and photographed as enlargements. Different brain parts are delineated on these prints as shown in Figure 13.11, cut out, weighed, and summed up, so, that a total brain size results as paper weight. This is compared with the actual brain weight (fresh brain weight) which had been obtained before fixation. A conversion factor is calculated taking into account the degree of photographic enlargement, the weight of photographic paper, the distances of slices and the specific gravity of fresh brain tissue (= 1.035). This factor is then needed to convert brain part sizes obtained in paper weight into actual volume sizes of brain tissue. This conversion is particularly important, since our experience shows that brains shrink at different rates in the process of fixation and histological treatment.

Shrinkage values of 32 % to 59 % have been obtained in rat, pig and tylopod brains (Kruska 1980). They seem to be independent of the species under investigation, the absolute brain size, and the sex and age of the animals as well as the fixation fluid and the time of storage. For this reason, in general, measurements of surface, nuclear diameter and lamination thickness in histological tissue produce no clear-cut results, since we cannot control for shrinkage in comparative investigations. However, the converted volume sizes of certain parts of the brain at best represent actual sizes and can be contrasted in the same way as total brain sizes of wild ancestral versus domestic mammals (Kruska and Stephan 1973).

Brains of several species were treated in this way: wild rats versus laboratory rats of the Wistar strain (Kruska 1975 b, 1975 c; Kruska and Schott 1977), European boar versus domestic pig (Kruska 1970, 1972, 1973 b; Kruska and Stephan 1973), wild versus domestic sheep (Ebinger 1974, 1975 a, 1975 b), guanaco versus llama and alpaca (Kruska 1980), polecat versus ferret (Schumacher 1963) and wolf versus poodle (Schleifenbaum 1973). We first compared the medulla oblongata, the cerebellum, the mesencephalon, the diencephalon and the telencephalon. Later we determined the volume sizes of telencephalic structures: the corpus striatum, septal region, allocortex and neocortex, and occasionally also of grey and white matter separately. This yielded a big range of different reduction values due to domestication for the brain parts considered.

Among the five fundamental parts of the brain, the telencephalon consistently shows the highest degree of size reduction in all the species investigated (Figure 13.12). We must also mention the cortex, since this part of the forebrain is especially involved, even though the investigated species are cerebralized to different degrees (Kruska 1980). It appears that, in general, phylogenetically younger parts of the brain are particularly reduced in size under the influence of domestication. Contrasting phylogenetic events with those of domestication, Röhrs (1985, p.547) concluded, "In domestication this (phylogenetic) process has been reversed: we may speak of regressive evolution."

On the other hand, we know from quantitative neuroanatomical studies that the enlargement or decrease of certain parts of the brain is consistent with special adaptations of species to different environments. Regressive as well as progressive changes occurred during phylogeny with respect to the size and differentiation of sense organs and related brain parts, as well as with other nuclear masses of central nervous circuits.

Adaptations to life in water, for example, led to reduction or even disappearance of the olfactory bulbs in cetacean species. Adaptations to a more or less subterranean life can be related to reductions in the optic system, arboreal life is related to the enlargement of the cerebellum and so on. These changes can be found in various taxonomic groups of mammals. Since they are often rather uniform at different levels of cerebralization, we can also ask whether there are any typical effects of domestication on certain parts of the central nervous circuits.

Some comparative quantitative studies of wild versus domestic mammals have included such functional aspects. For example, in the process of domestication of the European boar into domestic pig, receptor cells in the eye's retina (Wigger 1939) as well as in the olfactory epithelium (Güntherschulze 1979) are reduced in number to a degree comparable to the decrease in size of associated brain structures (Kruska 1980).

Another example concerns the allocortex, a special basal and medial region of the telencephalon. The architectonically distinguishable areas of this part of the mammal forebrain serve mainly in two functional systems, the olfactory and the limbic (Stephan 1975). The olfactory system can be called directly sense-dependent, as olfactory inputs from the olfactory epithelium reach the brain through the olfactory bulb and move along different pathways into the secondary olfactory centers of the allocortex.

The limbic system, in contrast, is mostly independent of the senses. Its telencephalic centers consist mainly of the hippocampus formation, but the septum, schizocortex and certain amygdaloid nuclear masses must also be included in this system. Of course, the hippocampus region receives afferent impulses from several, and perhaps even all, sensory systems, but as a whole it mostly acts endogeneously. According to present knowledge the hippocampus formation plays an important functional role in several behavioral complexes, such as emotionally guided behavior, but also in learning ability and memory. Among others, emotional reactions, aggression and other affective functions, attention as well as motivational and activating functions seem to be guided, controlled and regulated by the hippocampal region.

The various areas of these systems are also smaller in size in domestic forms than in wild living ancestors. The olfactory archicortical structures of these brain parts are also reduced to different degrees in the species investigated (Table 13.2) but to a much lesser degree than the total brain. This can be seen in Figure 13.13, where the degree of size decrease in olfactory structures is scaled to the level of decrease in total brain size. I have also drawn an orientation line which assumes a decrease in olfactory structures similar to total brain decreases due to domestication. In reality, olfactory structures lie below this line.

Surprisingly, comparable effects can be shown in rats, sheep, pigs and poodles. Their plotted points lie at nearly identical distances from the orientation line. This means that comparable effects obviously occur in species differently adapted to special modes of life, at different levels of cerebralization and with different decreases in total brain size caused by domestication. South American tylopods are the only exception, because the olfactory centers of the llama and the alpaca are hardly affected when compared with the guanaco.

With respect to the limbic system, in the species studied (Table 13.2), the hippocampal region by itself and the sum of several telencephalic limbic structures decrease in size in response to demands of domestication. In a comparative scaling in Figure 13.14, the picture in the limbic system is different from that for olfactory structures. Its regions are reduced in size to a higher degree than the brain

Table 13.2. Percentage changes in volume of brain structures in selected functional systems under domestication relative to the wild "ancestral" (from Kruska 1980).

	Rat	*Lama*	Sheep	Pig	Poodle
Olfactory structures	-5.9	-4.2	-21.7	-31.1	-32.7
Limbic structures	-9.6	-6.4	-35.2	-40.7	-34.4
Hippocampal formation	-11.6	-3.3	-41.0	-43.5	-42.0
Visual structures	-4.3		-25.9	-41.0	
Optic tract	+23.5		-20.6	-48.9	
Lateral geniculate	-15.7		-25.4	-38.6	
Striate area	-11.6		-30.2	-41.3	
Superior colliculi	-2.7		-12.1	-31.8	
"Motor" structures					
Cerebellum	-10.3	-12.4	-15.8	-27.3	-32.1
Corpus striatum	-10.9	-9.2	-20.8	-28.9	-27.1

Note. "Rat": Wistar strain laboratory rat; *"Lama"*: llama and alpaca.

as a whole. Here also tylopods are the outstanding exceptions, with almost no size effects differentiating the wild from the domestic form. It is worth mentioning here that the hippocampal region in sheep, pig and poodle, that is, in the so-called highly domesticated forms, is reduced to the remarkable degree of over 40 % in comparison to wild living counterparts. These are very impressive reduction values, which exceed even the relative decrease in neocortex, at least in these species.

We must also take into account the special cytoarchitectonical arrangement of nerve elements in this region. It is characterized by such a dense packing of neurons that counting perikarya is very difficult or even impossible. The high reduction values that go along with this extreme density of neuron packing probably indicate a tremendous functional decrease in the hippocampal region.

Because of the functional importance of this part of the limbic system these size changes may be indicative of a very special effect of domestication. Attenuation of aggressive behavior and diminution of endogeneously produced temptation or drive are fundamental attributes of the domestication process. Long term experiments involving several generations of silver foxes demonstrate the importance of selective breeding in domestication when seeking certain behavioral patterns (Belyaev 1969, 1979, 1980). Destabilizing effects led to the creation of aggressive individuals but resulted also in silver foxes with almost dog-like behavior. Especially these last mentioned foxes showed other morphological and physiological attributes characteristic of many domestic mammals.

Although brains of these foxes have not been investigated quantitatively, we can relate this attenuation of aggressive behavior to the decrease in size of limbic brain structures. At the least, such quantitative functional changes enable man to keep and handle large mammals without danger and related problems. Consciously or unconsciously man must have sought these changes, especially during the initial phase of domestication. Undoubtedly they were an important selective gain for domesticated mammals. A wolf, for example, will, roughly speaking, always remain an aggressive carnivore and will never behave like a docile dog, not even after having been tamed or habituated. At least this seems to be true for other highly domesticated species intensively bred by man and needed for special purposes.

In contrast, the South American tylopods llama and alpaca are kept in relatively natural conditions. Several factors of natural selection remain in force, and artificial selection by man is of minor importance. Sometimes the domestic forms even interbreed with the guanaco. The relatively small changes in hippocampal size from wild to domestic tylopods can be seen as related to the special wild-life-like conditions of these domestic forms.

Comparative investigations on the brains of wild living caribou (*Rangifer tarandus*) versus domesticated reindeer should be of interest in this connection, since many conditions of life in the wild can be looked at in the domesticated form. Unfortunately, information is lacking.

Other brain regions that serve in a functional system have, however, been investigated in certain species, too; for example centers of the sense-dependent visual system (Ebinger 1975 a; Kruska 1972; Kruska and Schott 1977). In this system, the tractus opticus, the lateral geniculate body, the striate area and the superior colliculi must be considered as the principal functional regions of the mammalian brain. In general these are also reduced in size, at least from the wild Norwegian rat to the laboratory rat, from the wild to the domestic sheep and from the European boar to the pig. They also show higher levels of size decrease than do total brains and olfactory structures (Table 13.2).

Surprisingly, however, the optic tract of rats increased in size due to domestication. Here it is important to note that only albino laboratory rats of the Wistar strain were used for the comparison with wild counterparts. This strain is known to possess heavier eyes than wild rats (Ebinger 1972; Wang 1927), while totally pigmented laboratory rats of the DA strain possess eyes of the same weight as wild rats do, independently of body size (Kruska, unpubl. material). Moreover, Wistar rats have heavier eye muscles (McCrady 1934, Robinson 1965) and larger diameters of n. oculomotorius, n. trochlearis, and n. abducens, i.e. the nerves of the eye muscles (McCrady 1934). Hence the unusual size increase in the volume of the optic tract from wild to this special laboratory rat strain most probably can be related to albinism.

Nuclear masses of the visual system are also reduced in size in Wistar rats, which may indicate a functional decrease due to domestication. To date there have been no studies on this problem with totally pigmented or hooded laboratory rat strains.

Sheep and pigs, on the other hand, show rather uniform size changes of nuclear regions to a certain degree, especially when contrasted with total brain size decrease (Figure 13.15). But there are also mutualities when we arrange the reduction values from low to high. Here evolutionary implications for this functional system become evident because the phylogenetically younger centers, the striate area and the lateral geniculate have diminished in size to a higher degree than the superior colliculus. This structure represents the phylogenetically old, established optical tectum in mammalian brains.

In this connection we must discuss a functional dualism within the mammalian visual system as it reflects neuroanatomical organization. The superior colliculus is mainly responsible for simple object localization, and it plays an elementary role in controlling head movements and correcting bodily adjustment during locomotion (Abplanalp 1971; Diamond and Hall 1969; Hassler 1965, 1966; Schneider 1967, 1969). These elementary functions are obviously only slightly reduced in domestic forms. The geniculo-striate system is involved in higher visual functions: The striate cortex and the surrounding neocortical areas account mainly for pattern discrimination. These functions of identification and judgement of objects seem to have decreased to greater extents under domestication.

Information on changes in the intensity of motor functions is also of interest. Here calculations on the cerebellum and the corpus striatum are applicable, because these bodies are responsible for motor coordination and regulation. They have also decreased in size due to domestication (Table 13.2). Although there is no uniformity in size decrease in the two brain regions when we compare different species, the reductions are generally smaller than in the total brain, at least in the more highly cerebralized artiodactylean and carnivore species (Figure 13.16). In rats, however, these brain parts are greatly diminished. The neocortical motor region, the area gigantopyramidalis, was also investigated in sheep (Ebinger 1975b). Again evolutionary implications become evident here as this region has been reduced in size to an extreme degree (Figure 13.16).

To summarize, let me state that while comparable decreases in size of certain brain parts due to domestication were found in the investigation of nuclear masses of certain functional circuits, the species-specific evolutionary level of a given system is always involved. Visual structures are reduced to a greater extent than olfactory and motor nuclear masses. For domestic mammals size reduction of limbic structures seems to be especially typical.

Comparison of Behavior

Wild and domestic forms also show differences in general behavior and in certain behavioral patterns. This has been investigated from special points of view in captive wolves versus poodles (Zimen 1971), European boar versus pig (Gundlach 1968; Briedermann 1971; Hafez et al. 1962; Reiher 1969), wild cavy versus guinea pig (Stahnke 1987) and in great detail in Norway rats versus laboratory strains (Richter 1949, 1954; Boice 1970, 1972). The investigations showed that no new behavioral patterns have developed due to domestication, but, with the use of ethological methods, qualitative as well as quantitative changes were found in reproductive behavior, nest-building behavior, mother-young interactions, social behavior, agonistic behavior, spontaneous locomotor activity, sensoric ability and so on.

In most cases a reduced quantity and quality of certain behavior patterns can be observed in the domestic forms. Very often dissociations of action chains known from the wild species occur due to domestication (Lorenz 1959). From the point of view of comparative quantitative neuroanatomy these behavioral changes are very probably connected in some way to size changes in brain regions and changes in brain proportions between wild and domestic animals, although no counts of neurons, dendritic branches or synapses in certain centers have been done to date.

But there is, of course, a problem in comparing neuroanatomical with behavioral changes observed by ethologists. The problem is the brain itself, and in particular its integrative, associative and coordinative functioning. Even in a rather simple action chain many histologically distinguishable brain parts are always involved in different ways, different qualities and different quantities. In most cases we don't even recognize these ways, or else we can't follow their complex branching.

In general, however, we may relate the reduction of brain substance to a decrease in behavioral patterns. Surprisingly, in domestic forms not only hypotrophied but also hypertrophied behavioral elements have been found, although no brain region is larger in size in comparison to wild species. Hypertrophied behavior in domestic forms undoubtedly occurs in sexuality. Experimental studies on rats and cats (Richter 1954; Schreiner and King 1956; Green et al. 1957) have elucidated the inhibitory character of certain allocortical structures, such as the piriform cortex and some deeper parts of the amygdaloid complex. Removal as well as destruction of these regions resulted in hypersexual behavior of the animals tested. It seems then that the size decrease in these nuclear masses due to domestication may be responsible for a behavioral increase.

In addition, domestic forms may show greater learning ability and even memory capacity in certain behavioral tests and experiments. This has been pointed out for laboratory rats as compared to wild rats by Boice (1970, 1972). At first, these results seem surprising, but there may be an explanation for this effect as well. As we all know, working experimentally with wild individuals often presents problems even after taming. These animals are constantly attentive and aware, most probably as a consequence of their larger sensory brain structures, greater sensory capacity and special brain construction. Because of that they may not concentrate on the particular tasks of test procedures. In contrast, domestic forms may act rather relaxed, most probably because their sensory and attentive behavior is depressed as a consequence of reduced brain size and changed brain proportions. If this turns out to be true, then larger brains seem to be a handicap rather than an advantage in special environments or altered situations.

Conclusions

The domestication of mammals has led to species-specific, gradually different decreases of total brain size and the size of basic parts of the brain. Mammals show changes in brain proportions from the wild to the domestic way of living. These changes are believed to have been caused by artificial selection because brain size is mainly genetically determined and man created domestic stock. Interbreeding of wolves with poodles, for example, has resulted in individuals with intermediate brain sizes in the first generation, while further interbreeding in the second generation led to offspring with a variety of brain sizes, including some with the weight of wolves and others of poodles (Weidemann 1970). Brain sizes of mules as a result of artificial interbreeding between horses and donkeys seem to reflect Mendelian rules to some extent (Kruska 1973).

Changes in brain size and proportions due to domestication must have been aimed at by man. They are intraspecific changes. Compared with interspecific events caused by natural selection during the phylogenetic evolution of species, they occur within shorter time intervals and are of minor dimensions. In most cases, however, phylogenetically younger brain parts are affected to a greater degree than phylogenetically old, established structures. For this reason these changes can serve as a model of regressive evolution on the species level.

On the other hand, domestication started with wild animals at different levels of cerebralization and resulted in gradually different yet unique effects with respect to nuclear masses of functional circuits. Size decreases in these brain regions in connection with changes of behavioral patterns seem to characterize particularly the domestic state and the intensity of domestication. Thus changes in brain proportions from wild ancestral to domestic form must be seen as intraspecific adaptations to the special "ecological niche" of domestication on different levels of cerebralization.

These results indicate that research on domestic animals can contribute to the elucidation of the phenomenon of changeability in brain size, brain proportions and brain functioning as well as of behavior.

Questions

Q. Are we domesticated animals?

A. This question has been dealt with in a lot of publications throughout the history of science. Maybe you as a philosopher are involved to a higher degree than I am, as this is more a question of definition. In my opinion and in that of some colleagues domestication is a kind of symbiosis between man and animals. But in this special symbiosis man plays a special, active role, whereas animals are mainly passively involved. So, when considering the activities of man, we can only discuss self-domestication. But, following the definition above, this may only occur when some human groups influence other human groups by selective breeding over many generations. Concerning brain size and brain proportions in humans no changes have been observed for long periods of time within a large variety, as far as I know. So, I think we are not domesticated or self-domesticated in this sense. But we have learnt to live together in special societies which might imply that we may be denaturalized to a certain degree.

Q. Are domestic mammals stupid as commonly believed?

A. I think stupidity is the opposite of intelligence, and the answer to your question depends on our human definition of both. If we take sensory capacity as a measure in an intelligence test, then certainly domestic mammals may be stupid compared with their wild living counterparts. But as soon as we measure memory capacity, for example, then domestic animals appear to be more intelligent. Maybe you noticed that even in a meeting intended for the discussion of the evolutionary biology of intelligence I mentioned neither stupidity nor intelligence. This was done on purpose, because in comparative neuroanatomy, in my view, no parameters can be found to evaluate either of these. In my opinion a domestic mammal is as well adapted to the domestic mode of living as a wild species is adapted to its special ecological niche. A wild animal may be more or less intelligent in its environment but stupid in domestication. Likewise a domestic animal may be intelligent in the domestic environment but stupid in the wild. In this sense I would answer your question with, no, they are not.

Q. I have heard that domestic mammals have been running wild again. Were brains of those animals investigated and what are the results ?

A. Yes, this has happened occasionally in Australia, the Americas, the Galapagos Islands and several other places. The running wild of domestic stock is viewed as the opposite of domestication and is called feralization. Brains of some species were investigated in small numbers. They showed no increase. That means that an enriched environment did not lead to increasing brain size, which, again, is in contrast to some experimental findings in rats. But most of these feralized mammals, such as pigs from the Galapagos Islands, may have been living in wild environments for a relatively short time. On the other hand, dingos from Australia and the so-called Hallstrom dogs from New Guinea are considered to be descendants of early domestic dogs. They have been around since at least 3.000 or at most 8.600 B.C. Recent dingoes and Hallstrom dogs have brain sizes equal to recent dog races of comparable body weight. So the effects on brain size due to domestication seem to be irreversible just like evolutionary changes according to the rule of Dollo.

Q. Are there race specific differences in brain size within domestic mammals?

A. Yes, there are small differences as has been shown in rats and cats, as well as in farm mink, according to our preliminary results. A large amount of material is needed for such investigations but unfortunately is seldom available. For dogs, however, this doesn't hold true. In this case we have the needed amount of material, but there seem to have been methodological problems in some recent investigations. Sometimes the material under study is evaluated without enough basic information. For example, a dog may seem to be a pure breed of a race and is evaluated accordingly when weights of brain and body are taken. But there is no certainty that the animal is a pure breed. That is why the use of reliable collections is so important. Unfortunately this is not always considered, and no accurate results can be obtained. However, there may be small differences between dog races, because they were bred for special sensory and psychologic performances. In the studies referred to here, only pure breeds of poodle were investigated.

References

Abplanalp P (1971) The neuroanatomical organization of the visual system in the tree shrew. Folia primatol 16:1-34

Apfelbach R, Kruska D (1979) Zur postnatalen Hirnentwicklung beim Frettchen *Mustela putorius f. furo* (Mustelidae; Mammalia). Z Säugetierkunde 44:127-131

Bauchot R, Stephan H (1966) Donnees nouvelles sur l'encephalisation des insectivores et des prosimiens. Mammalia 30:160-196

Belyaev DK (1969) Domestication of animals. Sci Journ 50:47-53

Belyaev DK (1979) Destabilizing selection as a factor in domestication. J of Heredity 70:301-308

Belyaev DK (1980) Destabilisierende Selektion als Evolutionsfaktor. Arch Tierzucht 23:59-63

Boice R (1970) The effect of domestication on avoidance learning in the Norway rat. Psychon Sci 18:13-14

Boice R (1972) Some behavioral tests of domestication in Norway rats. Behaviour 42:198-231

Briedermann L (1971) Ermittlungen zur Aktivitätsperiodik des Mitteleuropäischen Wildschweins (*Sus s.scrofa L.*). Zool Garten NF 40:302-327

Bronson RT (1979) Brain weight - body weight scaling in breeds of dogs and cats. Brain Beh Evol 16:227-236

Darwin Ch (1873) Das Variieren der Thiere und Pflanzen im Zustande der Domestication. Bde I und II. Schweitzerbart'sche Verlagshandlung, Stuttgart

Diamond IT, Hall WC (1969) Evolution of neocortex. Science 164:251-262

Dubois E (1914) Die gesetzmässige Beziehung von Gehirnmasse zu Körpergrösse bei den Wirbeltieren. Z Morph Anthrop 18:323-350

Ebinger P (1972) Vergleichend-quantitative Untersuchungen an Wild-und Laborratten. Z Tierzüchtg Züchtungsbiol 89:34-57

Ebinger P (1974) A cytoarchitectonic volumetric comparison of brains in wild and domestic sheep. Z Anat Entwickl- Gesch 144:268-302

Ebinger P (1975a) Quantitative investigations of visual brain structures in wild and domestic sheep. Anat Embryol 146:313-323

Ebinger P (1975b) A cytoarchitectonic volumetric comparison of the area gigantopyramidalis in wild and domestic sheep. Anat Embryol 147:167-175

Ebinger P, Macedo H, Röhrs M (1984) Hirngrössenänderungen vom Wild- zum Hausmeerschweinchen. Z zool Syst Evolut-forsch 22:77-80

Espenkötter E (1982) Vergleichende quantitative Untersuchungen an Iltissen und Frettchen. Dissertation thesis Veterinary Hochschule Hannover

Fischer CJ (1973) Vergleichende quantitative Untersuchungen an Wildkaninchen und Hauskaninchen. Dissertation thesis Veterinary Hochschule Hannover

Frick H, Nord HJ (1963) Domestikation und Hirngewicht. Anat Anz 113:307-316

Gittleman JL (1986) Carnivore brain size, behavioral ecology and phylogeny. J of Mammalogy 67:23-26

Green JD, Clemente CD, de Groot J (1957) Rhinencephalic lesions and behavior in cats. J comp Neurol 108:505-536

Güntherschulze J (1979) Studien zur Kenntnis der Regio olfactoria von Wild- und Hausschwein (*Sus scrofa scrofa* L.1768 und *Sus scrofa f.*domestica). Zool Anz 202:256-279

Gundlach H (1968) Brutfürsorge, Verhaltensontogenese und Tagesperiodik beim Europäischen Wildschwein (*Sus scrofa* L.). Z Tierpsychologie 25:955-995

Hafez ESE, Sumption LJ, Jakway JS (1962) The behaviour of swine. In: Hafez ESE (ed) The behaviour of domestic animals. Bailliere, Tindall and Cox, London, p 334-369

Harvey PH,Bennett PM (1983) Brain size, energetics, ecology, and life history patterns. News and Views 306:314-315

Hassler R (1965) Die zentralen Systeme des Sehens. Ber Zus-kunft Dtsch ophthalm Ges 66:229-251

Hassler R (1966) Comparative anatomy of the central visual systems in day-and night-active primates. In: Hassler R, Stephan H (eds) Evolution of the forebrain. Thieme, Stuttgart, 419-434

Herre W, Thiede M (1965) Studien an Gehirnen südamerikanischer Tylopoden. Zool Jb (Abt Anat) 81:155-176

Jerison HJ (1973) Evolution of the brain and intelligence. Academic Press, New York-London

Kappers CUA, Huber GC, Crosby EC (1967) The comparative anatomy of the nervous system of vertebrates, including man. Vol. I-III Hafner, New York

Kruska D (1970) Vergleichend cytoarchitektonische Untersuchungen an Gehirnen von Wild-und Hausschweinen. Z Anat Entwickl-Gesch 131:291-324

Kruska D (1972) Volumenvergleich optischer Hirnzentren bei Wild-und Hausschweinen. Z Anat Entwickl-Gesch 138:265-282

Kruska D (1973 a) Cerebralisation, Hirnevolution und domestikationsbedingte Hirngrössenänderungen innerhalb der Ordnung Perissodactyla Owen, 1848 und ein Vergleich mit der Ordnung Artiodactyla Owen, 1848. Z zool Syst Evol-Forsch 11:81-103

Kruska D (1973 b) Domestikationsbedingte Grössenänderungen verschiedener Hirnstrukturen bei Schweinen. In: Matolcsi J (Hrsg) Domestikationsforschung und Geschichte der Haustiere. Akademiai Kiado, Budapest, S 135-140

Kruska D (1975 a) Über die postnatale Hirnentwicklung bei *Procyon cancrivorus* (Procyonidae; Mammalia). Z Säugetierkunde 40:243-256

Kruska D (1975 b) Vergleichend-quantitative Untersuchungen an den Gehirnen von Wander-und Laborratten. I. Volumenvergleich des Gesamthirns und der klassischen Hirnteile. J Hirnforsch 16:469-483

Kruska D (1975 c) Vergleichend-quantitative Untersuchungen an den Gehirnen von Wander-und Laborratten. II. Volumenvergleich allokortikaler Hirnzentren. J Hirnforsch 16:485-496

Kruska D (1977) Über die postnatale Hirnentwicklung beim Farmnerz *Mustela vison f.*dom. (Mustelidae;Mammalia). Z Säugetierkunde 42:240-255

Kruska D (1979) Vergleichende Untersuchungen an den Schädeln von subadulten und adulten Farmnerzen, *Mustela vison f.* dom. (Mustelidae;Carnivora). Z Säugetierkunde 44:360-375

Kruska D (1980) Domestikationsbedingte Hirngrössenänderungen bei Säugetieren. Z zool Syst Evol-Forsch 18:161-195

Kruska D (1982) Hirngrössenänderungen bei Tylopoden während der Stammesgeschichte und in der Domestikation. Verh Dtsch Zool Ges 1982:173-183

Kruska D (1987) How fast can total brain size change in mammals? J Hirnforsch 28: 59-70

Kruska D, Röhrs M (1974) Comparative-quantitative investigations on brains of feral pigs from the Galapagos Islands and of European domestic pigs. Z Anat Entwickl-Gesch 144:61-73

Kruska D, Schott M (1977) Vergleichend-quantitative Untersuchungen an den Gehirnen von Wander-und Laborratten. III. Volumenvergleich optischer Hirnstrukturen. J Hirnforsch 18:59-67

Kruska D, Stephan H (1973) Volumenvergleich allokortikaler Hirnzentren bei Wild-und Hausschweinen. Acta anat 84:387-415

Lapique L (1908) Le poids encephalique en fonction du poids corporel entre individus d'une meme espece. Bull Mem Soc Anthrop Paris, 249-271

Lorenz K (1959) Psychologie und Stammesgeschichte. In: Heberer G (Hrsg) Evolution der Organismen. Fischer, Stuttgart, 2. erw. Aufl. S. 131-170

Martin RD (1981) Relative brain size and basal metabolic rate in terrestrial vertebrates. Nature 293:57-60

Martin RD, Harvey PH (1985) Brain size allometry. Ontogeny and phylogeny. In: Jungers WL (ed) Size and Scaling in Primate Biology. Plenum Press, New York, 147-172

Mason IL (1984) Evolution of domesticated animals, Longman, London New York

McCrady E (1934) The motor nerves of the eye in albino and gray Norway rats. J comp Neurol 59:285-300

Reiher EG (1969) Sinnesphysiologische und lernpsychologische Untersuchungen an Schweinen. Forma Functio 1:353-404

Rempe U (1962) Über einige statistische Hilfsmittel moderner zoologisch-systematischer Untersuchungen. Zool Anz 169:93-140

Richter CP (1949) The use of the wild Norway rat for psychiatric research. J of Nervous and Mental Disease 110:379-386

Richter CP (1954) The effect of domestication and selection on the behavior of the Norway rat. J of the Nat Canc Inst 15:727-738

Robinson R (1965) Genetics of the Norway rat. Pergamon Press, Oxford

Röhrs M (1955) Vergleichende Untersuchungen an Wild-und Hauskatzen. Zool Anz 155:53-69

Röhrs M (1985) Cephalization, neocorticalization and the effects of domestication on brains of mammals. In: Duncker H-R, Fleischer G (eds) Functional morphology in vertebrates. Fischer, Stuttgart New York, p 544-547

Röhrs M, Ebinger P (1978) Die Beurteilung von Hirngrössenunterschieden zwischen Wild-und Haustieren. Z zool Syst Evolut-forsch 16:1-14

Röhrs M, Kruska D (1969) Der Einfluss der Domestikation auf das Zentralnervensystem und Verhalten von Schweinen. Dtsch tierärztl Wschr 76:514-518

Schleifenbaum Ch (1973) Untersuchungen zur postnatalen Ontogenese des Gehirns von Grosspudeln und Wölfen. Z Anat Entwickl-Gesch 141:179-205

Schneider GE (1967) Contrasting visuomotor functions of tectum and cortex in the golden hamster. Psychol Forsch 31:52-62

Schneider GE (1969) Two visual systems. Science 163:895-902

Schreiner L, King A (1956) Rhinencephalon and behavior. Amer J Physiol 184:486-490

Schultz W (1969) Zur Kenntnis des Hallstromhundes (*Canis hallstromi* Throughton, 1957). Zool Anz 183:47-72

Schumacher M (1963) Quantitative Untersuchungen an Gehirnen mitteleuropäischer Musteliden. J Hirnforsch 6:137-163

Stahnke A (1987) Verhaltensunterschiede zwischen Wild-und Hausmeerschweinchen. Z Säugetierkunde 52:294-307

Starck D (1962) Die Evolution des Säugetiergehirns. Steiner, Wiesbaden

Starck D (1982) Vergleichende Anatomie der Wirbeltiere. Springer, Berlin, Heidelberg, New York, Vol 3

Stephan H (1960) Methodische Studien über den quantitativen Vergleich architektonischer Struktureinheiten des Gehirns. Z wiss Zool 164:143-172

Stephan H (1975) Allocortex. Springer, Heidelberg New York

Stephan H, Bauchot R, Andy OJ (1970) Data on size of the brain and of various brain parts in insectivores and primates. In: Noback ChR, Montagna W (eds) Advances in Primatology Vol I. Appleton-Century-Crofts, New York, 289-297

Stephan H, Frahm HD, Stephan M (1982) Comparison of brain structure volumes in Insectivora and Primates. J Hirnforsch 23:375-389

Stephan H, Baron G, Frahm HD, Stephan M (1986) Grössenvergleiche an Gehirnen und Hirnstrukturen von Säugern. Z mikrosk-anat Forsch (Leipzig) 100:189-212

Wang CC (1927) On the postnatal growth in the area of the optic nerve in Albino and in gray Norway rats. J Comp Neurol 43:201-220

Weidemann W (1970) Die Beziehung von Hirngewicht und Körpergewicht bei Wölfen und Pudeln sowie deren Kreuzungsgenerationen N_1 und N_2. Z Säugetierkunde 35:238-247

Wigger H (1939) Vergleichende Untersuchungen am Auge von Wild-und Hausschwein unter besonderer Berücksichtigung der Retina. Z Morph Ökol Tiere 36:1-20

Wirz K (1950) Studien über die Cerebralisation: zur quantitativen Bestimmung der Rangordnung bei Säugetieren. Acta anat 9:134-196

Zeuner FE (1963) A history of domesticated animals, Hutchinson, London

Zimen E (1971) Wölfe und Königspudel. Vergleichende Verhaltensbeobachtungen, Piper, München

Appendix 13.A. Wild mammalian species (left) and related domestic forms (right)

LAGOMORPHA

Wild		Domestic
Oryctolagus cuniculus	-	*Oryctolagus cuniculus* f. dom.
(Old world rabbit)	-	(Domestic rabbit)

RODENTIA

Wild		Domestic
Mesocricetus auratus	-	*Mesocricetus auratus* f. dom.
(Golden hamster)	-	(Laboratory hamster)
Rattus norvegicus	-	*Rattus norvegicus* f. dom.
(Norway rat)	-	(Laboratory rat)
Mus musculus	-	*Mus musculus* f. dom.
(House mouse)	-	(Laboratory mouse)
Cavia aperea	-	*Cavia aperea* f. porcellus
(Wild cavy)	-	(Domestic guinea - pig)
Chinchilla veliger	-	*Chinchilla veliger* f. dom.
(Wild chinchilla)	-	(Farm chinchilla)
Myocastor coypus	-	*Myocastor coypus* f. dom.
(Coypus)	-	(Farm nutria)

CARNIVORA

Wild		Domestic
Canis lupus	-	*Canis lupus* f. familiaris
(Gray wolf)	-	(Dog)
Alopex lagopus	-	*Alopex lagopus* f. dom.
(Arctic fox)	-	(Blue fox)
Vulpes vulpes	-	*Vulpes vulpes* f. dom.
(Red fox)	-	(Silver fox)
Procyon lotor	-	*Procyon lotor* f. dom.
(Raccoon)	-	(Farm raccoon)
Mustela putorius	-	*Mustela putorius* f. furo
(European polecat)	-	(Ferret)
Mustela vison	-	*Mustela vison* f. dom.
(American mink)	-	(Ranch mink)
Felis silvestris	-	*Felis silvestris* f. catus
(Wild cat)	-	(Domestic cat)

Appendix 13.A (Cont.)

PERISSODACTYLA

Equus przewalskii	-	*Equus przewalskii* f. caballus
(Przevalsky' horse)	-	(Domestic horse)
Equus africanus	-	*Equus africanus* f. asinus
(Wild ass)	-	(Donkey)

ARTIODACTYLA

Sus scrofa	-	*Sus scrofa* f. dom.
(European wild boar)	-	(Domestic pig)
Lama guanacoe	-	*Lama guanacoe* f. glama
(Guanaco)	-	(Llama, Alpaca)
Camelus ferus	-	*Camelus ferus* f. bactriana
(Wild bactrian camel)	-	(Domestic bactrian camel)
Rangifer tarandus	-	*Rangifer tarandus* f. dom.
(Wild reindeer)	-	(Domestic reindeer)
Bubalis arnee	-	*Bubalis arnee* f. bubalis
(Arni)	-	(Domestic water buffalo)
Bos primigenius	-	*Bos primigenius* f. taurus
(Aurochs)	-	(Domestic cattle)
Bos mutus	-	*Bos mutus* f. grunniens
(Wild yak)	-	(Domestic yak)
Bos javanicus	-	*Bos javanicus* f. dom.
(Banteng)	-	(Bali cattle)
Bos gaurus	-	*Bos gaurus* f. frontalis
(Gaur)	-	(Gayal)
Capra aegagrus	-	*Capra aegagrus* f. hircus
(Bezoar goat)	-	(Domestic goat)
Ovis ammon	-	*Ovis ammon* f. aries
(Mouflon)	-	(Domestic sheep)

Appendix 13.B. Cerebralization indices (CI) in species of fissipede carnivores, symbols as in Fig. 13.9; average body and brain weights in grams.

Species	Symbol	Body	Brain	CI
Arctoidea				
Ailuridae				
Ailuropoda melanoleuca	Am	90 350	256.80	112
Ailurus fulgens	Af	3 844	44.76	114
Procyonidae				
Bassariscus astutus	Ba	1 200	19.25	94
Procyon lotor	Pl	4 524	40.71	95
Procyon cancrivorus	Pc	4 553	69.93	162
Nasua nasua	Nn	3 414	36.93	101
Nasua rufa	Nr	3 175	34.00	96
Potos flavus	Pf	2 426	32.84	108
Mustelidae				
Mustela nivalis	Mn	47	1.82	55
Mustela erminea	Me	170	4.51	66
Mustela nigripes	Ms	449	7.80	66
Mustela putorius	Mp	834	9.22	55
Martes martes	Mm	1 000	18.76	102
Martes foina	Mf	1 477	21.79	95
Eira barbara	Eb	2 668	47.05	147
Galictis cuja	Gc	1 190	18.80	92
Galictis vittata	Gv	1 328	17.97	83
Meles meles	Ml	12 042	61.42	83
Helictis personata	Hp	1 250	15.00	72
Mephitis mephitis	Mh	3 289	10.90	30
Conepatus chinga	Cc	1 587	18.66	78
Conepatus humboldti	Ch	1 294	12.62	58
Lutra lutra	Ll	5 080	73.38	160
Pteronura brasiliensis	Pb	18 000	115.50	124

Appendix 13.B (cont.)

Species	Symbol	Body	Brain	CI
Ursidae				
Ursus horribilis	Uh	146 270	301.65	100
Ursus torquatus	Ut	69 860	269.00	135
Ursus arctos	Ua	197 000	407.00	114
Ursus americanus	Um	100 000	238.00	98
Melursus ursinus	Mu	136 080	267.00	92
Helarctos malayanus	Hm	30 005	354.00	285
Thalarctos maritimus	Tm	249 630	508.40	125
Cynofeloidea				
Canidae				
Canis lupus	Cl	26 950	139.25	119
Canis latrans	Cs	10 885	89.12	127
Canis aureus	Cu	8 576	62.61	102
Canis adustus	Cd	6 155	50.81	99
Canis mesomelas	Cm	7 396	55.17	97
Vulpes vulpes	Vv	5 757	51.16	104
Vulpes fulva	Vf	4 625	33.35	122
Vulpes velox	Vx	2 758	33.35	102
Alopex lagopus	Al	3 193	41.52	117
Urocyon cinereoargenteus	Uc	2 308	33.99	115
Otocyon megalotis	Om	3 194	28.22	80
Dusicyon gracilis	Dg	2 487	36.63	119
Dusicyon gymnocercus	Dy	3 390	41.50	113
Dusicyon culpaeus	Dc	6 234	57.68	112
Dusicyon griseus	Dr	4 324	42.47	125
Cerdocyon thous	Ct	5 600	49.50	102
Chrysocyon brachyurus	Cb	22 000	115.00	110
Lycaon pictus	Lp	23 150	134.65	126
Cuon alpinus	Cp	8 680	98.50	159
Fennecus cerda	Fc	1 014	20.71	111
Felidae				
Felis silvestris	Fs	3 706	39.93	104
Lynx canadensis	Lc	14 969	69.50	83
Lynx rufus	Lr	6 350	65.00	125
Caracal caracal	Ca	7 690	51.90	90
Leptailurus serval	Ls	8 373	55.79	92

Appendix 13.B (cont.)

Species	Symbol	Body	Brain	CI
Leopardus pardalis	Le	8 618	62.70	102
Herpailurus yagouarondi	Hy	2 485	40.94	133
Panthera pardus	Pp	26 425	139.15	120
Panthera leo	Po	114 020	237.35	91
Panthera tigris	Pt	124 570	255.70	93
Panthera onca	Pn	24 005	151.73	139
Acinonyx jubatus	Aj	40 000	130.00	89
Herpestoidea				
Viverridae				
Genetta genetta	Gg	1 500	17.50	76
Genetta servalina	Gs	1 295	16.00	75
Genetta tigrina	Gt	1 270	15.55	74
Viverra tangalunga	Vt	3 130	30.20	86
Viverra civetta	Vc	8 500	42.10	69
Paradoxurus hermaphroditus	Pm	3 314	22.65	63
Herpestidae				
Suricata suricatta	Ss	590	11.45	83
Herpestes javanicus	Hj	321	7.00	72
Herpestes smithi	Hs	1 205	14.60	71
Herpestes galera	Hg	2 675	28.80	90
Herpestes cafer	Hc	1 764	24.10	95
Mungos mungo	Mg	1 523	10.90	47
Mungos obscurus	Mo	865	9.99	59
Crossarchus zebra	Cz	930	10.80	61
Ichneumia albicauda	Ia	2 843	25.15	76
Cynictis penicillata	Cp	462	10.00	83
Cryptoprocta ferox	Cf	4 656	36.88	84
Hyaenidae				
Proteles cristatus	Pa	4 503	36.35	85
Hyaena hyaena	Hh	25 815	89.07	78
Crocuta crocuta	Co	55 120	171.60	98

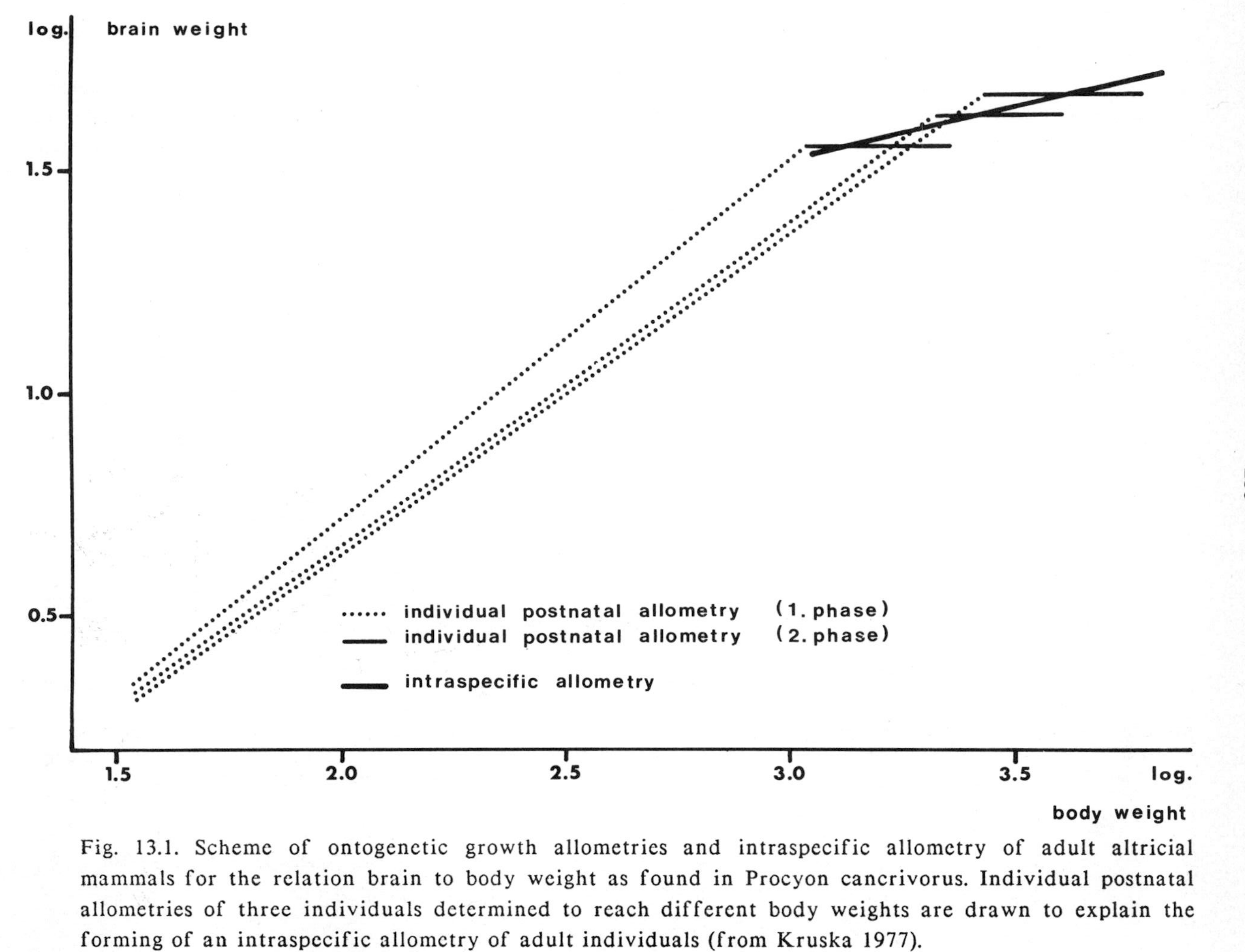

Fig. 13.1. Scheme of ontogenetic growth allometries and intraspecific allometry of adult altricial mammals for the relation brain to body weight as found in Procyon cancrivorus. Individual postnatal allometries of three individuals determined to reach different body weights are drawn to explain the forming of an intraspecific allometry of adult individuals (from Kruska 1977).

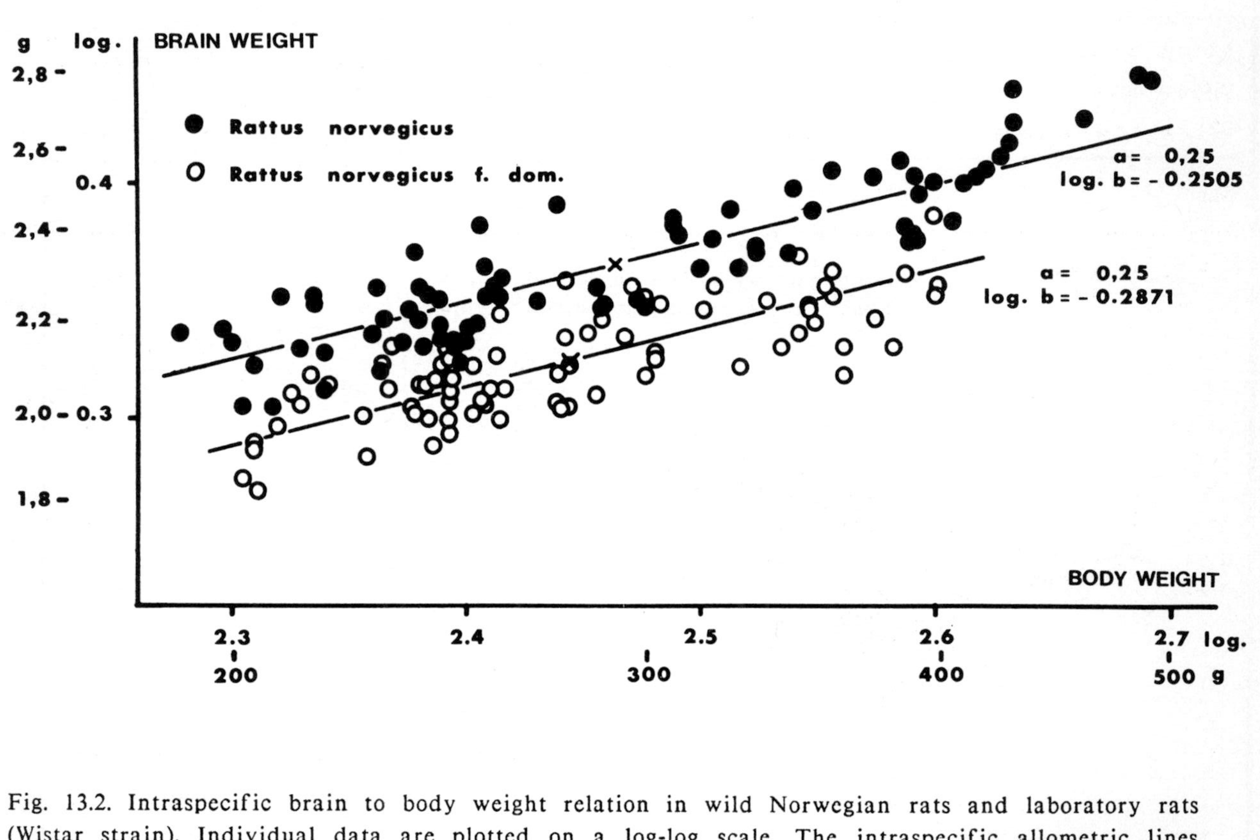

Fig. 13.2. Intraspecific brain to body weight relation in wild Norwegian rats and laboratory rats (Wistar strain). Individual data are plotted on a log-log scale. The intraspecific allometric lines indicate medium brain size at a given body size within both groups (from Kruska 1975 b).

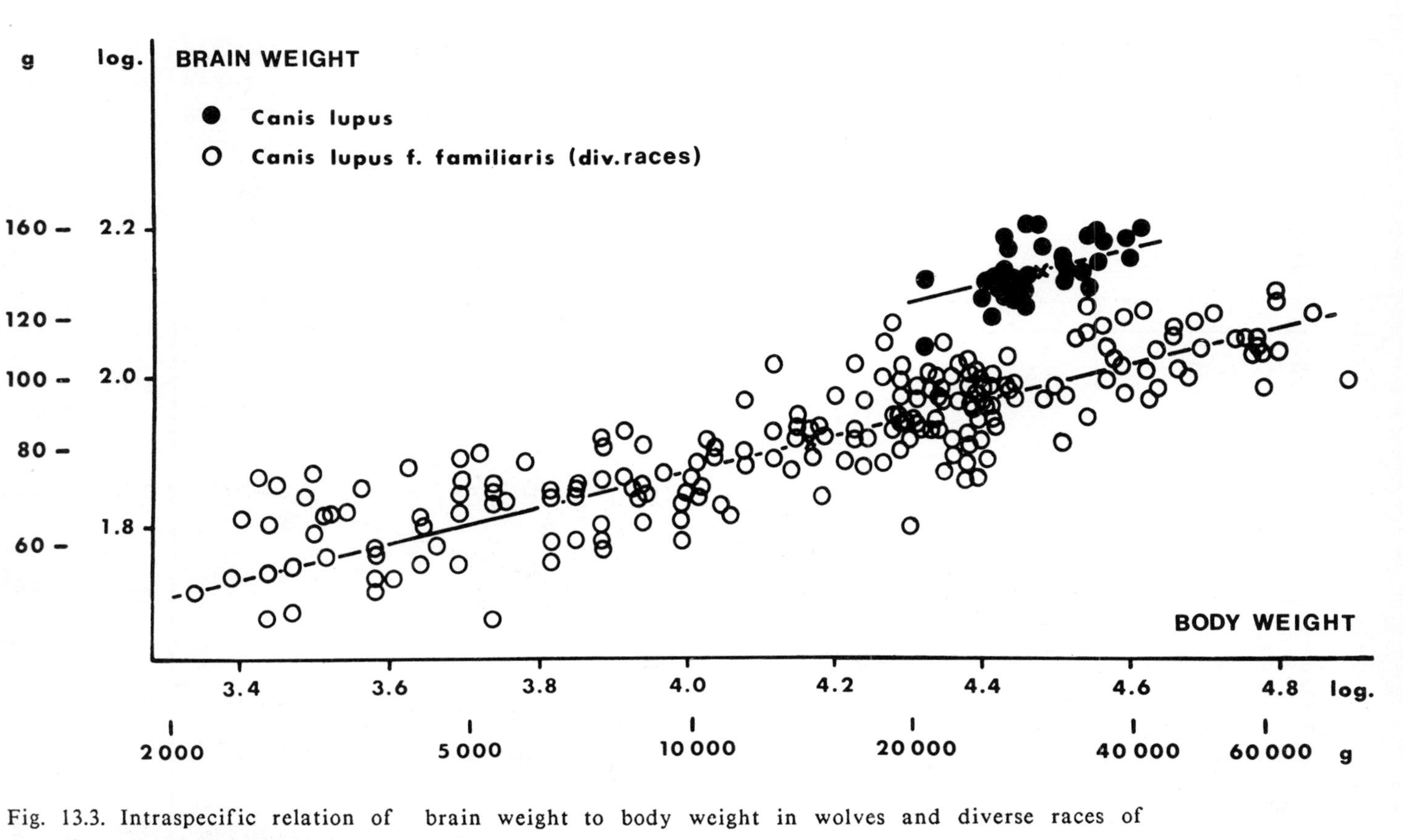

Fig. 13.3. Intraspecific relation of brain weight to body weight in wolves and diverse races of dogs (from Schultz 1969).

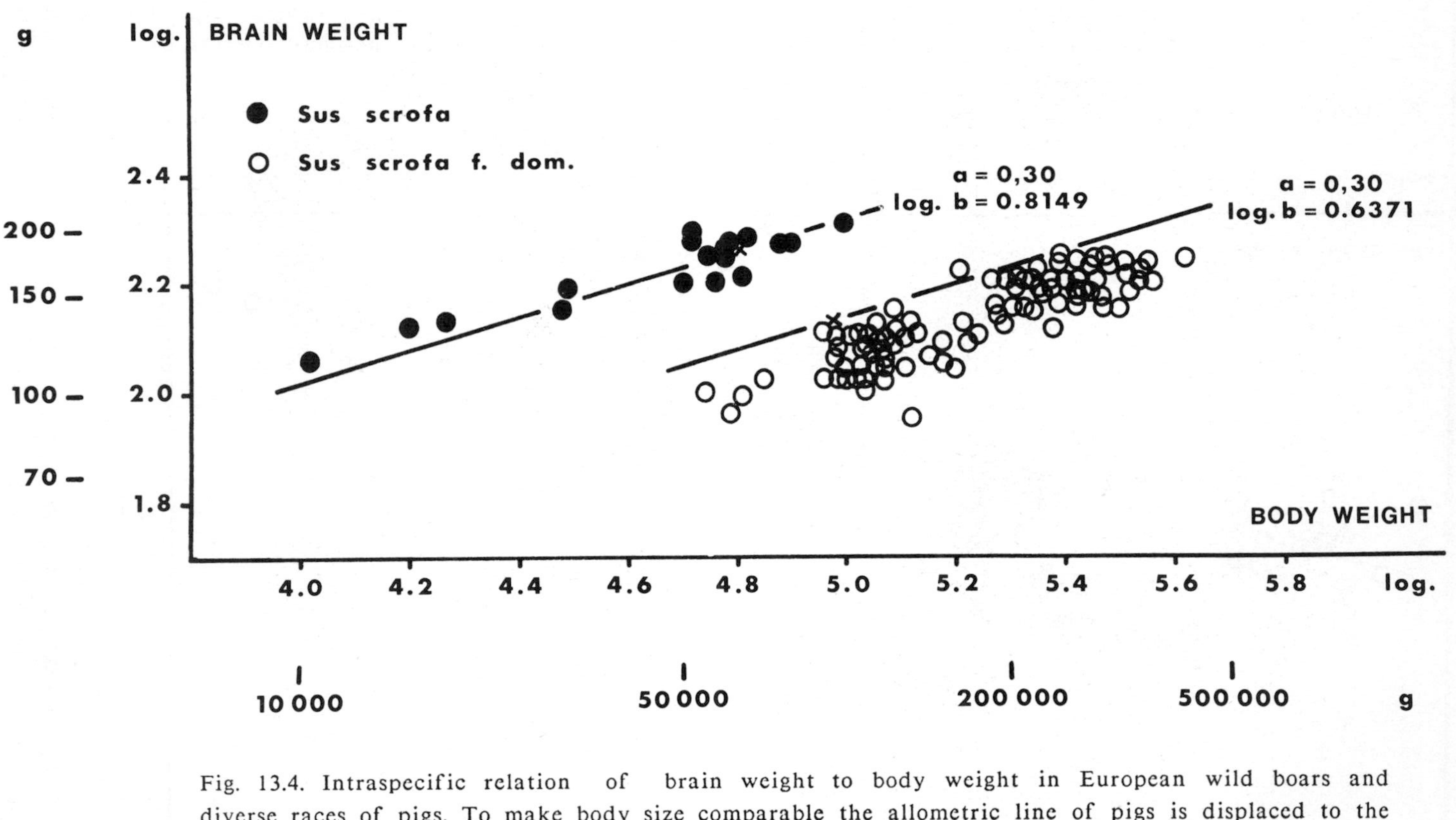

Fig. 13.4. Intraspecific relation of brain weight to body weight in European wild boars and diverse races of pigs. To make body size comparable the allometric line of pigs is displaced to the left as fat is taken into account in the domestic form (from Kruska 1970).

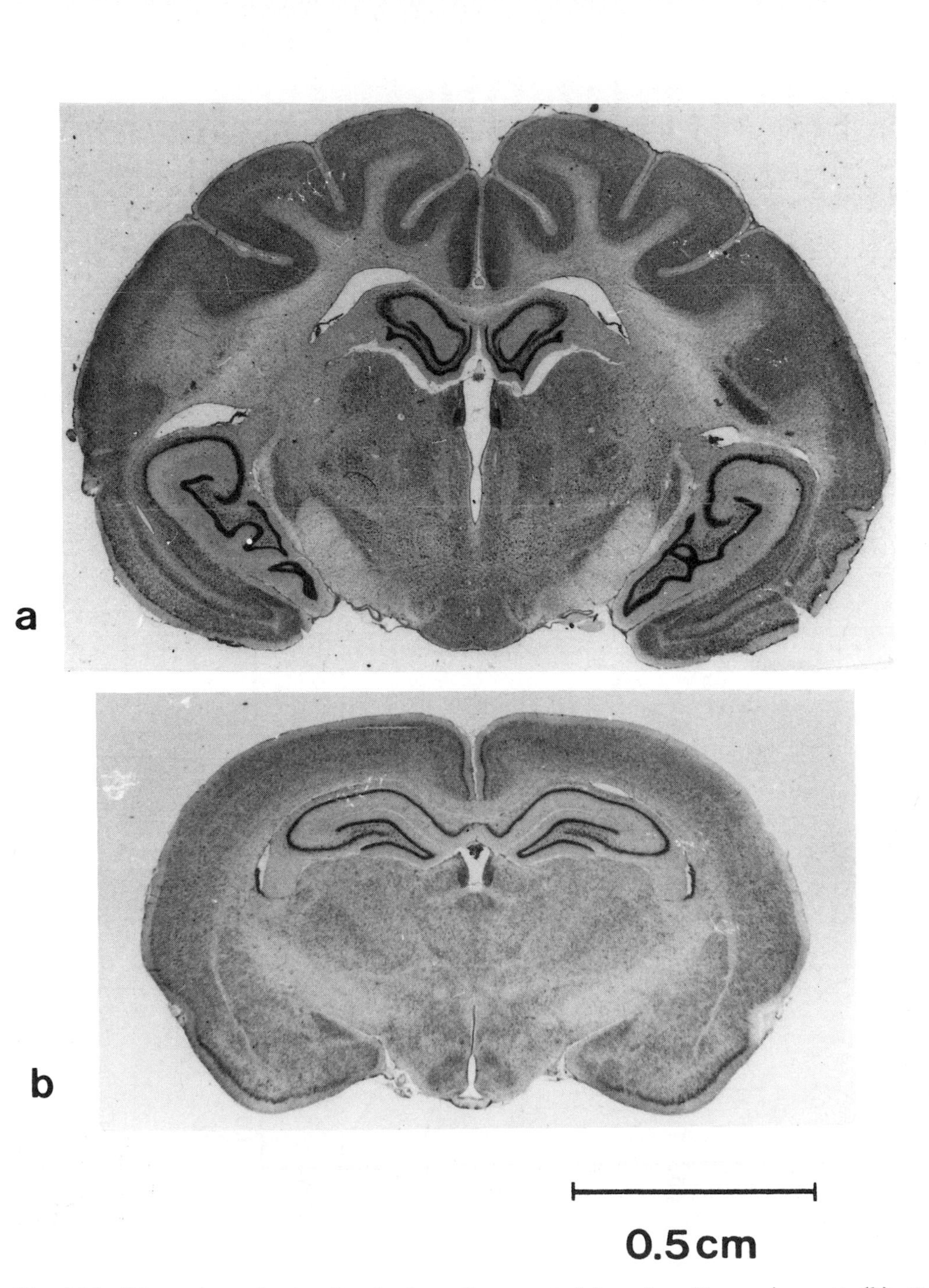

Fig. 13.5. Slices through the forebrains of a stoat (a) and a Norwegian rat (b) at comparable regions of the habenular nuclei to show differences in neocortex size and differentiation of mammals at nearly the same body weight.

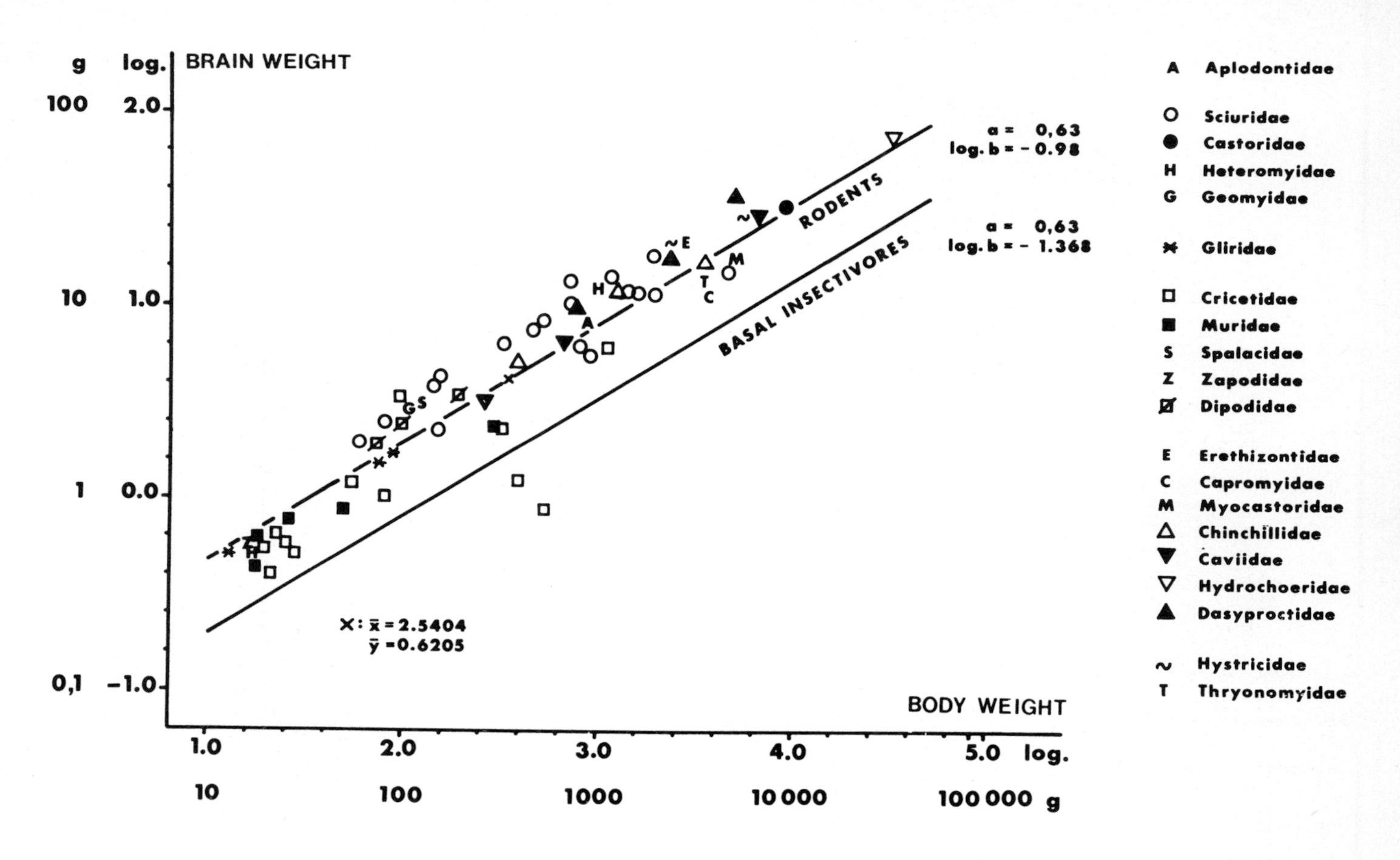

Fig. 13.6. Interspecific relation of brain weight to body weight in the order Rodentia. Average species-specific data of 64 species of diverse families are plotted in a double log. scale. The interspecific allometric line indicates medium cerebralisation level of this order. The allometric line of Basal Insectivora is drawn for comparison (from Kruska 1980).

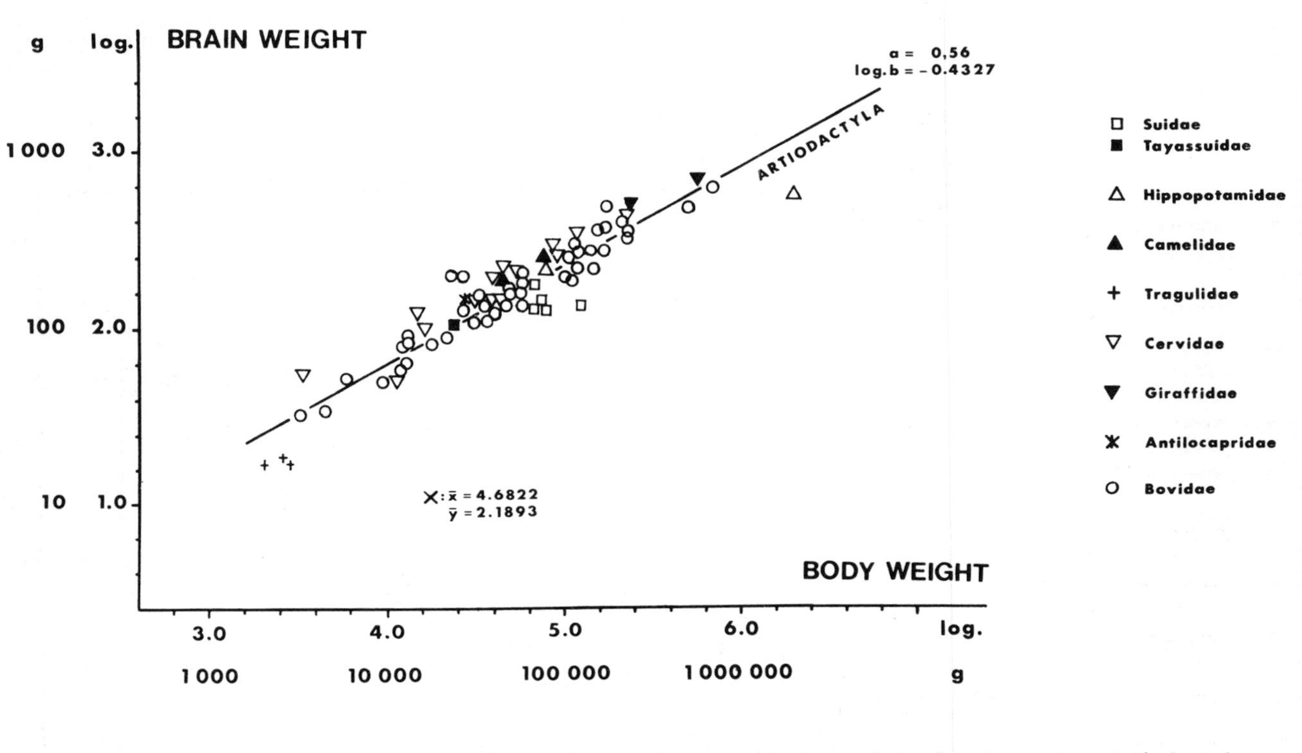

Fig. 13.7. Interspecific relation of brain weight to body weight in the order Artiodactyla based on species-specific data of 74 species (from Kruska 1980).

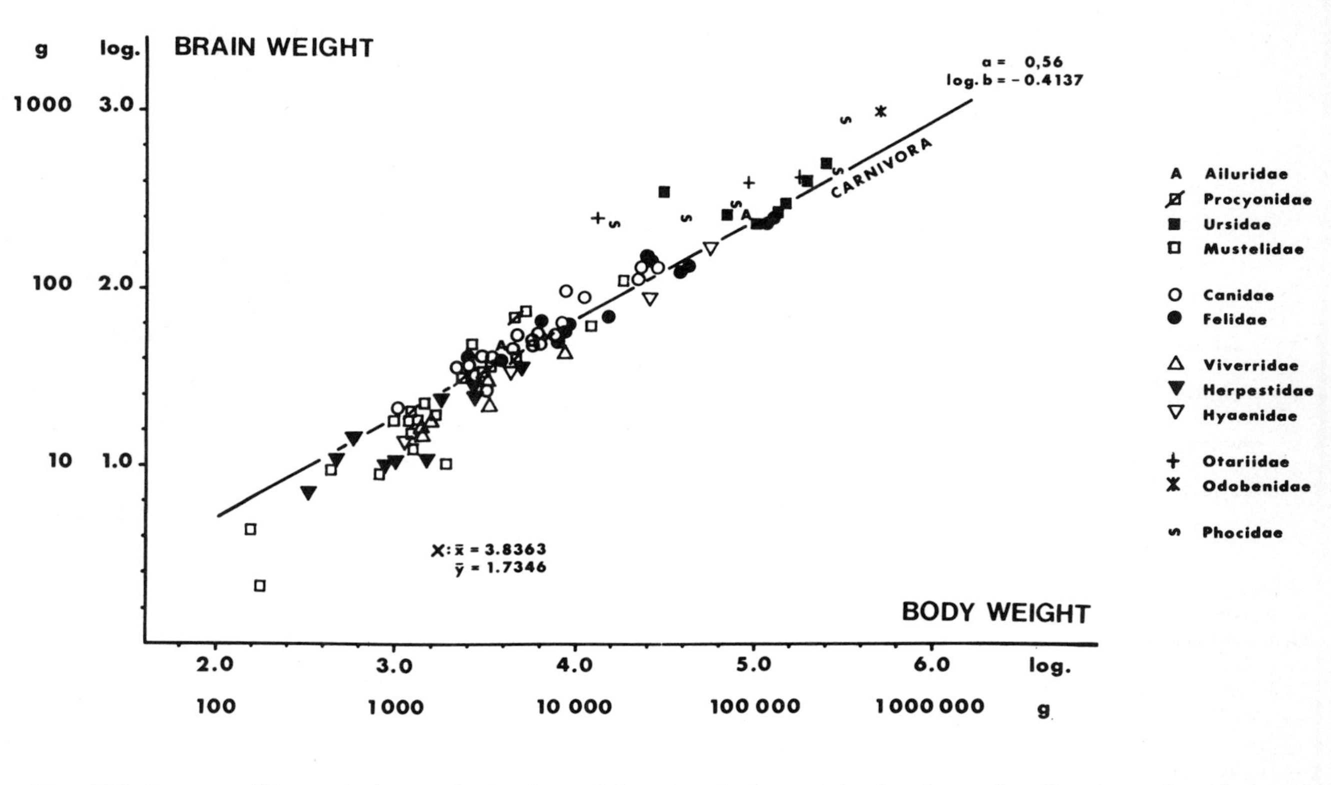

Fig. 13.8. Interspecific relation of brain weight to body weight in the order Carnivora based on species-specific data of 93 species (from Kruska 1980).

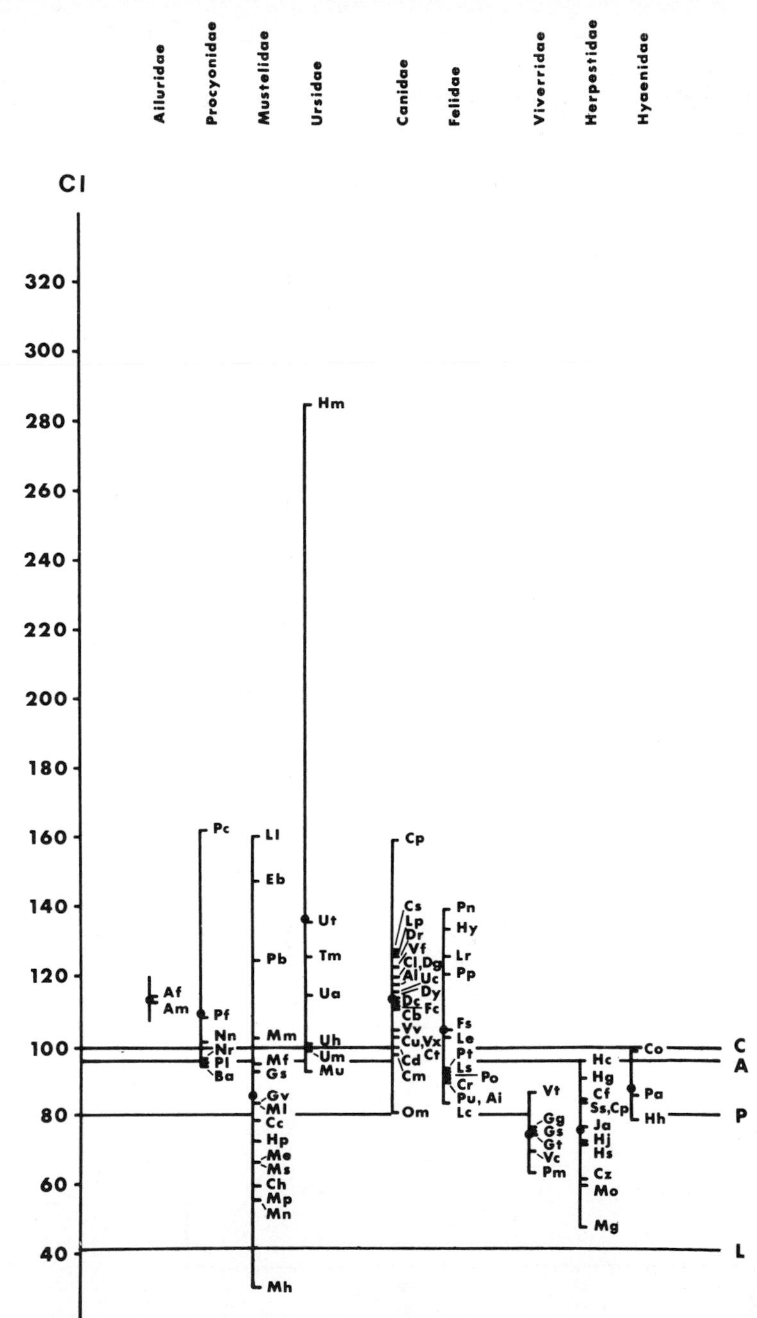

Fig. 13.9. Scaling of cerebralisation indices (CI) of several fissiped carnivore species within their families. The CI were calculated as percentual deviations of species' brain sizes from medium carnivor brain size at given body weights. Medium level of cerebralisations of Carnivora (C), Artiodactyla (A), Perissodactyla (P), and Lagomorpha (L) are also indicated. For abbreviations of the species' names see Appendix 13.B.

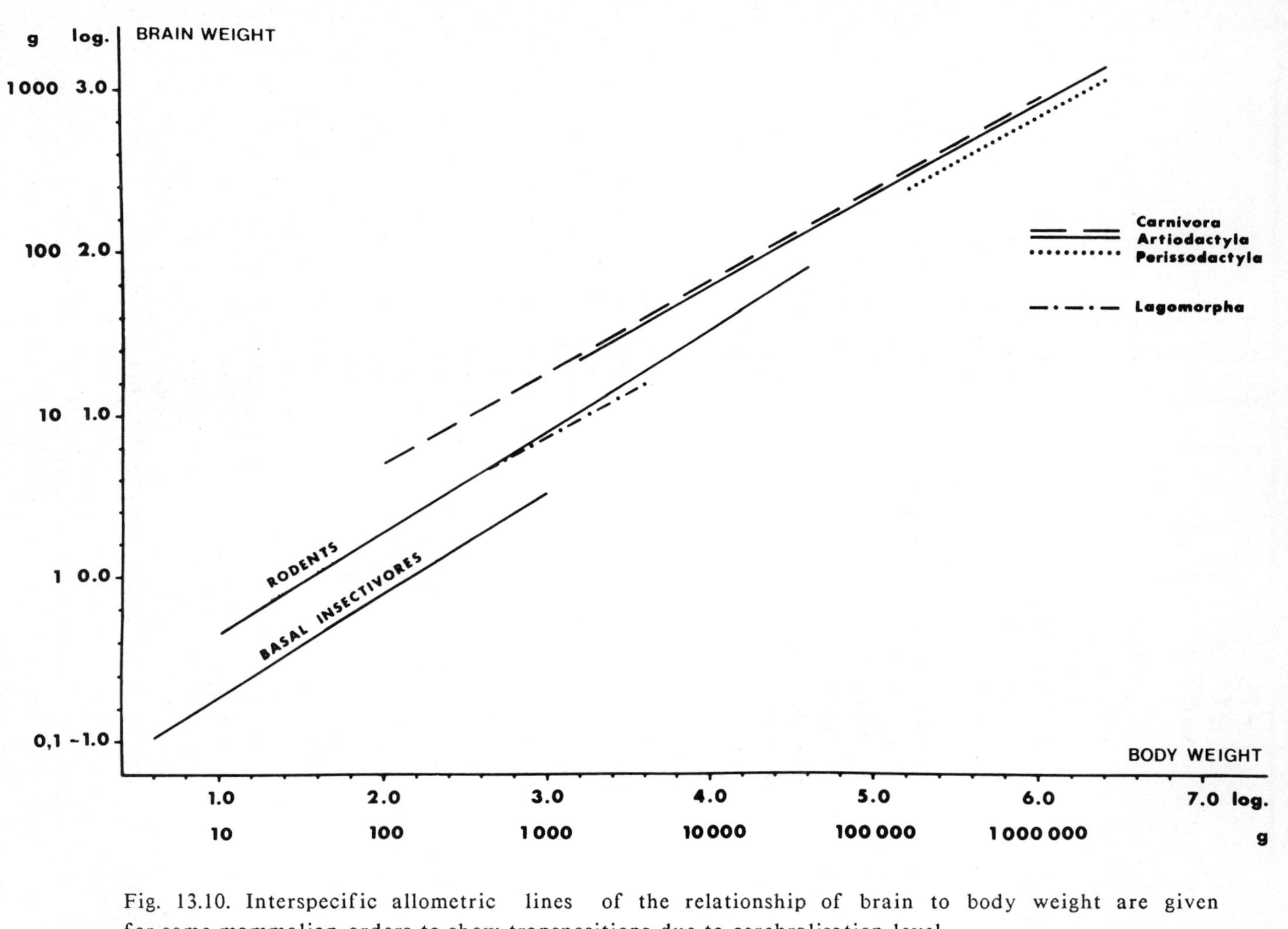

Fig. 13.10. Interspecific allometric lines of the relationship of brain to body weight are given for some mammalian orders to show transpositions due to cerebralisation level.

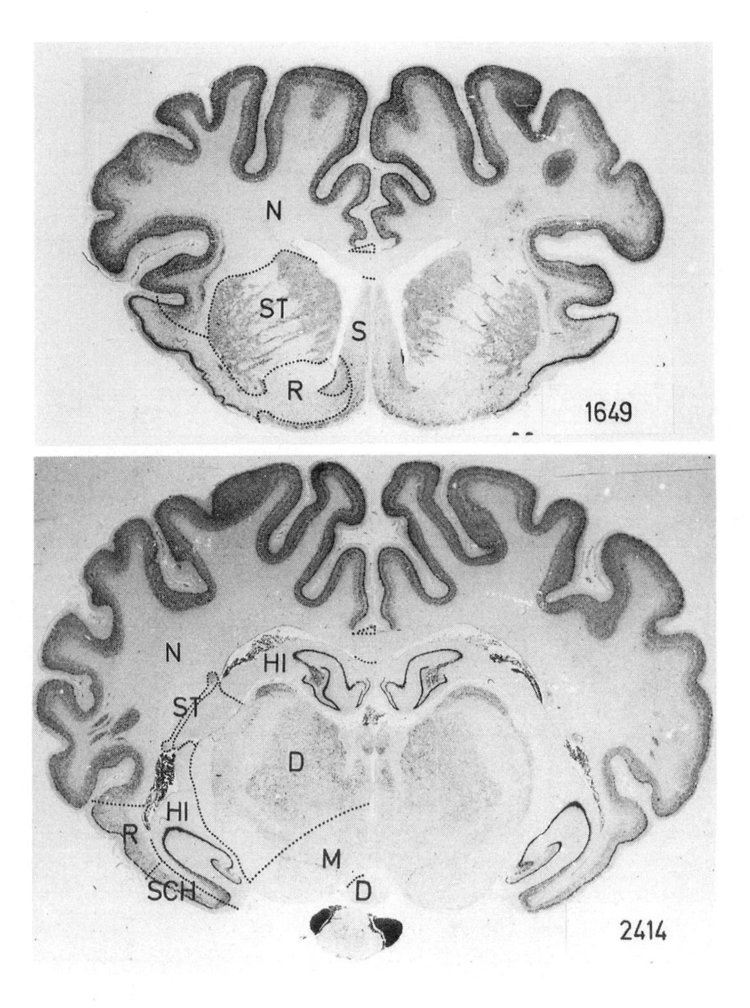

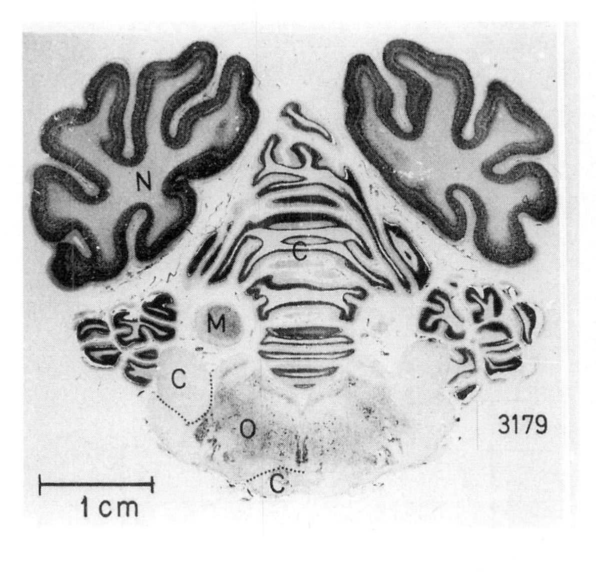

Fig. 13.11. Photographs of sections through the brain of a European wild boar at different regions and demonstration of brain region delimitation. C - cerebellum; D - diencephalon; HI - hippocampus formation; M - mesencephalon; N - neocortex; O - medulla oblongata; R - olfactory allocortex; S - septal region; ST - corpus striatum; SCH - schizo - cortex

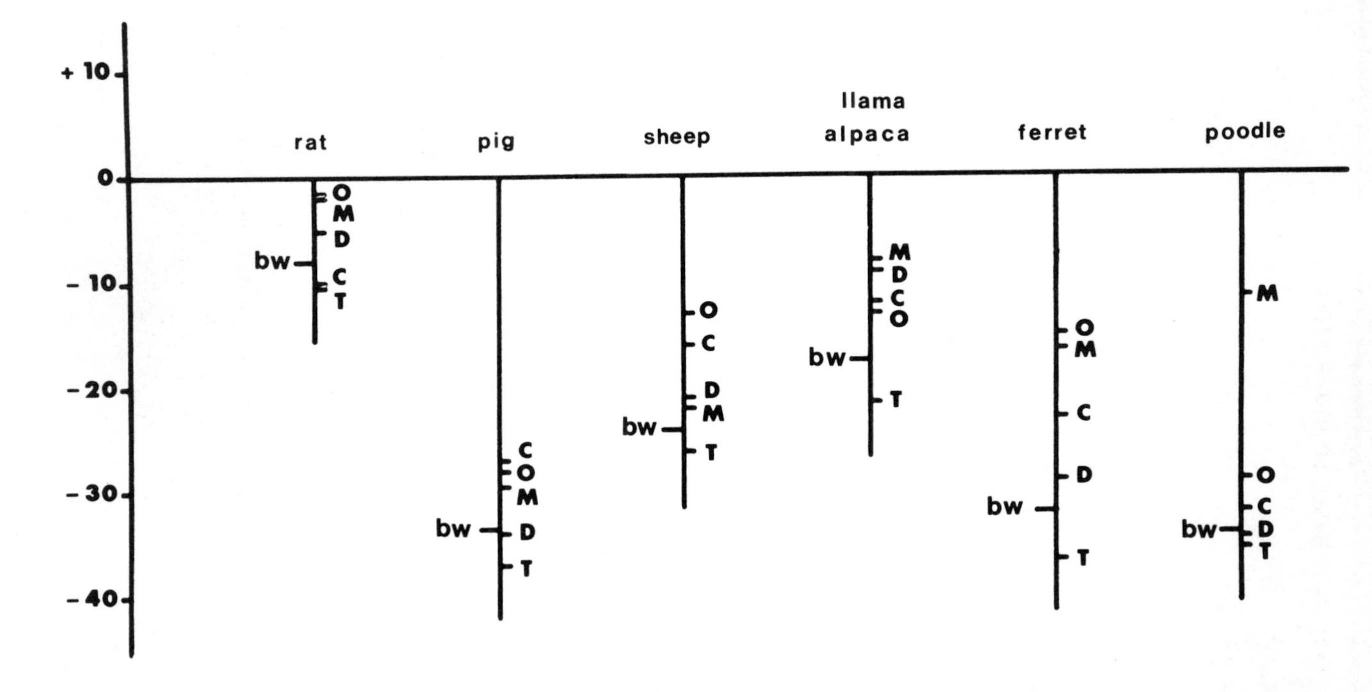

Fig. 13.12. Scaling of size reduction values of total brain size (bw) and volumes of fundamental brain parts due to domestication for several species. C - cerebellum, D - diencephalon, M - mesencephalon, O - medulla oblongata, T - telencephalon (from Kruska 1980).

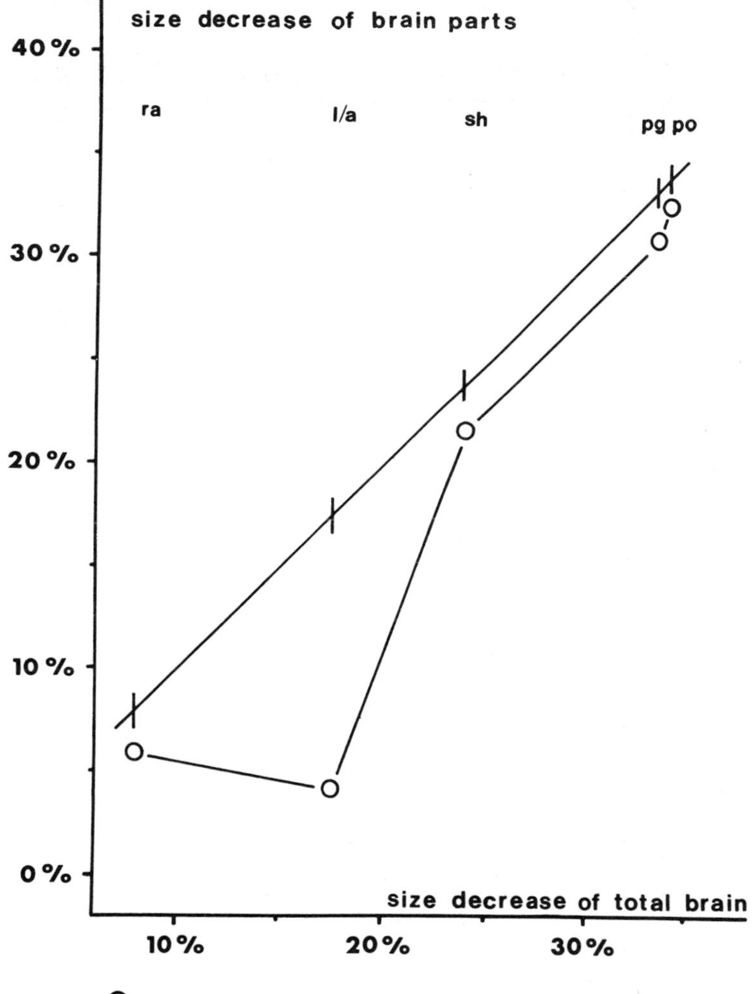

Fig. 13.13. Relation between size decrease of olfactory structures and total brain due to domestication in several species. By assuming that brain part sizes have decreased to the same extent as total brain volume, we can draw the uninterrupted line. This line serves as an orientation line. ra - laboratory rat; l/a - llama and alpaca; sh - sheep; pg - pig; po - poodle (from Kruska 1980).

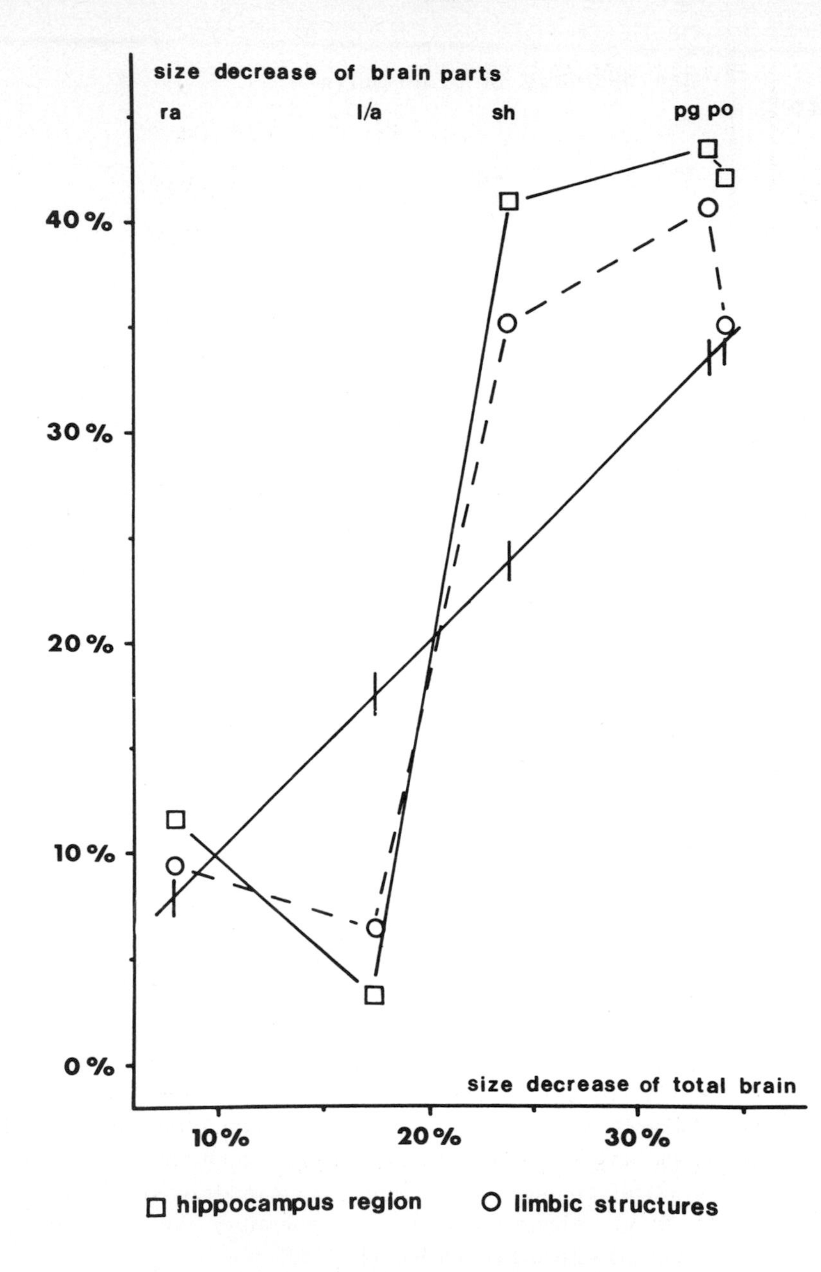

Fig. 13.14. Relation between size decrease of limbic structures and total brain due to domestication in several species. Orientation line and abbreviations as in fig. 13.13 (from Kruska 1980).

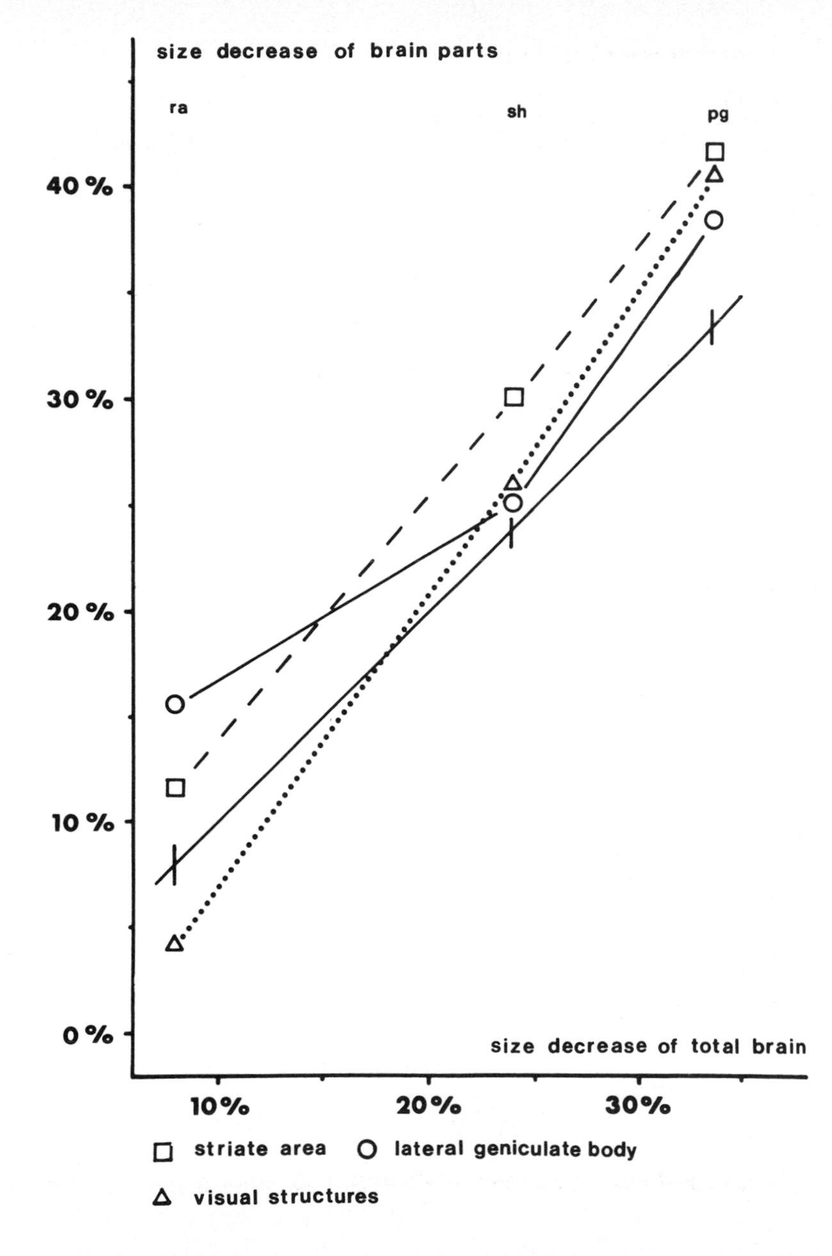

Fig. 13.15. Relation between size decrease of visual structures and total brain due to domestication in several species. Orientation line and abbreviations as in fig. 13.13. (from Kruska 1980)

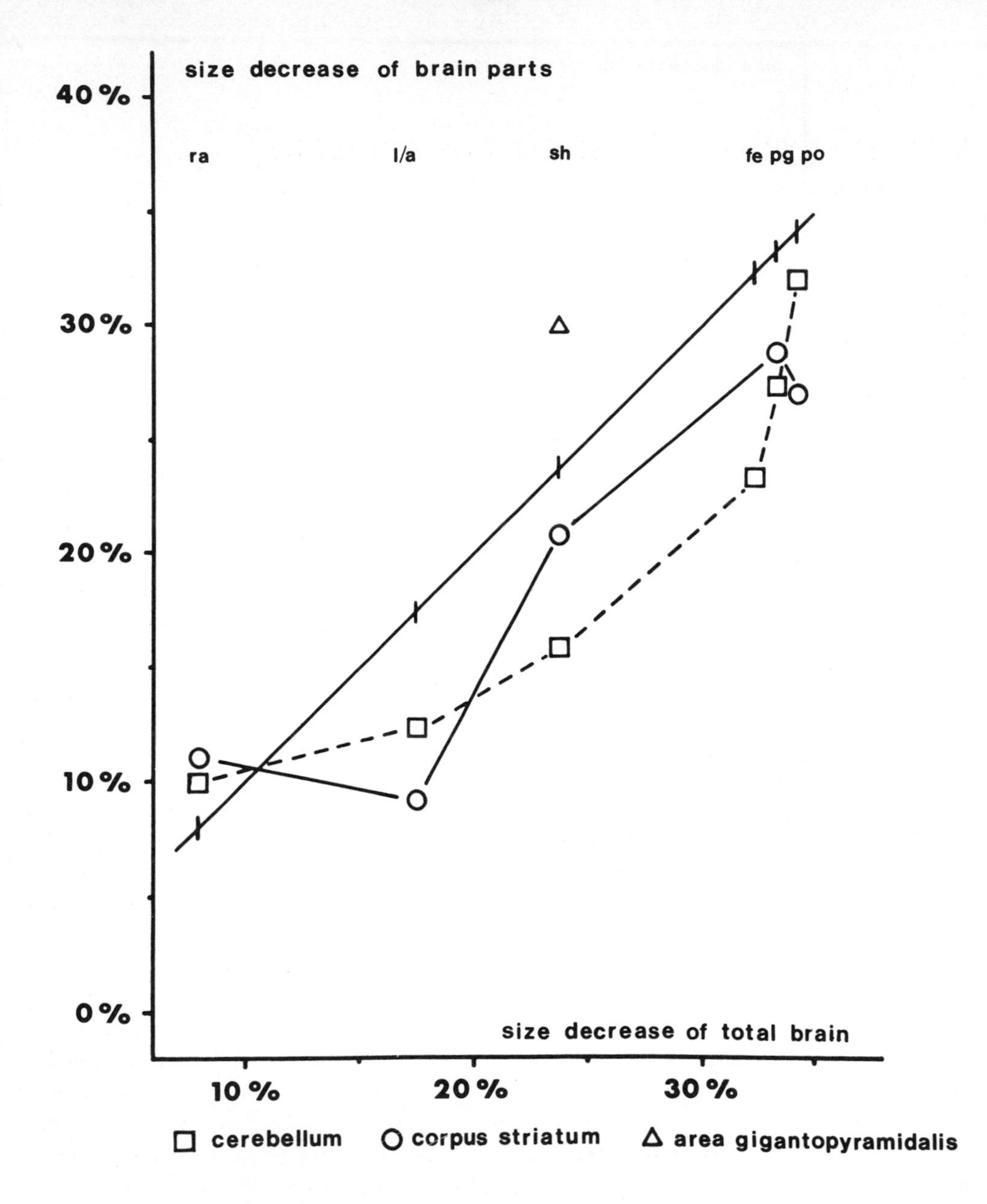

Fig. 13.16. Relation between size decrease of motoric structures and total brain due to domestication in several species. Orientation line and abbreviations as in fig. 13.13 (from Kruska 1980).

VERTEBRATE-INVERTEBRATE COMPARISONS

M. E. Bitterman
Bekesy Laboratory of Neurobiology
University of Hawaii
Honolulu, Hawaii 96822

Given the wider province of this symposium, I should say at once that my concern here is only with (non-human) animals, and not with intelligence in general but only with learning, which students of animal intelligence have taken to be its fundamental component and to the analysis of which their experiments have been directed almost exclusively since the turn of the century. As to whether anything like creative as distinct from purely reproductive intelligence need be attributed to their animals, there was substantial disagreement among early investigators; the influential Thorndike (1911), for one, thought not. Experienced subjects, especially primates, sometimes solved problems in ways that suggested "observation of essential features or relations" rather than "random actions and the selection of profitable acts" (Yerkes 1927, pp. 277-278), which alone did not require the assumption of more advanced capabilities or provide any satisfactory indication of what they might be. Clearly prerequisite to such solutions, however, was the long exercise of reproductive intelligence, and that, it soon became evident, was where the analysis should begin.

Another disclaimer is in order at the outset, which is to say that what we know about learning in animals comes primarily from intensive work with a small number of vertebrate species. Early investigators looked first to the "higher" animals for the evidence of mental continuity demanded by Darwin's theory, and later, as comparative experiments pointed to the generality of the laws of learning, there was little incentive to turn away from the familiar species that seemed so well suited to the discovery of those laws. Since the fashion of recent years among ethologists and the comparative psychologists influenced by them has been to denigrate what is called "general process theory," it should be emphasized that there do in fact seem to be important commonalities in the learning of vertebrates. We now have quite a long list of phenomena that may reasonably be assumed on the basis of the taxonomic diversity of the species in which they are found (if not the sheer number of species) to be general phenomena of vertebrate learning and understandable in terms of general principles (Bitterman and Woodard 1976). A variety of those phenomena will be described here. To say that there seem to be important commonalities in the learning of vertebrates is not, of course, to deny divergence, however difficult it may be to demonstrate (Bitterman 1975a; Terrace 1984), and there will be occasion

NATO ASI Series, Vol. G17
Intelligence and Evolutionary Biology
Edited by H. J. Jerison and I. Jerison

here also to examine some evidence of divergence.

About learning in invertebrates, relatively little is known. Although many species have been studied, the work has on the whole been rather primitive by vertebrate standards, and only with few species has it gone much beyond the question of whether they are capable of learning at all (Corning et al. 1973 1975; Sahley 1984). The principal exceptions when I set out some years ago to try to remedy the deficiency were octopuses and honeybees, whose performance in various learning situations had been reported to be so like that of vertebrates as to suggest the operation of common principles (Menzel and Erber 1978; Sanders 1975; Sutherland and Mackintosh 1971; Wells 1973), but there was not much useful information even about those animals. The octopus experiments, often quite sophisticated in purpose and design, could not be taken seriously because of unsuitable laboratory conditions and poor behavioral technique (Bitterman 1975b). The work with honeybees, mostly by zoologists interested in foraging who had come upon questions about learning with which they were not educated to deal, was highly idiosyncratic, clouded by looseness of conception and inadequate control of important variables. It served nevertheless to advertise the possibilities of the animal, setting the stage for the more conventional work that followed.

Now we have a substantial number of analytical experiments with honeybees that are patterned directly after work with vertebrates, the results of which -- contrary to my own expectations (Bitterman 1975a) -- do indeed closely resemble those for vertebrates. As interesting perhaps as the wide range of similarities in the performance of honeybees and vertebrates, whose common ancestor lived more than half-a-billion years ago and probably did not have much brain at all, is the total absence of novelty: Not a single phenomenon has appeared in the work with honeybees that is unknown to students of vertebrate learning. My purpose here is first to try to give some picture of the nature and scope of the new results, and then to say what I think they mean.

Appetitive Conditioning in Free-Flying Honeybees

Extinction, frustration, contrast.

Much of our work on learning in honeybees has been done with one or another variation of a training technique developed many years ago by von Frisch (1914) for the study of sensory capacities. Individual foragers fly back and forth between the hive and the shelf of an open laboratory window, where they feed to repletion from a drop of sucrose presented on a distinctive target -- a disposable petri dish which can be labeled with a disk of color, or an odor, or both. An association between the target and the sucrose develops in the course of these visits, and in one of our earliest experiments (Couvillon and Bitterman 1980) we set out to measure it in terms of what in the vertebrate literature is called "resistance to extinction." After 1, 2, 3, 6, or 12 visits to sucrose, an animal returning to the target from the hive found, not sucrose, but water (which was unacceptable and which could be distinguished from the sucrose only by taste), and its persistence in responding to the target during a 10-min test period was measured. When the animal made contact with the water, it would leave the target immediately, land again after a brief interval, leave again, return again, and so forth, the intervals between successive contacts with the target increasing progressively until the animal stayed away altogether. The pattern is apparent in the extinction curves of the five groups, which are plotted in Figure 14.1 in terms of mean cumulative frequency of response to the target over successive 30-sec intervals. The curves show expected increases in resistance to extinction (persistence) as the number of training visits increased from 1-6, followed at 12 visits by an unexpected -- but quite familiar -- decline.

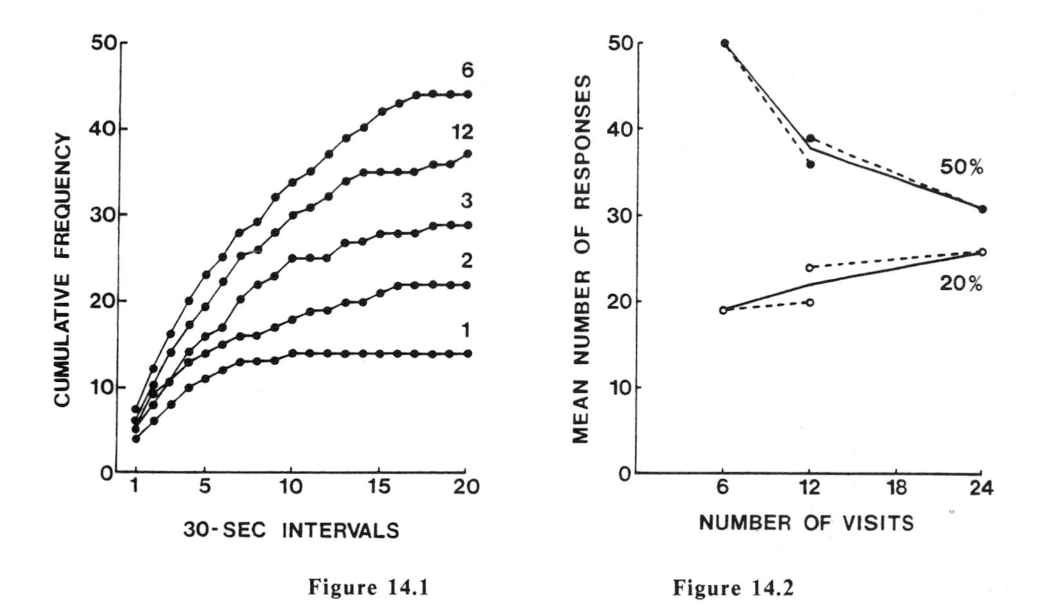

Figure 14.1 Figure 14.2

What we seemed to have come upon here was the paradoxical "overlearning-extinction effect" known from work with rats (North and Stimmel 1960). The effect was replicated immediately with new 6-and 12-visit groups, and later in several other experiments (Couvillon and Bitterman 1984) in which it was found to depend, as it does in rats (Ison and Cook 1964), on magnitude of reinforcement. The interaction with magnitude is shown in Figure 14.2, which is plotted in terms of the mean number of responses in extinction by eight groups of honeybees after 6, 12, or 24 training visits to 20% or 50% sucrose solution. Resistance to extinction declined as the number of training visits increased for the 50% but not for the 20% animals.

Two alternative interpretations of these results suggested themselves. The most intriguing was that we had found in honeybees something akin to what was known as "frustration" in rats (Amsel 1962); that is, the performance of our animals in extinction might have been disrupted by an emotional reaction to unrealized expectation of food that became very strong in the course of prolonged training with 50% sucrose. A more mundane possibility was that what we were seeing was only the dependence of resistance to extinction on hunger, another familiar vertebrate phenomenon (Perin 1942); since sucrose is taken up continuously from the social stomach, the resistance to extinction of the 50% animals may decline in the course of prolonged training simply because they become less hungry. In appetitive experiments with vertebrates, the hunger of various groups of animals that are differently rewarded in training is easily equated, but for free-flying honeybees, which come to the laboratory of their own accord, special procedures are required.

The hunger interpretation is ruled out by the results of an experiment (Shinoda and Bitterman 1987) in which three groups of bees were extinguished on a distinctive target, Group 6 after 6 visits to that target, Group 18 after 18 visits to that target, and Group 6-12 after 6 visits to that target which were interspersed among 12 visits to a distinctly different target. As shown in Figure 14.3, which is plotted in terms of mean cumulative frequency of response over successive 30-sec intervals, we found a target-specific overlearning-extinction effect; resistance to extinction was less in Group 18 than in Group 6-12 despite the fact that the total number of feedings was the same for both. It is tempting to conclude from such results that the effect of nonreinforcement on performance is the same in honeybees as in rats, where it seems not only to weaken the expectation of reinforcement but also, at least when expectation is strong, to generate some sort of competing reaction.

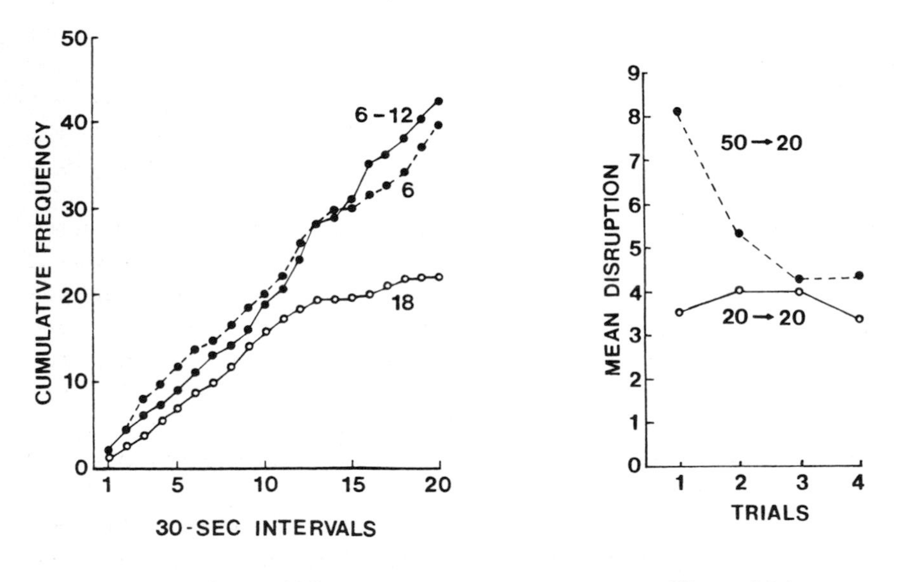

Figure 14.3

Figure 14.4

More direct evidence that the performance of honeybees may be disrupted by unrealized expectation of reinforcement is provided by an experiment on "successive negative incentive contrast." It has long been known from work with rats and monkeys (Tolman 1932) that the acceptability of a less-preferred food is reduced by anticipation of a preferred food, and the same is true of honeybees (Couvillon and Bitterman 1984). Free-flying foragers were permitted a series of visits to 50% sucrose on one target (A) alternating with visits to 20% sucrose on a distinctively different target (B), after which there were four visits to 20% sucrose, either on target A for one group (the 50-20 group) or on B for a second (20-20) group. On all visits, the acceptability of the sucrose was measured (inversely) in terms of the frequency with which the animals broke off contact with it (disruption) during feeding to repletion. The rate of disruption is normally somewhat higher for the 20% than for the 50% concentration. As Figure 14.4 shows, the 20% sucrose was even less acceptable for a time (mean disruption was greater) when it was presented on the target that before had contained 50% sucrose (50-20). Since the two groups had exactly the same experience prior to the tests, their adaptation and nutritive levels were the same, which seems to leave us only with unrealized expectation as a key to the behavior of the 50-20 animals.

Compound conditioning.

The target-specific overlearning-extinction results shown in Figure 14.3 indicate that extinction tests may be used to inquire into what has been learned about the specific properties of a target. In the simplest case, some feature of the target is changed, and its associative value is estimated from the effect on response. The overlearning-extinction results indicate also, and experiment confirms (Couvillon and Bitterman 1980), that such tests become less sensitive as the number of training visits becomes large (reducing response even to unaltered targets), for which reason the number of training visits is kept small. Extinction tests have been used in a series of experiments on compound conditioning which again have produced a variety of familiar vertebrate phenomena, these reflecting the ways in which the components of a compound stimulus may interact with each other.

To a considerable extent, as Pavlov (1927) found in his work with dogs, the interaction is summative (Couvillon and Bitterman 1980): Honeybees respond more to a compound of two components (a color and an odor) that have been separately reinforced (presented with sucrose) than to either component alone ("summation of excitation"); respond less to a compound of two components that have been separately nonreinforced (presented with water) than to either alone ("summation of inhibition"); and less to a compound of one separately-reinforced and one separately-nonreinforced component than to the separately reinforced component alone ("conditioned inhibition"). Often, however, the interaction is more complex.

For one thing, as Pavlov discovered, "compound-unique" properties may be generated. Shown in Figure 14.5 are the results of an experiment in which the training target was labeled with an orange disk together with the scent of jasmine on some visits and a yellow disk together with the scent of violet on others, after which the subjects were extinguished with a set of four targets presented simultaneously - two labeled with the training compounds and two with new compounds produced by recombination of the training components (orange-violet and yellow-jasmine). Honeybees responded more to the old compounds than to the new ones (Couvillon and Bitterman 1982). In as yet unpublished experiments by a different method developed in previous work on the determinants of choice (Couvillon and Bitterman 1985a 1986 1987), we have been able to demonstrate "conditional discrimination" in honeybees, a phenomenon known also, for example, in goldfish (Bitterman 1984a). On each visit, two targets were presented, one containing sucrose and the other containing water, with the animal free at once to correct an erroneous choice. In one problem, green rather than blue was reinforced when both targets were scented with peppermint, and the opposite when both were scented with geraniol; in another, peppermint rather than geraniol was reinforced when both targets were green, and the opposite when both were blue; and so forth. Figure 14.6, which is plotted in terms of the proportion of animals choosing correctly on each visit, shows that these problems are difficult, but that the animals are capable of significantly better-than-chance performance. Although conditional choice is conveniently described in quasi-conceptual terms, it can be explained simply in terms of the discrimination of compound-unique properties.

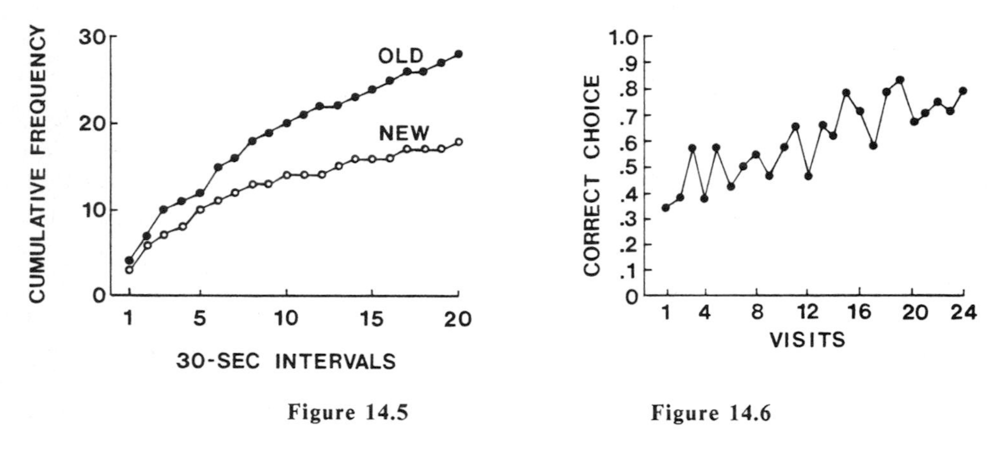

Figure 14.5 **Figure 14.6**

In his work with dogs, Pavlov also discovered a phenomenon called "overshadowing," which is that the presence of one stimulus may interfere with the conditioning of another. The effect has since been found in a variety of vertebrates, including carp (Tennant and Bitterman 1975), and in honeybees. In one such experiment (Couvillon and Bitterman 1982), a group of bees was trained with an orange-jasmine target (OJ+), while for another group the training target was labeled with an orange disk alone (O-), after which both groups were extinguished with two targets presented simultaneously -- one labeled with an orange disk (O-) and the other unlabeled (G-) -- to measure what had been learned about the color. As Figure 14.7 shows, the second group clearly preferred the orange target, but the first group did not -- in Pavlov's terms, color was overshadowed by odor. Pavlov suggested that overshadowing could be understood in terms of stimulus-intensity ("stronger" stimuli overshadowing "weaker" ones), but the matter is not quite so simple.

In honeybees, as well as in vertebrates such as rats and rabbits (Wagner et al. 1968), what may be thought of as information-value seems to play an important role. Consider the following experiment with honeybees (Couvillon et al. 1983) that was patterned directly after the work with vertebrates: One group of animals (the "true discrimination" or TD group) was trained to discriminate between two orange targets, one scented with jasmine and reinforced, the other scented with violet and nonreinforced. A second group (the "pseudodiscrimination" or PD group) was trained with the same targets equally often reinforced and nonreinforced, that is, jasmine did not reliably predict reinforcement. When both groups were tested with an unscented orange target, there was, as Figure 14.8 shows, less response in the TD group despite the fact that orange had been equally-often reinforced for both. Color was overshadowed by odor more when the odor predicted food than when it did not.

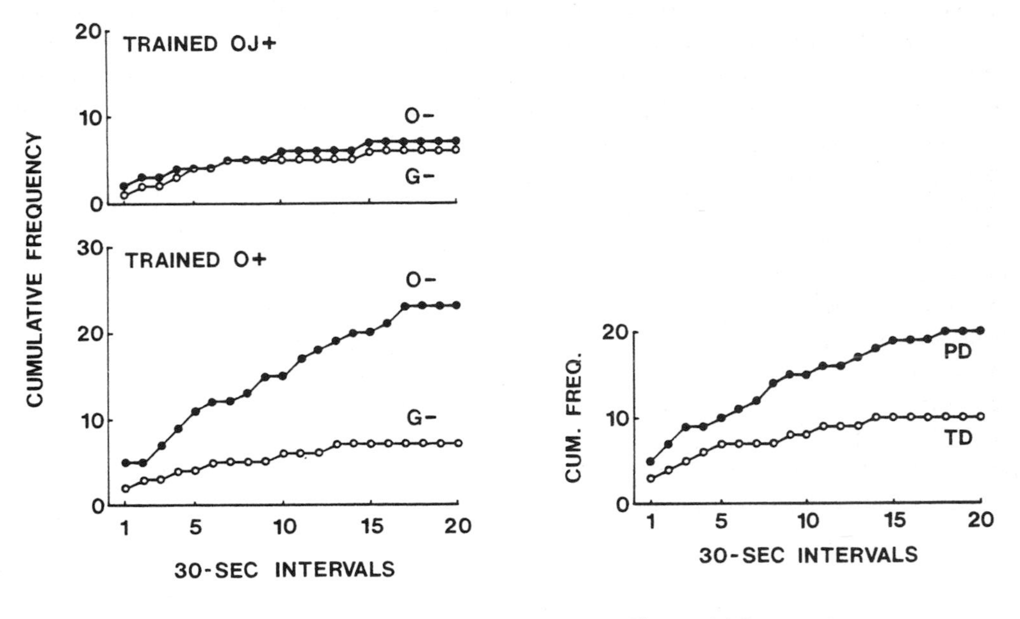

Figure 14.7

Figure 14.8

In recent years, an effect precisely the opposite of overshadowing has been discovered in rats: the presence of one component may in certain circumstances enhance or "potentiate" the conditioning of another (Durlach and Rescorla 1980). This phenomenon has been also been found in an experiment with honeybees (Couvillon and Bitterman 1982). One group of foragers was trained with an orange-jasmine target (OJ+), while for a second group the training target was labeled only with the scent of jasmine (J+), after which both groups were given a choice in extinction between a target labeled with jasmine (J-) and an unlabeled target (G-). As Figure 14.9 shows, both groups responded more to the the jasmine, but the preference was significantly greater in the group trained with orange-jasmine -- odor was potentiated by color. An interesting feature of these results is that the preference for J after training with J alone was substantially less that the preference for O after training with O alone (compare Figure 14.7); that is, although J overshadows O in training with the compound, it is separately less-conditionable than O. There is evidence also from recent work with vertebrates (Bouton et al. 1986) that -- contrary to Pavlov's intuitively-reasonable impression -- weakly conditionable stimuli tend to be potentiated and strongly-conditionable stimuli to be overshadowed.

Compound-uniqueness and potentiation have been explained in terms of "within-compound association," a phenomenon recently discovered in rats (Rescorla and Cunningham 1978) that appears also in goldfish (Bitterman 1984a). In the first stage of an experiment with honeybees (Couvillon and Bitterman, 1982), the animals were fed sometimes on an orange-jasmine target and sometimes on a target labeled with a yellow disk and the scent of lemon. In the second stage, the animals were trained to discriminate between two targets labeled with the odors alone (say, the jasmine containing sucrose and the lemon-scent containing water), which established a preference for jasmine. In the third stage, there was an extinction test with a pair of unscented orange and yellow targets in which the animals showed a clear preference for orange. If lemon-scent rather than jasmine contained sucrose in the second stage, the preference in the third stage was for yellow.

The results are plotted in the lower portion of Figure 14.10, where the color paired in the first stage with the positive odor of the second stage is designated "with +" and the alternative "with -." If color was differentially reinforced in the second stage, there was a corresponding preference for odor in the third stage. Plotted in the upper portion of Figure 14.10 are the results of a like experiment, in the first stage of which response to the compounds was nonreinforced, the animals being fed instead on an unlabeled target. The results show that in honeybees -- as in vertebrates (Rescorla and Durlach 1981) -- the components may be associated with each other even while the compound is nonreinforced.

A "blocking" experiment with honeybees (Couvillon et al. 1983) that was patterned directly after work with rats (Kamin 1969) is especially interesting because it did produce what seemed, at least for a time, to be a unique result. The

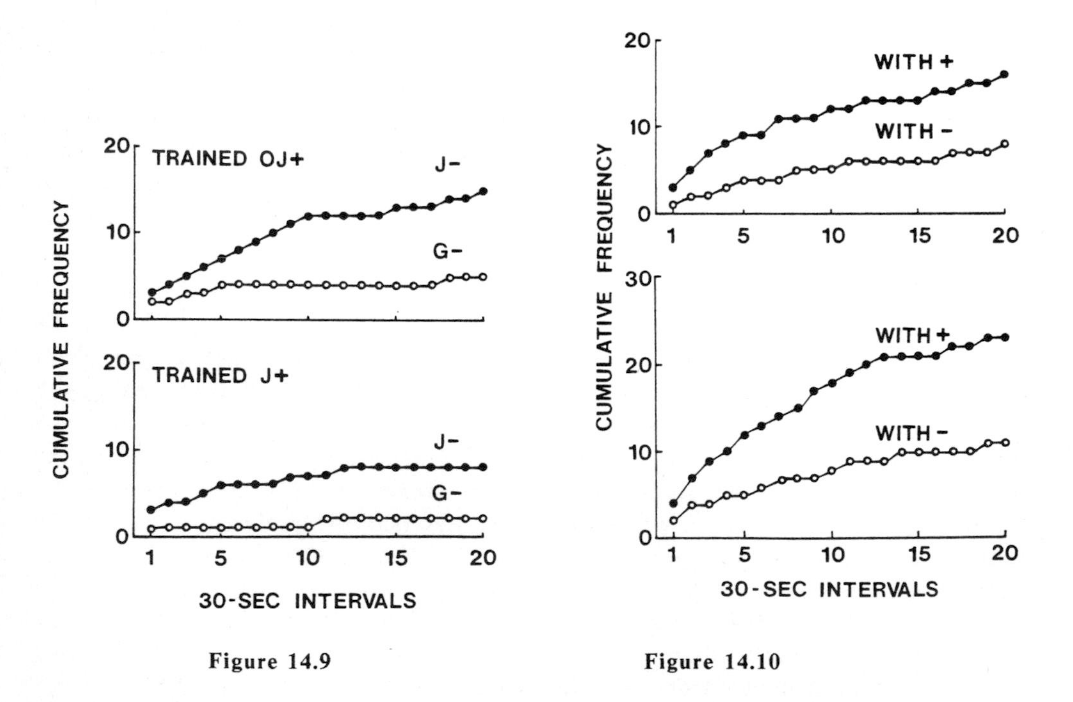

Figure 14.9

Figure 14.10

blocking group was fed first on a target labeled with only one component of an orange-jasmine compound, and then on a target labeled with the compound, after which it was tested with the other component. For the comparison group, the first two stages of training were given in reverse order (training with the compound before training with one of the components). The common result obtained in work with vertebrates -- and in a recent experiment with snails as well (Sahley et al. 1984) -- is that prior training with one component reduces (blocks) learning about the other in subsequent compound training (the comparison group responds more in the test than the blocking group), but, as Figure 14.11 shows, what we found instead in honeybees was "negative blocking" -- more response in the blocking group. Soon afterward, however, negative blocking appeared in rats (Dickinson et al. 1983), and then, in an experiment with a different technique considered below, we did find evidence of blocking in honeybees. Depending on circumstances which are as yet incompletely defined, rats may show overshadowing or potentiation, blocking or negative blocking, and the same is true of honeybees. The mechanisms of compound conditioning are perhaps no less complicated in honeybees than in rats.

Appetitive Conditioning in Temporarily-Confined Honeybees

In the hope of achieving better control of important training variables than is possible with free-flying honeybees, we have used situations to which foragers come as usual of their own accord, but in which they are confined temporarily for an automated series of training trials. In his doctoral work, Sigurdson (1981a 1981b) developed several such situations, one of which is diagrammed in Figure 14.12. The two doors in the tunnel leading to the outside together with the neighboring photodetectors (PS) monitor and control the comings and goings of the subject. The conditioned stimuli are colored lights which transilluminate a Plexiglas response-tube and the Plexiglas disk surrounding it; sucrose solution can be pumped in measured amounts into the bottom of the response-tube. Dipping into the tube is recorded with a photodetector in the manner of Grossmann (1973), and approach to the tube

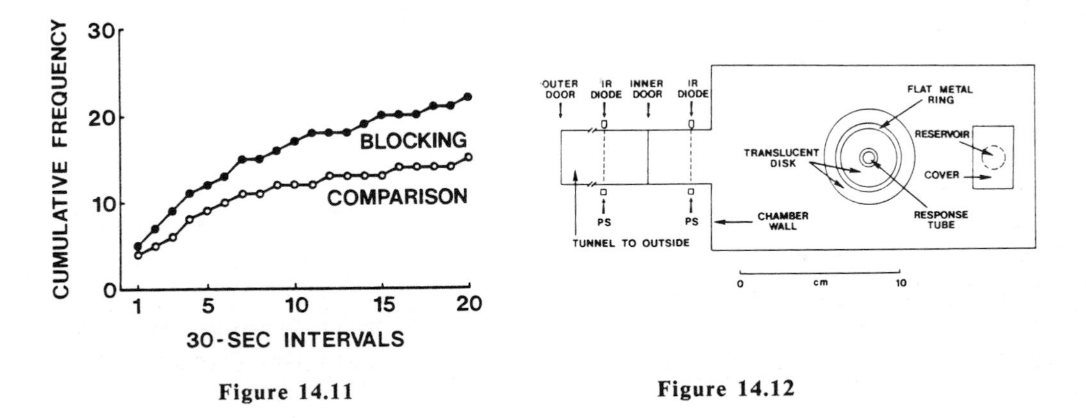

Figure 14.11 **Figure 14.12**

also is recorded (with an encircling metal ring and a capacitance device). At the end of each programmed series of trials, the solenoid-operated cover of a reservoir of sucrose solution is lifted, the animal completes the filling of its crop, and only then is permitted to leave for the hive, from which it returns for another series of trials a few minutes later. An advantage of this procedure is that there is complete control of the location of the animal during the presentation of stimuli. Furthermore, there can be repeated trials on each visit, some reinforced and others not, with the intertrial interval determined entirely by the experimenter; the unreplete animal is not free to forage elsewhere during a series of trials.

One of a variety of vertebrate learning phenomena demonstrated in this situation is the "partial reinforcement extinction effect," which has long been known in rats (Skinner 1938) and many other vertebrates. Figure 14.13 shows the performance of two groups of honeybees trained in 15 trials per visit. Each trial began after a variable intertrial interval in darkness with illumination of the response-tube, and on the first six (pretraining) visits for both groups each dipping response was rewarded with 2 microliters of sucrose solution. On the next 14 (training) visits, the treatment of the consistently reinforced (CRF) group continued as before, but the animals of the partially reinforced (PRF) group were rewarded only on half the trials in quasi-random sequence; on a nonreinforced trial the dipping hole remained illuminated for 8 sec whether or not it was entered. Then there were 15 extinction visits on each of which the animals of both groups had 15 nonreinforced trials. As shown in Figure 14.13, which is plotted in terms of mean ln (natural log) latency of response, there was rapid acquisition in both groups, then more rapid responding in the consistently reinforced group as compared with the partially reinforced group, and, finally, more resistance to extinction in the partially reinforced group. The differences are small but highly reliable, and the results might easily be mistaken for those obtained in a host of comparable experiments with pigeons, rats, and goldfish.

Pretraining a honeybee to dip for food into an illuminated tube is much like training rats or pigeons to insert their heads into the illuminated aperture of a feeder, and we can build on it in the same way, as, for example, in "autoshaping" experiments (Brown and Jenkins 1968). If the illumination of the dipping tube with white light, which signals the availability of food, is preceded by an 8-sec period of blue light, honeybees quickly come to dip in response to the blue light despite the fact that food is neither given for response to blue nor contingent upon response to blue. If among such trials we intersperse 8-sec presentations of green light which are not followed by white light and food, the animals continue to respond to blue but not to green, which shows that the response to blue is not simply a generalization effect.

Now we ask a question like that asked about classical appetitive conditioning in vertebrates, which is whether the honeybees respond to blue because blue is associated with food or because *response* to blue is followed by food ("adventitiously rewarded"). The answer (as in vertebrates) is provided by an "omission" experiment (Sheffield 1965). Naive animals are trained as before except that the blue light is followed by white light and food only when they do *not* dip in response to the blue; if they do, the blue light is simply turned off (as the green always is) 8 sec after onset. Mere approach to the dipping tube, which also is monitored, has no effect on the presentation of food. The results plotted in Figure 14.14 show powerful control of dipping by the omission contingency. The animals learn not to dip in response to the blue light (S+) despite the fact that, at asymptote -- precisely because they usually do not dip -- the blue light is almost always followed by white light and food. What the animals do on S+ trials is to approach the tube, wait near it for the light to turn white, and only then dip for food. On S-trials (with green light), there is a good deal of general activity during which the approach-detector may be triggered from time to time, but there is no organized approach response. Experiments with vertebrates show that some conditioned appetitive responses, such as insertion of the head into the aperture of a feeder by rats, are strongly suppressed by omission training, while others, such as rearing by rats and key-pecking by pigeons, are not (Holland 1979; Woodard et al. 1974), and the same is true of honeybees. As will be shown later, conditioned extension of the proboscis in response to a signal for food is not suppressed by omission training.

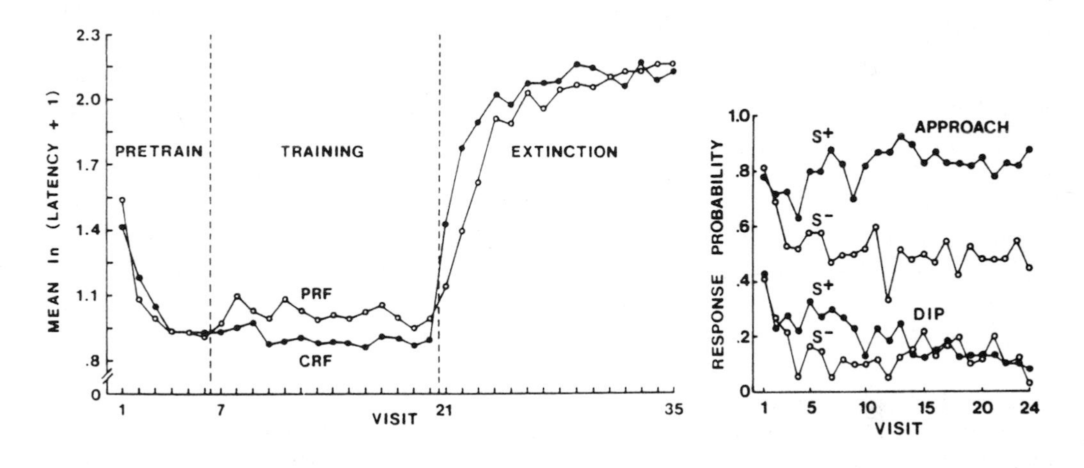

Figure 14.13 **Figure 14.14**

In honeybees as in the vertebrate species commonly used in learning experiments, high levels of performance can be generated by small and infrequent reinforcement (Skinner 1938). The extent to which the performance of honeybees can be controlled by a few minuscule drops of sucrose solution has been shown for the dipping response by Grossmann (1973). Here it is strikingly illustrated by some results obtained in another of Sigurdson's situations, this one designed to measure and to instrumentalize consummatory responding. The technique is to detect contacts of the proboscis with the aperture of a syringe needle through which sucrose is delivered (each contact having to be broken before another can be registered). Shown in Figure 14.15 is the cumulative record of a bee responding at the rate of about 100/min on an "FR-80 schedule" (2 microliters of sucrose solution for every 80 responses); the recording pen steps upward with each response, and the diagonal slashes indicate reinforcements. This animal was trained first with every response reinforced, and then the number of responses required for reinforcement was increased progressively in successive visits. The flat portions of the record indicate where the animal returned to the hive, as here it was free to do whenever it chose. Figure 14.15 also shows the record of a bee responding at the rate of about

50/min on an "FI-40 sec schedule." (Two microliters of sucrose solution could be earned no more often than once every 40 sec; additional responses during the 40-sec interval were not reinforced.) These records look very much like those obtained in comparable experiments not only with rats but with such divergent vertebrate species as goldfish (Holmes and Bitterman 1969) and painted turtles (Pert and Bitterman 1969), and also with octopuses (Henderson et al. 1975).

Diagrammed in Figure 14.16 is a situation designed by Sigurdson for the study of choice in temporarily-confined honeybees. It contains two vertically-oriented screens on which colored lights are projected from the rear, with a horizontal response-tube at the base of each. So that the animal will always be at the same point -- removed from the two stimuli and approximately equidistant from them -- when it makes a choice, a common vertebrate strategy is used, which is that the animal itself turns on the stimuli. To do so, it must activate the photodetector inside the inner door, which illuminates both screens until the animal makes a choice. If the positive stimulus is chosen, sucrose solution is delivered; if not, the stimuli are turned off without reinforcement and the animal earns the opportunity to correct itself by returning to the inner door.

In this situation, honeybees show a variety of familiar vertebrate phenomena, one of which is "progressive improvement in habit reversal" (Dufort et al. 1954). Figure 14.17 plots the number of errors made by a single forager in each of a series of reversals, with 10 trials per visit and 8 visits (80 trials) per reversal. On the first eight visits (reversal 0), choice of the left panel was reinforced; on visits 9-16 (reversal 1), the right panel; on visits 17-24 (reversal 2), the left again; and so forth. Reversal of choice was difficult at first, but it became progressively easier as the training continued.

Another phenomenon is "probability matching," a close correspondence between choice-ratio and reinforcement-ratio, which Sigurdson found in 70:30 color problems (problems in which one color was reinforced independently of its position on a random 70% of trials, and the other on the remaining trials). This phenomenon is known from work with such diverse vertebrates as African mouthbreeders (Behrend and Bitterman 1961), painted turtles (Kirk and Bitterman 1965), and pigeons (Bullock and Bitterman 1962). Both progressive improvement in habit reversal and probability matching have been found also in cockroaches (Longo 1964).

Appetitive Conditioning in Harnessed Bees

A simpler and more direct way to control the location of a bee during an experiment is to harness it in a small tube as shown in Figure 14.18. Under these circumstances, extension of the proboscis, reflexly elicited by sucrose to the antennae, is rapidly conditioned to a contiguous olfactory stimulus that is ineffective at the outset. Although known for some time (Kuwabara 1957), the phenomenon was not analyzed properly until recently (Bitterman et al. 1983). The first step in such an analysis is, of course, to ask if the new response to the odor does in fact depend on the pairing with sucrose, which is to say that it is necessary to control for effects of the conditioned and unconditioned stimuli (odor and sucrose) independently of their temporal relation. A common way to do so is to use an "explicitly un-

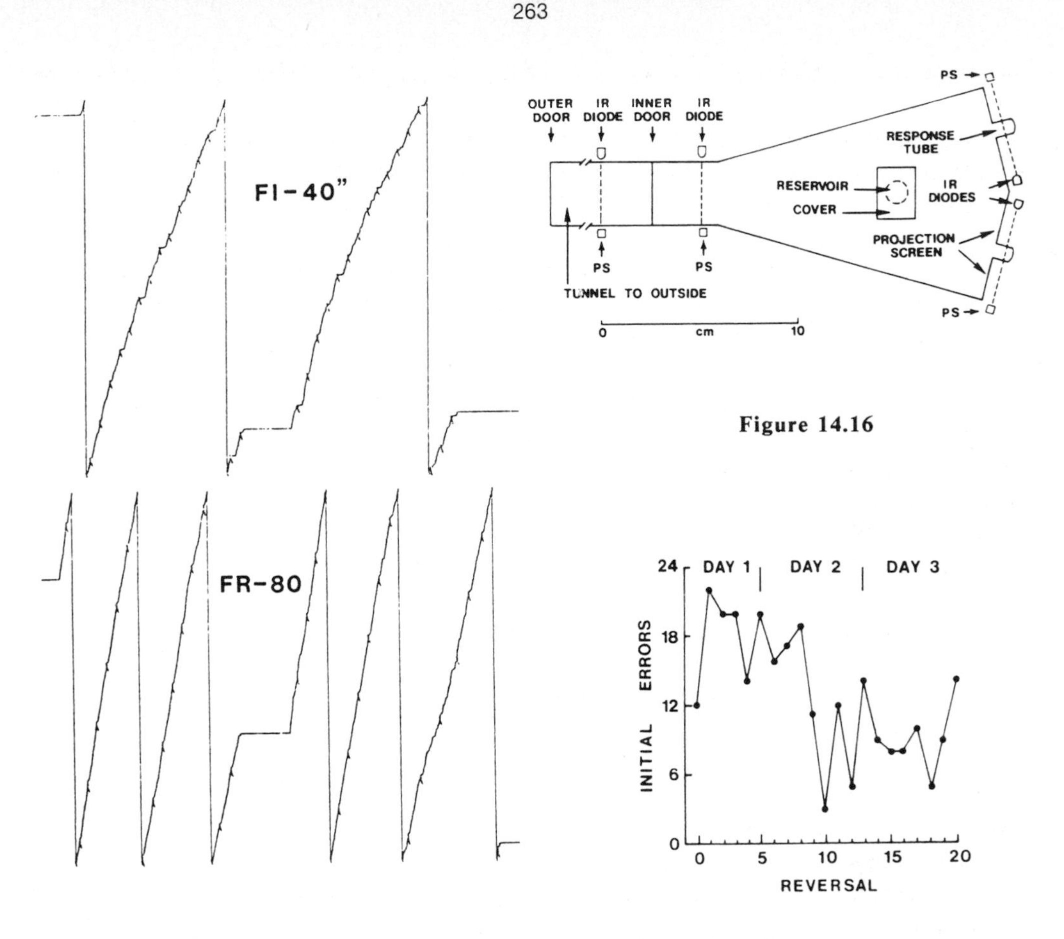

Figure 14.16

Figure 14.15

Figure 14.17

paired" group which has the same number of exposures to the two stimuli as a paired group, but in quasi-random sequence and never contiguous. The results of such an experiment are plotted in Figure 14.19 in terms of the proportion of animals in each group that responded with proboscis-extension on each presentation of the odor. Acquisition in the paired group (P) proved to be very rapid and, given the performance of the unpaired group (U), clearly dependent on pairing.

The role of adventitious response-reinforcer contiguity in proboscis-extension conditioning was examined by the omission technique. As in the dipping experiment already described, two different stimuli (here two odors) were used, one (S+) which was followed by sucrose *except* when the animal responded to it, and another (S-) which never was followed by sucrose. Figure 14.20 shows that response to S+ was acquired rapidly and persisted at a high level despite the fact that it prevented the delivery of sucrose. With the animals responding to S+ on three trials in four on the average, it was paired with sucrose only on one trial in four, but Pavlov found long ago that salivation in dogs could readily be conditioned with such low probabilities of reinforcement. From the virtual absence of response to S-, it can be concluded that response to S+ was in fact due to the occasional pairings with

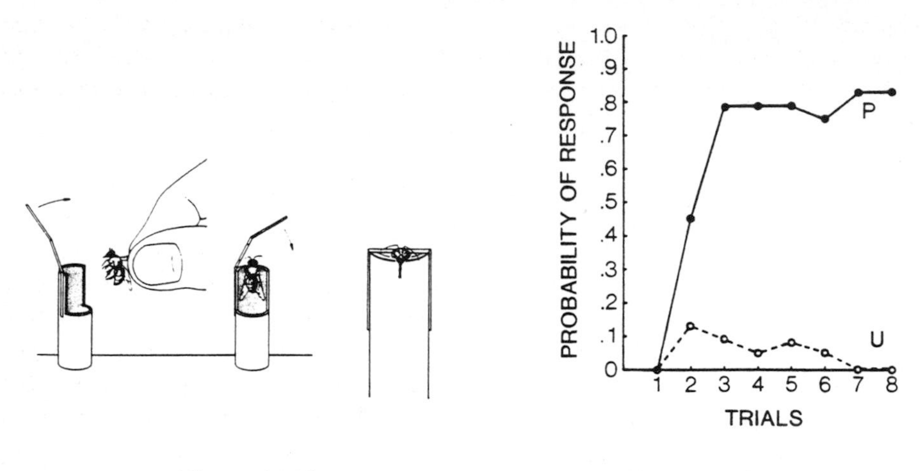

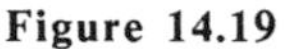

Figure 14.18 **Figure 14.19**

sucrose. These results closely resemble those of an experiment on key-pecking in pigeons (Woodard et al. 1974).

If reinforcement is terminated altogether, of course, the proboscis-extension response extinguishes and, as is true of vertebrates, more rapidly in "massed" than in "spaced" training (that is, more rapidly when the intertrial interval is short than when it is long). A response extinguished in massed trials typically reappears after a period of rest -- Pavlov's "spontaneous recovery" -- which Figure 14.21 shows in honeybees. The data plotted are for a series of 10 extinction trials given at intervals of 1 min except that the interval between trials 5 and 6 was 35 min.

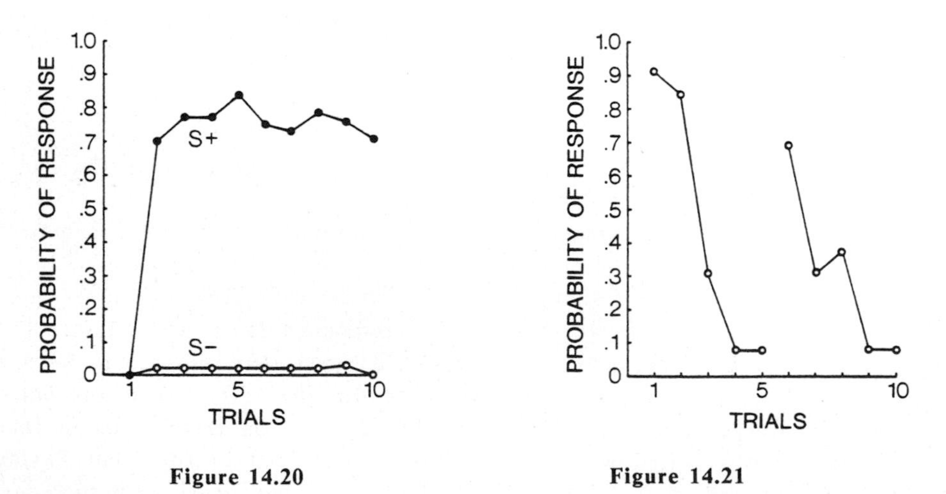

Figure 14.20 **Figure 14.21**

It may be useful to describe one other well-known phenomenon of vertebrate learning which was demonstrated in this work on proboscis-extension conditioning in honeybees, and which has been found also in recent work with snails (Sahley et al. 1984). Three groups of honeybees were trained in a two-stage experiment, the design of which was taken directly from contemporary work on appetitive conditioning in rats (Rescorla 1977). In the first stage, there were four pairings of an odor (the "first-order" conditioned stimulus) with sucrose for the animals of Groups PP and PU, which conditioned rapidly; for Group UP, there were four unpaired presentations of the stimuli. In the second stage, there were eight pairings of a different odor (the "second-order" conditioned stimulus) with the first odor for Groups PP and UP, while for Group PU the two odors were unpaired. Figure 14.22 shows the responses of Groups PP and PU to the two odors in the second stage of the experiment. Both continued to respond to the first-order stimulus, which for both had been followed by sucrose in the first stage, but response to the second-order stimulus developed only in Group PP, for which it was followed by the first-order stimulus -- Pavlov's "second-order conditioning." Group UP responded to neither of the stimuli; since the first-order stimulus was not itself conditioned in the first stage, it could not support conditioning of the second-order stimulus in the second stage. As noted before (Menzel and Bitterman 1983), the orderly and meaningful results obtained in experiments with harnessed bees, and their striking correspondence to the vertebrate results where comparison is possible, may be especially helpful in allaying frequently-voiced concerns about the appropriateness of studying animal intelligence under "unnatural" laboratory conditions.

Aversive Conditioning in Confined and Free-Flying Honeybees

Much of what is known about vertebrate learning is based on training with aversive reinforcement, both classical and instrumental -- experiments on punishment, escape, and avoidance -- and a strong case can be made for the essential similarity of the associative processes in appetitive and aversive situations, with different rules only for translating the associations into behavior (Mackintosh 1983). Figure 14.23 diagrams the analogue of a common vertebrate apparatus used by Abramson (1986) in his recent doctoral research on aversive conditioning to demonstrate an array of vertebrate phenomena in honeybees. It is a small Plexiglas shuttlebox in which the movement of the animal from one compartment to the other (under the hurdle) is monitored by photocells. The aversive stimulus is air from a bottle containing filter paper saturated with formic acid, which is substituted by a solenoid valve for the normal flow of pure moist air in from under the screen floor and out through the roof of the chamber. The basal level of activity shown by a naive honeybee in this apparatus can be modified rapidly (increased or decreased) by response-contingent presentations of the aversive stimulus, with a yoked animal trained concurrently in an identical apparatus to control for the effects of the aversive stimulus *per se*, a technique commonly employed in vertebrate experiments. The control animal is stimulated along with the "master" animal, independently of its own behavior.

If the master is punished whenever it responds, its frequency of responding declines more rapidly than that of the yoked animal (which sometimes is punished adventitiously). If, by responding, the master animal can escape from the aversive stimulus (which the computer turns on occasionally), its latency of response to the

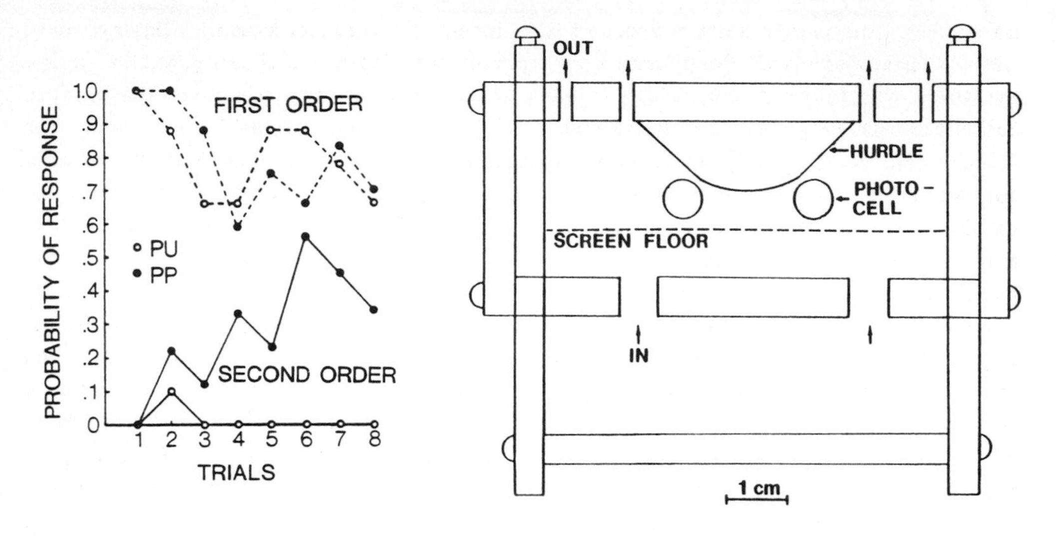

Figure 14.22

Figure 14.23

stimulus declines rapidly as compared with that of the yoked animal. The results of an escape experiment are shown in Figure 14.24, which is plotted in terms of the probability of prior response by yoked animals. To begin with, the yoked animals were as likely as the masters to respond first when the aversive stimulus was presented, but as training continued the proportion of trials on which the yoked animals responded first progressively declined.

Honeybees perform like vertebrates also in unsignaled or "Sidman" avoidance training (Sidman 1953). Figure 14.25 shows the results of an experiment in which a puff of the aversive odor was scheduled by a recycling 30-sec timer that was reset by each response of the master animal. At the outset the yoked animals happened to be more active than the masters, but the rate of responding by the masters increased and that of the yokes declined as (thanks to the behavior of the masters) the fre-

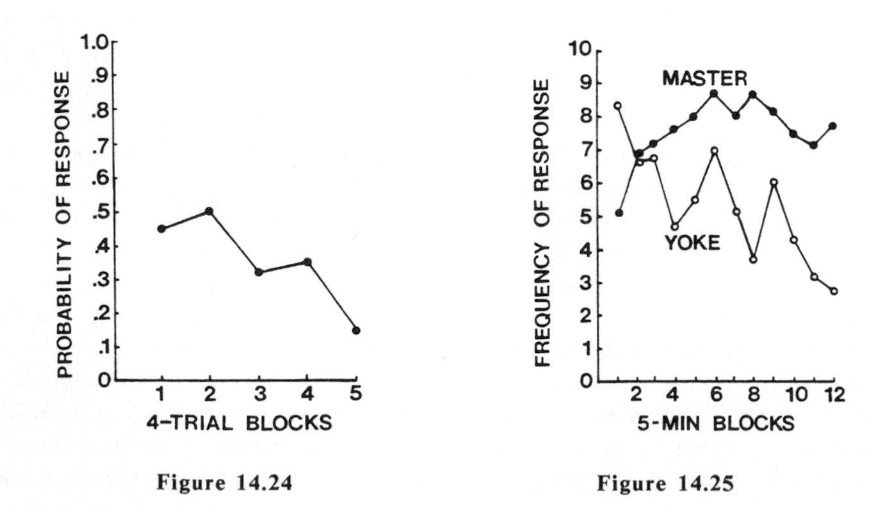

Figure 14.24

Figure 14.25

quency of aversive stimulation declined. A common (and rather complicated) explanation of analogous vertebrate results is that stimuli negatively correlated with an aversive reinforcer (in this case, feedback from the avoidance response) reduce fear and so reward the response that produces them (Weisman and Litner 1969).

Avoidance conditioning has been studied also in free-flying foragers, which come of their own accord to the shelf of an open laboratory window where they feed on the special target diagrammed in Figure 14.26. The food cup is so designed that a brief pulse of shock (the unconditioned stimulus) can be delivered while the animal's proboscis is in contact with sucrose solution. The conditioned stimulus, presented as the animal feeds, is low-amplitude vibration of the target or a stream of air from above, which typically antedates the scheduled shock by 5 sec. The response of the animal, both to the shock from the outset and to the signal after a few pairings, is to fly up from the target, following which the animal soon settles down again and continues to feed to repletion. As in the appetitive conditioning of proboscis-extension, response to a stimulus paired with shock can be shown to depend on the pairing; control animals, for which signal and shock are explicitly unpaired, hardly respond at all to the signal.

In recent experiments patterned closely after work with vertebrates, we have studied the effects on subsequent conditioning of unreinforced exposures of the to-be-conditioned stimulus alone and of unsignaled presentations of the unconditioned stimulus (Abramson and Bitterman 1986a 1986b), again with results strikingly like those for vertebrates. In a variety of vertebrate species, unreinforced preexposure of the conditioned stimulus has been found to retard acquisition (Lubow 1973). This phenomenon, which is known as "latent inhibition," occurs also in honeybees.

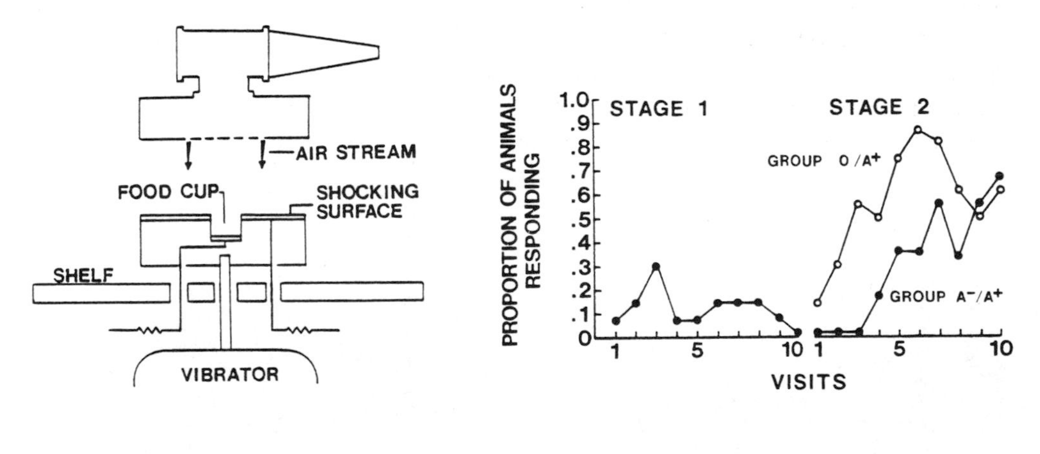

Figure 14.26

Figure 14.27

Figure 14.27 shows the results of an experiment in which the to-be-conditioned stimulus (A) was presented to Group A-/A+ once on each of the 10 visits in the first stage of training, while the animals of Group O/A+ were allowed to feed undisturbed. In the second stage, when a signal-shock pairing was scheduled once on each visit for both groups, acquisition in Group A-/A+ was found to be retarded. That the retardation was not due simply to habituation or loss of attention to A could be shown by a summation experiment, the results of which are plotted in Figure 14.28.

In the first stage of the training, there were 10 visits on each of which, for Groups A-/B+/B- and A-/B+/AB-, there was an unreinforced presentation of stimulus A, while the animals of Group 0/B+/AB- were undisturbed. In the second stage of training, there were five visits on each of which, for all groups, a pairing of stimulus B with shock was scheduled. On each visit of the third stage, there was an unreinforced presentation of A and B together for Groups A-/B+/AB- and 0/B+/AB-, and an unreinforced presentation of B alone for Group A-/B+/B-. The left-hand portion of Figure 14.28 shows rapid acquisition of response to B by all three groups in the second stage (in honeybees, as in vertebrates, latent inhibition is stimulus-specific). The right-hand portion shows that A, after unreinforced exposure in the first stage, actively suppressed response to B when presented with it in the third stage. It should be said at once that this is not the modal outcome of vertebrate experiments, which have tended on the whole to support an explanation of latent inhibition in terms of attentional decrement (Reiss and Wagner 1972), but there is an experiment with rats that points clearly, as does ours with honeybees, to inhibition (Kremer 1972). A possibility to be considered is that there are two opposing effects of preexposing the to-be-conditioned stimulus, with one or the other dominant under as yet undefined circumstances.

Vertebrate conditioning experiments show that acquisition is retarded by prior experience, not only with the conditioned stimulus alone, but also with the unconditioned stimulus alone -- the "US-preexposure effect" (Randich and LoLordo 1979). The most widely supported explanation is that contextual stimuli are conditioned during unsignaled presentations of the unconditioned stimulus and then block conditioning of the signal in subsequent paired trials. When the unconditioned stimulus is encountered only on paired trials, contextual stimuli are not conditioned (according to the theory) because they are overshadowed by the signal. The results for honeybees fit the theory nicely (Abramson and Bitterman 1986b).

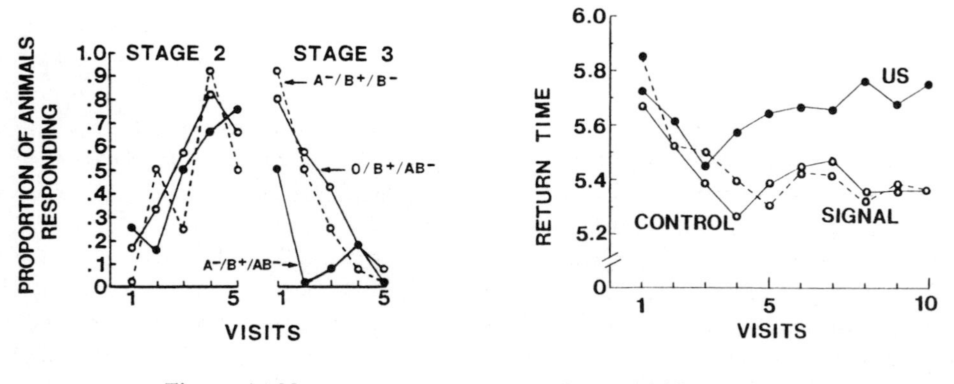

Figure 14.28

Figure 14.29

In the first stage of one of our experiments there were 10 visits, on each of which a brief, unsignaled, shock was given to the animals of a US group, while those of a control group fed undisturbed. Evidence of aversive contextual conditioning in the US group is shown in Figure 14.29, which is plotted in terms of the mean natural log return times for the two groups. Since the animals come to the target of their own accord (shuttling back and forth between the hive and the laboratory window), the time between leaving the target on one visit and landing again on the next (return time) affords a good measure of the attractiveness of the target, which may be enhanced by association with food and reduced by association with shock. Plotted also in Figure 14.29 are the return times of a group trained with signalled shock which give no indication of a developing aversion to the context.

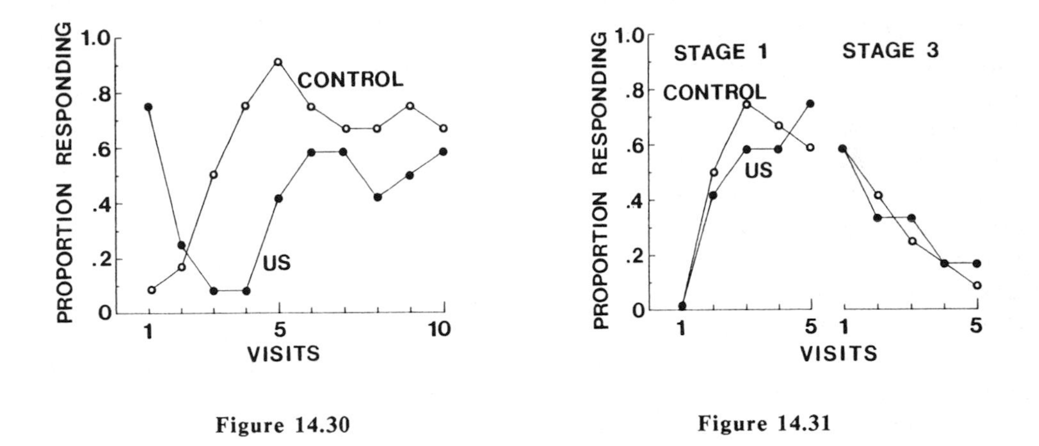

Figure 14.30

Figure 14.31

In the second stage of the experiment, there were 10 visits on each of which a pairing of the signal with shock was scheduled, and, as Figure 14.30 shows, the control animals conditioned normally. The responses of the US animals to the first presentation of the signal suggested that they had been sensitized by their previous experience with shock, but even so their conditioning was retarded, and it was interesting to note that, as the training with signaled shock continued, their return times fell to the control level. Another similarity between the honeybee results and those for rats (Randich and Ross 1984) and pigeons (Grau and Rescorla 1984) is that exposure to unsignaled shock which *follows* training with signaled shock conditions the context but does not interfere with the established response to the signal. Shown in Figure 14.31 is the performance of two groups of honeybees in the first stage of an experiment during which both were conditioned, and in the third stage of the experiment during which both were tested with the signal alone. In the second stage, there were 10 visits with unsignaled shock for the US animals (whose return times increased significantly), but not for the control animals (whose return times remained the same). As in vertebrates, then, conditioning the context interferes with the conditioning of the signal as distinct from response to the signal after it has been conditioned -- an associative rather than a performance effect.

Implications

What can these results mean? For one thing, they mean that it is reasonable to ask vertebrate questions about invertebrates, and perhaps that there is no other kind, although the answers may be different, and the mechanisms as well, even where the answers are the same. While the most parsimonious assumption with respect to phenomena common to vertebrates of widely divergent species is that they are mediated by common mechanisms, we must proceed more cautiously here. Menzel (1983 p. 510) considers the possibility that the mechanisms of learning in honeybees and vertebrates are homologous at both the "the cellular and network levels," which is difficult to understand since the networks seem to have evolved independently. The little we know of the common ancestor suggests that any homologies are simple indeed, restricted to the synaptic mechanisms long assumed to register (as Menzel puts it) the temporal relationships among events that reflect the causal structure of the world. While some of the common phenomena may be independent of network architecture -- except as it must permit the confluence of disparate events (thus determining *what* can be learned) -- most probably are not. Our discovery of incentive contrast in honeybees (Figure 14.4) in fact clearly suggests convergence.

The assertion that everything we know about learning in honeybees is familiar from work with vertebrates should not be taken to mean that all of the phenomena discovered in honeybees are phenomena common to all vertebrates. A case in point is provided by successive negative incentive contrast, which is omnipresent in mammals but absent in descendants of certain older vertebrate lines (Flaherty 1982). Some illustrative data obtained in comparative experiments of simple but powerful design are shown in Figure 14.32, data for rats in the left-hand portion and for goldfish in the right-hand portion.

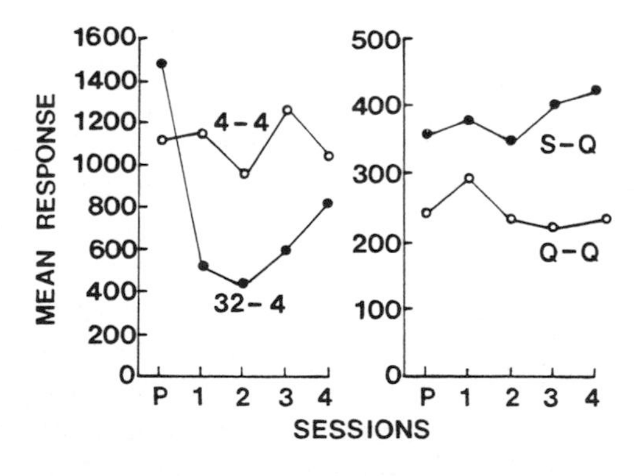

Figure 14.32

In the first stage of the experiment with rats (Flaherty et al. 1983), there was access for 5 min each day to a sucrose solution -- 32% for one group, 4% for another -- and consummatory responses were recorded. The last day of training is indicated on the abscissa as P. In the second stage, when the concentration for both groups was 4%, the rats previously trained with 32% sucrose (32-4) drank substantially less for several days than those previously trained with 4% (4-4). The right-hand portion of Figure 14.32 shows the results of an analogous experiment with goldfish plotted in terms of the number of consummatory responses in a period of 6 min. The animals were trained in the first stage either with a standard liquid diet (S) or with the same diet adulterated by the addition of quinine sulfate (Q), and in the second stage both were tested with the adulterated diet (Couvillon and Bitterman 1985b). Here the outcome was different than for the rats, the goldfish previously trained with the preferred food (S-Q) continuing to respond *more* than the control animals trained from the outset with the adulterated diet (Q-Q).

It should be noted that these results are quite representative of those obtained in a wide range of successive-contrast experiments with both species; there is even an as-yet-unpublished quinine-adulteration experiment with rats which shows a contrast effect at least as large or larger than the effect obtained with the two sucrose concentrations (personal communication from C. F. Flaherty). A good deal of related evidence from work with animals of other classes -- amphibians, reptiles, and birds -- and with other so-called "paradoxical reward effects" supports the idea that we have here a significant divergence in vertebrate learning and some indication of its nature. The indication, as I see it, is that the control of behavior by remembered reward and nonreward is not a general vertebrate characteristic, but one which appeared in a common reptilian ancestor of birds and mammals (Bitterman 1984b). In the older lines, behavior seems to be influenced by its consequences only in a more direct (perhaps Thorndikian) way.

If the mechanisms of incentive contrast in honeybees and vertebrates have indeed evolved independently, further analysis soon should reveal some functional differences on the assumption that convergence to the point of identity or even of seriously confusing similarity is unlikely in elaborately polygenic behavioral systems (Simpson 1964). Such analysis has already begun. In one recent series of experiments (Ammon et al. 1986), for example, we were unable to demonstrate that the disturbance produced by unrealized anticipation of reward can be conditioned and counterconditioned in honeybees as it is in rats. On the same assumption, even many of the phenomena which do seem to be general phenomena of vertebrate learning and which as yet have been little more than demonstrated in honeybees may turn out in the course of further analysis to be convergent. Convergence is important, of course, not only as a source of information about the range of feasible neural solutions to ubiquitous problems of adjustment, but in helping to define broader features of those problems which may hitherto have been obscured by undue preoccupation with relatively minor ecological details -- if, indeed, adaptive pressures always must be assigned a central role (Dumont and Ralston 1986; Gould 1982).

Interested as are many others in whether the results obtained in recent physiological work with various invertebrate preparations (mostly gastropod molluscs) will "generalize to vertebrates" (p. 243), Sahley, Rudy, and Gelperin (1984) argue from a few vertebrate phenomena found in *Limax* for common behavioral principles, and from common principles for common cellular mechanisms (which, they say, may or may not be homologous). Unfortunately, there is no easy road from learning phenomena to unifying principles (as there was not from falling apples and orbiting planets to Newton's laws of motion). After many years of work with vertebrates, we still have no general agreement on such fundamental matters, for example, as the necessary and sufficient conditions for the formation of an association between two stimuli. Nor is there an easy road from principles to mechanisms. While different principles imply different mechanisms, common principles do not necessarily imply common mechanisms; given the laws of physics and the properties of available materials, there may, as Pantin (1951) wrote, be only so many ways to build a bridge, but there are more than one.

Although the study of learning in honeybees has progressed rapidly in recent years, we still know little about it by vertebrate standards. There are many other vertebrate paradigms which it might be useful to explore, and as the scope of the inquiry is widened there is reason also to increase its depth, few of the resemblances to vertebrates having yet been probed very deeply. At the same time, it would be useful to begin to look carefully at other invertebrates for evidence not only of commonality, but also of divergence. It is not unlikely that many of the general phenomena of vertebrate learning that are also found in honeybees could be shown by their absence in other invertebrates to be convergent. For physiological purposes, it would be useful to find invertebrates with richer behavioral resources than the various species now commonly taken as models (Alkon and Farley 1984; Carew and Sahley 1986) and accessible brains as well; optimistic talk about honeybees notwithstanding (Menzel 1983), eight or nine hundred thousand neurons packed into a volume of less than a cubic millimeter (Mobbs 1982) will not generally be regarded as an attractive preparation. Unfortunately, there are few promising candidates -- readily available in large numbers, easily maintained in the laboratory, and with suitable sensory, motor, and motivational properties. The technical problems already well in hand (Henderson et al. 1975; Walker et al. 1970), my own inclination is to turn again to octopuses.

Acknowledgment

The work on learning in honeybees reviewed here was supported by grants BNS 79-05876 and 83-17051 from the U.S. National Science Foundation.

References

Abramson CI (1986) Aversive conditioning in honeybees *(Apis mellifera).* J Comp Psychol 100:106-116

Abramson CI, Bitterman ME (1986a) Latent inhibition in honeybees. Anim Learn Behav 14:184-189

Abramson CI, Bitterman ME (1986b) The US-preexposure effect in honeybees. Anim Learn Behav 14:374-379

Alkon DL, Farley J (eds) (1984) Primary neural substrates of learning and behavioral change. Cambridge University Press, Cambridge

Ammon D, Abramson CI, Bitterman ME (1986) Partial reinforcement and resistance to extinction in honeybees. Anim Learn Behav 14:232-240

Amsel A (1962) Frustrative nonreward in partial reinforcement and discrimination learning. Psychol Rev 69:306-328

Behrend ER, Bitterman ME (1961) Probability-matching in the fish. Amer J Psychol 74:542-551

Bitterman ME (1967) Learning in animals. In: Helson H, Bevan W (eds) Contemporary approaches to psychology. Van Nostrand, New York, p 139

Bitterman ME (1975a) The comparative analysis of learning. Science 188:699- 709.

Bitterman ME (1975b) Critical commentary. In: Corning WC, Dyal JA, Willows, AOD (eds) Invertebrate Learning, vol 3. Plenum, New York, p 139

Bitterman ME (1984a) Migration and learning in fishes. In: McCleave JD, Arnold, GP, Dodson, JJ, Neill, WH (eds) Mechanisms of migration in fishes. Plenum, New York, p 397.

Bitterman ME (1984b) Learning in man and other animals. In: Sarris V, Parducci, A (eds) Perspectives in psychological experimentation. Lawrence Erlbaum, Hillsdale, New Jersey, p 59

Bitterman ME, Woodard WT (1976) Vertebrate learning: Common processes. In: Masterton, RB, Bitterman, ME, Campbell CBG, Hotton N (eds) Evolution of brain and behavior in vertebrates. Lawrence Erlbaum, Hillsdale, New Jersey, p 169.

Bitterman ME, Menzel R, Fietz A, Schafer S (1983) Classical conditioning of proboscis extension in honeybees *(Apis mellifera).* J Comp Psychol 97:107-119

Bouton, ME, Jones DL, McPhillips, SA, Swartzentruber, D (1986) Potentiation and overshadowing in aversion learning: Role of method of odor presentation, the distal-proximal cue distinction, and the conditionability of odor. Learn. Motiv. 17:115-138

Brown PL, Jenkins, HM (1969) Autoshaping of the pigeon's key-peck. J Exp Anal Behav 11:1-8

Bullock DH, Bitterman ME (1962) Probability-matching in the pigeon. Amer J Psychol 75:634-639

Carew TJ, Sahley CL (1986) Invertebrate learning and memory: From behavior to molecules. Ann Rev Neurosci 9:435-488

Corning WC, Dyal JA, Willows AOD (eds) (1973, 1975) Invertebrate learning, vols 1-3. Plenum, New York

Couvillon PA, Bitterman ME (1980). Some phenomena of associative learning in honeybees. J Comp Physiol Psychol 94:878-885.

Couvillon PA, Bitterman ME (1982). Compound conditioning in honeybees. J Comp Physiol Psychol 96:192-199

Couvillon PA, Bitterman ME (1984) The overlearning-extinction effect and successive negative contrast in honeybees. J Comp Psychol 98:100-109

Couvillon PA, Bitterman ME (1985a) Analysis of choice in honeybees. Anim Learn Behav 13:246-252

Couvillon PA, Bitterman ME (1985b) Effect of experience with a preferred food on consummatory responding for a less preferred food in goldfish. Anim Learn Behav 13:433-438

Couvillon PA, Bitterman ME (1986) Performance of honeybees in reversal and ambiguous-cue problems: Tests of a choice model. Anim Learn Behav 14:225-231

Couvillon, PA, Bitterman ME (1987) Discrimination of color-odor compounds by honeybees: Tests of a continuity model. Anim Learn Behav 15:218-227

Couvillon PA, Klosterhalfen S, Bitterman ME (1983). Analysis of overshadowing in honeybees. J Comp Psychol 97:154-166

Dickinson A, Nicholas DJ, Mackintosh NJ (1983) A re-examination of one-trial blocking in conditioned suppression. Q J Exp Psychol 35:67-79

Dufort RM, Guttman N, Kimble, GA (1954) One-trial discrimination reversal in the white rat. J Comp Physiol Psychol 47:248-247

Dumont JPC, Robertson RM (1986) Neuronal circuits: An evolutionary perspective. Science 233:849-853

Durlach PJ, Rescorla RA (1980) Potentiation rather than overshadowing in flavor aversion learning. J Exp P: Anim Behav Proc 6:175-187

Flaherty CF (1982) Incentive contrast: A review of behavioral changes following shifts in reward. Anim Learn Behav 10:409-440

Flaherty, CF, Becker HC, Checke S (1983) Repeated successive contrast in consummatory behavior with repeated shifts in sucrose concentration. Anim Learn Behav 11:407-414

Frisch K von (1914) Der Farbensinn und Formensinn der Biene. Zool Jahrb 35:1-188

Gould SJ (1982) Darwinism and the expansion of evolutionary theory. Science 216:380-387

Grau JW, Rescorla RA (1984) Role of context in autoshaping. J Exp Psychol: Anim Behav Proc 10:324-332

Grossmann KE (1973) Continuous, fixed-ratio, and fixed-interval reinforcement in honey bees. J Exp Anal Behav 20:105-109

Henderson TB, Woodard WT, Bitterman ME (1975) Measurement of consummatory behavior in octopuses. Behav Res Meth Instr 7:265-266

Holland PC (1979) Effects of omission contingencies on various components of Pavlovian conditioned responding in rats. J Exp Psychol: Anim Behav Proc 5:178-193

Holmes NK, Bitterman ME (1969) Measurement of consummatory behavior in the fish. J Exp Anal Behav 12:39-41.

Ison JR, Cook PE (1964) Extinction performance as a function of incentive magnitude and number of acquisition trials. Psychon Sci 1:245-246

Kamin LJ (1969) 'Attention-like' processes in classical conditioning. In: Jones MR (ed) Miami symposium on the prediction of behavior: Aversive stimulation. University of Miami Press, Miami, p 9

Kirk KL, Bitterman ME (1965) Probability-learning by the turtle. Science 148:1484-1485

Kremer EF (1972) Properties of a preexposed stimulus. Psychon Sci 27:45-47

Kuwabara M (1957) Bildung des bedingten Reflexes von Pavlovs Typus bei der Honigbiene, *Apis mellifera*. J Fac Sci Hokkaido Univ (Zool) 13:458-464

Longo N (1964) Probability-learning and habit-reversal in the cockroach. Amer J. Psychol 77:49-51

Lubow RE (1973) Latent inhibition. Psychol Bull 79:398-407

Mackintosh NJ (1983) Conditioning and associative learning. Clarendon Press, Oxford

Menzel R (1983) Neurobiology of learning and memory: The honeybee as a model system. Naturwiss 70:504-511

Menzel R., Bitterman ME (1983) Learning by honeybees in an unnatural situation. In Huber F, Markl L (eds) Behavioral physiology and neuroethology. Springer, Heidelberg, p 206

Menzel R, Erber J (1978) Learning and memory in bees. Sci Amer 239:102-110

Mobbs PG (1982) The brain of the honeybee, *Apis mellifera.* I. The connections and spatial organization of the mushroom bodies. Phil Trans Roy Soc Lond (B) 298:309-354

North AJ, Stimmel DT (1960) Extinction of an instrumental response following a large number of reinforcements. Psychol Rep 6:227-234

Pantin CFA (1951) Organic design. Adv Sci 8:138-150

Pavlov IP (1927) Conditioned reflexes. Oxford University Press, Oxford

Perin CT (1942) Behavior potentiality as a joint function of the amount of training and the degree of hunger at the time of extinction. J Exp Psychol 30:93-113

Pert A, Bitterman ME (1969) A technique for the study of consummatory behavior and instrumental conditioning in the turtle. Amer Psychol 24:258-261

Randich A, LoLordo VM (1979) Associative and nonassociative theories of the UCS pre-exposure phenomenon: Implications for Pavlovian conditioning. Psychol Bull 86:523-548

Randich A, Ross RT (1984) Mechanisms of blocking by contextual stimuli. Learn Motiv 15:106-117

Reiss S, Wagner, AR (1972) CS habituation produces a "latent inhibition effect" but no active "conditioned inhibition." Learn Motiv 3:237-245

Rescorla RA (1977) In: Davis H, Hurwitz HMB (eds) Operant-Pavlovian interactions. Lawrence Erlbaum, Hillsdale, New Jersey, p 133

Rescorla RA, Cunningham CL (1978) Within-compound flavor associations. J Exp Psychol: Anim Behav Proc 4:267-275

Rescorla RA, Durlach, PJ (1981) Within-event learning in Pavlovian conditioning. In: Spear, NE, Miller, RR (eds) Information processing in animals: Memory mechanisms. Lawrence Erlbaum, Hillsdale, New Jersey, p 81

Sahley CL (1984) Behavior theory and invertebrate learning. In: Marler P, Terrace HS (eds) The biology of learning. Springer, Berlin, p 181

Sahley CL, Rudy JW, Gelperin A (1984) Associative learning in a mollusk: A comparative analysis. In: Alkon DL, Farley J (eds) Primary neural substrates of learning and behavioral change. Cambridge University Press, Cambridge, p 243

Sanders GD (1975) The cephalopods. In: Corning WC, Dyal JA, Willows AOD (eds), Invertebrate Learning, vol 3. Plenum, New York, p 1

Sheffield FD (1965) Relation between classical conditioning and instrumental learning. In: Prokasy WF (ed) Classical conditioning: A symposium. Appleton-Century-Crofts, New York, p 302

Shinoda A, Bitterman ME (1987) Analysis of the overlearning-extinction effect in honeybees. Anim Learn Behav 15:93-96

Sidman M (1953) Avoidance conditioning with brief shock and no exteroceptive warning signal. Science 118:157-158

Sigurdson JE (1981a) Automated discrete-trials techniques of appetitive conditioning in honey bees. Behav Res Meth Instr 13:1-10

Sigurdson JE (1981b) Measurement of consummatory behavior in honeybees. Behav Res Meth Instr 13:308-310

Simpson GG (1964) Organisms and molecules in evolution. Science 146:1535-1538

Skinner BF (1938) The behavior of organisms. Appleton-Century-Crofts, New York

Sutherland NS, Mackintosh NJ (1971) Mechanisms of animal discrimination learning. Academic Press, New York

Tennant WA, Bitterman ME (1975) Blocking and overshadowing in two species of fish. J Exp Psychol: Anim Behav Proc 1:22-29

Terrace HS (1984) Animal learning, ethology, and biological constraints. In: Marler P, Terrace HS (eds) The biology of learning. Springer, Berlin, p 15

Thorndike EL (1911) Animal intelligence. Macmillan, New York

Tolman EC (1932) Purposive behavior in animals and men. Century, New York

Wagner, AR, Logan FA, Haberlandt, K, Price T (1968) Stimulus selection in animal discrimination learning. J Exp Psychol 76:171-180

Walker JJ, Longo N, Bitterman ME (1970) The octopus in the laboratory: Handling, maintenance, training. Behav Res Meth Instr 2:15-18

Weisman RG, Litner JS (1969) Positive conditioned reinforcement of Sidman avoidance behavior in rats. J Comp Physiol Psychol 68:597-603

Wells PH (1973) Honey bees. In: Corning WC, Dyal JA, Willows AOD (eds) Invertebrate learning, vol 2. Plenum, New York, p 173

Woodard WT, Ballinger JC, Bitterman ME (1974) Autoshaping: Further study of "negative automaintenance." J Exp Anal Behav 22:47-51

Yerkes RM (1927) The mind of a gorilla. Genet Psychol Monogr 2:1-193

SPECIES-SPECIFIC DIFFERENCES IN LEARNING

Marco D. Poli
Institute of Psychology, School of Medicine, University of Milan

From the beginnings of his written history, man has shown a deep interest in understanding "animal intelligence" and in trying to make sense of the many instances of apparently "intelligent" behaviors exhibited by many animals; quite obviously the species he was more interested in were usually either the domestic ones or those, such as the apes and the monkeys, appallingly similar to himself. Although formal philosophical theories on this topic can be dated back at least to Aristotle, the most influential developments of ideas on the subject are much more recent.

As in most other fields of biological thought, also in the study of the "mind" of animals Darwin's theory of evolution by natural selection marked a sharp change of perspective. Before Darwin, the dominant view had been the one espoused by Rene Descartes, stating that the gift of reason was absolutely unique to man, while all animals were to be viewed as automata, whose behavior was controlled only by instinct and by automatic reflexes, which allowed a certain degree of adaptation to the requirements of the environment. Darwin's ideas on this topic were quite different. In *The Descent of Man,* he explicitly stated that "animals possess some power of reasoning," and, while acknowledging that "the difference between the mind of the lowest man and that of the highest animal is immense," he also wrote that he was nevertheless convinced that the difference "certainly is one of degree, and not of kind. We have seen that the senses and intuitions, the various emotions and faculties, such as love, memory, attention, curiosity, imitation, reason, etc., of which man boasts, may be found in an incipient, or even sometimes in a well-developed condition in the lower animals." With his theory of natural selection, Darwin effectively laid the bases on which most subsequent research on animal behavior and intelligence would be built. Particularly relevant in this perspective are not only *The Expression of the Emotions in Animals and Man* (1872) and *The Formation of Vegetable Mould through the Action of Worms, with Observations on their Habits* (1881), but also the botanical works, such as *On the Various Contrivances by which Orchids are Fertilized by Insects* (1862), *Insectivorous Plants* (1875) and *The Power of Movement in Plants* (1880).

However, although he collected a large amount of material on this topic, Darwin never found the time to study in real depth the problem of animal intelligence and of its evolution and to publish in a systematic work his views on these matters; in due course, he passed all the material he possessed to the young George Romanes, who was to become one of last century's best known - and possibly most discredited - students of animal intelligence. In fact, Romanes' book *Animal Intelligence* (1882),

NATO ASI Series, Vol. G17
Intelligence and Evolutionary Biology
Edited by H. J. Jerison and I. Jerison

partly based on the notes received from Darwin, presented a large collection of data, mostly anecdotal, attempting to show that many animals, even of fairly low phylogenetic level, were indeed able of "intelligent" behavior. Romanes' work is nowadays usually mentioned only with an innuendo, and is completely dismissed by most students as one of the most typical examples of two of the worst possible sins in comparative research: anthropomorphism and careless collection of data. However, in fact Romanes did analyze and criticize some of the data he reported but, unwisely, he did so only in his second book, *Mental Evolution in Animals* (1884), whose principal aim was to develop an integrated theory of the mental abilities of animals, based on the data presented in *Animal Intelligence.* Since most people only read the first volume, which as a matter of fact is just an uncritical collection of anecdotal data, they completely dismissed him from the ranks of reliable scientists.

Some of the first severe and documented criticisms of Romanes' easy-going approach came from Conway Lloyd Morgan, who, especially in his *Introduction to Comparative Psychology* (1894), submitted some of Romanes' ideas about animal intelligence to a strict analysis, going so far as to test some of them under controlled conditions and to show experimentally that at least some of his arguments were untenable and that many of the behaviors that Romanes had hastily attributed to "intelligence" could be more economically explained in simpler terms without postulating any hypothetical cognitive process.

It is most likely that Morgan's ideas influenced E.L. Thorndike in his work on animal learning. Thorndike was the first psychologist to attempt a systematic experimental study of animal intelligence. In 1911 he published his seminal book *Animal Intelligence*, in which he described a series of tests he had carefully devised to study the learning abilities of animals of different species and to analyse how they solved problems. Thorndike's research is too well known to detail here. Suffice it to say that the results of his experiments convinced him that "intelligence," or better, "learning capabilities" were present in animals, and that their level was roughly comparable to the phylogenetic level of the species (this equation between "intelligence" and "learning ability," although quite arbitrary, has not been seriously challenged thereafter). Further, Thorndike was convinced by his experiments that animals only learned through a gradual process of Trial and Error; in his view, no other learning mechanism had any relevance. Neither complex cognitive abilities nor imitation learning had any place in his system.

Before undertaking the task of discussing some aspects of more recent research on animal intelligence, it would seem obviously essential to attempt at least a working definition of the problem under study. This, however, is very difficult. In fact, simply stating that, in principle, animals could also possess different degrees of intelligence without attempting a cogent definition of what "intelligence" actually is can't appreciably further our knowledge of the subject. Moreover, it could be argued that no significant and universally accepted progress in this area could ever be obtained in the absence of a satisfactory definition of the concept of animal intelligence. Unfortunately, defining animal intelligence has proved to be at least as difficult and elusive as defining human intelligence.

Even leaving aside for the moment the problem of definition, the comparison of the "intelligence" of different species is not an easy task. First, when studying nonhuman cognition, we tend to focus on processes that we regard as typical of our own intelligence, without paying much attention to the different adaptations of each species. Second, the speed and ease with which an animal learns a specific task seem to depend as much on intelligence as on sensory, perceptual and motor capabilities, and the several factors at play seem to be almost impossible to disentangle. Moreover, attempts to compare the intelligence of different animals often assume that they differ only in "general" intelligence, that is, in an abstraction supposedly identifiable in strict laboratory conditions. However, it appears that at least some animals do possess special abilities directed towards the behavioral solution of problems they are liable to face in their natural habitat and that, unless well identified beforehand, these abilities may often elude the most carefully planned laboratory studies. It is quite difficult to imagine laboratory research capable of revealing phenomena such as the social transmission of acquired habits exhibited by Japanese monkeys (Itani, 1965; Kawai, 1965). Taking into account these and other arguments, a number of researchers, stressing that behavior is subject to the same evolutionary principles as any other character of the organism, and reviving Nissen's (1958) dictum that "it seems just as logical and possible that an adaptive behavior should give selective value to a related structural character as that an adaptive anatomic feature should lend selective advantage to behavior which incorporates or exploits that structure", have tried to reintroduce in the psychology of animal learning the biological ideas it had for such a long time neglected.

Although the difficulty of satisfactorily defining animal intelligence has until now defied the inventiveness of researchers, they have tried to measure this capability operationally, much as has been done in the measurement of human intelligence, where most so-called "intelligence tests" are in fact such only in an operational sense. To that end, they have attempted to find tests equivalent to human intelligence tests which could be used in animal psychology. Following in Thorndike's footprints, most researchers have reasoned that undoubtedly one of the main observable consequences of intelligence is the ability to learn about environmental stimuli in a variety of situations of different complexity and to make use in further occasions of the experiences gathered in the past, that is, to appreciate relationships between events and to exploit such knowledge in subsequent appropriate situations. It has, therefore, although possibly not entirely correctly, become quite usual more or less explicitly to equate intelligence with the ability to learn. This assumption led to the identification of a wide variety of learning situations of intuitively different difficulty. Animals of different species were then tested in these situations, and attempts were made finally to arrange them on a "scale" of intelligence.

Even if this mode of procedure is accepted, it would be necessary both to identify a common criterion by which to compare different species and to agree on a reference criterion to which to compare the performance levels shown by different species in a common test. In humans, this reference criterion is usually success in school or in work, but no comparable criterion is available for animals. Some students, notably comparative psychologists of the school founded by Schneirla, advocated the use of phylogenetic scales as a reference criterion, but this encounters many difficulties. Since they often appear to confound inherently heterogeneous criteria and phenomena, such scales are often quite dubious themselves. For example, they usually rank monkeys higher than cats, which in turn rank higher than rats. It has been argued that, evolutionarily speaking, this is incorrect, since primates, carnivores and rodents all diverged at about the same time from the ancestral placental mammals (for a full discussion of this topic in a behavioral context, see for example Hodos and Campbell, 1969 and Hodos, 1982). This is not to suggest that phylogenetic scales are completely useless and that their use misses any justification, but it points out that they do in fact suffer from fairly severe limitations.

However, if the simple equation "intelligence = learning capability" is accepted at least operationally, a wide variety of tests can be devised, at least in principle. The simplest tests of an animal's ability to appreciate relationships between events are those in which the subject is required to learn that one event follows another in time, or that an action is followed by a particular consequence. This is the case with classical and instrumental conditioning respectively. The conditions can be made more difficult by introducing the need to discriminate between stimuli, or to respond according to some temporal scheme. However, it can be argued that, on the whole, research on simple conditioning situations, both Pavlovian and instrumental, has so far failed reliably to demonstrate differences in performance between different species. Apparently, in fishes, birds, mammals and even primates there is a substantial uniformity in the capability of solving the kind of problems involved in both simple conditioning situations, and such differences as have been identified appear to be more of degree than of kind (but see Bitterman, 1975, 1976). More fruitful has been the study of some more complex experimental paradigms, and especially of those in which animals are required to learn some more general rules, as, for example, in a series of problems sharing some aspects of the solution, so that the solution of a new problem can be facilitated by the abstraction of a generally valid rule from previous experience with problems of the same kind.

Other useful experimental paradigms include reversal tasks (in which either the position of the reward or the meaning of the discriminative stimuli is reversed without giving to the animal any indication of the change), conditional tasks (in which the animal is required to learn to choose sometimes one stimulus and sometimes another one, with an external signal indicating which one is correct), and matching to sample and oddity tasks (in which the subject has to choose respectively the stimulus of a series which matches a reference stimulus or the one that is at odds with it). In all these situations, to solve the problem the animal has to discover the general rule governing the experiment, mastering quite difficult "concepts." Some examples of these procedures will be discussed further on.

Some of the most comprehensive analyses of comparative studies of this kind have been presented by Bitterman (1972, 1975, 1976) and Bitterman and Woodward (1976). Recently, an ambitious but on the whole disappointing attempt to perform a systematic comparison of different species in a variety of learning tasks has been published by Angermeier (1984).

Biological Factors in Learning

During most of the first sixty years of this century, experimental psychologists were convinced that they had found in the study of learning theories the most fruitful path towards the understanding of animal learning - and therefore also of the intellectual capacities of the different species. Most of the great investigators of animal learning, from Pavlov to Thorndike, to Guthrie, to Hull, to Tolman, to Skinner, etc. attempted, explicitly or implicitly, to formulate universal and omni-comprehensive laws, general enough to incorporate every possible instance of learning, independent of the species under study and from the stimuli, responses, and reinforcements used. Only very few of these theoreticians, with the exceptions of Thorndike and Tolman, ever explicitly admitted even the faintest possibility that the background of biological adaptations possessed by each species could influence its learning capacities. Given these premises, the choice of the species to be studied and of all other aspects of the experimental situation was determined almost exclusively by factors like the ease of obtaining a continuing and economical supply of experimental subjects or the possibility of automatically recording the chosen aspects of behavior, and almost no true comparison of different species was usually performed (see for example Beach, 1950, and Lockardt, 1971). Since the mid-'60's, confidence in the all-embracing explanatory potentialities of traditional learning theories began to wane as a consequence of the gathering from a variety of experimental and observational studies of a sizeable mass of data not easily explained in terms of such theories. A progressively growing body of research has by now been collected apparently contradicting some aspects of these inflexible axioms and showing that learning laws do have at least some exceptions. Many observations performed both in the field and in the laboratory point out to the necessity of allowing a much larger influence than previously thought to the effects of interspecific differences in learning processes. This applies to phenomena as different as taste aversions and song learning, that cannot be fully understood without recognizing the all-embracing role of species-specific and state-dependent unlearned predispositions functionally adapted to the association of particular specific classes of stimuli. In other words, apparently some of the characteristics of associative learning do vary in synchrony with the adaptative needs of the species.

In fact, dissatisfaction with established learning theories came from many different directions, and to some extent originated within the theories themselves. The study of phenomena such as autoshaping, selective attention, preferential learning of certain responses amongst the many choices possible, conditioned learning of taste aversions, the solution of complex problems, etc. showed the impossibility for any single set of theoretical principles to encompass all the available experimental data. This undermined what had been one of the central points of most learning theories, namely, the attempt to create an organic and comprehensive theoretical system capable of explaining in terms of a limited number of basic principles every conceivable case of learning, in man as well as in all other animal species.

Further difficulties for conditioning theories arose from the outside, mainly in connection with the gradual gathering of a large amount of reliable and systematically collected ethological data that undermined the generality of the learning principles identified in laboratory research and the possibility of extending them to natural life conditions. Examples of these data, which, however, have no space to discuss here, relate to Imprinting, to the social or "cultural" transmission of information within a species, to song learning in many bird species, etc.

The phrases, "constraints on learning" and "biological boundaries of learning," have, therefore, come to indicate a heterogeneous body of phenomena, all of which represent on some account or another exceptions to accepted laws of learning. However, to speak of "constraints" or "boundaries" can be misleading since in some cases these phenomena consist in the "facilitation" rather than "limitation" of learning.

It is important to emphasize right now that one of the possible and most interesting explanations of these phenomena is that they are due to the action of biological mechanisms facilitating or constraining a certain response in a given experimental situation. If in the past all experimental research on animal learning was based on the assumption that the study of the behavior of laboratory animals under artificial and well controlled conditions was the best way to pinpoint and understand the basic principles underlying learning, and that these principles were of general applicability and importance, some ethologists and biologically-oriented psychologists have come to question this view, stressing in its place the adaptive significance of learning. In their view, the mechanisms responsible for learning have evolved like all other characters of the organisms and should therefore be viewed as specific adaptations contributing to the survival and reproduction of a given species in a particular ecological niche. This view has long been held by ethologists but more recently has gained a wider acceptance also amongst psychologists. For example, it has been claimed that research of the kind that will be exemplified further on has shown that "our principles of learning no longer have a claim to universality ... learning depends in very important ways upon the kind of animal that is being considered, the kind of behavior that is required of it, and the kind of situation in which the behavior occurs." (Bolles, 1975).

It is well known that an assumption common to most learning theories is that different stimuli, different responses and different reinforcements are essentially equivalent, and therefore interchangeable. However, it is enough to use experimental situations slightly different from the more common ones to show that the rule has many exceptions. For example, operant learning has been studied in the chaffinch *Fringilla coelebs* (Stevenson, 1967; 1969; Kling and Stevenson-Hinde, 1977) using as reinforcement either the commonly used presentation of food or the recorded song of a conspecific. Both stimuli could be shown to possess reinforcing properties, and in fact it was possible to obtain conditioning with both of them. However, there was a marked response-reinforcer selectivity. The birds could easily be conditioned to peck a key to get food, or to jump on a perch to listen to the recorded conspecific song, but it was by all means impossible to get them to peck to listen to the song or to jump on the perch to get food. Moreover, it is necessary to point out that this is not an isolated observation based on some badly controlled stimulus, but is just one of the many possible examples that can be drawn from a by now rich body of research. A further illustration can be drawn from a series of experiments showing that a reliable behavioral contrast effect, similar to the one described in pigeons by Reynolds (1961), can also be obtained in rats licking a "dry" tube for water but not in rats pressing a lever for water (Poli and Motta, 1979; Poli, Prato Previde and Gerli, 1984).

Among the earliest observations of phenomena difficult to explain within the established laws of learning were the well known ones by M. and K. Breland (1961) who described under the label "Misbehavior" a number of observations they had made during their quite unconventional attempts to train for commercial purposes a large number of animals belonging to a wide variety of species. More specifically, they reported a number of instances in which, after having shown signs of efficient learning of a conditioned response, animals engaged in behaviors which either delayed or cancelled the planned reward they would otherwise have received after performing the correct response. The Brelands thought that these behaviors interfering with the conditioned response could be explained in terms of "Instinctive Drift," that is by a progressive shift towards some biologically preadapted response appropriate to the experimental situation. However, these observations apparently were not well controlled and were therefore considered by most researchers as purely anecdotal and unreliable. Only quite recently has the Brelands' work stimulated more tightly controlled experimental studies (see for example Boakes, Poli, Lockwood and Goodall, 1978, and Timberlake, Wahl and King, 1982).

The evidence supporting the hypothesis that learning laws are not adequate to explain the results of all learning experiments can be divided into two broad categories. First, there are cases, such as imprinting or taste aversion learning, in which learning occurs with particular ease in spite of the fact that, according to the traditional laws of conditioning, under such circumstances it should be difficult or even impossible. Second, there are cases in which learning does not occur, or occurs with great difficulty, although the experimental paradigm is apparently suitable for the development of conditioning.

Imprinting and imprinting-like phenomena in birds and in mammals are some of the best known instances of exceedingly fast and easy learning. Possibly taste aversion learning makes an even clearer example, since it is quite easy to draw a parallel between the procedures used in these experiments and the ones used in classical conditioning.

One of the most widely cited experiments in this area has been performed by Garcia and Koelling (1966), who trained the first of two groups of rats to drink "flashing and noisy" water (made so by a circuit activated by the contact of the animal's body, which started a flashing light and a metronome), while the second group received a solution of saccharine in water. For half of the subjects of each group the drinking response was followed by an electric shock and for the remaining half by a short 54 Roentgen irradiation with X-rays or by the administration of a sublethal dose of lithium chloride. When the animals recovered from the effects of the treatment, they were tested with the same stimuli used during training. It was clear that poisoning and X-ray exposition selectively inhibited only the tendency to drink sweet, but not flashing and noisy, water, while electric shock also had selective effects, but of a different kind, preventing the drinking of flashing and noisy, but not sweet, water.

Similar conclusions have been reached by Wilcoxon, Dragoin and Kraal (1971), who studied rats and quails in an experimental situation conceptually similar to the one used by Garcia and Koelling. All subjects drank water coloured in blue and made bitter by a small amount of chloridric acid; after 30 minutes all animals were injected intraperitoneally with cyclophosphamide. A test performed some days later showed that quails selectively avoided the blue color, while rats avoided the bitter taste. On the whole, these experiments, like the many others subsequently performed on similar lines, show unequivocal instances of selectivity in the associations between responses and reinforcements. Thorough reviews of taste aversion learning have been presented by Rozin (1976), Milgram, Krames and Alloway (1977), Domjan (1981) and Braveman and Bronstein (1985). It is also worth noting that these experiments have stimulated some interesting medical applications, as in the attempt of reducing anorexia in terminal cancer patients (see for example Braveman and Bronstein, 1985).

The psychological literature also offers a good number of examples of situations in which learning does not occur, or is very slow and unstable, although the experimental paradigm used is apparently adequate for easy and stable conditioning to develop. Avoidance learning is possibly the clearest cut such case.

The law of effect predicts that a reinforcer will increase the probability of the behavior that produces it. The law is assumed to be valid for all behaviors, all reinforcers and all response-reinforcer combinations. Therefore, there should be no particular difficulty in training a pigeon to peck a key to avoid an electric shock. Although pigeons can easily be trained to peck a key to obtain food, or to run to one side of the cage to avoid shock, this has proved to be very difficult. The few successful experiments obtained conditioning only with the greatest difficulty and were constrained by the use of peculiar training techniques. Moreover, the resulting behavior was very vulnerable to interference (see, for example, Hineline and Rachlin, 1969). The same difficulty arises with rats. The relevance of the general characteristics of the responses used in avoidance learning situations has been pointed out by Bolles (1970, 1975), who, moving from the assumption that most animals possess in their behavioral repertoire a set of innate or precociously acquired defense responses appropriate to their ecological niche, offered an interesting analysis in terms of evolutionary theory. In Bolles' view, in order really to understand learning of avoidance and escape responses, it would be necessary to know how each particular species reacts to aversive situations in the wild. In laboratory conditions, conditioned responses similar to these species-specific defense reactions (SSDR) would be very easily learned, while responses incompatible with them would be almost impossible to learn. Later on, Bolles himself suggested some qualifications of the SSDR concept (Bolles, 1978). This hypothesis is consistent with ethological and evolutionary concepts. As pointed out by Fantino and Logan (1979), to accept Bolles' analysis does not mean to believe that all SSDRs are innate, but only that the flexibility allowed in the acquisition of avoidance responses is not infinite, but is limited to a well-defined class of behaviors. Moreover, this biologically-oriented explanation is at a different level of, but not necessarily incompatible with, other explanations more deeply rooted in the tradition of learning theories. An interesting analysis of the constraints on avoidance conditioning in the rat, with interesting implications for both taste aversion and avoidance experiments has recently been published by LoLordo and Jacobs (1983).

Response selectivity is not limited to avoidance learning. Examples of selectivity in positive reward situations are the already quoted experiments on chaffinches by Stevenson (1967, 1969) or the study by Shettleworth (1973), in which hamsters were trained to perform some naturally occurring behaviors such as scrabbling and washing the face to obtain food rewards. While some responses, such as scrabbling, were easily conditioned, others, such as washing the face, were not. Shettleworth argues that this result is probably to be viewed in relation to the different meaning and context of the two responses in natural conditions. Further examples are presented and discussed in Shettleworth (1974, 1983), Hinde and Stevenson-Hinde (1973) and LoLordo (1979).

Of the many possible explanations of these results, one, based on a re-evaluation of the role played by classical conditioning in all learning situations, even those usually thought of as typically instrumental, is particularly popular among animal psychologists, possibly because it does not require the introduction of any new principle into the accepted laws of learning.

As an example, it might be useful to discuss a possible explanation of food aversion experiments (e.g. Garcia and Koelling, 1966) in terms of classical conditioning. First of all, it is necessary to point out the peculiarities of this kind of experiment in comparison with more typical classical conditioning studies. The fact that conditioning occurs in just one trial is interesting but not unique. There are other instances of one-trial learning, especially in experiments using as the US some intensely aversive event. Also not unique is the fact that the interval between the CS and the US is longer than the few fractions of a second usually recommended. It has been shown that, contrary to what is reported by most manuals, the length of the CS-US interval can vary within a fairly wide range according to the specific experimental situation under study. It is, however, fair to say that no other examples are known of reliable conditioning with CS-US intervals as long as the ones used in taste aversion studies. Moreover, it has been shown that even in taste aversion experiments there is an orderly relationship between the length of the interval and the effectiveness of the conditioning, a result in agreement with those of standard experiments on Pavlovian conditioning. The one really significant difference with the standard situations is the strong selectivity in the formation of associations between specific kinds of cue (taste cues in the case of rats, visual cues in the case of quails) and illness. Therefore, these experiments point out the existence of strong preferential associations between particular categories of stimuli. These preferences can be explained in at least two ways. First, in terms of earlier experiences. See, for example, Mackintosh's (1973) demonstration that rats given uncorrelated presentations of a CS and a US (a tone and an electric shock respectively) can learn that the two events are not related and, although remaining capable of forming an association between the CS and some other US, can show difficulties in conditioning when later on the two events are deliberately paired. This result could be due to the fact that when rats serve as subjects in an experiment, they already possess a lifetime's experience of taste cues, external cues and internal states of illness or well-being; in this experience, taste cues will have consistently been correlated with some interior state, while the other cues will not. Under these conditions, it would be plausible to expect some transfer of previous informal learning to the conditioning situation.

A second possible explanation would be in terms of genetically transmitted biological adaptations: the animals better able to associate new taste stimuli to later gastric illness would have become more selective in their feeding habits, would have had better chances of survival and would therefore have given birth to a larger progeny, showing a better biological fitness and transmitting to their descendants the genetic information facilitating such selective associations. This explanation, rooted in ethological thought, can easily be extended to the selective associations exhibited by rats in Garcia's experiments. Feeding behavior is obviously not a neutral behavior "constructed" in the laboratory, but is based on a long and complex series of both past individual experiences and biological adaptations gathered during the phylogenesis of the species. In the wild, rats apparently select - or avoid - food mainly through gustative and olfactory cues, while auditory and visual ones are probably much less important. This is particularly evident in neophobia, the tendency to avoid new objects or new foods, that in the wild is for this species a most important survival strategy. It is therefore not surprising that also laboratory rats should show a strong tendency to use gustative or olfactory stimuli to choose food, selectively associating them with certain consequences and not others. On the other hand, most birds have quite weak olfaction and taste, and therefore tend to select food mostly on the basis of visual cues. This is in agreement with Wilcoxon experimental findings. On this topic, see also Kalat (1977) and some of the papers collected in the recent volume edited by Braveman and Bronstein (1985).

Generally speaking, in Pavlovian conditioning the pairing of a CS and a US will transfer to the former stimulus some of the properties of the latter (this was noted very early by Pavlov himself, who emphasized the role of "stimulus substitution" in conditioning). Therefore, a subject will tend to approach and contact a CS previously associated with a food US and to move away from a CS associated with a noxious event such as an electric shock. The topography of these approach and withdrawal responses will be very similar to the unconditioned response naturally elicited by the US. When the instrumental learning procedure contains Pavlovian elements, these elements will suffice to superimpose an unintended Pavlovian conditioning upon the instrumental response, making it easier to learn and enhancing its strength. An example is the very common experiment on pigeons where the required response is key-pecking: The bird pecking a lighted key for food reinforcement constantly sees the light just before getting the food so that the key can become an effective CS for the food US. This phenomenon, known as "autoshaping", has been first described by Brown and Jenkins (1968), and has aroused much interest (for recent discussions of the issue, see Hearst and Jenkins, 1974, Gamzu and Schwartz, 1976, and Locurto, Terrace and Gibbon, 1981). In fact, it has been argued that what really put learning theories in difficulty were not so much the inconsistencies, perplexing as they were, but the evident contradictions implicit in Brown and Jenkins' (1968) discovery of autoshaping, namely that Pavlovian contingencies held a faster and more stable conditioning of the pigeon's keypeck than did traditional hand-shaping procedures. Moreover, subsequently Williams and Williams (1969) demonstrated that such autoshaped contingencies maintained responding even when every keypeck cancelled reinforcement. Their experimental paradigm ruled out the possibility of instrumental explanations, such as adventitious reinforcement. The fact that animals go on responding even when the only consequence of their behavior is the removal of an effective reinforcement confirms the power of Pavlovian contingencies over instrumental ones (see also Hearst, 1978, and Mackintosh, 1983). Autoshaping has proved to be one of the more interesting fields of research to emerge in recent years: among the more recent contributions to the issue, in addition to the already cited volume by Locurto, Terrace and Gibbon (1981), see Kaplan's (1984) interesting experimental analysis of the local context hypothesis originally introduced by Kaplan and Hearst (1982), which stresses that conditioning to contextual cues is an important component of the modulation of the strength of trace conditioning, and also the study by Timberlake and Lucas (1985), which, although dealing explicitly with "superstitious" behavior, has also interesting implications for a biologically-oriented analysis of autoshaping. In fact, these authors conclude a painstaking analysis of the factors in play in superstitious behavior under periodic delivery of food by stating that it probably develops from components of species-typical patterns of appetitive behavior related to feeding.

An explanation on the same general lines is also available for those cases where conditioning is exceedingly difficult. Learning will be difficult when the required instrumental response is incompatible with the CR elicited by the concurrent classical conditioning. The animal will find it difficult to approach and peck a key to avoid shock when the same key, being associated with shock, becomes a CS eliciting the CR of withdrawing. In fact, the idea that Pavlovian and operant conditioning may involve common selective learning processes is nowadays progressively spreading.

These explanations are of course only tentative. What can be said with some degree of confidence is only that the phenomenon of selective associations is by now a robust result, but is still waiting for a conclusive and comprehensive explanation.

It is, however, imperative to be very careful not to throw away the baby with the tub water. Learning theories have not been shown either wrong or irrelevant. On the contrary, they do still possess an important heuristic value and are far from dead. The undeniable difficulties they are facing at present don't mean that the attention given to them in the past has been misplaced. Learning is today as in the past one of the most interesting and lively topics of psychology - and not only of comparative psychology. What is needed, and still is largely missing, is a synthesis of the more recent data and of the ethological evidence with the accepted knowledge and their integration in the body of learning theories. Unfortunately, only a few and timid steps have so far been made in this direction.

Some Aspects of Animal Cognition

Another effect of the discovery of some qualifications to the universality of learning theories has been the growing interest in a cognitive approach to the study of animal behavior. In fact, after a long interval of almost complete obsolescence, in recent years a number of systematic attempts have been made to explore the "cognitive" processes of animals in a comparative frame of reference. Although still far from conclusive, the results of these studies are beginning to take a coherent shape and to suggest new and stimulating directions of research. However, we are still very far from a general consensus on animal consciousness. It has sometimes been stated that self-awareness, the animal's ability to abstract and to form a conceptual framework of its environment so that it can perceive itself and its actions in relation to the environment, is a *conditio sine qua non* for speaking of consciousness. Nevertheless, this still is a very vague definition, and in fact it has been argued that it is impossible to conceive a single experiment or set of experiments which could conclusively prove the existence of self-awareness in animals lower than the great apes. This is very intriguing, but no easy way out of the impasse is easily available. At the Dahlem workshop held in March 1981, Griffin (1982) urged the participants to devise experiments that would not only examine animal cognition more thoroughly, but also provide evidence regarding animal experience. The results of such an effort, although stimulating, haven't been too satisfactory.

In fact, to study animal awareness it would first of all be necessary to agree both on in what respects it is useful to extend cognitive concepts to animal psychology (a fine discussion of the advantages and limitations of such an extension and of the use of cognitive terminology has been forwarded by Honig, 1978), and on the meaningfulness of asserting any knowledge of an animal's experience.

It has been argued that, to keep the concept of mind in a scientific context, it is necessary, first, to identify and define instances of mind, and establish a set of procedures and empirical markers with some degree of consistency; second, to show that the concept of mind will serve to more efficiently integrate and organize existing information; and, third, to demonstrate that the formulation permits the derivation of specific, testable predictions about the presence or absence of mind and its influences on behavior (Gallup, 1982). These recommendations, however, are very general, and do contain neither reliable nor simple formulas for deciding if and when we are allowed to use cognitive terms when dealing with animals (but see the discussion in Griffin, 1982). Unfortunately, by now it should be clear that there is no easy answer to such questions since, notwithstanding the very large amount of recent research in cognitive psychology, we still are very far from any convincing and reliable answer, not even in the context of human psychology.

A very interesting and much debated point is whether animals are capable to make use of abstract concepts in categorizing tasks. For example, with regard to the concept of number, both laboratory experiments and naturalistic observations have shown that rats can use a variety of external and temporal cues as predictors of danger and safety in avoidance and suppression situations (see Imada and Nageishi, 1982). Recently, attempts have also been made to demonstrate, amongst other supposed higher cognitive capabilities, the presence in lower species of the ability to count or, more precisely, the possession of the concept of "number." For example, Davis and Memmott (1982) presented a well documented review of counting behavior in animals, concluding that successful demonstrations of this capacity were most likely under relatively extreme experimental conditions, in which alternative predictors of food or safety were unavailable. In fact, Davis, Memmott and Hurwitz (1975) found that, when three unsignaled shocks were invariably presented superimposed upon a bar-pressing session, rats came to behave as though they had learned, "If three shocks, then no more shocks," that is, bar-pressing was accelerated after the third shock (of course, such a result can also be accounted for within the framework of conditioning theory, without the need of postulating any higher cognitive capacity). More recently Davis and Memmott (1982) were able to confirm reliably the previous findings concluding that rats could count three signaled shocks (but not three unsignaled shocks) superimposed upon bar-pressing.

However, the replication of these findings is not completely straightforward. Results in agreement with those of Davis and Memmott (1978) have been reported by Fernandes and Church (1982), who, having observed that most previous experiments allowed a confusion to be made between counting and timing of stimuli, trained rats to press a lever after a two-sound signal and a different lever after a four-sound signal. Later tests demonstrated that apparently rats could learn to discriminate by number alone, without other stimulus components. However, Imada, Shuku and Moriya (1983), having unsuccessfully tried to replicate Davis and Memmott's findings in a study in which special care was taken not to allow the confounding of the temporal cues with the frequency cues, were unable to detect any indication of counting ability in rats.

Research on the formation and use of number concepts has naturally not been limited to rats, but some primate studies are now also available. Although human mathematical abilities seem to derive from cognitive processes strictly related to the ones involved in linguistic abilities, there is surprisingly little evidence for mathematical abilities in apes. While in the past most studies on ape intelligence have therefore dealt with the question if apes possessed the capacity for human language, attention is now also given to the question of whether they possess numerical capabilities. In order to determine if nonhuman primates possess some type of number-concept appreciation, Ruby (1984) studied the ability of adolescent rhesus monkeys to form a concept of the third object in a row of three, four, five, six or seven objects. The results, in agreement with those of other studies of the number concept in nonhuman primates, lent further support to the idea that these animals do in fact possess at least some type of number appreciation. For example, Matsuzawa (1985) trained on an intermittent reinforcement schedule for correct answers a 5-years-old female chimpanzee to use Arabic numerals to name the number of items in a display, showing that the primate was able to master the numbers from 1 to 6 and to name the number, colour and object of 300 types of samples. Most of these studies, however, have been subjected to stringent criticisms. For example, a critical analysis of Matsuzawa's experiment has been put forth by McGonigle (1985), who argues that the explanation of these results doesn't necessarily require reference to cognition, suggesting a simpler explanation in terms of a rudimentary counting mechanism known as "subitizing."

A further interesting point has been dealt with in studies attempting to demonstrate experimentally that under some conditions animals are able to form abstract concepts, such as the "same/different" concept, which some believe to be based on the possession of cognitive capacities usually not attributed to nonhumans. Zentall, Edwards and Hogan (1984), for example, have demonstrated that pigeons can show a significant degree of "same/different" concept transfer. This kind of research usually involves matching or oddity-to-sample procedures to train the rules of "same" and "different" respectively. Pecks to a sample stimulus result in the presentation of two comparison stimuli, one having the same stimulus value as the sample, and the other different from it. In the matching situation, responses to the comparison stimulus similar to the sample are reinforced, while responses to the different alternative are not; in the oddity situation the reverse is true. Using this procedure, Zentall and Hogan (1974) trained pigeons on either a matching-to-sample or an oddity-from-sample task. When novel stimuli were substituted for the training stimuli, pigeons for which the task remained constant (matching to matching, or oddity to oddity) transferred to the new stimuli at a higher level of performance and acquired the transfer task at a faster rate than did pigeons for which the task was changed.

Similar results were also obtained by Zentall and Hogan (1978) in a study in which pigeons were trained to respond to a single key, either half of which could be independently illuminated. For half of the birds, responses were reinforced when the two halves matched, whereas they were not reinforced when they mismatched; the contingencies were reversed for the remaining birds. New stimuli were then substituted for the training stimuli, and once again birds for which the task remained the same performed at a higher level than did those for which the task was reversed. These results are not at all unique but have been confirmed in a variety of situations. For example, Pisacreta and Witt (1983) showed that pigeons were able to learn to observe five lighted keys and to peck on each trial the key that displayed a color different from the other four. Moreover, the performance was not disrupted when a novel color was introduced, and the birds were required to peck two non matching keys during each trial. Similarly, Wright, Santiago, Urcuioli and Sands (1984) reported successful learning of a same/different task in monkeys and in pigeons. Finally, Herrnstein and de Villiers (1980) have shown in a systematic series of experiments that pigeons can extract from color pictures of fishes the abstract category "fish." Moreover (see Edwards, Jagielo, and Zentall, 1983), pigeons seem to show significant transfer of the "same/different" concept under conditions analogous to those used by Premack (1976) who, by training chimpanzees to place a token representing "same" between pairs of objects that were alike (e.g. A-A) and another token representing "different" between pairs of objects that were unlike (e.g. A-B), was able to show that these primates can learn to use arbitrary symbols to represent the relational rules or concepts "same" and "different." In fact, after having learned the task, Premack's chimpanzees showed to be capable of appropriately choosing between the tokens when novel pairs of objects (e.g. C-C or C-D) were presented.

Although fashionable, cognitivistic interpretations of animal problem solving have not been unchallenged. Many psychologists of behavioristic background have attempted systematic alternative analyses of the results. For example, Epstein, Kirshnit, Lanza and Rubin (1984) performed a number of Koehler-like experiments on pigeons, showing that these birds did in fact solve the problems presented to them in a fundamentally chimpanzee-like way. However, in the opinion of these authors, the analysis of the data and the systematic variation of the animals' experiences indicate that most likely the new response was acquired by trial and error. For the solution of Koehler's problems, certain specific previous experiences are required. In Epstein's experiment, the birds were separately trained to two discrete responses: to push a cardboard box towards a green spot painted on the floor just below a banana hanging from the roof, out of the reach of the bird (all behavior directed towards the box in absence of the green spot was deliberately extinguished), and to climb on the box to peck the banana when it was placed under the fruit. Although box-directed behaviors in presence of the banana were extinguished, all birds developed in a very short time a complex response of pushing the box under the banana and then climbing on it to peck the fruit. Epstein et al. (1984) presented a suggestive analysis of the functions of the different previously learned components of the responses in the formation of this "insightful" behavior.

Moreover, many studies (see Epstein, Lanza and Skinner, 1980, 1981; Lanza, Starr and Skinner, 1982) have simulated in pigeons a variety of complex cognitive processes, showing not only that such behavior can be explained by pointing to fortuitous contingencies of reinforcement, but also that it can be reliably produced by arranging the necessary contingencies.

An explanation in non-cognitivistic terms of another behavior supposed to require the development of higher cognitive capabilities has been forwarded by Boakes and Gaertner (1977) in an analysis of the development of simple communication mechanism in pigeons. These authors, after discussing in term of autoshaping some of the studies on dolphins purported to demonstrate the spontaneous development of a form of communication, proceed to present an experiment on pigeons conceptually equivalent to the studies performed on dolphins, in which the different behaviors of the "sender" in presence of two discriminative stimuli serve as a cue for the "receiver" to autoshape the correct responses. The data show quite clearly that although apparently the two birds learned how to "communicate," in fact no higher cognitive ability was needed, the simplest explanation of the results being in terms of well known learning phenomena. This result has recently been confirmed by Millard, Deutsch-Klein and Glendon (1981) and by Lubinski and MacCorquodale (1984).

Of course, hypotheses such as Epstein's are fully supported by Skinner (1984) who, discussing his views on behavioral evolution, notes that in the evolution of any character (and, of course, this applies also to behavioral characters), it is important not only to point out its present survival value, but also to try to reconstruct earlier stages of it, which, to evolve, should also have had survival value. The current survival value of reflexes and of the released patterns of behavior studied by ethologists may be clear, but what about the construction of plausible sequences through which they could have evolved, with survival value at every stage? Skinner argues against the usefulness of referring to "cognitive" processes to explain animal behavior, pointing out that the complexities of natural selection over very long periods of time might suffice to produce extremely complex behaviors, without any need of postulating any hypothetical process. Apparently extremely complex behaviors, such as eel migration, bee dancing etc., are usually satisfactorily explained without any reference to cognition simply in terms of a series of small but advantageous modifications of preceding behavioral responses.

Quite naturally, not everybody agrees with this kind of analysis. For example, Gallup (1983) maintains that the experiments by Epstein, Lanza and Skinner (1981), demonstrating the possibility of establishing in the pigeon through operant conditioning a series of responses similar to the ones used in self-recognition tests (see further on) are not convincing confutations of studies showing the presence of higher cognitive abilities in animals, such as his own studies on self-awareness in chimpanzees and orangutans. However, the arguments offered in his dismissal of Epstein's papers are not very convincing.

Gallup's (1970, 1982; Suarez and Gallup, 1981) own studies on self-recognition in primates have aroused much interest because of the ingenuity of the techniques he devised. After a short experience with mirrors, higher primates learn to use the mirror as a mirror -- to respond to themselves. In inexperienced chimpanzees, the first response to a mirror was invariably to respond as if they were seeing another chimpanzee. However, after about two days, this "social" tendency would disappear, to be replaced by the beginning of a self-directed orientation, that is, by the tendency to use the mirror to respond to themselves, using it for autogrooming, inspecting visually inaccessible parts of the body and experimenting with reflections. Later on, a formal test was introduced, consisting in painting a not visible part of the face of the animal while he was unconscious and in controlling thereafter his reaction to the reflected image. All chimpanzees who already had a sufficient experience with mirrors immediately after awakening began to explore the painted spot using the reflected image as a reference, while none of the subjects who didn't have any preceding experience with the mirror did so.

These findings have been replicated in a number of other experiments on chimpanzees and other species, but only orangutans were as successful as the chimpanzees. In fact, with the exception of humans, chimpanzees, and orangutans all primates have so far failed to recognize themselves, even after very long exposures to mirrors. (Many species have been tested, including the gorilla [Suarez and Gallup, 1981].) The real surprise in this respect - a challenge to all interpretations - comes from the lack of self-recognition instances in gorillas, which are biologically very closely related to humans and chimpanzees.

One possible explanation, forwarded by Gallup, is that the species unable to use the mirror in this kind of experimental setting are lacking an essential cognitive capability: a sense of self. The argument is roughly as follows: An organism which is aware of itself is an organism which is aware of its own existence. Since introspection presupposes self-awareness, evidence of mind must be evidence of introspection. It can be argued that introspection presupposes language, but presumably this is not strictly necessary. Premack (1972) concluded that language did not provide the chimpanzee Sarah with any new concept, but merely allowed her to express what she already knew.

Still, the interpretations of Gallup's data have been criticized on many accounts. First, the definition of consciousness offered by Gallup, who claims that there are strong links between the ability of distinguishing "self" from "non-self" and self-consciousness, is extremely general and by no means impervious to criticisms. For example, immunologists assume that the immune system is capable of distinguishing between "self" and "non-self" ("Flawed lymphocytes can turn up with an inability to distinguishing between self and non self, and replication of these can bring down the entire structure with the devastating diseases of autoimmunity" [L. Thomas, *The Youngest Science,* 1983, p. 85]). If Gallup's argument is to be turned on end, we should conclude that lymphocytes are capable of self-awareness and are therefore conscious!

However, Gallup also argues that, since introspection presupposes self-awareness, evidence of mind must be considered as evidence of introspection. In his view, knowledge of self provides the possibility for an inferential knowledge of others: If I know that I exist and that I have some experiences, I can suppose that other conspecifics also have similar experiences to mine, and I can purposefully try to deceive or otherwise to influence them (Gallup, 1983). "Deception," or "looking guilty" would then be considered as reliable cues of the possession of mind. The logical cogency of this argument is, however, quite weak, since it doesn't offer any strict parameter by which to judge to which animals one should grant the possession of mind. Gallup argues that chimps do possess mind because they are capable of purposefully deceiving, or because they look "guilty" after having performed some forbidden behavior. However, any owner of a dog or a cat would eagerly testify to his own pet behaving "guiltily" or "deceptively"; on these matters it is very difficult to avoid superficial anthropomorphisms. These caveats notwithstanding, it is fair to say that there are some experimental data that seem reliably to confirm that chimpanzees can deliberately try to deceive. Premack and Woodruff (1978) and Woodruff and Premack (1979) have shown that apparently chimpanzees were able to make inferences and attributions about mental states in humans using a situation in which both humans and chimpanzees were involved in a dyadic search of rewards. When the members of the dyad were asked to compete, rather than to cooperate, for the goal, chimpanzees would selectively withhold information or even provide misinformation about the location of the reward to mislead their opponent.

Some penetrating criticisms of the procedure used in self-recognition studies have been published. See, for example, the lucid analysis forwarded by Anderson (1984); also, the demonstration by Epstein, Lanza and Skinner (1981) of the possibility of establishing in the pigeon through operant conditioning a series of responses similar to the ones used in self-recognition tests is, notwithstanding Gallup's (1983) dismissal, worth some further consideration.

A number of further demonstrations of supposed higher cognitive capabilities in animals have, of course, been presented, but, however suggestive some of them may be, on the whole the data are as yet not sufficient to discard explanatory interpretations in terms of simpler processes.

References

Anderson JR (1984) Monkeys with mirrors: some questions for primate psychology. Intern J Primat 5: 81-98

Angermeier WF (1984) The Evolution of Operant Learning and Memory. Karger, Basel

Beach FA (1950) The snark was a Boojum. Amer Psychol 5:115-124

Bitterman ME (1972) Comparative studies of the role of inhibition in reversal learning. In: Boakes RA, Halliday MS (eds.) Inhibition and Learning. Academic Press, London

Bitterman ME (1975) The comparative analysis of learning. Science, 88: 699-709

Bitterman ME (1976) Issues in the comparative psychology of learning. In: Masterton RB, Bitterman ME, Campbell CBG, Hotton N (eds.) Evolution of Brain and Behavior in Vertebrates. Lawrence Erlbaum, Hillsdale

Bitterman ME, LoLordo VM, Overmier JB, Rashotte, ME (1979) Animal Learning: Survey and Analysis. Plenum Press, New York

Bitterman ME, Woodward WT (1976) Vertebrate learning: common processes. In: Masterton RB, Bitterman ME, Campbell CBG, Hotton, N (eds.) Evolution of Brain and Behavior in Vertebrates. Lawrence Erlbaum, Hillsdale

Boakes RA, Gaertner I (1977) The development of a simple form of communication. Quart J. Exper Psychol 29: 561-575

Boakes RA, Poli MD, Lockwood MJ, Goodall G (1978) A study of misbehavior: token reinforcement in the rat. J Exper Anal Behav 29: 115-134

Bolles RC (1970) Species-specific defense reactions and avoidance learning. Psychol Rev 77: 32-48

Bolles RC (1975) Learning Theory. Holt, Rinehart & Winston, New York

Bolles RC (1978) The role of stimulus learning in defensive behavior. In: Hulse SH, Fowler H, Honig WK (eds.) Cognitive Processes in Animal Behavior. Lawrence Erlbaum, Hillsdale

Braveman NS, Bronstein P (eds.) (1985) Experimental Assessment and Clinical Applications of Conditioned Food Aversions. Ann NY Acad Science vol. 443

Breland K, Breland M (1961) The Misbehavior of organisms. Amer Psychol, 16: 681-684

Brown, Jenkins HM (1968) Auto-shaping of the pigeon's key peck. J Exper Anal Behav 11: 1-8

Davis H, Memmott J (1982) Counting behavior in animals: a critical evaluation. Psychol Bull 92: 547-571

Davis H, Memmott J, Hurwitz HMB (1975) Autocontingencies: a model for subtle behavioral control. J Exper Psychol: Gener, 104 169-188

Domjan M Ingestional aversion learning: unique and general processes.(1981) In: Rosenblatt JJ, Hinde RA, Beer C, Busnel MC (eds.) Advances in the Study of Behavior, vol. 11. Academic Press, New York

Edwards CA, Jagielo JA, Zentall TR (1983) "Same/different" symbol use by pigeons. Anim Learn Behav 11: 349-355

Epstein R., Kirshnit CE, Lanza RP, Rubin LC (1984) "Insight in the pigeon: antecedents and determinants of an intelligent performance. Nature 308: 61-62

Epstein R, Lanza RP, Skinner BF (1980) Symbolic communication between two pigeons (*Columba livia domestica*) Science 207: 543-545

Epstein R, Lanza RP, Skinner BF (1981) "Self-awareness" in the pigeon. Science 212: 695-696

Fantino E and Logan CA (1979) The Experimental Analysis of Behavior. Freeman, San Francisco

Fernandes DM and Church RM (1982), Discrimination of the number of sequential events by rats Anim Learn Behav, 10/2, 171-176

Gallup GG (1970) Chimpanzees: self-recognition. Science 167: 86-87

Gallup GG (1982) Self-awareness and the emergence of mind in primates Amer J Primat 2: 237-248

Gallup GG (1983) Toward a comparative psychology of mind. In: Mellgren RL (ed.) Animal Cognition and Behavior. North Holland, New York

Gamzu E, Schwartz B (1976) The maintenance of key-pecking by stimulus-contingent and response-independent food presentation. J Exper Anal Behav 19: 65-72

Garcia J, Koelling RA (1966) Relation of cue to consequence in avoidance learning. Psychon Sci 4: 123-124

Griffin DR (ed.) (1982) Animal Mind - Human Mind. Springer, Berlin

Hearst E (1978) Stimulus relationships and feature selection in learning and behavior. In: Hulse SH, Fowler H, Honig WK Cognitive Processes in Animal Behavior. Lawrence Erlbaum, Hillsdale

Hearst E, Jenkins HM (1974) Sign-tracking: the stimulus-reinforcer relation and directed action. Psychonomic Society Press, Austin, TX

Herrnstein RJ, de Villiers P.A. (1980) Fish as a natural category for people and pigeons. The Psychology of Learning and Motivation, vol. 14. Academic Press, New York

Hinde RA, Stevenson-Hinde J (eds.) (1973) Constraints on Learning. Academic Press, London

Hineline PN, Rachlin H (1969) Escape and avoidance of shock by pigeons pecking a key. J Exper Anal Behav 12: 533-538

Hodos W (1982) Some perspectives on the evolution of intelligence and the brain. In: Griffin DR) Animal Mind - Human Mind. Springer, Berlin

Hodos W, Campbell CBG (1969) *Scala Naturae*: why there is no theory in comparative psychology. Psychol Rev 76:337-350

Honig WK (1978) On the conceptual nature of cognitive terms: an initial essay. In: Hulse SH, Fowler H, Honig WK Cognitive Processes in Animal Behavior. Lawrence Erlbaum, Hillsdale

Imada H, Nageishi Y. (1982) The concept of uncertainty in animal experiments using aversive stimulation. Psychol Bull 91: 171-176

Imada H, Shuku H, Moriya M (1983) Can a rat count? Anim Learn Behav 11: 396-400

Itani J (1965) On the acquisition and propagation of a new food habit in the troop of Japanese monkeys at Takasakiyama. In: Imanishi K, Altmann SA (eds.) Japanese Monkeys. Emory University Press, Atlanta

Kalat JW (1977) Biological significance in food aversion learning. In: Milgram NW, Krames L, Alloway TM (eds.) Food Aversion Learning. Plenum, New York

Kaplan PS (1984) Bridging temporal gaps between CS and US in autoshaping: a test of a local context hypothesis. Anim Learn Behav 12: 142-148

Kaplan PS, Hearst E (1982) Bridging temporal gaps between CS and US in autoshaping: insertion of other stimuli before, during, and after CS J Exper Psychol: Anim Behav Proc 8: 187-203

Kawai M (1965) Newly acquired precultural behavior of the natural troop of Japanese monkeys on Koshima Islet. Primates 8: 1-30

Kling JW, Stevenson-Hinde J (1977) Reinforcement, extinction and spontaneous recovery of key-pecking in chaffinches. Anim Behav, 23: 424-429

Lanza RP, Starr J, Skinner BF (1982) "Lying" in the pigeon. J Exper Anal Behav 38: 201-203

Lockardt RB (1971) Reflections on the fall of comparative psychology: is there a message for us all? Amer Psychol 1: 168-179

Locurto CM, Terrace HS,Gibbon J (eds.) (1981) Autoshaping and conditioning theory. Academic Press, New York

LoLordo VM Constraints on learning. (1979) In: Bitterman ME, LoLordo VM, Overmier JB, Rashotte ME Animal Learning: Survey and Analysis. Plenum Press, New York

LoLordo VM, Jacobs WJ (1983) Constraints on aversive conditioning in the rat: some theoretical accounts. In: Zeiler MD, Harzem P (eds.) Biological Factors in Learning. Wiley, Chichester

Lubinski D, MacCorquodale K (1984) "Symbolic communication" between two pigeons (Columbia Livia) without unconditioned reinforcement. J Comp Psychol 98: 372-386

Mackintosh NJ (1973) Stimulus selection: learning to ignore stimuli that predict no change in reinforcement. In: Hinde RA, Stevenson-Hinde J (eds.) Constraints on Learning. Academic Press, London

Mackintosh NJ (1983) Conditioning and Associative Learning. Clarendon Press, Oxford

Matsuzawa T (1985) Use of numbers by a chimpanzee. Nature 315: 57-59

McGonigle B (1985) Can apes learn to count? Nature 315: 16-17

Milgram NW, Krames L, Alloway TM (eds.) (1977) Food Aversion Learning. Plenum, New York

Millard WJ, Deutsch-Klein N, Glendon FM (1981) "Communication" in the pigeon. Behav Anal Lett 11: 305-315

Morgan CL (1894) An Introduction to Comparative Psychology. Scott, London

Pisacreta R, Witt K (1983) Same-different discrimination in the pigeon. Bull Psychon Soc 21: 411-414

Poli M, Motta M (1979) Analisi sperimentale delle relazioni tra Behavioral Contrast e risposta consumatoria. Ricerche di Psicol 9: 69-87

Poli M, Prato Previde E, Gerli M (1984) Contrasto comportamentale nel ratto con due tipi di risposta. Ricerche Psicol 14: 2-27

Premack D (1972) Teaching language to an ape. Scient Amer 227: 92-99

Premack D (1976) Intelligence in ape and man. Lawrence Erlbaum, Hillsdale

Premack D, Woodruff G (1978) Does the chimpanzee have a theory of mind? Behav Brain Sciences 4: 515-526

Reynolds GS (1961) Behavioral Contrast. J Exper Anal Behav 457-471

Romanes GJ (1882) Animal Intelligence. Kegan, Paul, Trench & Co, London

Romanes GJ (1884) Mental Evolution in Animals. Appleton & Co., New York

Rozin P (1976) The selection of foods by rats, humans, and other animals. In: Rosenblatt JJ, Hinde RA, Shaw E, Beer C (eds.) Advances in the Study of Behavior, vol. 6. Academic Press, New York

Ruby LM (1984) An investigation of number-concept appreciation in a rhesus monkey. Primates 25: 236-242

Shettleworth SJ (1973) Food reinforcement and the organization of behavior in golden hamsters. In: Hinde RA, Stevenson-Hinde J (eds.) Constraints on Learning. Academic Press, London

Shettleworth SJ (1974) Constraints on learning. In: Rosenblatt JJ, Hinde RA, Shaw E, Beer C (eds.) Advances in the Study of Behavior, vol. 4. Academic Press, New York

Shettleworth SJ (1983) Function and mechanisms in learning. In: Zeiler MD, Harzem P (eds.) Biological Factors in Learning. Wiley, Chichester

Skinner BF (1984) The evolution of behavior. J Exper Anal Behav 41: 217-221

Stevenson J (1967) Reinforcing effects of chaffinch song. Anim Behav, 15: 427-432

Stevenson J (1969) Song as a reinforcer. In: Hinde RA (ed.) Bird Vocalizations. Cambridge University Press, Cambridge

Suarez SD, Gallup GG (1981) Self-recognition in chimpanzees and orangutans, but not gorillas. J Hum Evol 10: 175-188

Timberlake W, Lucas, GA (1985) The basis of superstitious behavior: chance contingency, stimulus substitution, or appetitive behavior? J Exper Anal Behav 44: 391-396

Timberlake W, Wahl G, King D (1982) Stimulus and response contingencies in the misbehavior of rats. J Exper Psychol: Anim Behav Proc 8: 62-85

Wilcoxon HC, Dragoin WB, Kraal PA (1971) Illness-induced aversions in rat and quail: relative salience of gustatory and visual cues. Science 171: 826-828

Williams DR, Williams H (1969) Auto-Maintenance in the pigeon: sustained pecking despite contingent non-reinforcement. J Exper Anal Behav 12: 511-520

Woodruff G, Premack D (1979) Intentional communication in the chimpanzee: the development of deception. Cognition 7: 333-362

Wright AA, Santiago HC, Urcuioli PJ, Sands SF (1984) Monkey and pigeon acquisition of same/different concept using pictorial stimuli. In: Commons ML, Herrnstein RJ, Wagner AR (eds.) Quantitative Analyses of Behavior, vol. 4. Ballinger, Cambridge MA

Zentall TR, Edwards CE, Hogan DE (1984) Pigeons' use of identity. In: Commons ML, Herrnstein RJ, Wagner AR (eds) Quantitative analyses of behavior. Acquisition II. Ballinger, Cambridge MA

Zentall TR, Hogan DE (1974) Abstract concept learning in the pigeon. J Exper Psychol 102: 393-398

Zentall, TR, Hogan DE (1978) Same/different concept learning in the pigeon: the effect of negative instances and prior adaptation to the transfer stimuli. J Exper Anal Behav 30: 177-186

CONTRIBUTION OF THE GENETICAL AND NEURAL MEMORY TO ANIMAL INTELLIGENCE

V. Csanyi
Department of Behavior Genetics
L. EötvösUniversity of Budapest
Javorka S.-u 14, GöD
H-2131, Hungary

Mechanisms that adjust the internal "parameters" of organisms to random or temporary changes in the environment can be found at every level of living organization (Plotkin and Odling-Smee 1981). At the most basic level, these mechanisms are in gene expression and in the epigenesis of the phenotype. This basic level of organization may be thought of as "genetic memory;" much animal performance involves mechanisms at that level and is based mostly or entirely on gene expression.

Mayr described the behavior of the Yucca moth, which illustrates such a memory: "The female moth collects a ball of pollen from several flowers, then finds a flower suitable for ovipositing. After depositing her egg in the soft tissue of the ovary, by means of a lancelike ovipositor, she pollinates the flower by pushing the pollen to the bottom of the funnel-shaped opening of the pistil. This permits the larva to feed on some of the developing seeds of the fertilized flower, and yet guarantees the development of enough seeds in the nonparasitized sectors of the fruit to permit the Yucca plant abundant reproduction" (Mayr, 1976, p. 31). The behavior is a complex "biotechnology", carried out by the moth on the basis of inherited knowledge encoded in the genome (Morgan 1896). Functionally the underlying process is equivalent to the retrieval of some stored memory traces.

Acquiring and changing genetic "memory traces" are relatively slow processes. Acquisition of new essential knowledge of the type that governs the behavior of the Yucca moth may require millions of generations. Animals must also cope with rapid, more or less predictable, changes in their environment. The nervous system is the special organ of adaptation that evolved to adjust the organism to rapid changes (Figure 16.1). Genetic and neural memory interact and supplement each other. Genetic and neural mechanisms are interconnected and fine tune animal behavior to a complex and sometimes disordered environment. In this paper I develop a functional description of how these two mechanisms contribute to behavioral adaptation.

1. Models of the environment in the animal brain

It was shown brilliantly by MacKay (1951-52) that the essence of the adaptation performed by the nervous system is *goal directed behavior*, and neither vitalistic nor

NATO ASI Series, Vol. G17
Intelligence and Evolutionary Biology
Edited by H.J. Jerison and I. Jerison

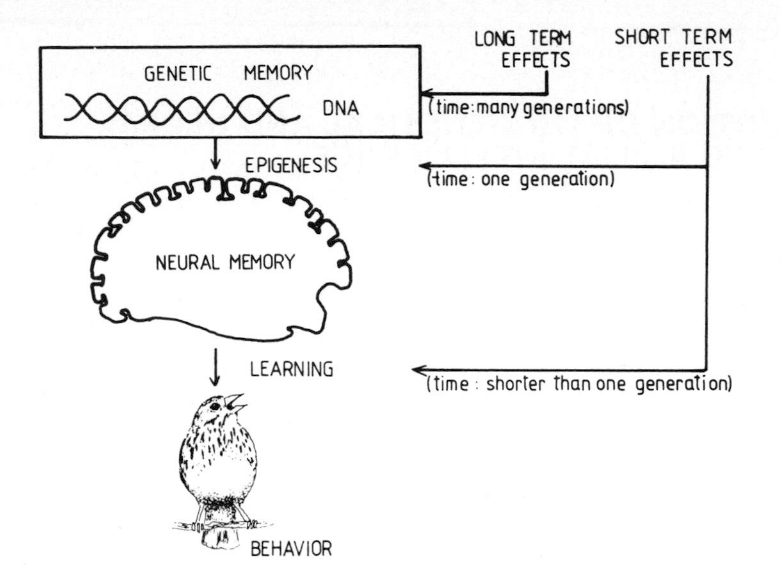

Figure 16.1. Role of genetic and neural memory in adaptive processes of the animals

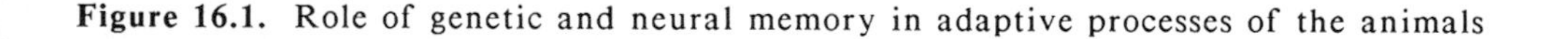

anthropomorphic concepts are required to explain it. Goal directed behavior, or generally speaking, intelligence, has a simple cybernetic explanation if we consider the organism as a system with a number of feedback loops, capable of changing its own parameters.

Definition of goal "X" pursued by system "A" would be as follows. Let "Y" represent "A" and its present environment, and let "X" equal "A" in a state in which it has reached its goal. Then "A" shows goal-directed behavior if it performs internal or external movement that minimizes the difference between "X" and "Y". This definition of goal directed behavior is valid not only for living organisms but for artificial cybernetic constructions. It follows from the definition, that a system capable of showing goal-directed behavior has to possess certain mechanisms. It is necessary that the system be able to distinguish between states "X" and "Y", that is, to have some kind of *recognition sub-system*, which, in case of artificial systems, can be very simple (e.g. a heat sensor of a thermostat). Furthermore it is necessary that it be able *to change its own state*, based upon this recognition process. Finally it is necessary that the system carry *internal representations* of the possible goals in order to recognize differences and change.

Subsystems of recognition (perception) and change (behavior) are well-known in living organisms. Internal representations are made possible by mechanisms of genetic and neural memory. The question of the organization of internal representation is, however, more complex. The adaptive "goal" of an organism is to survive in its surroundings, and the internal representation must, therefore, be of an environment that includes the organism itself.

The internal representation of the environment can be considered as a construction, which in essence is a model (Craik 1943), more precisely, a *dynamic model of the environment.* I mean "model" in the systems science sense: a model of a complex system is always a simpler system whose components and their interactions correspond in some way to components and some interactions of the complex system (Mesarovic 1964). Model building is always a simplification involving a special identification between two different systems, of which one is the model and the other is the system being modelled. The effects of operations on the model are used to predict the behavior of the system being modelled.

The most important biological function of the animal brain is the construction of such a dynamic model of the environment. This model includes environmental factors and interactions most important for the survival and reproduction of the animals, the continuous maintenance and operation of the model, and the use of the obtained data for *predictions* in the interest of the survival and reproduction of the animal.

Mackay's very useful concept of cognition is as the *ability to construct models.* It is obvious that all animals, including humans, belong to a common class, because all nervous systems should be able to model their environment, and in that sense all animals have some kind of cognition. This view takes differences as well as similarities into account, because it is possible to model a complex system very simply but also in more complicated ways.

Simple models can predict some simple but important things for adaptation. For example, an animal may "know" that dark periods are followed by daylight. This knowledge is in the form of a *model* of the dark-light rhythm of the environment that establishes the diurnal cycle in the nervous system of animals. The mechanisms are basically genetic. Different species may be active during the dark or during the light period, but all have the inherited "genetic memory" of the periodicity of this environment. Both genetic and neural memory contribute to the final performance.

More complex phenomena can also be modelled. A wolf's brain presumably models not only diurnal cycles but the behavior of prey, of fellow members of the pack, order of dominance within the pack, effects of past events, and many other things. Ethological studies have demonstrated higher mental functions in animals (Griffin 1984, Denett 1983, Epstein et al. 1984), and such functions are also evident in electrophysiological experiments (John 1972).

When the neuronal network is relatively simple, its model of the external world may be quite primitive. The structure of the model could involve only the spatiotemporal activity of a few excited neural elements. In higher organisms with more extensive neuronal networks, very elaborate models of the external world may be established. Fish, amphibians, birds and mammals are all capable of storing essential parameters of an encountered image in memory and of acting appropriately in a later situation in the absence of the relevant environmental stimuli (Beritashvili 1971). Rats can memorize maps of complex mazes and, by comparing the actual stimuli with the internalized map, they can orient themselves precisely in physical space (O' Keefe and Nadel 1974, Olton and Samuelson 1976).

According to MacKay, the neural model is not only a simple projection, but a kind of complex reconstruction, which also contains *instructions of the possible behavior of the organism* in response to the stimuli of the external world (MacKay 1951-52, 1965). The animal's activity is not simply organized by responses to the external stimuli, but also by expectations and analyses of situations based on the internal analysis of the model formed in the nervous system (Gallistel 1980). The model may be thought of as a program of sequences of events, which can be used by the brain as an internal reference against which sequences of external events may be compared. Information analyzed in this way probably includes that involved in eliciting fear (Hebb 1946), orientation (Sokolov 1960), attack and defence (Archer 1976), and avoidance of predators (Csanyi 1985a, 1985b, 1985c).

In higher animals, the formation of an environmental model may be assumed to involve the internal representation of the animal itself, which would include the emergence of self-consciousness in apes (Gallup 1970) and in man (Anderson 1984).

Ideas of modelling appeared in behavioral science more than half a century ago (Krechevsky 1932, Tolman 1932), but they were submerged by other trends in behaviorism. More recently, ethological studies provided further evidence of the existence of internal models and images, and of their important role in the behavior of animals, for example in the formation of prey-predator relationships (Tinbergen 1960, Croze 1970, Humphrey, 1978)

2. Basic units of the environmental model: the *concept*

The modelling activity of the brain manifests itself in the development of higher organization above the level of neurons. Milner (1976,1977), following Hebb (1949), assumed the emergence of structures based on intercellular connections, which would correspond to the stimulus, to the drive, to the response, or even to the set of possible consequences of possible responses. I refer to these associative structures as *concepts* (italicized to indicate this technical usage). The confluence of *concepts* forms higher active neural networks that determine the behavior of the animal in a given situation, and these neural networks represent the modelling activity of the brain.

Neurophysiological experiments have been carried out to prove the existence of *concepts* (Stuss and Picton 1978). The connections between motor functions and cortical models were examined in detail. Analysis of such experiments indicated that cortical processes are linked less to a motor activity itself than to its purpose and consequences (MacKay 1966, Evarts 1967, Grastyan et al. 1978). Watanabe (1979) found the activity of certain groups of neurons in the cat cerebral cortex to be correlated only with the context and content of the stimulus and not to its physical parameters. The connections between cortical areas and motor activity are quite loose (Pribram 1976), and it is presumably through models produced in the brain that finely organized motor activity is generated and controlled in fine detail.

Although beyond the scope of this paper, the study of modelling by the brain at the neurophysiological level is a rapidly developing field in neurobiology. Two important points should be made. First, the modelling cannot be a simple kind of photographic representation. It is the work of a complex, hierarchically ordered and abstractly organized information system. A mass of information from the periphery is concentrated and interpreted by the brain, and only selected and edited parts, or excerpts, pass to the next levels of hierarchy where the analysis may be repeated and reorganized several times (Jerison 1978, 1985). Second, with respect to basic processes, vertebrate (including mammalian) brains are only marginally different from one another. And even the differences among animal species in higher processes may be more quantitative than qualitative (Griffin 1976, Macphail 1982, Menzel and Johnson 1976, Sperry 1976).

3. Structure of a *concept*

What is the correlation between components of the environment and the structure of a *concept*? A *concept* is assumed to contain smaller, genetically determined elementary behavior programs called *instructions* (Mackay 1951-52). These are necessarily unchanged during a lifetime but they must have some genetic variability to have evolved. Cloak (1975) has suggested that these instructions should be assumed to be connected with and activated by a definite environmental stimulus-cue. Since *concepts* are learnable they must contain other smaller programs, motivational mechanisms, and programs to perform various simple logical operations (if, or, no, etc. switches) have to have some connections to the *concept*.

Let us continue with the help of an analogy to the structure of nervous systems, even the simplest of which are characterized by a triple division. Separate receptor neurons interact with stimuli, motor neurons create or mediate behavior instructions, and interneurons between receptors and motor neurons serve as a reference for the activity of the three as a whole. An analogous triple division of a *concept* is into parts related to cues, behavioral instruction, and referential structures. This division is correlated with the three necessary conditions formulated by MacKay for goal directed behavior by any cybernetic device (Figure 16.2). The *concept* defined in this way is the basic functional unit of the higher nervous system and can be used to describe any kind of behavior act.

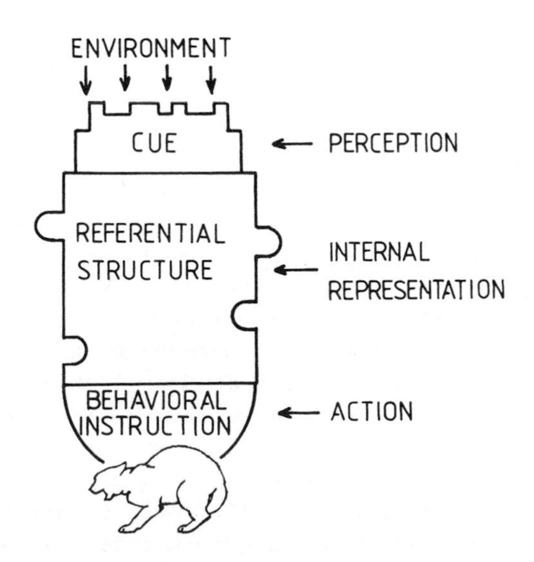

Figure 16.2. Functional structure of the *concept*

Sensitization and habituation in simple animals are good test cases. In the course of sensitization, the excited state of receptor cells (cue), transmitted to the interneurons (referential structure), not only evokes an immediate reaction (behavioral instruction) within the nervous system, but also produces a lasting state of excitation in the interneurons which is reflected in their referential role. These cells react to a second appearance of the stimulus with a more immediate, more direct and a more effective reaction. The mechanism of habituation is similar.

The three functional components of behavior are present in higher nervous systems, but the complexity of the parts is increased enormously and functional overlaps have developed. A higher animal is not stimulated by a simply defined stimulus; rather it finds itself in a "situation". It searches for food, is chased or chases others, etc. Lights, sounds, and mechanical vibrations signal concrete situations: e. g., the presence of a predator or a prey. In a higher animal, stimulus-components of a particular situation elicit no isolated responses. There is more to the behavior than reflexes, taxes, or fixed action patterns. The animal is in action: it escapes, attacks, or is just thinking of what to do. The reference (*referential structure*) is not simply a state of some neurons, but memory traces and their mutual interactions, which create experience and a general view of environment. After an animal makes a decision, which is a hierarchically organized process (Dawkins 1974), and performs its behavior, it produces an altered environment that includes changes both of the animal itself and its surroundings. The altered environment might present new cues, and a new round of decision-making is initiated. In the next sections the components of this triple division are reviewed in more detail, with emphasis on genetic and ontogenetic determinants.

3.1. Perception-cue

The sensory apparatus is mainly genetically determined. Rather than simply sensing and measuring one-dimensional stimuli, sensory organs of higher animals have complex interpretation mechanisms that mediate relations of the external world to the central nervous system. The retina of the frog's eye, for example, is itself a complex nervous system that extracts important and directly useful information from the environment. Retinal activity of higher animals like cats or monkeys contributes less to the processing of information, primarily mediating changes and contrasts. The interpretation of these occurs at higher centers of the brain (Arbib 1972).

It is not too much to assume that information processed by a perceptual system is divided into elementary units, such as Julesz's (1981) "textrons" or "features" of linearity or obliqueness of contours. Genetic determination prevails in these small units, while higher constructions may be learned. The perception of contours and lines is inherited, that of the cube is learned. Learning processes continuously influence perception, and this is manifested, for example, as selective attention to specific "search images" when an animal is looking for specific food, prey, etc.

Key-stimuli and innate releasing mechanisms defined by ethologists may also be considered as parts of the perceptual process (Lorenz 1981). A certain configuration of stimuli, including several key-stimuli, may induce adaptive responses without any previous experience. It can be shown that it is not the complete configuration of stimuli that influences the behavior of an animal but that the key-stimuli act separately; the behavior is a sum of their effects. After several repetitions, additional learning forms a kind of Gestalt from the participant stimuli and other features of the situation. Releasing mechanisms are relatively primitive forms of response, because stimulus and response are not yet separated, and perception is immediately followed by action without previous consultation with the reference neurons. Relative simplicity is also shown by the fact that releasing mechanisms play a dominant role in lower animals, like insects and other invertebrates. They are occasionally found in higher animals, mostly in case of fish and birds, but very often learning may modify their effects.

There are some properties of the physical environment, namely space, objects, light, etc., which are especially important for all higher species. Visual perceptual units such as lines, contours, angles, and movements are generally perceived by animals. The complex consisting of key-stimulus and releasing mechanism has a similar informative function, but the message is species-specific and is mostly in the genetic memory. It mediates information from objects, like parents, progeny, food, enemies, etc., relevant for a species. Just as in the visual perception process, only the most essential part of the information in the environment is represented by the key-stimulus. Influence of epigenetic mechanisms on perception was shown by the famous experiments of Hubel and Wiesel (1974). They found that line perception in young cats is bound to the activity of certain groups of neurons, and these are able to develop normally only if the young animal is appropriately stimulated visually. If such visual stimulation does not occur at a certain early age, then later the animals are unable to perceive horizontal or vertical lines.

3.2. Referential mechanisms

Environmental information reaches the central nervous system after being processed by receptors, and this processing continues in the higher centers. All mechanisms located in the central nervous system and dealing with further interpretation and processing of information are regarded functionally as *referential mechanisms* or referential structures. A part of these referential mechanisms is genetically determined. Constraints resulting from the morphology of the animal, its primary motivational systems, inherited fears, phobias and preferences, mechanisms of the internal rhythms, hormones, all belong here (Hinde and Stevenson-Hinde 1973). Another part of the referential structure is formed during ontogenesis: learned behavior, plans, intentions, and especially memory belong here (Miller et al. 1960).

The referential structures of the animal brain embody all mental factors that participate in processing information mediated by perception, until decision to select the relevant action is taken. Referential structures are species-specific. Traits of given species assign certain acceptor points or fields, and memory traces bound to these play special roles in the life of the animal. In other parts of the referential structures, memory traces can be built up only with great difficulties or not at all, i.e., the animal is unable to fix certain associations. This may be very well illustrated by the concept of "preparedness" used in the physiology of learning. Rats are unable to associate the immediately perceived features of a food with its toxic effects. However association is very quick (one trial is enough) between the food's taste and its toxicity. It is just the opposite in some other species: quail, for example, are able to associate color and shape with toxicity but unable to do so between taste and toxicity (Garcia et al. 1966). The more developed a species is, the greater are its possibilities to position a memory trace in its referential structures but this placing process depends not only on the number of experiences or other parameters of the learning process but on the genetically determined factors of the referential structures.

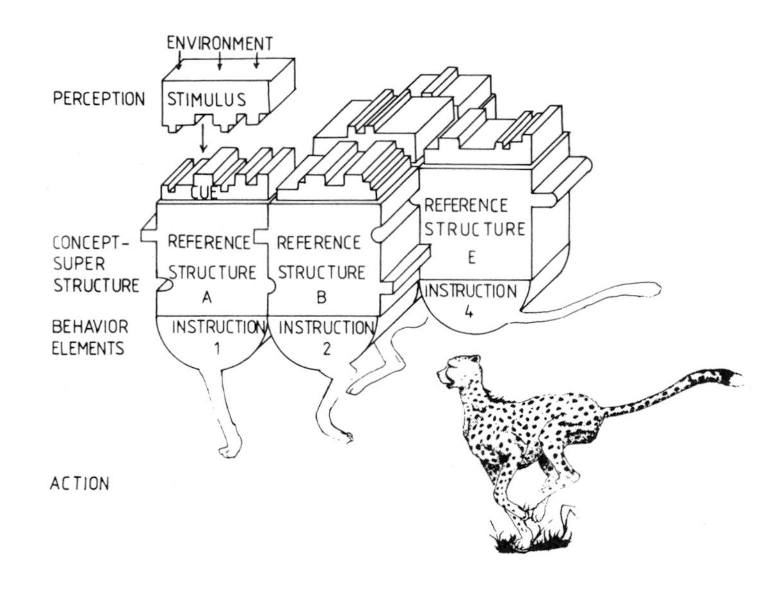

Figure 16.3. Concept-superstructures and animal action

Referential structures may serve to connect various *concepts* into structures of higher complexity (Figure 16.3). Higher animals and man have very complex referential structures and the number of *concepts* connected by these referential structures may be enormous. Thus a higher organization, a new superstructure, practically a system in itself, creates the mental "personality" of the animal, which varies according to the sequence of acquisition of new information, thus producing highly variable individual behavior. Genetic and environmental effects expressed in these higher superstructures are mutually complementary. Without environmental effects, as in the case of an isolated, deprived animal, *concept* structures remain very simple, because only the combination of actions of an active animal with the responses of the environment can build various basic *concepts* into higher structures. But the rules of this construction are determined genetically. We put a greater emphasis on genetic constraints to compensate for the opposite emphasis of the behaviorist tradition in emphasizing learning.

Learning has been considered as a general process, but recent results urge us to change the old slogan that 'everything can be taught and everybody can learn', to 'everything can be taught to somebody, and everybody can learn something'. Learning may be viewed as a special means of adaptation, an ability evolved by animals in the course of evolution which can only be properly studied by considering the close and dynamic relationship between the behavior repertoire of a given species and its environment. The various forms of learning may be considered as parts of an ethogram in the same way as the genetically better defined, less plastic behavior systems. An ethologically meaningful description of the behavior of a species can only be achieved by considering the characteristic features of its learning process and their adaptive significance, as well as considering their functional relationships with other elements of the ethogram. On the other hand it is quite possible that at lower organizational levels than behavior, i.e., neural or synaptical levels, common mechanisms exist for the various forms of associative learning. A number of articles discussing these contrasting approaches to learning and behavior by psychology and ethology suggest the need to work out a unified, ecologically oriented direction in behavior research (Johnston 1982, Domjan and Galef 1983).

An important feature of the referential structure is that its organization can be transformed according to certain rules. Memory traces can be combined, transferred and new solutions of problems, i.e., new *concepts*, may emerge. These transformations are active processes and are based on the feature that the referential structures can accept information not only from the outside but from internal processes as well. When the brain prepares a plan, the external information just initiates the process, and future action is built up by transformation from referential structures already existing. Thus there is an active internal construction mechanism in the brain making the behavior of the higher animals so varied, planned and conscious.

Summing up, the referential structures of animal brain are used to build up a complex dynamic *concept* system, embodying hierarchically organized and mutually interacting structures, which can manifest itself as a systemic unity, a kind of "world view", but of course not in human terms. It is this which governs the general behavior of an animal and its relations to its environment.

3.3. Behavioral instructions, action

As we have already noted, all animals possess numerous genetically determined behavioral instructions, responses like reflexes, taxes, fixed action patterns, innate aversions and preferences. The responses are closely connected to environmental stimuli, i.e., to special cues. If an object approaches the eyes of a human or animal, they respond with a defensive reaction, with a reflex, and there is no need to consult complex referential mechanisms. In the same way releasers and key stimuli directly activate various inherited behavior patterns in lower animals. But there is a tendency in evolution for the generalization of responses, that is, cue-components not directly connected with the primary stimulus-response unit may contribute to the activation of a response, possibly through *concept* superstructures. For example, fixed action patterns can be elicited or inhibited by conditioning. Males of the jungle fowl cannot be conditioned to perform their courtship display for food rewards, but they can readily be conditioned when access to a female is the reinforcement (Kruijt 1964). This is a mental substitution: the features of the closed door of the conditioning apparatus are substituted for the features of the females (cues). A certain mental rigidity of this species is shown by the males' inability to associate the courtship dance with a food reward. In other words, they are unable to do a particular mental transformation. The substitution possibilities of the various cues are species specific.

More possibilities for neural transformations were created by the evolutionary process. In the phenomena of ritualization (Eibl-Eibesfeldt 1970), stimulus-control of inherited elements of behavior is altered genetically to enable a completely different group of stimuli to control the ritual. Ritualization is a clear example of the substitution of cues, or in some cases of instructions, in genetic memory.

The feed-back mechanism that Miller, Galanter, and Pribram (1960) described as the TOTE (Test-Operate-Test-Exit) and which determines some goal-directed behavior would also be considered as largely genetically determined.

Inherited responses are often complemented by learned instructions through the referential structures. Animals are able to store engrams of complex behavior sequences in their *concept* complexes, and under appropriate conditions these constructions direct their behavior. There is evidence of the existence of such internal representations of learned motor responses both in man and animals (Terzuolo and Viviani 1979). Acquisition of skills is based on *concepts* containing these representations. The motor sequences may be very rigid and stereotypic, or they may be somewhat flexible and subject to internal transformations.

Learned motor patterns such as these have an important role in various behaviors of animals. They permit the much faster serial responses than would be possible with ordinary proprioceptive feedback mechanisms. That is, the brain gives orders in advance that are necessary for the next future motor response. When a monkey or a man picks up a well-known object, an apple for example, the object is not dropped, because the muscles of the arms use just the right amount of force needed to keep an apple in hand. We might think that the brain received information about the weight of the apple through the proprioceptors located in the arm, and that this information served as a control in maintaining the right muscle tone. It is easy to show that this is not the case, by presenting the subject with an apple containing a hidden lead weight. The hands move downward, and the apple may be dropped, as the unexpected weight remains uncompensated. As the arm moves down, appropriate feedback begins and with some delay muscles assume their correct tone. Electrophysiological experiments have shown that when objects are offered to and taken by monkeys the brain gives its calculated order several ten msec in advance and the feed-back mechanism makes only later adjustments if necessary (Terzuolo and Viviani 1979).

So far we have been speaking of responses, but an isolated response is very rare. Animals perform highly complex *actions*. An animal does not respond to the tone of a bell but it wants to go out from a cage, or, if hungry, it wants food. The *action* is an organized response built up from units of simple responses.

Finally I would emphasize that *concepts* are not static but dynamic structures. They are in "parallel" (rather than "serial") mode (Arbib 1972), and participate in hierarchically organized decision mechanisms (Dawkins 1974). The decisions are with respect to many possible responses, or subroutines (Gould and Gould 1982). The state of the environment may change from moment to moment, and the actions and their consequences, whether successful or not, are immediately represented in the referential structures of the brain and influences further events.

4. Evolution of *concept* structures

4.1. Concepts *based upon genetic memory*

Considering the evolution of animals we can recognize that the construction ability of the animal brain is increasingly sophisticated as we advance on the evolutionary tree. Even primitive nervous systems are capable of recognition and selection of appropriate responses, as illustrated by von Uexkull's description of the behavioral functional cycle of the tick (von Uexkull 1921). Mated females climb bushes and wait until a mammal passes by. When they perceive the odor of butyric acid, which is secreted by the skin glands of all mammals, they release their grip and drop. If they should strike something warm, they begin to search for an area of skin and suck blood to feed. The simple *concept* generated by the tick nervous system determined most probably exclusively by the genetic memory, consists of the cue: receptors of the butyric acid, referential structure: developmental stage of the post-mating period, and behavior instruction: to drop. In this case the simple *concept* that directs the behavior of the animal is determined solely by genetic memory.

4.2. Concepts involving both genetic and neural memory

As the referential structures of the animal become more complex, more complex actions appear. This is illustrated by the predator recognition and avoidance learning of paradise fish (Csanyi 1985a, 1985b, 1985c).

We studied avoidance behavior of paradise fish in the presence of a living pike, a living goldfish, and various fish-like dummies. They were in semi-natural environments or aquatic shuttle-boxes. Detailed description of the experiments and their analysis are in the above-cited articles.

Our ethological observations may be summarized as follows: A hungry pike attacks the paradise fish, and, depending on the relative size of the two animals, this can be quickly fatal to the prey. If the encounter is between a satiated pike and a paradise fish, there is a tendency in naive paradise fish to approach and explore the pike at the first meeting. This exploratory activity quickly fades during habituation trials but it does not cease entirely. If a paradise fish is chased by a hungry pike once or twice, a permanent behavioral change occurs. The approach tendency disappears and is replaced by active avoidance and escape behavior. When a satiated or a hungry pike was a companion in an aquatic two-way shuttle box, paradise fish showed fast avoidance learning in the presence of a hungry pike and showed no observable learning in the presence of a satiated one.

To determine whether the predator's features were really necessary for attracting the paradise fish we observed the behavioral reaction of paradise fish to non-predatory goldfish. Approach tendency was very high in the first meeting and quickly waned thereafter. Contrary to the experiment with the pike, escape behavior did not increase during the trials.

We then searched for specific cues by which paradise fish recognize and distinguish predatory species from the harmless ones. We found that a mild electric shock applied in the presence of a goldfish for 30 seconds permanently changed the behavior of paradise fish. It showed significantly less approach behavior in all subsequent meetings. The same shock treatment did not change the behavior of experimental fish towards conspecifics.

To summarize the results of these experiments:

1. Naive paradise fish tend to approach and explore a satiated predator or a harmless other fish at the first meeting.
2. This exploratory activity quickly fades during habituation trials.
3. If the paradise fish are chased by the predator or if subjected to mild electric shock in the presence of a harmless fish of another species, then a permanent behavioral change occurs. At the next meeting active avoidance movements are emitted in the presence of the previous companion. In appropriate experiments avoidance learning can be demonstrated.
4. Shock in the presence of conspecifics does not results in avoidance behavior.

To determine the cues that directed the avoidance learning in the paradise fish we performed experiments using fish-like dummies in one of the compartments of the shuttlebox and applied mild electric shock whenever the paradise fish entered into this compartment. The most important feature eliciting avoidance behavior was having two eyes. Only dummies with two eye-like spots provided appropriate cues. Dummies with one spot or no spots failed as cues. Dummies or living goldfish alone without shock punishment also failed to induce avoidance behavior.

To account for the main features of paradise fish avoidance behavior I have developed a new model, the *Interacting Learning Hypothesis* (ILH) (Csanyi 1985a, 1985b). It is based upon the assumption that the brain actively models its environment. The central feature of this hypothesis is the *species-specific key stimulus* (SSKS). Appearance of such stimuli in the environment of the animal results in an identification process which starts with immediate orientation and approach and results in learning (Figure 16.4). If the carrier of the key stimuli is harmless, then the exploratory behavior slowly diminishes by simple habituation. If the carrier of the SSKS interacts with the paradise fish and causes fear or pain, then a species-specific defense reaction (Bolles 1971) is activated, and the exact features of the carrier are fixed in the memory. If the same stimulus carrier returns, the defense reaction is elicited by its *representation* in memory without exploratory behavior. We call this hypothesis interactive learning because learning is the result of the paradise fish's active search for the SSKS carriers.

In terms of the modelling function of the brain: the brain of the paradise fish is assumed to be able to build up *concepts* which direct its avoidance behaviour. The cue part of the *concept* is some kind of "wired in" representation of the key stimuli (two eye spots), the *behavioral instructions* are expressed in the approach

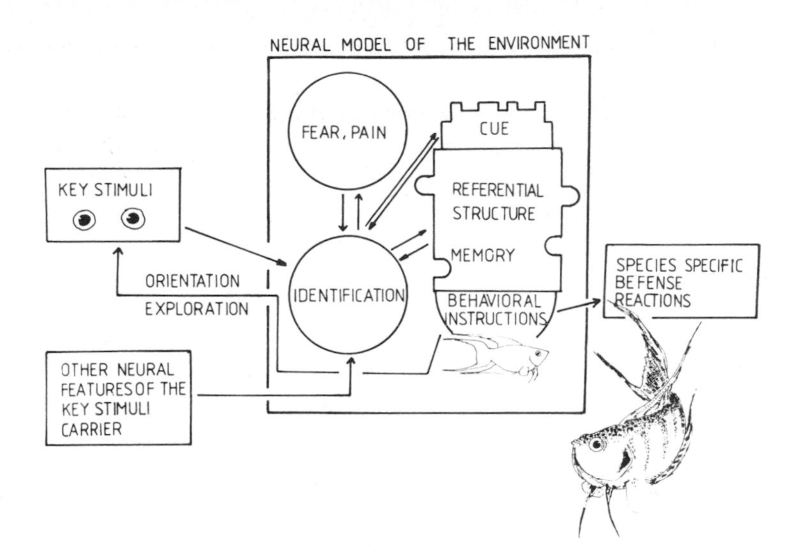

Figure 16.4. Sketch of the Interacting Learning Hypotheses

and escape behavior actions, and the *referential structure* is built up during some of the first encounters. The referential structures after perceiving and recognizing the cue can direct the selection of the most appropriate behavioral action depending on previous experience.

The paradise fish lives in small ponds and streams where the number of the possible predator species is certainly limited. The occasional meeting with one of them results in an SSKS - directed learning process which creates a network of representations of the appropriate predator in the brain that remains active for the rest of the animal's life. (We have some data that one encounter with a living fish results in a memory trace in the paradise fish lasting for at least three months.) This network built up from various *concepts* can be regarded as a *dynamic model* of the environment. The dynamics of this model result in *avoidance behavior* by the paradise fish if and only if it is necessary, that is, in the case of appearance of the predator. The ability to rely on this model may be thought of as an ability to predict future contingencies.

4.3. Concepts of 'insight' in higher animals.

W. Kohler in his famous reports of 'insight' behavior in the chimpanzee described several cases of intelligent tool-using of captive animals. For example, his chimpanzees could put boxes on top of one another in order to reach a banana suspended from the ceiling (Kohler 1921). From Kohler's description it is evident that the behavior sequences were not learned by trial and error. A chimpanzee would sit and look around at the box, at the place under the banana, and at the banana, until the solution had been found. These experiments have frequently been cited as illustrating the highest intelligence shown by animals, although later research suggested that the chimpanzees could not solve problems of this sort if they did not first have certain experiences.

Quite recently Epstein and his coworkers (Epstein et al., 1984) repeated these experiments with pigeons, much simpler animals than apes. Pigeons that acquired the relevant skills by operant conditioning were able to solve the Kohler-problem in a remarkably chimpanzee-like fashion. However, they had to be trained to make three independent behaviors: a. directional pushing of a paper-box, b. climbing onto the box, and c. peck the "banana", a small banana shaped object (this final act was supposed to control the nearby food-magazine). In the crucial experiments the banana was attached to the ceiling of a large round arena in which an appropriate paper box was placed on the floor, far from the banana. The performance of the properly trained birds were very similar. At first each pigeon appeared to be confused. It stretched and turned beneath the banana, looked back and forth from the banana to box, and so on. Then each bird began rather suddenly to push the box in the direction of the banana. Each subject stopped pushing in the appropriate place, climbed and pecked the banana. With various control procedures it was found that omitting any of the three above-mentioned training procedures resulted in failure. It is also important to note that these pretraining procedures did not include training to push the box towards the banana, nor was pushing the box reinforced in the presence of the banana. Consequently the successful performance must be regarded as genuinely novel action.

In this elegant experiment formation of separate *concepts* is revealed, which in the final test were united into one complex structure by their referential mechanisms resulting in an appropriate action pattern. The pigeon's brain is already able to perform mental transformations on its concepts. It is able to figure out the results of possible behaviors and to build up proper environmental models solely by mental operations on previously acquired concepts. This is a process which, in ordinary language is called thinking.

5. Replicative memory-model of the brain

Modelling by the brain occurs in the formation of higher organizations above the neuronal level. In our earlier studies we considered the epigenetic development of the animal brain as an evolutionary process with evolving *concept* components, and we termed it "neural evolution." We treated it in the framework of an autogenetic, replicative model (Csanyi 1978, 1980, 1982, 1985d, 1986). In the organization of the biosphere hierarchically nested replicative networks of various components, i.e., organisms, can be found. Animals as components of these replicative networks are needed to process replicative information of their own replication, so the animal brain has evolved a capacity for problem solving. The fuzzy set of problem solving algorithms that a given species need to have for its survival and reproduction may be called "animal intelligence". In the previous sections we discussed the *concept* as the basic functional unit of animal problem solving. In the next sections we discuss some evidences that *concepts* are able to evolve and that concept-superstructures actually are evolving component-systems.

Based on neurophysiological data and theories, *concepts* are assumed to develop in the brain from the effect of external stimuli through a group-degenerative selection of neurons and reentrant signals, (Edelman 1977). Each cycle of the reentrant signal mechanism would correspond to a replicative cycle. So concept development is essentially a replicative process.

5.1. Selection, mutation and recombination of concepts

Selection is assumed to be an inherent feature of the replicative memory model. The idea of some kind of selective mechanism in mental processes has already been proposed on an intuitive basis (Pringle 1951, Pattee 1965, Darlington 1972, Campbell 1974, Glassman 1977). Dawkins (1976) has even suggested the multiplication of memory traces analogous to the multiplication of genes. The replication of *concepts* is assumed to take place under the continuous controlling effect of various motivational mechanisms and of the environment in a "generate-test-generate" fashion. *Concepts* having adaptive value for an animal would enter the multiplication cycle more often, and therefore be produced in greater numbers than irrelevant ones or those that are harmful. Only a finite number of neurons is available for these activities, and *concepts* would be in constant competition for excitable elements. Differential reproduction of *concepts*, being the essence of selection, would, therefore, have to take place. It is easy to see that "mutational" changes of *concepts* should also occur. Selection of elements of the concept-component is a group-degenerative process, namely it should often happen that in a replicative cycle an element with new properties is incorporated into a *concept*. It follows from the selective mechanism that there would be direct competition among the different variants. It is conceivable that recombination of simultaneously replicating concepts could also occur, thus producing new, even more adaptive forms.

5.2. Autogenesis of concepts

It is generally accepted that the nervous system of simpler animals is largely "hard wired", that is a majority of neuronal connections are genetically determined, and the final structure of most *concepts* can, therefore, be traced back to the genetic memory. At the same time even the most primitive organisms show habituation and sensitization, indicating a possibility for emergence of simple *concepts*.

In our study of the various autogenetic processes, (Csanyi 1982, 1985d, Csanyi and Kampis 1985) we have found that replicative organization emerges when the dynamics of the components of a given organizational level of a system are influenced and controlled by constraints that can be interpreted on a higher level. In the case of the nervous system this higher level is the level of overt behavior. For an analysis of the organization of the nervous system, data must be provided by the behavioral sciences. Behavior is the mode by which the animal secures the integrity, maintenance and reproduction of its organism. Behavior is the output of the brain and at the same time it is the "interface" that relays the effects of environment to the animal. The body structure of the animal and its niche represent constraints and potentials that also influence the brain of an animal. These constraints determine the boundary conditions under which the behavior of an animal is optimal and adaptive with respect to survival and reproduction. This, however, does not mean that all behavioral patterns are *a priori* adaptive, but we must search for these and provide solid evidence for their adaptive value.

If the brain's environmental model constructed from concept-components has indeed a replicative organization, then we must find the factors that control it at the level of behavior. There are two such important characteristics of animal behavior: cyclicity and variability.

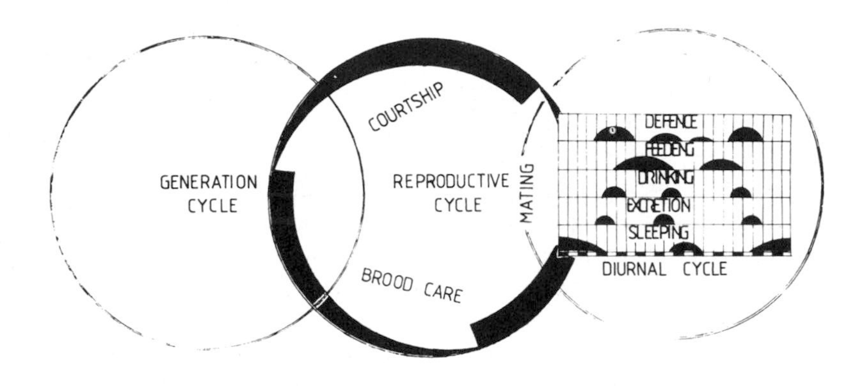

Figure 16.5. Interconnected cyclic processes in the animal life

1. *Behavior is cyclic.* Maintenance of life and reproduction are based on hierarchically organized cyclic processes (Figure 16.5). In addition to daily feeding and resting cycles, there are seasonal cycles of reproduction and migration, and, encompassing these, the great cycle of changing generations. This cyclic organization is the basis of the type of cyclic activity in which *concepts* generate behavior units. Cyclicity is the condition that provides the mechanism for changing *concepts*, ensuring their recurrent activity. This cyclic organization is the exact analogue of the reproductive cycles of a molecular autogenetic system, and also of the replicative cycles of cellular and organismic autogenesis (Csanyi 1985d, 1986). *Concepts* not capable of cyclic activity will be selected against and will not participate in a replicative organization. All other *concepts* can exist, provided that their cyclic activation and the resulting behavior does not interrupt the great cycle necessary for survival and reproduction. Replicative selection of concept-components is provided by this organization. It is a creative process, since all concepts that are not harmful in the above sense can survive whatever behavior they activate. Practically, their survival depends also on competitive selection in two senses. There is a competition among *concepts*, as reinforcement mechanisms of the brain favor *concepts* that lead to adaptive behavior in the given environment. On the other hand, in the case of strong competition among individuals, those will survive who can generate better (more adaptive) *concepts*. This latter type of selection influences the concept-generating mechanisms of the animal brain at the

genetic level of organization.

2. *Behavior is variable.* Behavior, like any other phenotypic character, can be traced to the genetic constitution of the animal, and all forms of behavior are remarkably variable. There are two distinctive factors creating this variability of behavior. One depends on variation in genetic constitution. The other is the learning process. To have evolved under natural selection, behavior mechanisms must have been genetically variable. Indeed a wide range of genetic variability of behavior is known in behavior genetics (Fuller and Thompson 1978, Fuller and Simmel 1983, Csanyi 1985a,b,c; Csanyi and Gervai 1986, Gervai and Csanyi 1985). On the other hand, the biological function of different types of learning is to provide an even greater "ontogenetic" variability beyond the level of genetic organization. The main selective factor of the sequence of behavioral patterns develops through learning and shapes the most adaptive forms of behavior in a given situation. The idea that there is a close analogy between the learning process and biological evolution is not new, and it appears frequently in studies of *animals behavior* (Pringle 1951, Skinner 1966, Pask 1972, Cloak 1975). The existence of a selective mechanism based upon learning has been elaborately analysed by Staddon and Simmelhag (1971).

If *concepts* are replicative, as has been assumed, then selection operating in the replicative process that is the main factor in creating ontogenetic variability of behavior. The essential process of learning according to the present framework is the selection of *concepts* generated in and by the brain and in that way the construction of the dynamic concept-superstructures of the brain's environmental model.

References

Anderson JR (1984) The Development of Self-Recognition: A Review. Dev Psychobiol **17**:35-49

Archer J (1976) The Organization of Aggression and Fear in Vertebrates. In: Bateson PPG, Klopfer P (Eds) Perspective in Ethology. Vol 2 p 231-199. Plenum, New York

Arbib MA (1972) The Metaphorical brain. Wiley, New York

Beritashvili IS (1971) Vertebrate Memory, Characteristics and Origin. Plenum Press, New York

CampbellDT (1974) Evolutionary Epistemology. In: Schilpp PA (ed) The Philosophy of Karl Popper. Open Court Publ, Lasalle Illinois

Cloak FT Jr (1975) Is Cultural Ethology Possible? Human Ecology **3**:161-182

Craik KJW (1943) The Nature of Explanation. Cambridge Univ Press, Cambridge UK

Croze H (1970) Searching Image in Carrion Crows. Z Tierpsychologie **5**:1-86

Csanyi V (1978) Az evolucio altalanos elmElete. Fizikai Szemle **28**:401-443

Csanyi V (1980) The General Theory of Evolution. Acta Biol. Hung Acad Sci **31**:409-434

Csanyi V (1982) General Theory of Evolution. Publ House Hung Acad Sci, Budapest

Csanyi V (1985a) Ethological Analysis of Predator Avoidance by the Paradise Fish *(Macropodus opercularis)*: I. Recognition and Learning of Predators. Behaviour **92**:227-240

Csanyi V (1985b) Ethological Analysis of Predator Avoidance by the Paradise Fish *(Macropodus opercularis)*: II. Key Stimuli in Avoidance Learning. Anim Learn Behav **14**:101-109

Csanyi V (1985c) How is the Brain Modelling its Environment? A Case Study by the Paradise Fish. In: G Montalenti, G Tecce (eds), Variability and Behavioral Evolution, Proceedings, Accademia Nazionale dei Lincei, Roma (1983 Quaderno No 259

Csanyi V (1985d) Autogenesis; Evolution of Selforganizing Systems. In: Aubin JP Saari D, Sigmund K (eds) "Dynamics of Macrosystems" Proceedings Laxenburg, Austria, pp 253-268, Springer, Berlin

Csanyi V (1986) Evolutionary Systems: A general Theory. California Univ Press (submitted), pp 330

Csanyi V, Gervai J (1986) Behavior-Genetic Analysis of the Paradise Fish *(Macropodus opercularis)*: II Passive Avoidance Conditioning in Recombinant Inbred Strains. Behav Genet **16**:553-557

Csanyi V, Kampis Gy (1985) Autogenesis: The Evolution of Replicative Systems. J Theor Biol **114**:303-321

Csanyi V, Toth P, Altbacher V, Doka A, & Gervai J (1985a): Behavior Elements of the Paradise Fish *(Macropodus opercularis)*: I Regularities of Defensive Behavior. Acta Biol Hung **36**:93-114

Csanyi V, Toth P, Altbacher V, Doka A, & Gervai J (1985a): Behavior Elements of the Paradise Fish *(Macropodus opercularis)*: II A Functional Analysis. Acta Biol Hung **36**:115-130

Darlington, PI (1972) Nonmathematical Concepts of Selection, Evolutionary Energy and Levels of Evolution. Proc Natl Acad Sci, USA **69**:1239-1243

Dawkins R (1974) Some Descriptive and Explanatory Stochastic Models of Decision Making. In: McFarland, DJ (ed), Motivational Control System Analysis. Academic Press, London

Dawkins R. (1976) The Selfish Gene. Oxford Univ Press, New York

Denett DC (1983) Intentional Systems in Cognitive Ethology: The "Panglossian Paradigm" Defended. Behav Brain Sci **6**:343-390

Domjan M, Galef BG Jr (1983) Biological Constraints on Instrumental and Classical Conditioning: Retrospect and Prospect. Anim Learn Behav **11**:151-161

Edelman GM (1977) Group Degenerate Selection and Phasic Reentrant Signaling: A Theory of Higher Brain Function. In: Schmitt FO (ed) The Neurosciences: Fourth Study Program. The MIT Press, Cambridge

Eibl-Eibesfeldt I (1970) Ethology, the Biology of Behavior. Holt Rinehart Winston, New York

Epstein R, Kirshnit CE, Lanza RP, Rubin LC (1984) "Insight" in the Pigeon: Antecedents and Determinants of an Intelligent Performance. Nature **303**:61-62

Evarts EV (1967) Representation of Movements and Muscles by Pyramid Tract Neurons of the Precentral Motor Cortex. In: Yahr MD, Purpura DP (eds) Neurophysiological Basis of Normal and Abnormal Motor Activity pp 215-251, Raven Press, New York

Fuller JL, Thompson WR (1978) Foundations of Behavior Genetics. Mosby, Saint Louis

Fuller JL (1983) Behavior Genetics: Principles and Applications. Erlbaum, Hillsdale

Gallistel CR (1980) The Organization of Action. A New Synthesis. Lawrence Erl baum, Hillsdale pp 335-360

Gallup GG (1970) Chimpanzees, Self-recognition. Science **167**:86-87

Garcia J, Ervin FR, Koelling R (1966) Learning with Prolonged Delay of Reinforcement. Psychonomic Science **5**:121-122

Gervai J, Csanyi V (1985) Behavior-Genetic Analysis of the Paradise Fish *(Macropodus opercularis)*: I Characterization of the Behavioral Responses of Inbred Strains in Novel Environments. A Factor Analysis. Behav. Genet **15**:503-519

Glassman RB (1977) How Can So Little Brain Hold So Much Knowledge? Psychol Rec 2:393-415

Gould JL, Gould CG (1982) The Insect Mind: Physics or Metaphysics? In: Griffin DR (ed) Animal Mind- Human Mind. Springer, Berlin

Grastyan E, John ER, Bartlett I (1978) Evoked Response Correlate of Symbol and Significance. Science **201**:168-171

Griffin DR (1984) Animal Thinking. Harvard Univ Press, Cambridge

Hebb DO (1946) On the Nature of Fear. Psychol Rev **53**:259-276

HebbDO (1949) The Organization of Behavior: A Neuropsychological Theory. Wiley, New York

Hinde RA, Stevenson-Hinde J (1973) Constraints on Learning. Academic Press, New York

Hubel DH, Wiesel TN (1974) Sequence Regularity and Geometry of Orientation Columns in the Monkey Striate Cortex. J Comp Neurol **158**:267-294

Humphrey NK (1978) Nature's Psychologists. New Scientist 78:9OO-9O3

Jerison HJ (1973) Evolution of the Brain and Intelligence. Academic Press, New York

Jerison HJ (1985) Animal Intelligence as Encephalization. Phil Trans R Soc Lond B **308**:21-35

John ER (1972) Switchboard Versus Statistical Theories of Learning and Memory. Sciences **172**:850-864

Johnston TD (1982) Selective Cost and Benefits in the Evolution of Learning. Adv Study Behav **12**:65-106

Julesz B (1980) Textons, the Elements of Texture Perception and Their Interactions. Nature **290**:91-97

Kohler W (1921) Intelligenzprüfungen an Menschenaffen. Berlin

Krechevsky I (1932) Hypotheses in Rats. Psychological Review **39**:516-532

Kruijt JP (1964) Ontogeny of Social Behavior in Burmese Red Jungle Fowl. Behavior Suppl **12**:1-201

Lorenz K (1981) The Foundations of Ethology. Springer, New York

Mackay DM (1951-52) Mindlike Behavior of Artefacts. Brit J Phil Sci **2**:105-121

Mackay DM (1966) Cerebral Organization and the Conscious Control of Action. In: Eccles, J (ed) Brain and Conscious Experience. pp 422-445, Springer-Verlag, New York

Macphail EM (1982) Brain and Intelligence in Vertebrates. Clarendon, Oxford

Mayr E (1976) Evolution and Diversity of Life. Belknap, Cambridge, Mass. and London

Menzel EW, Johnson MK (1976) Communication and Cognitive Organization in Human and Other Animals. Ann NY Acad Sci **280**:131-142

Mesarovic MD (ed) (1964) Views on General Systems Theory. John Wiley, New York, p 5

Miller GA, Galanter E, Pribram KH (1960) Plans and Structure of Behavior. Holt Rinehart Winston, New York

Milner PM (1976) Models of Motivation and Reinforcement. In: Wanguing A, Rolls ET (eds), Brain Stimulation Reward. p.543-556

Milner PM (1977) Theories of Reinforcement, Drive, and Motivation. In: Iversen L, Iversen S, Synder SH (eds) Handbook of Psychopharmacology, 7:181-202 Plenum Press, New York

Morgan CL (1896) Habit and Instinct. Edward Arnold, London

O'Keefe J, Nadel L (1974) Maps in the Brain. New Scientist **62**:749-751

Olton DS, Samuelson RJ (1976) Remembrance of Places Passed: Spatial Memory in rats. J Exp Psych: Anim Behavior Proc **2**:97-116

Pask G (1972) Learning Strategies, Memories and Individuals. In:Robinson WH, Knight DE(eds) Cybernetics Artificial Intelligence and Ecology, Proc.Fourth Ann Symp Am Soc Cybernetics, Spartan Books, New York Washington

Pattee HH (1965) The Recognition of Hereditary Order in Primitive Chemical Synthesis. In Fox SW (ed) The Origin of Prebiological Systems Academic Press, New York

Plotkin HC, Odling-Smee FJ (1981) A Multiple-level Model of Evolution and its Implications for Sociobiology. Behav Brain Sci **4**:225-268

Pribram KH (1976) Problems Concerning the Structure of Consciousness. In: Globus GG, Maxwell G, Savodnik I (eds), Consciousness and the Brain. Plenum Press, New York

Pringle JWS (1951) On the Parallel Between Learning and Evolution. Behaviour **3** :174-214

Skinner BF (1966) The Phylogeny and Ontogeny of Behavior. Science **153**:1205-1213

Sokolov EN (1960) Neuronal Models and The Orienting Reflex. In: Brasier MAB (ed) Central Nervous System and Behavior Macy Foundations, New York

Sperry RW (1976) Mental Phenomena as Causal Determinant of Brain Function. In: Globus GG, Maxwell G, Savodnik I (eds), Consciousness and the Brain. p 163-179 Plenum Press, New York

Staddon JER, Simmelhag VL (1971) The "Superstition" Experiment: Reexamination of Its Implication for the Principles of Adaptive Behavior. Psych Rev **78**:3-43

Stuss DT, Picton WT (1978) Neurophysiological Correlates of Human Concept Formation. Behav Biology **23**:135-162

Terzulo CA, Viviani P (1978) The Central Representation of Learned Motor Patterns. In: Talbot RE, Humphrey DR (eds) Posture and Movement, Raven, New York

Tinbergen N (1960) The Natural Control of Insects in Pine Woods Arch Neder Zool **13**: 265-379

Tolman EC (1932) Purposive Behavior in Animals and Man. Appleton-Century-Crofts, New York

Uexkull J von (1921) Umwelt and Innenwelt der Tiere 2nd ed Berlin

Watanabe M (1979) Prefrontal Unit Activity and Delayed Conditional Discrimination In: Agranoff B, Tsuhada V (eds), Memory and Learning, Raven Press, New York

ANIMAL LANGUAGE RESEARCH: MARINE MAMMALS RE-ENTER THE CONTROVERSY

Ronald J. Schusterman
Institute of Marine Sciences
University of California
Santa Cruz, California 95064
and California State University
Hayward, California 94542

Robert Gisiner
Institute of Marine Sciences
University of California
Santa Cruz, California 95064

A horse named Clever Hans, who lived in Germany at the turn of the century, drew world-wide attention because he could apparently talk and solve arithmetical problems. By tapping with his front leg the horse could not only do arithmetic, but could also combine letters into words, words into sentences, and thus express his thoughts. What his questioners were unaware of was that Clever Hans used a simple "go" or "no-go" set of cues. The "go" cue was a slight leaning forward by the questioner, the "no-go" cue was an inadvertent straightening-up by the questioner when the correct number of taps had been reached (Pfungst, 1911; Sebeok and Rosenthal, 1981). Every time the correct number of taps was given the horse received a food reward. There are two lessons to be learned from the Clever Hans Effect, as it has come to be known. The first lesson is an obvious one; take precautions to prevent inadvertent cueing of subjects in psychological experiments. The second, and related, lesson is that one should interpret animal behavior parsimoniously rather than otherwise, i.e. use simple and straightforward assumptions and explanations of behavior rather than invoking uncalled-for complex processes.

However, as several recent authors have pointed out (e.g. Griffin, 1984) simple explanations and assumptions should not obfuscate or deny the existence of complex processes. Two examples from the animal problem-solving literature serve to illustrate this point. In the first example, a recent experiment on size transposition by a California sea lion (*Zalophus californianus*), it was shown that the animal learned about stimulus relations as well as about particular instances reflecting those relations (Schusterman and Krieger, 1986). The authors had to invoke both a cognitive (search image) as well as a mechanistic (stimulus generalization) concept in order to interpret their data. The second example comes from Köhler's "insight" learning by chimpanzees (Köhler, 1925). Although the phenomenon cannot be explained by a simple stimulus generalization mechanism, in which transfer is based on a *perceptual* similarity, it is partially explainable *conceptually* in terms of "rule learning" or learning set (Harlow, 1949). In this case, chimpanzees, by means of transfer of training, transcend the perceptual characteristics of the stimuli and respond with an abstract rule or strategy (Schusterman, 1962). Epstein, Kirshnit, Lanza,

NATO ASI Series, Vol. G17
Intelligence and Evolutionary Biology
Edited by H. J. Jerison and I. Jerison

and Rubin (1984) have recently shown that experience with each component of a complex task leading to a spontaneous interconnection of the components may provide a different interpretation of "insightful" problem solving.

This paper will address the issue of appropriate parsimonious explanations for the learning of symbols and syntax by two Atlantic bottlenose dolphins (*Tursiops truncatus*) and three California sea lions. We will address a few of the issues recently raised by language comprehension projects in which one female dolphin named Akeakamai and one female sea lion named Rocky have been signaled gesturally to perform actions on objects (Herman, in press; Herman, 1986; Herman, Richards, and Wolz, 1984; Schusterman and Krieger, 1984; Schusterman and Krieger, 1986). Objects, properties and locations of objects, and actions were assigned unique symbols which could be combined to generate a limited set of commands.

Brief Outline of the Animal Language Research (ALR) Controversy, with Special Emphasis on Marine Mammals

Many of us who are now comparative psychologists, ethologists and sociobiologists were, as children, thrilled, intrigued and inspired by Hugh Lofting's fictional country doctor, John Doolittle, who could talk with the animals. As noted by Lorenz (1952), the goal of talking with animals appears an ancient one and extends back at least to biblical times in the legend of King Solomon talking the language of "beasts and of fowl and of creeping things, and of fishes" (I Kings IV.33). Today we seem to be realizing our earlier dreams by finding numerous kinds of mental attributes that "dumb" animals share with linguistically sophisticated humans (Roitblat, 1987). Right from the start, however, contemporary ALR (as it has been called by Hoban, 1986) was considered controversial (Wood, 1973). The crux of the controversy appears to be centered around the "all-or-none" question of animals (particularly anthropoid apes) being capable of communicating linguistically. Furthermore, Premack (1986) and Hoban (1986) suggest that language is not a unitary phenomenon, probably did not evolve solely from call systems of nonhuman primates and does not function exclusively as a communication system. They contend, rather, that the antecedents of human language are multifaceted and include some of the following "higher order mental processes:" (a) storing networks of percepts to form concepts, (b) using symbols to refer to objects and events, and (c) organizing events into a serial order. These and other linguistic abilities frequently depend on higher order conditioning that allows associations to be formed between the signals themselves, producing "logical worlds . . . built of concatenated stimulus events" (Hollis, 1984).

Currently many investigators in ALR, despite earlier controversy, agree that anthropoid apes, unlike most five-year-old children, are not linguistically competent enough to produce or comprehend an intelligible sentence. However, with training they are capable of semantic communication using multisign sequences (Miles, 1983). They can communicate this way with their teachers and/or with one another in either Pidgin Sign English (Fouts, Fouts and Schoenfeld, 1984), or in an artificial language consisting of visual symbols (Savage-Rumbaugh, 1986). With few exceptions, the earlier all-or-none focus of ALR on grammatical competence has been abandoned in favor of attempts to describe and understand the emergence of symbolic communication capabilities in animals as a function of species differences, training and developmental variables. Attempts to determine the limits of artificial language acquisition in animals are of interest because they may enable us to pinpoint and analyze precursors of human language ability in the cognitive processing skills of nonhuman animals. What are some of the protolinguistic skills of animals, and are there some animal groups that are capable of acquiring a natural language?

The first failed attempts to teach spoken language to chimpanzees occurred during the 1930's and began as part of a series of cross-fostering studies aimed at describing the similarities and differences between the mental and physical development of human and chimpanzee infants (Kellogg, 1968). The breakthrough in ALR occurred in the late 1960's with two infant chimps (Washoe and Sarah) in programs initiated by Allan and Beatrice Gardner (1969) and by David Premack (Premack and Schwartz, 1966). Apes were considered logical choices for ALR based on a variety of homologous psychological traits shared by apes and humans, as a consequence of their relatively recent descent from common ancestors.

Lilly and Dolphins

Chimpanzees were not the only subjects of ALR in the 1960's. Several ALR programs which, in fact, preceded the ape investigations were inspired by John Lilly's (1961) attempts to teach Atlantic bottlenose dolphins to understand and speak English. Lilly's choice of dolphins for ALR was based on two interlocking criteria:

1. Bottlenose dolphins have a relatively large brain with a proportionally large neocortex. This can be expressed quantitatively as a high encephalization quotient or EQ (Jerison, 1973). According to Jerison, EQ is a measure of biological intelligence or information processing capacity. Whereas California sea lions and other pinnipeds have EQ's estimated to be between one and two, similar to those of terrestrial carnivores, the EQ's of dolphins* are more similar to chimpanzees, ranging between two and five (Eisenberg, 1981; Worthy and Hickie, 1986).

2. The bottlenose dolphin is a highly vocal and social creature and, according to Lilly, showed signs of possessing semanticity in its vocal communication with members of its own species. In addition, Lilly thought bottlenose dolphins were an exceptional species for ALR because they showed "kindliness" toward man and appeared to be capable of mimicking human voice sounds in an intelligible way (Lilly, 1961).

Although Lilly failed in his attempt to establish the English language as a direct means of communication between humans and bottlenose dolphins, his ideas stimulated several others to use dolphins in ALR (e.g. Bastian, 1967; Batteau and Markey, 1968; Lang and Smith, 1965). The follow-up work to Lilly's initial effort has been succinctly detailed and summarized by Wood (1973). We will briefly review the work of Batteau and his group since it and the ape work seem to have led rather di-

---

*We are speaking of dolphins as a group (relative to pinnipeds as a group). The EQ of the bottlenosed dolphin is about 2.8, at the middle lower end of the range. [EQ was computed with exponent (slope) = 0.755; constant = 0.0537. See Harvey, this volume, and the final "Afterthoughts" chapter. Ed.]

rectly to the resurgence of ALR with dolphins, particularly to research on "sentence comprehension" by dolphins (Herman, Richards, and Wolz, 1984).

Batteau's Failed Dolphin Language Studies

Batteau used Skinnerian shaping techniques to acquire a high level of stimulus control over three different "object-action" commands. The object-actions were "hit ball with pectoral flipper," "swim through hoop," and "retrieve bottle." Each of these object-actions was under the stimulus control of a whistle sound projected by an underwater speaker. The whistles were human vocal sounds transformed by Batteau's ingenious electronic devices. Additional whistle sounds controlled up to 15 separate behaviors; these included raising flukes, jumping, emitting sonar pulses, responding to their own "name," etc. With reference to Herman's later work, it is important to note that one of Batteau's dolphins, Maui, also learned to swim right or left in response to different signals and that she could mimic the sounds controlling many of the behaviors.

As most marine mammal trainers know, training dolphins to respond differentially to 15 or more distinct command signals is a far cry from demonstrating that a dolphin has even the basic rudiments of semantic comprehension. In fact, when probe trials were given by Batteau and Markey (1968) so that Maui was signalled to perform two behaviors in sequence without any new special training, the dolphin only performed the first object-action. This finding suggests that either Maui did not understand the task or that at least some dolphins may have difficulty with the grammatical rule of *recursion*. In addition, control trials showed that both dolphins had been conditioned to associate an object-action command signal with a specific location. For example, when the ball and hoop were reversed from their usual positions, the dolphins were confused and first went to the appropriately signalled object, but performed an incorrect action. Then they went to the inappropriate object and performed a correct action. It was also found that both dolphins were biased in their responses.

In Batteau's experiments the dolphins were trained to respond to holophrastic commands and not to separate elements of object and action, unlike the chimp language studies which were just beginning when Batteau met an untimely death by drowning. In the chimp language studies the investigators trained their animals to work with signs that could potentially be combined and recombined in such a way that each combination could convey a separate meaning.

Do Herman's Dolphins Comprehend Sentences?*

It is Herman and his colleagues (Herman, Richards, and Wolz, 1984) who, of all the investigators involved in ALR, have made the strongest claim to date that their dolphins have a "tacit knowledge of syntactic rules" and can comprehend literally thousands of novel sentences up to five words in length. The original claims by the

---

* [I have asked Professor Herman to review specialized issues raised here, and I discuss his comments in the "Afterthoughts" chapter (see also, Herman, 1987). Ed.]

Gardners and by Premack (Gardner and Gardner, 1969; Premack, 1976) that their chimpanzees combined symbols in grammatically competent fashion have since been shown to be exaggerated (Pettito and Seidenberg, 1979; Terrace, 1979).

The basic reasons Herman chose to use dolphins in ALR were the same as those stated by Lilly 15 years earlier: high EQ and a complex vocal communication system. The reason Herman has emphasized language comprehension in dolphins has been his belief that language comprehension and language production may develop as separate systems with comprehension emerging prior to production, and thus probably being more fundamental. For this reason, Herman (in press) has emphasized the study of receptive competencies in his ALR with dolphins. Support for this approach of stressing receptive competencies has recently come from the finding that Kanzi, a young pygmy chimpanzee, needed only exposure, not training, to attain symbolic skills comparable to those attained by common chimpanzees trained in the use and comprehension of symbols (Savage-Rumbaugh, Rumbaugh and McDonald, 1985). Savage-Rumbaugh and her associates think that Kanzi's learning of English merely by hearing it means that the capacity to comprehend speech is possibly an evolutionary precursor to its production, and that language comprehension, more than vocal speech production, is the "essence of language" (Savage-Rumbaugh et al., 1985).

Herman's two bottlenose dolphins, Akeakamai (Ake) and Phoenix, were taught a similar but not identical language in different modalities (an acoustic language for Phoenix and a gestural language for Ake). For purposes of exposition, and in order to make direct comparisons with our own parallel ALR on California sea lions, we will focus most of our comments on Herman's claims about Ake's linguistic competencies as reported in Herman, Richards, and Wolz (1984). In Ake's artificial language, relationships between objects were constructed by signing a goal item first, an item to be acted upon (transported) second, and finally the relational sign FETCH. Herman et al. (1984) referred to this relational sequence as an "inverse grammar."

Ake was trained in a gestural language in which signals were the movements of a trainer's arms and hands. Each object, position modifier and action was assigned a unique signal. The signals could then be combined and recombined following certain rules to produce a circumscribed but complete set of legal instructions. Thus, there was provision for substitution of items in object or action categories, making it possible to replace the object sign HOOP with the sign BASKET or the action sign UNDER with OVER. In addition to object and action signals Ake was given modifier signs referring to left and right and could be instructed to act on an object on her left, but only if there was a paired member on her right. Thus the categories of the dolphin Ake's gestural vocabulary included modifiers (LEFT, RIGHT), objects (e.g. WATER, BALL, PIPE, FRISBEE, etc.) and actions (TOSS, SPIT, TAIL-TOUCH, PECTORAL-TOUCH, etc.). Ake used two classes of objects; *non-transportable* objects that could be moved around the rim of the tank by the researchers (e.g. WATER, water jetted from a hose), and *transportable* objects that floated freely in the pool and could be moved by Ake (e.g. BALL). Ake also had two classes of actions; a relational action (e.g. FETCH, take a transportable object to another object) and direct actions (e.g. SPIT, spit a stream of water at an object). It is important to note that relational actions and direct actions were mutually exclusive categories. This was not the case in the sea lion Rocky's vocabulary where the identical FETCH sign functioned as a direct action or relational action, depending on the signs that preceded it.

At the time of Herman's publication of test data in 1984 (about six years after her training was begun) Ake's comprehension "vocabulary" included two modifier signs, eleven object signs, eight action signs and one relational sign (FETCH). Armed with this sparse vocabulary of 22 signs the dolphin Ake nevertheless had the potential to carry out over a thousand unique commands. An example of the longest sign sequence given to Ake might be WATER, RIGHT PIPE FETCH or RIGHT WATER, PIPE FETCH: these two different sets of instructions contain the same elements, but in a different order, giving each set of signs a different meaning. In the first sequence the referent WATER is the goal item and the referent RIGHT PIPE is the transported item, whereas in the second sequence the referent RIGHT WATER is the goal item and the referent PIPE is the transported item.

The sign sequences given to Ake formed two types of instructions. One type of instruction had the dolphin act on a single object and the other type had the dolphin perform an action relating two objects. By employing various discriminative learning techniques to teach Ake to 1) associate gestural signs with directions, objects, and actions and 2) discriminate between categories of signs by making Ake sensitive to the serial order or sequence of signs, Herman and his associates were able to demonstrate that Ake could respond appropriately to sentence-like commands containing as many as four signs. Moreover, by *generalizing* within categories of signs the dolphin could respond appropriately to a variety of novel commands. For example, Ake could respond correctly the first time she was given the instruction PIPE, RIGHT BALL FETCH ("take the ball on your right to the pipe") or RIGHT BALL, PIPE FETCH ("take the pipe to the ball on your right") after several experiences with commands like RIGHT BALL OVER ("jump over the ball on your right") and PIPE, BALL FETCH ("take the ball to the pipe").

Below are several statements by Herman et al. (1984) summarizing and interpreting the results of their tests of the linguistic competency of their dolphins, in particular Ake's competency on "sentence" comprehension in which the relational action FETCH is used:

> "In summary, the understanding of the function of object names as direct or indirect objects, and of how modifiers may be attached to object names, further illustrate the considerable sensitivity of dolphins to syntactic structure" (p. 199).
>
> "As in natural languages, tacit knowledge of the syntactic rules underlying the language was necessary for correct interpretation of the function of lexical items in the sentence, and for an understanding of the unique semantic proposition being expressed. This is most obvious for the inverse rules in Akeakamai's gestural language." (p. 203).
>
> "Within the nonlinear grammar, Akeakamai demonstrated her ability to assign and reassign functions to earlier words in a sentence, i.e., parse the sentence, on the basis of a succeeding word or words." (p.207)

As Herman himself has noted (1987), his ALR with dolphins was started following the heated criticisms of ALR with apes and thus was designed to avoid the methodological pitfalls of the ape language work. Nevertheless, the dolphin language work by Herman can and has been criticized from a conceptual standpoint. One can readily discern from the above quotes that the interpretation of the dolphin's performance by Herman et al. (1984) is in terms which are linguistically loaded (e.g. "grammar", "sentence", "lexical", "direct object", "indirect object", "semantic proposition"). In addition, these statements seem again to reflect an emphasis, as did earlier ALR, on grammatical competence as an all-or-none factor. This has led to rather sharp criticism from Premack (1986) on the grounds that this "flurry of linguistic terms is gratuitous" because an interpretation of the dolphins' performance requires no linguistic terms of any kind. According to Premack, "Herman et al. have shown two things: discrimination of temporal order and the learning of rules based on perceptual classes" (Premack, 1986, p. 26). The two rules used to generate all the instructions to the dolphins were: (1) (Modifier) + Object + Action, and (2) (Modifier) + Object A + (Modifier) + Object B + Action 2, where Action 2 was a signal instructing Ake to take Object B to Object A. Reversing the positions of the two object signals in the string reversed the meaning of the command. Although we agree that these competencies are the major ones that have been (more or less) demonstrated in Herman's dolphins, we disagree with Premack's dismissal of these competencies as unnecessary for demonstrating language learning. Rather, we believe that all of the cognitive abilities that the dolphins have demonstrated are necessary for language learning, but are perhaps not sufficient. (See Herman, in press, for a rebuttal of Premack's criticism of his ALR with dolphins.)

Hoban (1986) essentially agrees with many of Premack's criticisms of Herman's work and has added some herself. For example, we think that Hoban is correct in pointing out that Herman et al. have prematurely used the term "word" in their description of signs and their associated referents. Although it is true that the dolphins (as well as the sea lions) can correctly choose the ball or the pipe when given the BALL sign or PIPE sign, can they choose the correct signs for ball or pipe when they are presented with the referents ball and pipe? In other words, are the signs and their referents interchangeable as are human words and the things they signify? The answer is no. An even more advanced form of symbol use occurs when words or symbols refer to other symbols. There is no indication whatsoever that dolphins are, as yet, capable of using symbols or words to stand for one another. The referential quality of the dolphin symbols has not been as well established as that of the chimpanzee symbols used in ALR by Savage-Rumbaugh (1986).

In general, the arguments by both Premack and Hoban are that Herman's experiments, despite their methodological elegance, have not, as yet, demonstrated anything "even remotely reminiscent of linguistic skills in the behavior of the dolphins" (Hoban, 1986, p. 144). We agree that the results of Herman's ALR with dolphins have not been interpreted cautiously enough, i.e. the dolphin's language-like learning skills have not been conceptualized in accordance with Lloyd Morgan's Canon.

In the present paper it will be our contention that what has been taught to both the dolphins and the sea lions is indeed a language, albeit a pidgin or simplified language, consisting of three categories of signs arranged according to two types of rules which give the language its "openness" or "combinatorial productivity" (Miller, 1967). The syntax chiefly involves the temporal sequence or the serial order of signs. We will attempt to show that acquiring a *conditional sequential discrimination* is a key factor in the dolphin and sea lion being able to comprehend relationships encoded in gestural signs strung together to form sentence-like commands. We will support this hypothesis with data from a series of experiments with a California sea lion (Rocky) which directly parallel experiments conducted by the Herman group with a bottlenose dolphin (Ake). We will focus primarily on the most complex sign sequence form, the relational sequence. In the relational sequence dolphins and sea lions have learned a serial ordering rule that, in one form, can be expressed as "take the second designated object to the first designated object."

Methods

Sea Lions and Facilities

Our experiments with ALR have involved three California sea lions (*Zalophus californianus*): Bucky, a six-year-old male housed at Marine World/Africa USA in Redwood City, California (now located in Vallejo, California), Gertie, a five-year-old female housed at Long Marine Lab, University of California, Santa Cruz and Rocky, an eleven-year-old female also housed at Long Marine Lab. Since our research on relational sequences has gone furthest with Rocky, we will focus primarily on her performance in the experiments discussed in this paper. However, it should be noted that both Gertie and Bucky have also learned to carry out instructions relating two designated objects.

At Long Marine Lab, experiments were conducted in the sea lions' home pool, a 7.6m diameter by 1.8m deep circular concrete tank surrounded by a deck that is flush with the pool rim. During an experimental session, sea lions not being tested were placed in a separate adjoining pool area. The pools at Long Marine Lab are filled by a flow-through system that provides minimally-filtered seawater at ocean temperature (approximately 15 °C).

Basic Procedure

The basic experimental procedure, controls and training techniques for giving single-object sequences to Rocky have been detailed by Schusterman and Krieger (1984; 1986). We should point out that we have used many blind procedures to guard against unintended cueing by the signaler and other individuals involved in the experiment. These control procedures indicated that Rocky, as well as the other two sea lions, responded only to the gestural signs given by the signaler.

Table 17.1 lists Rocky's repertoire of gestural signs used in the current experiments. The signs are given by the signaler in a specific sequence (Table 17.2).

Figure 17.1. illustrates a typical relational sequence. A signal sequence started with Rocky "on station", her chin resting on the signaler's foot. The first sign and each subsequent sign in a sequence were cued by an observer who told the blindfolded signaler (via radio headphones) that Rocky was in position to receive the next sign. Rocky typically left station after an object signal, visually searched for the object and, after locating it, returned to station. She did not leave station or scan the pool area after modifier signals. This scanning search following the object sign is very informative regarding a sea lion's cognitive capabilities (see Schusterman and Krieger, 1986). Because the sea lion's eyes are aimed primarily forward and up it must essentially "point" with its head in order to bring the object into the field of focused, binocular vision. In comparison, bottlenose dolphins (like those used in Herman's ALR) have laterally placed eyes that can independently scan large areas, without their making head movements that might indicate what in particular they are looking at. These physical differences in visual perception can result in differences in training and signing. The dolphins received a rapid, uninterrupted signal sequence while, potentially at least, simultaneously scanning large areas of the pool environment with the eye not watching the signaler. Rocky, on the other hand, tended to interrupt attention to the signaler with sometimes prolonged (greater than 2 seconds) and methodical searches of the environment

Table 17.1. Current repertoire of gestural signs used with Rocky.

Objects	Modifiers	Actions
Transportable[1]	*Brightness*	*Direct*[4]
PIPE	BLACK	FETCH
BALL	WHITE	FLIPPER-TOUCH
RING	GRAY[3]	MOUTH
WATERWING	*Size*	OVER
CLOROX	SMALL	UNDER
CAR	LARGE	TAIL-TOUCH
DISC		*Relational*[5]
BAT		FETCH
CUBE		
FOOTBALL		
CONE		
Nontransportable[2]		
WATER		
PERSON		

Note: [1]Objects that can be moved by the sea lions.
[2]Objects movable by experimenters but not by sea lions.
[3]Not currently in use.
[4]Involves only one object.
[5]Involves two objects.

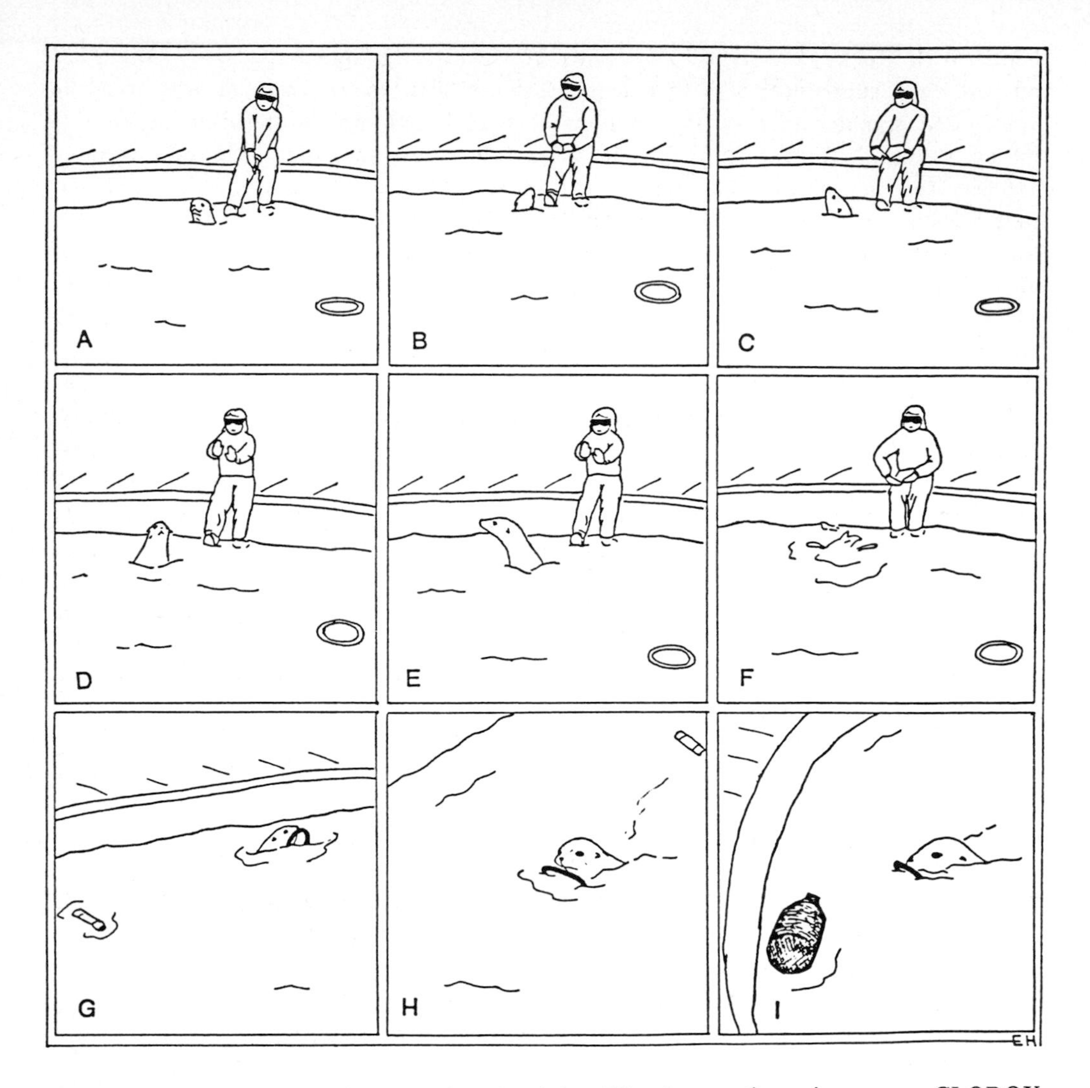

Figure 17.1. Example of a relational trial. The instruction given was CLOROX, BLACK RING FETCH. A) The signaler signs CLOROX, Rocky searches for the object. B) The signaler pauses between GI and TI sign sequences, Rocky remains at station. C) The TI modifier sign BLACK is given, Rocky turns slightly, but does not scan the pool area. D) The signaler gives the RING sign, in this frame Rocky is looking at the white ring. E) Rocky continues her search, she is now looking at the black ring (out of view in this frame). F) The action sign FETCH has been given and Rocky has been released (signaler lowers foot). G) Rocky has gone to the black ring and is starting to move it while scanning the pool for a goal item. H) Rocky is pushing the ring toward the clorox bottle (just out of view in this frame). I) Rocky is placing the black ring in contact with the clorox bottle which constitutes a successful response to the instruction CLOROX, BLACK RING FETCH.

Table 17.2. Sign sequences given to Rocky, with examples.

Sequence	Example
SINGLE OBJECT SEQUENCES	
Two Sign	
O+A	BALL OVER
Three Sign	
M+O+A	WHITE BAT TAIL-TOUCH
Four Sign	
M1+M2+O+A	LARGE BLACK CONE MOUTH
RELATIONAL SEQUENCES	
Three Sign	
GI+TI+A	BALL, RING FETCH
Four Sign	
M+GI+TI+A	WHITE CAR, DISC FETCH
GI+M+TI+A	WATER, SMALL CUBE FETCH
Five Sign	
M1+M2+GI+TI+A	BLACK SMALL CONE, BAT FETCH
GI+M1+M2+TI+A	CAR, WHITE SMALL BALL FETCH
M+GI+M+TI+A	LARGE CONE, BLACK RING FETCH
Six Sign	
M+GI+M1+M2+TI+A	WHITE DISC, SMALL BLACK CONE FETCH
M1+M2+GI+M+TI+A	WHITE LARGE CONE, SMALL BALL FETCH
Seven Sign	
M1+M2+GI+M1+M2+TI+A	SMALL WHITE CLOROX, BLACK SMALL CUBE FETCH

Note: O = object sign in a single object sequence.
GI = sign designating the goal item in a relational sequence.
TI = sign designating the transported item in a relational sequence.
M= modifier sign, modifies object sign following it.
A = action sign.

after each object signal, resulting in a much more extended signaling process. Following an action sign Rocky is usually (except on "no-go" trials) released from station by the signaler withdrawing her/his foot. The "no-go" trials help prevent the sea lion from anticipating release and thus not attending to the action sign. Occasionally, following a long, unproductive search for a specific object, Rocky returned to station and, following the action sign, she did not perform the indicated action, but remained at station when released by the signaler's foot-drop. The lack of a specific action following release was termed a *balk*.

Experimental Design

Experimental trials were inserted as probe trials in a semi-random manner into the standard daily training sessions, or "baseline." Each baseline session lasted about one hour and contained approximately seventy trials. The trials were divided into "sets" of four to eight trials. After each set the objects were replaced with a new group of objects. There were typically four to ten objects in the pool during each set (the mean was about six objects).

Approximately equal numbers of all single-object sequence forms (refer to Table 17. 2) were run in a semi-random fashion to maintain competency in all familiar sequence types. After the three-sign relational instruction had been trained it was included in the baseline sessions. Later, four-sign relationals were also added to the baseline sessions. This procedure maintained Rocky's competence in all sign sequence forms and provided relatively context-free baseline sessions (i.e. Rocky was not able to anticipate the type of trial she was about to be given).

Training the Relational Sequence

The principal topic of this paper is a special type of sign sequence referred to as the relational sequence, so-called because the outcome requires Rocky to fetch one object to a second, rather than to fetch a single object directly to the signaler. The order of the two object signs indicates the relation between them, that is, which object is the goal item (GI) and which is the transported item (TI). For Rocky, the first object sign indicates the GI and the second object sign indicates the TI. The terms GI and TI are solely descriptive and are not given a grammatical connotation or interpretation. In contrast, Herman et al. (1984) refer to the same sign sequence, when given to the dolphin Akeakamai, as an "inverse grammar" and use the grammatical terms indirect object (IO) for the first object sign and direct object (DO) for the second object sign. Our introducing superficially similar nomenclature may be initially confusing for some, but we believe it is important to free the performance of the relational action from grammatical terminology when such terminology is not needed and, indeed, contains surplus meaning.

The relational sequence was introduced to Rocky in April, 1985. Training began by teaching Rocky to fetch an object to something other than the signaler. A single object was placed in the pool and Rocky was released from station without being given any gestural signals. This elicited a standard response of fetching the object to the signaler: both Rocky and Gertie were normally required to periodically "gather up" the objects in the pool in order to have the session continue. A training target (a pool float at the end of a broom handle) was then introduced in the pool near the signaler. The target had been frequently used in other training situations and had a very strong association with reinforcement. After a few trials with the target Rocky had learned to transport the object in any direction to get it to the target, even away from the signaler.

In the next stage of training a second object was placed in the pool and the relational signal sequence was introduced. At this stage of the training process it started with a touch signal on Rocky's forehead indicating that this was a relational sequence, followed by the GI object signal, TI object signal, then the action signal. On the initial attempt Rocky made an intention movement to bring the TI to the GI, but then fetched the object to the signaler. In subsequent attempts the training target was used to guide Rocky to the GI and then removed just before she arrived. The target was quickly faded out of the procedure. The touch signal was soon dropped from the sequence because we felt it was distracting to Rocky. Her performance of the relational action did not change after the touch was discontinued. The context cue of two object signs was apparently sufficient to indicate a relational action.

After a few repetitions of the fully trained sequence with just two objects, we introduced alternating trials with other actions to eliminate context cues, then added one or two objects to provide a choice of destination objects. The entire training procedure took about two hours time distributed over three days. After this training, the simple three-sign relational sequence was incorporated into baseline sessions, intermixed with familiar single object sequences.

Experiments

Response bias for goal items on novel complex relationals. "Novel" sequences were sign combinations that had never been given to the subject. They were used in experiments to eliminate the potential of the animals producing rote responses to repeated familiar sequences. Since there were in fact thousands of unique instructions possible, it is unlikely that the animals were capable of organizing their response behavior in this way. Nevertheless, both we and the Herman group took the precaution of using novel sequences in most experiments. The relative effect of sequence novelty on performance is discussed further in the Results section.

Novel sign sequences were of two types: those that contained a combination of signs never before given to the animal, but of a sequence type that the animal had practiced (what Herman et al., 1984, refer to as "lexical novelty") or those that contained a novel number of signs. Herman et al. (1984) refer to the latter as "structurally novel" sequences, but that term is somewhat misleading since the only novelty was the addition of one (in Ake's case) or sometimes more (in Rocky's case) modifiers. Adding familiar modifiers to a familiar sequence to make it "novel" increased the complexity of the instructional sequence, but did not require truly novel response behavior, or novel integration of the information in the signs. Any other form of structural novelty (adding, subtracting, or rearranging signs) produced unusual responses that are discussed under the Anomalous Sequence headings of this paper.

In this experiment, novel sequences containing an unmodified or single modified GI and a double modified TI (e.g. WHITE WATERWING, SMALL WHITE BALL FETCH) were inserted as probes in standard baseline sets at the rate of two to four trials per session, for a total of 32 trials. Half (16) of the experimental trials had "positive" GI's and half had "nonpositive" GI's. The classification of an object as a positive or nonpositive GI refers to the probability of a correct response on a baseline relational with that object as the GI. For this experiment we arbitrarily designated all GI's with a greater than 0.5 probability of eliciting a correct response as being positive GI's (Table 17.3). For example, BAT was considered a positive GI because baseline trials with BAT as GI had a 0.81 probability of being correct based on performance of baseline trials immediately preceding the experiment.

CONE was a nonpositive GI because baseline relationals with CONE as GI had a probability of 0.23 of being correct. Table 17.3 also shows that the positive/nonpositive status of an object can change over time. Further investigation of this object-related response bias is in progress, but for our purposes in this paper this somewhat simplistic and arbitrary identification of positive and nonpositive goal items is in fact quite informative.

Each trial with a positive GI was replicated using a nonpositive GI. Both were run with the same number of objects in the pool, and the same TI. For example, the trial BLACK BAT, LARGE WHITE BALL FETCH, which includes a positive GI, was conducted under the same conditions as the trial BLACK CONE, LARGE WHITE BALL FETCH, which contains a nonpositive GI. In both trials the same number of objects were present in the pool, but two cones (a black and a white) had been substituted for two bats. The difference in performance on the two groups of novel complex relationals could therefore be attributed solely to the effect of the GI object.

Table 17.3. Rocky's response bias and consequent goal item (GI) object type.

Object (type)	Probability (and Number) of Correct Response Before Expt.	During Expt.	Status Change?
PIPE	0.25 (20)	0.13 (16)	N
BALL	0.13 (16)	0.00 (4)	N
RING (*)	0.56 (18)	0.50 (10)	N
WATERWING (*)	0.82 (11)	0.57 (7)	N
CLOROX BOTTLE (*)	0.67 (15)	0.63 (8)	N
CAR (*)	0.47 (15)	0.56 (16)	Y
DISC	0.18 (11)	0.50 (10)	Y
BAT (*)	0.81 (16)	0.44 (9)	Y
CUBE	0.11 (19)	0.05 (20)	N
FOOTBALL	0.33 (15)	0.50 (10)	Y
CONE	0.23 (13)	0.15 (13)	N
WATER (*)	0.57 (28)	0.88 (17)	N
PERSON (*)	0.44 (31)	0.76 (21)	Y

Note: (*) indicates positive GI object; i.e., Correct response probability > 0.50

Relationals to a nontransportable GI. The nontransportable objects WATER and PERSON were added to Rocky's repertoire after training on the relational behavior. Nontransportable objects differed from transportable objects in that they were always located on the rim of the pool and could not be moved by the sea lions. Transportable objects, on the other hand, floated in the pool and could be moved by the sea lions. Furthermore, wind and Rocky's movements caused the transportable objects to drift about the pool during a set, whereas nontransportable objects stayed in one place during an entire set, unless deliberately moved by the researchers. Nontransportable objects were usually moved only at the end of a set.

A limited number of three-sign relationals to nontransportable GI's were included in the baseline. Using our standard experimental procedure we also ran 42 novel relationals to nontransportable GI's (21 to WATER and 21 to PERSON). Table 17.3 indicates that nontransportable GI's were associated with a high probability of correct response.

Reversals. This experiment consisted of 54 paired *original* and *reversal* three-sign familiar relational sequences. The originals were run as probes in baseline sets until Rocky responded correctly and the response was reinforced. The reversal was run immediately after the reinforced original. In other words, if Rocky responded correctly to the sign sequence CAR, CONE FETCH we next gave her the sign sequence CONE, CAR FETCH.

A particular object was used as a GI in originals about as often as it was used as a TI to prevent the possibility of Rocky making a predictive association between certain GI objects and the possibility of a forthcoming reversal.

Modifier-reversals. This experiment was intended to illustrate a potential problem of modifier assignment in relational sign sequences. In this experiment a modifier sign in a novel complex relational command was "reversed" or switched from a GI modifier to a TI modifier or vice versa by changing its position in the second or "reversal" trial. The relative positions of the two object signs, however, remained unchanged: the GI in the first trial remained the GI in the second trial of the modifier-reversal pair. For example the sign sequence WHITE BALL, BAT FETCH is the modifier-reversal of BALL, WHITE BAT FETCH (the object signs remain in the same order). The modifier-reversal experiment was run in the same manner as the reversal experiment described in the preceding section, with one exception; the modifier-reversal trial was not run immediately after the successful, reinforced original because certain objects needed to be added or removed before it was possible to run the modifier-reversal trial. The modifier-reversal trial was given some time later, either a few trials later in the same set or as long as three days after the successful original.

Four modifier-reversal pairs were run with a single modified GI as the original and four pairs were run with a single modified TI as the original. In addition one pair was run with a double modified GI original and one pair with a double modified TI original, for a total of ten modifier-reversal pairs.

Anomalous sequences: Type I. The anomalous sequences in this experiment retained the double object signs indicative of a relational sequence, but had a different action sign substituted for the familiar FETCH action sign (the only action sign that had heretofore been associated with two object signs). An example of this kind of anomalous sequence is CUBE, DISC FLIPPER-TOUCH. This experiment was designed to exactly parallel an experiment with the dolphin Ake reported by Herman et al. (1984). Twelve novel anomalous sequences of this form were run as probe trials in baseline sessions, one or two per session. All responses were nonreinforced.

Anomalous sequences: Type II. This group of anomalous sequences either had their signs out of normal sequence, contained added signs, or were missing signs normally present. Fourteen novel anomalous sequences were run as probes, one or two probes per session. All responses were nonreinforced.

Calculating probabilities of correct response for relational fetch sequences. The method we used to calculate probabilities of chance correct responses to relational instructions differed from that used by Herman et al. (1984, Appendix A). Their calculations were based on the total number of potential "meaningful" sign combinations of that sequence form. According to their method, the likelihood of Rocky producing a correct chance response to a three-sign relational sequence would be the reciprocal of 130 or 0.0077 (refer to Table 17.4 for numbers of potential combinations in each sequence form). Some of the reasons for choosing the method we used over others are presented in detail at the end of the Results section. Here we present our basic assumptions as well as the mechanics of our method of calculating probabilities.

Table 17.4. Number of potential sign combinations in each sequence type in Rocky's repertoire.

Sequence Type	No. Combinations	
SINGLE OBJECT SEQUENCES		
O+A	76	
M+O+A	528	
M1+M2+O+A	240	
(TOTAL)		(844)
RELATIONAL SEQUENCES[1]		
Three Sign		
GI+TI+A	130	
Four Sign		
M+GI+TI+A	313	
GI+M+TI+A	376	
(TOTAL)		(689)
Five Sign		
M1+M2+GI+TI+A	384	
GI+M1+M2+TI+A	464	
M+GI+M+TI+A	956	
(TOTAL)		(1804)
Six Sign		
M+GI+M1+M2+TI+A	1192	
M1+M2+GI+M+TI+A	1192	
(TOTAL)		(2384)
Seven Sign		
M1+M2+GI+M1+M2+TI+A	1216	
GRAND TOTAL	7067	

[1]Matching object combinations, e.g., PIPE, PIPE FETCH, excluded.

After learning the basic relational task, the sea lion Rocky, and the dolphin Ake, erred almost solely on the goal item. They rarely failed to go to the correct TI or to take that TI to another object (correct action), but they often took the TI to an object other than the GI indicated by the signaler (see Tables 17.5, 17.6 and especially 17.7 for supporting data).

We therefore made the probability of a *completely* correct response dependent upon the number of objects in the pool available to serve as GI's. The probability that Rocky would go to the correct TI and take it to another object was nearly 1.0. Her selection of an object to serve as GI, if she was choosing at random, would be a function of the number of objects in the pool, less one (the object she was using as TI). If six objects were in the pool, five would be available for use as GI, and there would be a 1/5 (0.20) chance probability of taking the TI to the correct GI.

For a single trial, therefore, the probability of a chance correct response was calculated as 1/(x-1), where x equals the number of objects in the pool. For a group of trials, the estimated number of correct responses expected by chance was calculated by first dividing the trials into classes based on the number of objects in the pool and then multiplying the number of trials in each class by 1/(x-1). This method was derived empirically and is a conservative way to estimate chance performance to a relational instruction. That is, we slightly overestimate the likelihood of chance performance because our calculation did not include any other error factors, e.g. incorrect TI or incorrect action, etc., although these error factors contributed slightly to the total observed errors (see Table 17.7).

Results and Discussion

Baseline Relationals

We believe that our results with the sea lion Rocky are directly comparable to those of Herman et al. (1984) with the dolphin Ake. Ake's performance on familiar three-sign relationals was, like Rocky's, much lower than her performance on single object sequences of the same length. As Table 17.5 shows, the number of signs given in the instructions were relatively trivial compared to the type of instruction given: that is, whether it was a relational instruction or nonrelational. For example, compare Rocky's performance on M1+M2+O+A sequences (four signs) with her performance on GI+TI+A sequences (three signs).

Table 17.5. Rocky's performance on baseline sequences (data from September, 1986).

Sign Sequence	Prob. Correct	(C.R./TOTAL)
O+A	0.967	(349/361)
M+O+A	0.922	(403/437)
M1+M2+O+A	0.927	(268/289)
GI+TI+A[1]	0.436	(52/119)

[1]Includes only three-sign relationals to transportable GI's.

Table 17.6. Rocky's performance, in terms of correct responses (CR), on three-sign baseline relationals.

Time Period	Total Trials	Obs. CR	CR Prob.	Chance CR	Chance Prob.	chi^2	p
Jan-Feb '86	200	50	(0.25)	40	(0.20)	8.8	>0.10
May-Jun '86	200	82	(0.41)	46	(0.23)	30.2	<0.01
Sep-Oct '86	200	80	(0.40)	44	(0.22)	32.7	<0.01

Note: (d.f. = 4 on all chi^2 values)

Table 17.6 shows Rocky's performance values on the three-sign relational sequences given during baseline sessions from three different time periods: one period preceding the start of experiments (January-February, 1986), one period at the start of experiments (May-June, 1986) and one period at the end of the experiments (September-October, 1986). Rocky's performance prior to June, 1986 (as represented by data for January and February, 1986 in Table 17.6) was not significantly different from chance, based on our conservative method for calculating the probability of a chance correct response. Thereafter, her performance reliably exceeded chance values of probability.

Rocky's data indicate that, while she was responding at above chance levels (between 40 and 50% correct responses) on relationals, her performance on single object trials was almost perfect (exceeded 90% correct responses). Ake's data shows the same marked differences for probably the same reasons. We will attempt to demonstrate that Herman's interpretation of Ake's performance on relational instructions is not the most parsimonious explanation supported by his data and that both Ake's and Rocky's data better support a different interpretation.

First, almost all of Rocky's and Ake's errors on relational sequences were confined to the GI, the first sign or signs of the so-called "inverse grammar", those indicating the goal item (Herman's indirect object). Table 17.7 lists the proportion of total errors attributable to each element of the sign sequence. The table supports our viewpoint that when Ake and Rocky attempted to carry out a relational command they would either be correct and obtain a food reward or they would take the appropriate TI to the wrong GI and not be reinforced.

One factor that influenced Ake's ability to select the correct GI was the addition of a relative position (left/right) modifier to the GI (Herman et al., 1984). This apparently allowed Ake to remember the object's position, an easier mental task than remembering its identity encoded by the object sign. For Ake, the sign sequence form M+GI+TI+A resulted in a higher percentage of correct responses than the sequences GI+M+TI+A or GI+TI+A. The modifiers given to Rocky referred to object qualities (size/brightness) rather than object positions and there was, therefore, a decrement in performance on GI-modified relationals.

We found that factors not analyzed by Herman et al. influenced Rocky's performance of relational sequences but not on single-object sequences. These

Table 17.7. Proportion of total error attributable to each sign category of the three sign relational sequence (GI+TI+A).

	Total Trials	Total Errors	Probability of errors (by category)				
			GI[2]	TI[3]	A	Mult.[4]	Total
Ake[1]	53	18	0.89	0.055	0.0	0.055	1.00
Rocky[5]	200	120	0.975	0.0	0.0	0.025	1.00

Note: [1]data from Herman et al. (1984), Table 17.7, p. 170.
[2]GI = goal item.
[3]TI = transported item.
[4]Mult. = errors on more than one sign.
[5]Data for September, 1986.

factors may have had an effect on Ake's performance as well. They include:

1) The number of objects in the pool.
2) Bias for goal items.
3) Whether a goal item was fixed in space, that is whether it was a transportable or nontransportable object.
4) Whether the goal item and the transported item were reversed on successive trials.

The Effect of the Number of Objects in the Pool

If, as we hypothesize, Rocky had difficulty selecting a GI and was often selecting the GI object at random ("guessing") then her performance should vary with the number of objects available in the pool. Our data from the three-sign baseline relationals and from the reversal experiment support this hypothesis; the percentage of correct responses increased as the number of choice items decreased (Table 17.8).

Table 17.8 also shows that Rocky's performance on single object instructions did not vary as a function of the number of available objects. These results strongly support the notion that there were only two rules influencing Rocky's performance: one rule for integrating signs in order to act on a single object and a second rule for taking that object (TI) to one of several available GI's. As we have previously noted, Table 17.5 shows that the number of signs that are given in a sequence is relatively trivial. Rather, the critical factor is retaining information about the first designated object (GI) in a relational sequence. Despite the fact that Rocky produced an orientation to the appropriate object when the GI sign (the first object sign in the relational sequence) was given, information in the second object sign apparently interfered with memory for the first sign (retroactive interference).

Response Bias for Goal Items on Novel Complex Relationals

Rocky's performance on the 16 trials with positive GI objects was significantly

Table 17.8. Effect of number of objects on Rocky's probability of making a correct response (CR).

	Number of Objects in Pool						
	3	4	5	6	7	8	9+
RELATIONAL INSTRUC.							
Expected (relat.)[1]	0.50	0.33	0.25	0.20	0.17	0.14	<0.11
Baseline prob.[2]	---	0.41	0.41	0.37	0.29	0.14	0.0
Baseline (N)	(0)	(39)	(27)	(19)	(28)	(14)	(2)
Reversal prob.[3]	0.50	0.39	0.28				
Reversals (N)	(32)	(89)	(99)				
SINGLE-OBJ. INSTRUC.							
Expected (1 obj.)[4]	0.33	0.25	0.20	0.17	0.14	0.13	<0.11
Baseline prob.[5]	---	0.95	0.96	0.95	0.95	0.92	0.94
Baseline (N)	(0)	(42)	(181)	(306)	(306)	(153)	(97)

Note: [1]Expected = 1/(x-1), where x = the number of objects in the pool.
[2]Baseline prob. = 3-sign baseline relational trials, June, 1986
[3]Reversal prob. = familiar three-sign relational trials from the reversal experiment.
[4] Expected = 1/x, where x = the number of objects present in the pool.
[5] Baseline prob. = total single-object trials for June, 1986.

above chance (chi^2 = 19.0, d.f.=2, p<0.01) while her performance on the 16 trials with nonpositive GI objects remained at chance (chi^2 = 1.3, d.f.=2, p>0.1). The results are presented graphically in Figure 17.2.

These results show that carrying out a novel instruction relating two objects depends primarily on bias involving the goal item. Therefore, we believe that Herman et al. (1984) have misplaced their emphasis by stressing sequence novelty in their experiments. All of the trials in this experiment were novel combinations. Both their data and ours show that factors related to coding the GI in short-term memory, such as object number, position modifiers (Ake's LEFT/RIGHT modifier), or GI bias have a greater influence on performance than sequence novelty.

Relationals to a Nontransportable GI

When Rocky was given novel 3-, 4-, and 5-sign relationals to nontransportable goal items her performance was significantly above chance (Table 17.9). As previously noted in Table 17.3, nontransportable objects (WATER, PERSON) were highly positive GI's. In Figure 17.3 Rocky's performance on familiar three-sign relationals to nontransportable GI's is compared to her performance on three-sign relationals to transportable GI's and to performance expected by chance.

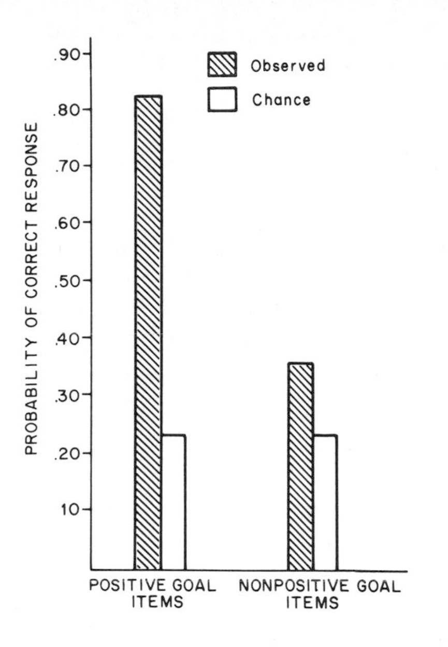

Figure 17.2. Relative probability of a correct response on novel complex relational instructions containing positive GI's versus those containing nonpositive GI's.

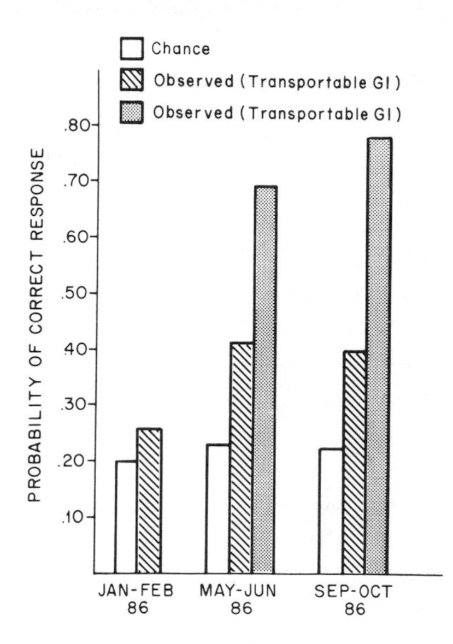

Figure 17.3. Rocky's performance on 3-sign baseline relationals to transportable and nontransportable GI's compared to predicted chance performance.

Table 17.9. Rocky's performance on novel complex relationals to nontransportable GI's (d.f.=2 for all chi^2 values).

Object	Total Trials	Correct (Obs.)	Correct (Exp.)	chi^2	p
WATER	21	14	4	29.5	<0.01
PERSON	21	9	4	7.0	<0.01
TOTAL	42	23	9	27.3	<0.01

In Table 17.10 (next page) numerical data from September 1986 are presented for familiar three-sign relationals to a nontransportable GI and to a transportable GI. The two sets of data are then combined to show how the relative proportion of relationals to positive destination objects (such as nontransportable objects) can alter the animal's overall performance level. Without breaking down the data on relational sequences into categories of GI's (such as nontransportable and transportable or positive and nonpositive) it is difficult to determine what factors influenced the sea lion Rocky or the dolphin Ake in arriving at a correct solution of the relational problem.

Table 17.10. Rocky's performance on familiar 3-sign relationals to transportable and nontransportable GI's.

	Probability of Correct Response		
	Jan-Feb '86	*May-Jun '86*	*Sep-Oct '86*
Transportable GI	0.25 (50/200)	0.41 (82/200)	0.40 (80/200)
Nontransport. GI	(see note)	0.69 (49/71)	0.77 (48/62)
TOTAL (Trans+Non)	0.25 (50/200)	0.48 (131/271)	0.49 (128/262)

Note: Nontransportable objects were introduced after February, 1986.

In the original report by Herman et al. (1984) there was no attempt at subdividing the GI into transportable and nontransportable categories. However, two years later Herman (1986) does note that there were fewer errors in Ake's relationals to a nontransportable GI (his "relocatable object"), but he still does not provide quantitative data on the relative error rates of relationals to transportable and nontransportable GI's, nor on the proportion of trials to nontransportable versus transportable GI's.

Reversals

The reversal experiment was designed to assess the relative effect of sign sequence and memory for object type on Rocky's performance of relational fetches. A feature of relational sequences that is not present in single object sequences is that elements can be transposed to create a sequence with new meaning. For example, the sequence BALL, RING FETCH can be changed to RING, BALL FETCH to make two legal, but different instructions. The sea lion can extract the correct information from the sequence only if it understands that sign order is important. If sign order is ignored the information in the signs themselves is not sufficient to yield an unambiguous message.

One might expect that Rocky's previous experience with sign transposition would predispose her to ignore sequence. Her two classes of modifier signs (size and brightness), when used in conjunction, are transposable, i.e. both BLACK LARGE CONE and LARGE BLACK CONE can be used to indicate the same object. Rocky accepted such transpositions from the time of their first introduction (Schusterman and Krieger,1986).

As Herman et al. (1984) pointed out, if the role of sign sequence in determining which object is to be the GI and which is to be the TI is not well understood or remembered by the animal then we might expect a number of errors in which the correct objects are used but in reverse relation (GI used as TI, TI used as GI). If this were the case, by reinforcing an original just before running its reverse, we might expect even more of these errors on the reversal due to the interference effect of the previous sequence. In fact, Rocky made no such errors on either the

Table 17.11. Rocky's performance on relational reversal pairs.

	Total Trials	Reversal Errors[1]
Original	166	0
Reversal	54	0

[1]If the instruction CUBE, RING FETCH is given, a reversal error response would be RING, CUBE FETCH.

original instruction or its subsequent reversal (Table 17.11).

If, on the other hand, the goal item in a relational sequence is not remembered then a preceding reinforced relational containing the same elements, albeit in a different order, might improve performance on the subsequent reversal by *priming* memory for the objects involved (Domjan and Burkhard, 1982). If, for example, the sequence PIPE, BALL FETCH is performed and reinforced, when the reverse sequence BALL, PIPE FETCH is given, Rocky would be primed to respond to the two objects, ball and pipe. Furthermore, Rocky will not have to choose the GI from two objects primed in memory since she is using one of the primed objects as the TI.

The other plausible priming scenario would yield the same effect. If, as we have previously hypothesized, Rocky has difficulty remembering the GI, but readily retains the TI of the original command in memory, then the GI of the reversal will still be primed. In simplified form we can think of the original command as the A-B sequence; if only the latter element, B, is primed, then on the reversal (B-A) the hard to remember element (the first) is the one that has been primed.

Table 17.12 shows that Rocky did perform considerably better on the reversal sequences than on the originals ($chi^2 = 3.4$, d.f.=1, $p<0.1$, one-tailed test). These results suggest that presenting signs referring to the same object on successive trials, despite their reversed order, enabled Rocky to "rehearse" or better attend to the signs and their associated referents. Additional experiments of this kind, especially one where the sequence was responded to incorrectly and followed by a reversed sign order should give us a much better understanding of the variables involved in what Herman et al. (1984) have called "comprehension of semantically reversible sentences."

Table 17.12. Correct responses to relational reversal pairs.

	Total Trials	Correct Responses	(%)
Originals	166	54	(32.5)
Reversals	54	25	(46.3)

Modifier Reversals

Herman et al. (1984) attach great significance to Ake's apparent ability correctly to assign a centrally placed modifier to the appropriate object in the sign sequence Object + Modifier + Object + Action. They attribute this ability to a "precedence rule" that indicated Ake understood "how modifiers may be attached to object names" and considered the correct assignment of the modifier to the appropriate object as another indicator of "the considerable sensitivity of the dolphins to syntactic structure." (Herman et al. 1984: p.199).

We, on the other hand, consider this a perfect example of the danger of applying gratuitous grammatically-rooted explanations to phenomena that admit a simpler explanation. In fact, the dolphin *cannot* misassign the modifier, because the objects in the pool will not allow it. If the sign sequence BALL, LEFT PIPE is given there must necessarily be only one ball, but two pipes in the pool. How could the animal express an incorrect assignment of the LEFT modifier sign to BALL if there were not two balls from which to choose?

With the understanding that context would not allow an inappropriate response of the form Herman et al. (1984) hypothesize, we ran ten novel modifier reversal pairs (Table 17.13). Remember that in these relational sequences "reversal" refers to the positional shift of a modifier from preceding the GI to preceding the TI or vice versa. Thus, the modifier-reversal of the instruction SMALL BALL, CUBE FETCH becomes BALL, SMALL CUBE FETCH.

Rocky's overall performance was significantly better than chance ($chi^2 = 16.5$, $d.f = 3$, $p<0.01$). There was no decrement of performance on modifier-reversals relative to originals as might be expected if modifier assignment on the original was confusing Rocky's modifier assignment on the reversal (Table 17. 13). We were also able to use Rocky's orienting behavior during the signal sequence to verify that Rocky was indeed incorporating the modifier into the orienting response for the appropriate object (see Figure 17.1 for an illustration of the orienting behavior).

Table 17.13. Rocky's performance on novel reversed-modifier pairs.

FORM OF ORIGINAL	Original	Reversal	Total
Four Sign Sequence			
M+GI+TI+A	0.57 (4/7)	0.75 (3/4)	0.64 (7/11)
GI+M+TI+A	0.31 (4/13)	0.25 (1/4)	0.29 (5/17)
Five Sign Sequence			
M1+M2+GI+TI+A	0.50 (1/2)	1.00 (1/1)	0.67 (2/3)
GI+M1+M2+TI+A	0.33 (1/3)	1.00 (1/1)	0.50 (2/4)
Total	0.40 (10/25)	0.60 (6/10)	0.46 (16/35)

Anomalous Sequences: Type I

Herman et al. (1984) called these sequences anomalous because the substitution of another action sign for the relational fetch action sign creates a sequence that cannot be carried out using the rules the animal had learned up to that point. The animal therefore had to "create" a response to the new sequence. Ake consistently (on eleven of twelve trials) performed a single object response to the correct second object using the correct action. The first object sign was either ignored or forgotten. Herman et al. (1984) interpreted this outcome as indicating that Ake processed but rejected the first sign and acted only on the second object plus action because it was the only legitimate instruction given her linguistic abilities. They consider the possibility of a recursive solution to the problem (performing the indicated action to each of the two objects in succession) and interpret Ake's failure to produce the recursive solution as a possible learning limitation in dolphins, generally.

However, within their linguistic model, there are more, equally consistent, potential outcomes. For example, Ake could have rejected the action sign as inappropriate and acted on the two object signs as implying a relational command, or she could have rejected the second object sign rather than the first. When tested on the same problem, Rocky performed almost exactly as Ake had. In 11 out of 12 anomalous sequences of the same form Rocky responded by performing the correct action to the second signed object (Table 17.14; $p<0.01$, binomial test). These results are virtually identical to those obtained from Ake. We interpret Ake's and Rocky's response to the anomalous sequence as we did the earlier finding that errors on relational constructions are most frequently GI errors. These findings are consistent with the notion that both Rocky and Ake use just two rules to comprehend all sets of instructions; namely, one rule to designate an object and another rule for bringing that object to another object.

Table 17.14. Rocky's responses to anomalous commands of the form GI+TI+A, where A is an action not associated with two object signs prior to the experiment.

Sign Sequence Given	Response
CUBE, BALL TAIL-TOUCH	BALL TAIL-TOUCH
CLOROX, FOOTBALL UNDER	FOOTBALL UNDER
CAR, BAT MOUTH	BAT MOUTH
BALL, BAT OVER	BAT OVER
DISC, BALL OVER	BALL OVER
FOOTBALL, WATERWING TAIL-TOUCH	WATERWING TAIL-TOUCH
PERSON, DISC MOUTH	DISC MOUTH
WATERWING, PIPE MOUTH	PIPE MOUTH
WATER, BAT UNDER	BAT UNDER
WATERWING, CAR FLIPPER	BLACK RING, CAR FETCH
RING, WATERWING OVER	WATERWING OVER
BALL, CUBE FLIPPER	CUBE FLIPPER

These rules require that the animals depend on sign sequence alone to assign the appropriate meaning to each sign. Syntax, in the human grammatical sense, can change the meaning of words based on types of structural dependency other than word order. In the simplified language used in these experiments the syntax is only serial order or sequence, where the signs are related to each other only in terms of sequence and never in any other way. In human language the sequence can be changed and its meaning still remain the same. For example, "The boy bit the dog" and "The dog was bitten by the boy" mean the same thing because it is always the dog that is acted upon by the boy, despite changes in the positions of the nouns.

Anomalous Sequences: Type II

Type II anomalous sequences include a variety of sequence types designed to assess how Rocky processed sign sequences. The sequences are listed in Table 17.15 and have been subdivided into:

1.) *transposed sign sequences* in which a familiar sign sequence has its component signs rearranged in a novel form,

2.) *omitted sign sequences* in which a sign type normally present has been deleted from the sequence, and

3.) *added sign sequences* in which a sign type is present in the sequence more times than usual.

Rocky's orienting responses are given in parentheses in Table 17.15, so that the type of sign given and the observed response can be easily compared. Since she gave a distinctive orienting behavior after receiving an object sign, and another distinctive stereotyped behavior when she received a modifier sign, an inappropriate response on her part was informative about the type of sign she was anticipating. A dash was used if Rocky did not move her head from station when a signal was given. Since Rocky shows no overt response to an action sign, it wasn't possible to tell whether Rocky was not responding to the sign or treating it as an action sign.

Her responses to these anomalous sequences indicate that Rocky relied primarily on sign order to organize the information conveyed by the signaler's gestures. The fact that Rocky balked on transposed sign sequences containing all the necessary elements indicates that she is sensitive to the ordering of the signs. When the standard sequence was violated her orienting responses sometimes indicated that she was anticipating a different sign. For example, when given an action sign after a modifier sign in the sequence O+M+A, she performed an object orientation on the action sign. Her previous experience would lead her to expect an object sign after a modifier sign. Likewise, when she was given the added-modifier sequence BLACK LARGE WHITE BALL TAIL-TOUCH she performed an object orientation in response to the third modifier sign, probably because she expected an object sign to follow two modifiers, rather than a third modifier sign. When she was given the sequence WHITE OVER and only a black and white ball were present, the information was sufficient for Rocky to make a choice, but she did not respond (she balked). This could be interpreted either as her treating the modifier sign as an abstract property that must be assigned to an object, or as sensitivity to serial order (i. e. modifier signs must be followed by object signs and action signs must be preceded by object signs).

Table 17.15. Rocky's responses to Type II anomalous sequences.

Sign Sequence	Response
TRANSPOSED SIGNS	
<u>O + M + A</u>	
CUBE SMALL MOUTH (O) (M) (O)	balk
BALL BLACK FLIPPER (O) (M) (-)	balk
<u>O + M + M + A</u>	
CUBE BLACK SMALL UNDER (O) (M) (M) (-)	balk
<u>M + O + M + A</u>	
LARGE BALL WHITE OVER (M) (O) (M) (-)	balk
<u>A + O</u>	
TAIL CAR (-) (-)	balk
<u>O + A + O</u>	
WATERWING FETCH CLOROX (O) (-) (-)	balk
<u>A + O + O</u>	
FETCH FOOTBALL WATERWING (-) (-) (-)	balk
OMITTED SIGNS	
<u>M + A</u>	
WHITE OVER (M) (-)	balk
LARGE UNDER (M) (O)	balk
<u>O + A</u>	
BALL OVER (O) (-)	WHITE BALL OVER
CLOROX UNDER (O) (-)	BLACK LARGE CLOROX OVER
ADDED SIGNS	
<u>M + M + M + O + A</u>	
BLACK LARGE WHITE BALL TAIL (M) (M) (O) (O) (-)	DISC, LARGE WHITE BALL FETCH
<u>O + A + A</u>	
CAR UNDER MOUTH (O) (-) (-)	CAR UNDER
CONE OVER TAIL-TOUCH (O) (-) (-)	CONE OVER

Key to symbols: O = object signal; (O) = typical orienting response to an object signal; M = modifier signal; (M) = typical response to a modifier signal; A = action signal; (-) = no discernible response, which usually occurred when Rocky was given an action signal.

Finally, in the sequence with an added action sign (O+A+A) only the action sign immediately following the object sign was performed. Such a response fits with the hypothesis that the sign sequence gives meaning to the sign. An action sign not preceded by an object sign was out of context in this sequential language and was ignored by Rocky just as the first object sign was ignored in the Anomalous Sequences I (refer to Table 17.14).

Although sign sequence appears to be the major factor guiding Rocky's interpretation of the instructions, there is evidence that she classifies signs into functional categories as well. For example, Rocky's behavior following action signs was the same regardless of the action sign's position in the sequence (i.e. she stopped orienting to all subsequent signs). As noted earlier, she accepted transpositions of modifiers, and did so the first time the double modifier was introduced (Schusterman, unpubl.). She did not accept transpositions of objects and actions, modifiers and actions, or modifiers and objects.

She also treated modifier signs differently from object and action signs in that she performed an action to a correct object type if a necessary modifier was omitted, but balked on a M+A sequence when an object sign was not needed. For example, if there were a large and a small ball in the pool (along with other object types), she responded to the sequence BALL FLIPPER by performing a BALL FLIPPER response to one of the balls, even though the signs given were not adequate to specify to which particular ball the action was to be directed. When a modifier and action were given (e.g. WHITE OVER) she did not perform the indicated action, even though, in this case, the information given her was sufficient to indicate the correct response object.

Calculating Probabilities of Chance Correct Responses

No matter how complex the task, there is some probability that the animal could produce a correct response by chance alone. Herman et al. (1984) recognized that "models of completely random choice are inappropriate, and not conservative, given what is known about the dolphin's responses." We disagree, however, with their contention that there is not sufficient information to choose a model for calculating the probability of a chance correct response based on performance of elements or groups of elements within the sign sequence (a "phrase structure" model, in their terminology).

We have shown that errors in relational sequences are almost always GI errors, and this is true for both Rocky and Ake. We have further shown that Rocky's performance on relationals varies with the number of objects available, unlike her performance on single-object sequences which does not vary with the number of objects present. We suspect that a similar analysis of Ake's data would yield a similar result. These data indicated to us that the relational action and object to be transported (TI) were not being selected by chance, and that the probability of completing the entire instruction correctly rested principally on the probability of Rocky going to the correct goal item (GI).

Factors that enabled Rocky to retain the goal item's identity in memory, such as positive object bias, object mobility (nontransportable objects), or successive trials with the same objects (reversals), affected Rocky's ability to perform relational instructions successfully. We believe that the same variables probably affected the dolphin Ake's performance on relational fetch sequences.

Sequence novelty, heavily emphasized by Herman et al. (1984), did not appear to affect Rocky's or Ake's performance significantly, when compared with performance of familiar sequences. Thus, once Rocky had learned the basic three-sign relational instruction, she was capable of successfully completing novel relational instructions of seven signs. Emphasizing sequence novelty and the potential number of unique sequences within a sequence type (e.g. three-sign relationals), as Herman et al. (1984) have done, does not reflect some of the most critical variables actually affecting performance of the animals.

Furthermore, their method of using the potential number of unique sequences within a type imposes some unrealistic assumptions about the animal's potential responses. First, this method assumes that all the objects are available to the animal in every trial (since they assume that the animal can produce any sequence within a type). However, all objects are *not* present for every trial. If there is no "fish" item present (see Table 17.1 in Herman et al., 1984) then all potential responses using the item "fish" are obviously impossible and should not be included in a calculation of the probability of a correct response. Their method also excludes responses outside the sequence type, yet such responses are quite possible. For example, if a three-sign relational instruction was given, but Ake used one of two hoops (left and right) as a GI, then her response would be considered a four-sign relational.

For these reasons we rejected the method used by Herman et al. (1984) to calculate the probability of a correct response to a relational instruction by chance and instead chose to make the probability of a chance correct response to such an instruction dependent on the number of goal item choices available.

Summary and Conclusions

The ALR effort, after almost three decades, has yet to find a nonhuman animal having all the attributes of human language. Most particularly, efforts at demonstrating syntactic competency based on symbol sequentiality have resulted in ambiguous results (Terrace, 1979). However, this paper suggests that with similar training and testing regimens two different types of marine mammals -- the high EQ bottlenose dolphin and the lower EQ California sea lion --show similar abilities cognitively to process a syntax consisting of ordered strings of signs relating two objects. We believe that such an explanation relies on learning and cognitive skills and not on linguistic skills as such.

Both sea mammals responded at better than chance levels to novel sets of instructions (conveyed by a trainer's gestural signs) by carrying out different behaviors depending on the serial order of the signs. Even with the same signs, differences in sequence convey differences in meaning, e.g. the sign sequence BALL, BLACK PIPE FETCH produces the instruction "take the black pipe to the ball" whereas the sign sequence BLACK BALL, PIPE FETCH produces the instruction "take the pipe to the black ball." This suggests that, by using a sequence rule, both marine mammals perceived that modifier categories depend on the object category while object categories remain distinct from one another except when followed by the relational term FETCH. Nevertheless, by using the relational sequence rule the sea lion Rocky was even capable of correctly carrying out novel commands containing as many as six or seven signs (e.g. BLACK SMALL FOOTBALL, LARGE BLACK CUBE FETCH; glossed as "take the large black cube to the black small football.")

We believe that sea lions, like dolphins, that have been conditioned to associate signs and objects, code things and dimensions imaginally and not in words or "grammatical terms". The grammatical thinking of animals is in the eye of the beholder. It is an error due to our own thinking in a formal grammar and therefore expressing the phenomena in such terms. The sea lion Rocky, like the dolphin Ake, apparently learned two rules and was able to apply them quite effectively under a variety of circumstances.

In one experiment we showed that if relational instructions between two objects were reversed immediately following a successful response then the likelihood of correct responses was higher on the reversals than on the original relational instructions. Thus Rocky was more likely to take the ball to the pipe when given the sequence PIPE, BALL FETCH than she was to take the pipe to the ball when she was given the immediately preceding instruction BALL, PIPE FETCH. We attribute better performance on *reversed* relationals than on *original* relationals to a *priming* effect. That is, since the first signed object is sometimes either ignored or forgotten, we believe that presentation of signs designating the same objects on successive trials (despite their reversed order) enables the sea lion to think actively about the signs and their associated referents.

Further experiments have corroborated the idea that a California sea lion can be trained to be as sensitive to the sequentiality of signs as dolphins. For example, if the standard sequence (Modifier) + Object + Action was changed to Object + (Modifier) + Action the sea lion Rocky would not even leave station. And when given a series of commands like PERSON, DISC MOUTH, Rocky, like the dolphin Akeakamai, mouthed the disc and ignored the person. These responses to unfamiliar, novel sequences are explicable in terms of just two learned rules:

1. If an object is designated by one, two or three signs (an object sign and up to two modifiers), then perform the designated action to that object.
2. If two objects are designated (again, by one to three signs each) and the action is FETCH, then take the second designated object to the first.

The animals' responses do not require the ability to treat the signs as syntactic elements in the full grammatical sense of that term.

What, then, are some of the mental tools needed for a simplified language? We conclude that the precursors of language are likely to be found in animals that are at least capable of combining the following learning and cognitive skills:

1. Paired associate learning or higher order conditioning.
2. Perceiving and categorizing objects and events into class and relational concepts -- each with their own subcategories (Thomas, 1980).
3. Acquiring conditional sequential discriminations.

Acknowledgments

This research was supported by contract N00014-85-K-0244 from the Office of Naval Research to RJS. RG's participation in this symposium was made possible by support from the Institute of Marine Sciences, University of California, Santa Cruz. We would like to acknowledge the invaluable contributions of Kathy Krieger, Michelle Jeffries, Evelyn Hanggi, Brigit Grimm, Faith Dunham, Howard Rhinehart, Rebecca Hardenbergh, and the many volunteers that participated in this research project. We would also like to thank Christine Johnson, Carolyn Ristau and Forrest G. Wood for critical reviews of an earlier draft of this manuscript.

References

Bastian J (1967) The transmission of arbitrary environmental information between bottlenose dolphins. In: Busnel RG (ed) Animal sonar systems, vol II. Laboratoire de Physiologie Acoustique, Jouy-en-Josas, France

Batteau DW, Markey PR (1967) Man/dolphin communication. Final report, contract N00123-67-1103, 15Dec1966-13Dec1967. US Naval Ordnance Test Station, China Lake, California

Domjan M, Burkhard B (1986) The principles of learning and behavior, 2nd edn. Brooks/Cole, Monterey California

Eisenberg JF (1981) The mammalian radiations. University of Chicago, Chicago Ill.

Epstein R, Kirshnit CE, Lanza RP, Rubin LC (1984) "Insight" in the pigeon: antecedents and determinants of intelligent performance. Nature 308:61-62

Fouts RS, Fouts D, Schoenfeld D (1984) Sign language conversational interaction between chimpanzees. Sign Lang Stud 42:1-12

Gardner RA, Gardner BT (1969) Teaching sign language to a chimpanzee. Science 165:664-667

Griffin DR (1984) Animal thinking. Harvard University Press, Cambridge

Harlow HF (1949) The formation of learning sets. Psych Rev 56: 51-65

Herman LM (1986) Cognition and language competencies of bottle-nosed dolphins. In: Schusterman RJ, Thomas JA, Wood FG (eds) Dolphin cognition and behavior: a comparative approach. LEA, Hillsdale NJ

Herman LM (1987) Receptive competencies of language-trained animals. In: Rosenblatt JS (ed) Advances in the study of behavior, Vol 17 pp 1-60. Academic Press, New York

Herman LM, Richards DG, Wolz JP (1984) Comprehension of sentences by bottlenosed dolphins. Cognition 16:129-219

Hoban E (1986) The promise of animal language research. PhD dissertation, University of Hawaii, Manoa

Hollis I (1984) Cause and function of animal learning processes. In: Marler P, Terrace H (eds) The biology of learning. Springer-Verlag, New York, p 357

Jerison HJ (1973) Evolution of the brain and intelligence. Academic Press, New York

Kellogg WN (1968) Communication and language in the home-reared chimpanzee. Science 182:423-427

Köhler W (1925) The mentality of apes. Harcourt, New York

Lang TG, Smith HAP (1965) Communication between dolphins in separate tanks by way of an electronic acoustic link. Science 150:1839-1844

Lilly JC (1961) Man and dolphin. Doubleday, New York

Lorenz K (1952) King Solomon's ring, new light on animal ways. Methuen, London

Miles HL (1983) Ape and language: the search for communicative competence. In: de Luce J, Wilder HT (eds) Language in primates. Springer-Verlag, New York, p 43

Miller GA (1967) The psychology of communication. Basic Books, New York

Petitto LA, Seidenberg MS (1979) On the evidence for linguistic abilities in signing apes. Br Lang 8:162-183

Pfungst, O (1911) Clever Hans: the horse of Mr. von Osten. Holt, Rinehart, and Winston, New York

Premack, D (1976) Intelligence in ape and man. LEA, Hillsdale NJ

Premack D (1986) Gavagai! or the future history of the animal language controversy. MIT Press, Cambridge

Premack D, Schwartz A (1966) Preparations for discussing behaviorism with chimpanzee. In: Smith FL, Miller GA (eds) The genesis of language. MIT Press, Cambridge

Roitblat HL (1987) Introduction to comparative cognition. WH Freeman, New York

Savage-Rumbaugh ES (1986) Ape language: from conditioned response to symbol. Columbia University Press, New York

Savage-Rumbaugh ES, Rumbaugh DM, McDonald K (1985) Language learning in two species of apes. Neurosci Biobehav Rev 9:653-665

Schusterman RJ (1962) Transfer effects of successive discrimination-reversal training in chimpanzees. Science 137:422-423

Schusterman RJ, Krieger K (1984) California sea lions are capable of semantic comprehension. Psych Record 34:3-23

Schusterman RJ, Krieger K (1986) Artificial language comprehension and size transposition by a California sea lion (*Zalophus californianus*). J Comp Psychol 100:348-355

Sebeok TA, Rosenthal R (eds)(1981) The Clever Hans phenomenon: communication with horses, whales, apes, and people. Ann NY Acad Sci, vol 364. New York Academy of Sciences, New York

Terrace HS (1979) Is problem solving language? J exp Anal Behav 31:161-175

Thomas RK (1980) Evolution of intelligence: an approach to its assessment. Br Behav Evol 17:454-472

Wood FG (1973) Marine mammals and man: the Navy's porpoises and sea lions. Robert B Luce, Washington DC

Worthy GAJ, Hickey JP (1986) Relative brain size in marine mammals. Am Natur 128:445-459

BASIC PROCESSES IN HUMAN INTELLIGENCE

Ian J. Deary
Department of Psychology
University of Edinburgh
7, George Square
Edinburgh EH8 9JZ

Tests of intelligence are unpopular with Western intellectuals today. Notoriously, the tests appear to discriminate against 'minorities' and against 'the working class'; and women are under-represented in the higher ranges of IQ. Moreover, the tests are often held to be 'circular': critics who make this charge are repeating, however unknowingly, the dictum of one of the few geniuses who have ever applied their talents to psychology - namely Edwin Boring, who was the first to remark, in 1923, that intelligence "is what the tests test."

Until recently it has not been possible to answer this objection definitively, even if one accepts the major discovery of the psychometricians that, given a large random sample of the population, when a group is tested on a variety of different mental tests the correlation matrix that results is almost entirely positive. This is the finding that led Spearman (1904) to propose that, due to differences in brain functioning, people had reliable individual differences in general intelligence (or g): yet this conclusion has not proved easy to substantiate. For, while Spearman's positive manifold is a common finding, its interpretation is not unproblematic.

Recent principal components analysis of the 11 very different subtests of the WAIS-R (Canavan, Dunn and McMillan, 1986) yielded a g factor which accounted for over 55% of the variance between subjects on the tests. In this study no subtest had a loading of less than 0.64 on the first, general factor. More generally, attempts to replace conventional IQ-style tests with non-g loaded mental ability tests have not been successful. Hooper, Hooper and Colbert (1984) reported a study involving subjects of different age groups who were administered standard psychometric intelligence tests alongside tests developed from Piaget's theory of formal operational thinking. The average correlation between Raven's Matrices and twelve formal operational reasoning tasks was 0.53. These two studies are merely illustrative of many similar efforts. Yet while these are not in dispute as typical findings, the problem lies in the reliance on specific statistical methodology.

It is possible to choose a particular factor-analytic method and, from the characteristics of the method, fashion one's own model of intelligence. Thus, followers of Spearman such as Burt (1909-10), Jensen (1980) and Eysenck (1982) have continued to extract the general factor using methods like principal components analysis. This method, though, tends to ensure a general factor, others will claim.

NATO ASI Series, Vol. G17
Intelligence and Evolutionary Biology
Edited by H. J. Jerison and I. Jerison

Loudest among these remains Guilford (1985) who, Nelson-like, has not 'seen any g' because he has chosen to implement factor analytic techniques that ignore g. Thus he has been able to construct a model of intelligence which contains 120 individual abilities although, as Eysenck (1979) pointed out, many of these are correlated and, if submitted to higher order analysis, will yield higher order factors.

How can such a strong empirical finding - the positive manifold of mental test correlation - give rise to such very different models? Boring (1923) had a point. Within the psychometric field there has been no resolution of the various general or multi-box models of intelligence. There is a need to find some external criterion or some antecedent variable that is related to performance on mental ability tests. Resolution may be coming in two ways.

First, the nature of factor analysis is changing. It is now possible actively to test hypothetical models of intelligence with techniques such as LISREL. Using this method Gustafsson (1984) administered 16 tests of intelligence to 1000 subjects and tested various models of intelligence for their goodness of fit to the data. Interestingly, he was able to give comfort to many previous theorists. Older models of the structure of human abilities appeared to be special cases of his hierarchical unifying model of intelligence. At the lowest level in the hierarchy of ability lay primary factors similar to those proposed by Thurstone (1938) and Guilford (1985). At a higher level there appeared fluid and crystallised intelligence, as hypothesised by Cattell (1963). At the peak of the hierarchy lay g, which was indistinguishable from the second order fluid intelligence factor. Models which had seemed irreconcilable for so long emerged as complementary rather than contradictory; in the past various authors have merely drawn attention to various parts of the larger model.

The Biology of Intelligence

Second, and more important, escape from circularity has come with the advent of an area of study known as the 'biology of intelligence.' But this term has its problems: it immediately alienates those who see this as an exercise in genetic reductionism in what appears to be an arbitrary and culture-biased set of puzzles (Gould, 1981; Rose, Lewontin and Kamin, 1984). To others the term is problematic because of its diversity, for it applies to no one area of study and to no one set of methods. In what follows I will examine the experimental evidence for one aspect of the biology of intelligence - namely, those 'basic' psychological processes that correlate with IQ test scores. But, before that, it is useful to catalogue the various approaches that have been taken by other researchers.

First, biology as a model. In Robert Sternberg's (1985) review of models of intelligence he included Piaget's work as an example of the biological approach. In this sense Piaget's contribution has been important and very different from that of the psychometricians. While Piaget's tests of conservation, etc. have provided mental tests with moderate g-loadings and much variance that is specific to the tests themselves (see Jensen, 1980), his work on the biology of knowledge growth has been much less studied. Knowledge accretion, in Piaget's account (1971, 1978, 1980) involves the brain acting in the way that other organs do: with a substrate (information from the world) and with products which are transforms of the substrate (schemata, formed by the processes of assimilation and accommodation). Piaget's writings on the growth of intelligence have a structure that is close to that of the evolutionary epistemologists (Wuketits, 1985; Campbell, 1974) and those AI workers whose first assumptions and constraints involve what is known about brain development (Edelman and Reeke, 1982). This form of the biology of knowledge might form a superstructure for psychometricians but it is not of immediate empirical concern.

Second, biology as race, genetics and heredity. It is this aspect of the psychometric endeavour that has coloured many others and has often prevented rational argument. The discovery of the Burt fraud (Hearnshaw, 1979) and the reactions to the work of Jensen (1969) and Eysenck (1973) by non-specialist writers like Kamin (1974), Gould (1981) and Rose, Kamin and Lewontin (1984) has made others wary of the field. Jensen argued recently (1985; there are many peer criticisms appended to the article) that the one standard deviation difference between US blacks and whites on IQ tests has its basis in the tests' g-loadings. Nevertheless, Jensen also reports that Blacks are superior to whites on other abilities such as memory. On the other hand, Mackintosh's (1986) research in this area has drawn attention to evidence that fails to support a genetic origin for mental ability differences between ethnic groups: Mackintosh implicates differences in the social circumstances of the different ethnic groups.

Third, biology as neurobiology. This endeavour involves searching for biological, often biochemical, correlates of intelligence test scores. Patient groups with impaired levels of cognition such as Alzheimer or Down syndrome patients are often used. Weiss (1984, 1986) has reviewed evidence that Down's syndrome involves excess peroxidation of neuronal membranes and a build up of the products of oxidation within the neuron. Both of these will impair information transfer. One enzyme involved in the prevention of excess oxidation is glutathione peroxidase (GSHPx). Sinet, Lejeune and Jerome (1979) argued that the level of this enzyme in Down's syndrome patients would be correlated with the level of accurate information transfer and that this in turn would be demonstrated by a correlation between the enzyme level and the patient's IQ. They reported a correlation of O.5 between GSHPx with IQ scores in 50 Down's syndrome patients. Uric acid is also involved in the prevention of lipid oxidation by free radicals and Inouye, Park and Asaka (1984) have reported a correlation of 0.334 ($p<.025$) between uric acid level and IQ. These workers have argued, from twin data, that the two traits (uric acid level and IQ) might have partial communality of gene loci.

Using Alzheimer patients and normal controls Chase, Fedio, Foster, Brooks, DiChiro and Mansi (1984) (1984) found a correlation of 0.68 between glucose use (as measured by positron emission tomography scanning of fluorodeoxyglucose F18) in the cerebral cortex and full scale IQ on the WAIS. Further evidence for the biological basis of IQ scores came from their localisation data: verbal IQ related best to glucose metabolism rate in the left temporal region ($r = 0.76$) and performance IQ correlated best with glucose metabolism in the right parietal region ($r = 0.70$). Also using Alzheimer patients, Deary, Hendrickson and Burns (1987) have demonstrated a correlation of 0.5 ($p < .02$) between serum calcium levels and cognitive scores assessed by the Mini Mental State (Folstein, Folstein and McHugh, 1975). These authors have argued that calcium has a central role in the maintenance of the integrity of the normal cytoskeleton and that low calcium levels impair information transfer.

Basic Processes in Intelligence

Fourth, biology as basic psychological processes. In this sense biology refers to psychological tests that index basic information-processing functions and constraints. The renewed interest in basic processes marks a return to the empirical efforts of Galton (1883), Spearman (1904) and Spearman's students such as Abelson (1911) and Carey (1914-15). The success of the Binet approach to intelligence testing - the estimation of general ability using a hotch-potch of mental tests with appropriate weighting before an average is taken - has had its drawbacks. In 1911 Abelson warned that the immediate practicality of the Binet test, (in education, mental handicap, job selection, etc.) which measured 'higher-level' processes like memory, reasoning and judgement, meant that psychometry would become divorced from experimental psychology. The worry was that the more the tests proliferated the less was becoming known about the psychological processes that underlay success in performing them. Although there has been a trickle of experimental studies hinting that perceptual speed, reaction time and sensory discrimination were reliable and moderately high correlates of IQ test performance (see Deary, 1986 and 1987 for reviews), until the last decade there has been no concerted research effort to discover what amount, if any, of the interindividual variance on IQ scores is attributable to individual differences in simpler psychological performance.

This revived effort has been driven by two ideas. First, the idea that the more intelligent person has some advantage in mental speed has always had some currency. Experimental evidence (see below), professional and lay opinions agree that intelligent people tend to be "quick-witted", "quick on the uptake" and "quick thinkers" (Sternberg, 1981). Second, the researchers in this field have implicitly or explicitly accepted that the solution of IQ test items involves many psychological processes and that these may be expressed as a hierarchy. Thus Gustafsson's hierarchy of psychometric intelligence is mirrored by an internal hierarchy of psychobiological processes. In this account (Jensen, 1985) IQ test items are solved by psychological metaprocesses. These metaprocesses are combinations of basic psychological processes and their orchestration and combination are affected by prior experience, education and coaching. Thus the efficiency of metaprocesses will have some imperfect correlation with the functioning of individual basic processes. Jensen (1985) suggests that these basic processes might include stimulus apprehension, iconic memory, stimulus encoding, short term memory, rehearsal of short term memory, memory scanning, retrieval from long term memory, mental rotation, response execution, etc. At an even more basic level these processes will share the performance constraints of a common neurology and thus their efficiency will be correlated to yield a biological general intelligence factor. Thus the efficiency of neuronal transmission (affected by inherited factors and environmental factors such as nutrition and exposure to neurotoxic agents) correlates with the basic psychological processes (affected by time of day, sedative drugs, etc.) which combine to form metaprocesses which solve informational problems. These problem solutions are factor analysed to yield ability clusters (verbal, visuo-spatial, etc) which correlate and yield a psychometric g factor. Therefore both biological g and psychometric g are hypothetical and not, as yet, able to be indexed directly.

Three more or less basic psychological tests have attracted much interest in the last decade as correlates of intelligence. All involve some form of mental speed and are: reaction time (RT); average evoked potentials (AEPs); and inspection time (IT).

Reaction time measures have a history of reliable but modest correlation with IQ scores (see Beck, 1933 for a review of the early studies) but the current research has been dominated by the work of Jensen using the Hick (1954) paradigm. In the Hick method subjects are positioned before a response panel and place their preferred index finger on a 'home' button. Surrounding the home button is a semicircle of 8 equidistant lights, each of which has a response button in front of it. The task involves waiting for a light to come on whereupon the subject must release the home button and, as quickly as possible, move to the button in front of the appropriate light and press it. This yields two measures: the time it takes for the subject to release the home button after the light has been switched on is termed the movement time (MT); and the time from the release of the home button until the button in front of the target light is depressed is called the response time (RT). Together they are called the reaction time. The Hick paradigm involves varying the number of lights from which the target light may be expected: Jensen uses 1,2,4 and 8 light sets. Thus reaction times to 0,1,2 and 3 bits of information may be assessed. Hick's law states that there is a linear increase in reaction time as the number of bits of information increases (i.e. $\log_2$ of the number of stimulus alternatives).

Jensen (1986) has collected 20 independent subject samples where the slope of Hick RT has been correlated with mental ability test scores. Twelve of the studies are from Jensen's laboratory (1055 subjects): of 22 correlations the N-weighted mean r is -0.091, SD = 0.109. Eight studies from other laboratories (503 subjects) have delivered 13 correlations: their N-weighted mean is -0.181, SD = 0.147. Jensen makes two important points about these correlations. First, despite their low correlation the significance of these results is at the 0.1% level. Second, many of the samples include university students whose ability range is less than half of the population range: this reduces the size of the correlations and Jensen estimates that a true population correlation between Hick RT slope and IQ would be in the region of -0.3.

Jensen (1986) has also reviewed the work on the correlation between IQ and the RT for the individual stimulus set sizes on the Hick paradigm. From 31 studies (1129 subjects) the N-weighted mean correlations are: -0.18 for 1 light (0 bits); -0.19 for 2 lights (1 bit); -0.22 for 4 lights (2 bits); and -0.23 for 8 lights (3 bits). Again, 19 of these studies are from Jensen's own laboratory. Two recent critical accounts of this work have appeared. Longstreth (1986) proposes that the large number of studies that come from the same laboratory or from the laboratories of Jensen's former students (especially P.A. Vernon) biases the results in some unspecified way. He also suspects that negative studies are less likely to be published hence elevating the reported mean correlation. Mackintosh (1986) has suggested that RT is not a basic process: he attributes the correlations to subjects' willingness to concentrate on a boring task. Frearson and Eysenck (1986), using 37 normal adults tested on Raven's APM, found an IQ-RT correlation of about -0.3 and an IQ-MT correlation of -0.45 regardless of the number of stimulus alternatives, contrary to Jensen's findings. When these authors made the Jensen task more cognitive by asking subjects to respond to one of three lights on the basis of its relative position the IQ-RT correlation increased to between -0.5 and -0.6. These results indicate that speed of RT and MT to even a single stimulus is significantly correlated to IQ but, obviously, if the task is made more difficult, the correlation will rise (because, presumably, extra basic psychological processes are being sampled).

Empirical and theoretical implications of the correlations between IQ scores and indices derived from AEPs to simple auditory stimuli have been investigated most recently by A.E. and D.E. Hendrickson (1980, 1982). Earlier work by Schucard and Horn (1972) and Ertl and Schaffer (1969) had demonstrated a low correlation between amplitude and latency measures on the AEP and psychometric intelligence. The Hendricksons have used the 'string length' of the AEP, a measure of AEP complexity. They argue that high IQ subjects are characterised by few errors in neuronal transmission while low IQ subjects make more errors on average. Thus, they claim, when a large number of evoked potentials to an identical stimulus are averaged for a high IQ subject the low error rate (over many neurons) will result in similar waveforms. For the low IQ subject the higher error rate will result in more variability in individual evoked potentials. Therefore the low IQ AEP loses much of the complexity of the individual evoked potentials that compose it. The Hendricksons measure complexity by a computer-run algorithm which, in essence, places a string over the AEP for a given epoch and measures its length.

Using the average of 90 EPs to 85 dB, 30 ms, 1000 Hz tones Blinkhorn and D.E. Hendrickson (1982) correlated 'string' measures with Raven's Advanced Progressive Matrices Scores on 34 students (17 male, 17 female). Post stimulus 512 ms strings correlated 0538 (p<.001) with APM scores. Unstimulated EEG strings correlated at 0.127 (n.s.) with APM scores. Hendrickson and Hendrickson (1982) reported a correlation of 0.72 in 219 schoolchildren. Haier, Robinson, Braden and Williams' (1983) replication found correlations between 0.13 and 0.50 (many were non-significant) although they did find that stimulus intensity was important. In a small study where the Hendricksons' methods were followed closely, Caryl and Fraser (1985) found a correlation of 0.72 between Alice Heim test scores and string length measures of AEP. Mackintosh (1986) failed to replicate this finding and suggests, again, that while the work of the Hendricksons is important, their results might be due to a willingness to comply with and to continue to concentrate upon a boring task.

Inspection time is the newest of the three basic measures. Although similar tasks were shown to be useful in discriminating subjects of different levels of ability many decades ago (Cattell, 1886; Burt 1909-10), the IT task as it is used today was first suggested formally by Vickers, Nettlebeck and Willson in 1972. The first study of the relationship between individuals' ITs and IQ scores was carried out in 1976 by Nettlebeck and Lally. They reported correlations of -0.89 and -0.92 between WAIS performance IQ and two estimates of inspection time. The study was small (10 subjects) and the IQ range was wide (from 47 to 119). Subsequent replications confirmed the finding (Lally and Nettlebeck, 1977; Brand, 1981), and by 1982 Brand and Deary reviewed the nine known studies of IT and IQ and found a median correlation of -0.8 in five studies where young adults of mean IQ around 100 had been tested with 'culture-fair' tests.

To perform the IT task requires very little active cognition. The IT is an estimate of a person's speed of intake of sensory information. Practically, the task, as originally conceived, involved subjects viewing two matte black vertical lines of markedly different lengths in a tachistoscope. The lines, after a brief exposure, were backward masked by thicker matte black lines to prevent further examination of the stimulus for information. The task of the subject was to state whether the long line was the right or the left hand member of the pair of lines. The measure taken (the subject's IT) is the exposure duration at which the subject is able to perform the discrimination to a pre-set level of accuracy (often around 90%). No reaction time is taken in the standard task and subjects are encouraged to respond at leisure. Significant correlations have been found, in the expected direction, in IT tasks involving 2, 3 and 4 lines, 2 lights, and animal names as stimuli. The correlation has been found in diverse subject groups: young children (Brand and Deary, 1982; Anderson, 1986); mentally handicapped and normal adult populations (Brand and Deary, 1982; Nettlebeck, 1986); and in undergraduate samples where ability range is very restricted (McKenzie and Bingham, 1985; Longstreth, Walsh, Alcorn, Szeszulski and Manis, 1986; Deary, Caryl, Egan and Wight, in preparation).

The basic nature of the IT phenomenon is supported by the fact that the correlation holds for auditory as well as visual processing speed. In the auditory IT task the subject has to identify the temporal order of two tones ('low-high' or 'high-low') of markedly different pitch (usually about 100 Hz) which are played for a brief time one after the other and are then backward masked either by white noise or by a warble that contains the frequencies of both of the stimulus tones (Deary, 1980, reported in Brand and Deary, 1982; Nettlebeck, Edwards and Vreugdenhil, 1986). The correlations between auditory and visual IT and IQ are similar: in a recent review Nettlebeck found that of 24 independent studies (including auditory and visual stimuli), 16 found a significant negative correlation between IT and IQ. The correlations ranged from +0.1 to -0.61 with a mean of -0.34. Many of the studies involved groups that were of restricted ability range (and all the studies involving mentally retarded subjects were excluded) and Nettlebeck estimated a true IT-IQ correlation of around 0.5 in non-retarded young adults.

In a short time the IT measure has attracted much interest. Not all of this interest has been helpful in advancing the psychological understanding of the measure. The early hopes that IT might provide a widely-acceptable culture-fair estimate of ability were clearly not realised (Brand and Deary, 1982; Nettlebeck, 1983; Mackintosh, 1981). However, the combination of the reliability and size of the IT-IQ relationship, the number of independent groups who have replicated it and the apparent simplicity of the task put IT in a unique position for the study of basic processes in intelligence. The following points are a series of problems and possibilities that the measure has thrown up in its short existence.

Individuals' strategies in performing the IT task have been examined (Mackenzie and Bingham, 1985; Mackenzie and Cumming, 1986; Egan, 1986; Fitzmaurice and Nolan, 1983). When the two lines task is performed using computer-drawn lines or on a series of light-emitting diodes as opposed to the tachistoscope lines then about half of an undergraduate sample will report an apparent-movement artefact between the stimulus lines and the backward mask and they report using this impression of movement to solve the items. Mackenzie and his colleagues have found that the IT-IQ correlation holds for the non-strategy-users but disappears in the strategy using group. Unpublished work in Edinburgh (Egan, personal communication) has found both the strategy-users and the non users to have a similar IT-IQ correlation in a study involving a group with normal mean and standard deviation for IQ. However, the strategy involved here seems to be an artefact of computer-presentation devices and has not been reported in the tachistoscopic visual or with the auditory form of the test.

A more general approach to the strategy argument is put forward by those who claim that the high IQ person has a better IT by virtue of his willingness to sustain attention in a boring task. A reply to this criticism may be found in the observation that individuals' results, no matter how slow the IT, show regular psychophysical curves. In other words, in the IT test, where exposure durations are varied unpredictably and are then, at the end, plotted as a duration versus % correct, one would be hard put to argue for an attentional hypothesis that explained how an 'unattending' testee managed to score 100% at a duration of, say, 80ms and then fell off regularly until he or she was responding at chance levels when the duration came to 65ms. The subjects' hypothesised fatigue or unwillingness to respond because of boredom does not in fact result in the expected random error pattern. Finally, there is another objection to the attentional hypothesis. In a commonly used psychophysical IT procedure, the PEST (Taylor and Creelman, 1967) adaptive staircase, the subject has a short run of trials at one duration that suddenly jumps to a faster duration. If the IT task is simply a matter of attention then the items to be solved at these sudden duration jumps (when the duration gets shorter) would be less well discriminated by the low IT subjects. A re-analysis of data in this department by myself and Egan has shown that low IT and low IQ subjects cope with unexpected shortenings of duration just as well as those with shorter IT and higher IQ.

The cross-modal study of IT has added weight to its purported basic nature. Also, the auditory-visual correlation has been studied to find if the two tasks have some common variance, for it is possible that their separate correlations with IQ scores are a result of independent strategies. The original claim of a correlation of near to unity between the two (Brand and Deary, 1982) was certainly dependent upon the inclusion of mentally retarded subjects and Nettlebeck, Edwards and Vreudgenhil (1986) and Deary, Caryl, Egan and Wight (in preparation) subsequently found a correlation of about 0.4 for university samples. Given the restricted ability range of this population this suggests that there is a considerable amount of shared variance between the auditory and visual processing speeds and that this might reflect some property of the CNS. However, the correlation of IQ and processing speed of tactile information (Edwards, 1984) suggests that the result will not generalise to this third sense. But, given the cognitive unimportance of tactile as opposed to visual and auditory information perhaps that is not surprising.

The initial strength of the IT-IQ finding lay in the fact that it could be replicated on different hardware using different stimuli (this, of course, is not a weakness of the phenomenon as Mackintosh (1986) suggests). In his comprehensive review Nettlebeck (1986) called for a concerted international effort to standardise the test. At present the IT is performed on tachistoscopes, computer screens and LEDs; experimenters have used lights, lines and sounds of different intensities and of different discriminabilities; and different psychophysical techniques have been used in its estimation (adaptive staircases and MCS methods as well as combinations of the two). In order that a subject's IT becomes more than an arbitrary number it will be necessary to standardise these many factors.

Meanwhile IT presents many advantages as a test of basic processing speed. It shows little practice effect (a reduction of 17-30% is typical over the first few sessions with very little improvement thereafter (Nettlebeck, 1986)) and may be used in repeated testing in pharmacological studies, time of day work, occupational testing and in longitudinal studies. It continues to remain one of the few correlates of general mental ability that has not been accused of some kind of bias. Unlike RT, IT has the appearance of a task that indexes one basic process: the speed of intake of sensory information. Unlike AEP, IT may be tested relatively quickly by workers using non-specialist equipment.

Moreover, IT offers the tantalising possibility of testing a human ability in animals such as rats, doves and non-human primates that are commonly used in the psychological laboratory. Indeed, perhaps that is the most important way forward for the IT measure in the biology of intelligence. Speculatively, it is possible to name three factors that might contribute to inter- and intra-species differences in what we call intelligence. Speed of information processing is an obvious candidate: faster intake speed (as indexed by IT) will lead to enhanced ability to make sense of the incoming flow of information and will allow more detailed study of time-limited information sources (like road traffic or the motion of prey). Second, the degree of corticalisation will allow more complex and higher level thinking: allowing an animal to plan ahead with more foresight, to make more use of stored information and to organise incoming information in more detailed indexes. Third, there might be a place for the ability to make fine discriminations with the senses (Deary, 1987): a sensory system that has a smaller JND increment will accrue more information from the environment and will make fewer stimulus confusions. Here, then, is a proposed triad for the study of the biology of intelligence, both intraspecies and interspecies: sensory sensitivity, information processing speed and relative size of cerebral cortex.

References

Abelson AR (1911) The measurement of mental ability of backward children. British Journal of Psychology, 4: 268-314

Anderson M (1986) Inspection Time and IQ in young children. Personality and Individual Differences, 7: 677-686

Beck LF (1933) The role of speed in intelligence. Psychological Bulletin, 30: 16 9-178

Blinkhorn SF, Hendrickson DE (1982) Average evoked responses and psychometric intelligence. Nature, 295: 596-597

Boring EG (1923) Intelligence as the tests test it. New Republic, 35: 35-37

Brand CR, Deary IJ (1982) Intelligence and 'Inspection Time'. In (HJ Eysenck ed) A Model for Intelligence. Springer Verlag

Burt C (1909-10) Experimental tests of general intelligence. British Journal of Psychology, 3: 94-177

Campbell DT (1974) Evolutionary epistemology. In (PA Schilpp, ed) The Philosophy of Karl Popper, vol 1, pp 413-436 Open Court, LaSalle

Canavan AGM, Dunn G, McMillan TM (1986) Principal components of the WAIS-R. British Journal of Clinical Psychology, 25: 81-85

Cattell J.McK (1886) The time taken up by cerebral operations. Mind, 11: 220-242, 377-392, 524-538

Cattell RB (1963) Theory of fluid and crystallised intelligence: A critical experiment. Journal of Educational Psychology, 54: 1-22

Carey N (1914-15) Factors in the mental processes of schoolchildren. British Journal of Psychology, 7: 453-490

Caryl PG, Fraser IC (1985) The Hendrickson 'string-length' measure and intelligence - a replication. Paper presented at the Psychophysiology Society - 2nd Scottish meeting (September 1985)

Chase TN, Fedio P, Foster NL, Brooks R, Dichiro G, Mansi L (1984) WAIS performance: Cortical localisation by fluorodeoxyglucose F18-positron emission tomography. Archives of Neurology, 41: 1244-1247

Deary IJ (1986) Inspection time: Discovery or rediscovery? Personality and Individual Differences, 7: 625-631

Deary IJ (1987) The nature of intelligence: Simplicity to complexity and back again. In (D Forshaw and M Shepherd, eds) The Maudsley Essay Series in the History of Psychiatry, vol 1 (In press)

Deary IJ, Hendrickson AE, Burns A (1987) Calcium and Alzheimer's disease: A finding and an aetiological hypothesis. Personality and Individual Differences, 8: (in press)

Edelman GM, Reeke GN Jr (1982) Selective networks capable of representative transformations, limited generalisations and associative memory. Proceedings of the National Academy of Sciences USA, 79: 2091-2095

Edwards C (1984) Inspection time in three sensory modalities and its relation to intelligence. BA (Hons) Thesis, University of Adelaide

Egan V (1986) Intelligence and inspection time: Do high-IQ subjects use cognitive strategies? Personality and Individual Differences, 7: 695-700

Eysenck HJ (1973) The Inequality of Man. Temple Smith

Eysenck HJ (1979) The Structure and Measurement of Intelligence. Springer Verlag

Eysenck HJ (ed) (1982) A Model for Intelligence. Springer Verlag

Fitzmaurice G, Nolan JM (1983) Inspection time: A misguided model for intelligence. Paper read to the Irish Psychological Society

Folstein MF, Folstein SE, McHugh PR (1975) 'Mini Mental State'. A practical method for grading the cognitive state of patients for the clinician. Journal of Psychiatric Research, 12: 189-198

Galton F (1883) Inquiries into Human Faculty. Dent

Gould SJ (1981) The Mismeasure of Man. Norton

Guilford JP (1985) The structure-of-intellect model. In (BB Wolman, ed) Handbook of Intelligence, pp 225-266. John Wiley Ltd

Gustaffson JE (1984) A unifying model for the structure of mental abilities. Intelligence, 8: 179-203

Haier RJ, Robinson DL, Braden W, Williams D (1983) Electrical potentials of the cerebral cortex and psychometric intelligence. Personality and Individual Differences, 4, 591-599

Hearnshaw L (1979) Cyril Burt: Psychologist. Cornell University Press.

Hendrickson AE, Hendrickson DE (1982) The biological basis of intelligence: Parts 1 and 2. In (HJ Eysenck, ed) A Model for Intelligence. Springer Verlag

Hendrickson DE, Hendrickson AE (1980) The biological basis of individual differences in intelligence. Personality and Individual Differences, 1: 1-33

Hick W (1954) On the rate of gain of information. Quarterly Journal of Experimental Psychology, 4: 11-26

Hooper FH, Hooper JO, Colbert KK (1984) Personality and memory correlates of intellectual functioning: Young adulthood to old age. Karger

Inouye E, Park KS, Asaka A (1984) Blood uric acid level and IQ: A study in twin families. Acta Genet Med Gemellol, 33: 237-242

Jensen AR (1969) How much can we boost IQ and scholastic achievement? Harvard Educational Review, 39: 1-123

Jensen AR (1980) Bias in Mental Testing. Methuen

Jensen AR (1985) The plasticity of 'intelligence' at different levels of analysis. In (J Lockhead, J Bishop and D Perkins, eds) Thinking: Progress in Research and Teaching. Franklin Institute Press

Jensen AR, Vernon PA (1986) Jensen's reaction-time studies: A reply to Longstreth Intelligence, 10: 153-179

Kamin LJ (1974) The Science and Politics of IQ. Lawrence Erlbaum Associates

Lally M, Nettlebeck T (1977) Intelligence, reaction time and inspection time. American Journal of Mental Deficiency, 82: 273-281

Longstreth LE (1986) The real and the unreal: A reply to Jensen and Vernon. Intelligence, 10: 181-191

Mackenzie B, Bingham E (1985) IQ, inspection time, and response strategies in a university population. Australian Journal of Psychology, 37: 257-268

Mackenzie B, Cumming S (1986) Inspection time and apparent motion. Personality and Individual Differences, 7: 721-729

Mackintosh NJ (1981) A new measure of intelligence? Nature, 289: 529-530

Mackintosh NJ (1986) The biology of intelligence? British Journal of Psychology, 77: 1-18

Nettlebeck T (1982) Inspection time: An index for intelligence? Quarterly Journal of Experimental Psychology, 34A: 299-312

Nettlebeck T (1986) Inspection time and intelligence. In (PA Vernon, ed) Speed of Information Processing and Intelligence. Ablex

Nettlebeck T, Edwards C, Vreugdenhil A (1986) Inspection time and IQ: Evidence for a mental speed-ability association. Personality and Individual Differences, 7: 633-641

Nettlebeck T, Lally M (1976) Inspection time and measured intelligence. British Journal of Psychology, 67: 17-22

Piaget J (1971) Biology and Knowledge. Edinburgh University Press

Piaget J (1978) Behaviour and Evolution. Routledge, Kegan and Paul

Piaget J (1980) Adaptation and Intelligence. University of Chicago Press

Rose S, Kamin LJ, Lewontin RC (1984) Not in Our Genes. Pelican

Schucard DW, Horn JL (1972) Evoked cortical potentials and measurement of human abilities. Journal of Comparative and Physiological Psychology, 78: 59- 68

Sinet PM, Lejeune J, Jerome H (1979) Trisomy 21 (Down's syndrome), glutathione peroxidase, hexose monophosphate shunt and IQ. Life Sciences, 24: 29-33

Spearman C (1904) "General intelligence," objectively determined and measured. American Journal of Psychology, 15: 201-293

Sternberg RJ, Conway BE, Ketron JL, Bernstein M (1981) People's conceptions of intelligence. Journal of Personality and Social Psychology, 41: 37-55

Sternberg RJ (1985) Human intelligence: The model is the message. Science, 230: 1111-1118

Thurstone LL (1938) Primary mental abilities. Psychometric Monographs, No 1. University of Chicago Press

Vickers D, Nettlebeck T, Willson RJ (1972) Perceptual indices of performance: The measurement of 'inspection time' and 'noise' in the visual system. Perception, 1: 263-295

Weiss V (1984) Psychometric intelligence correlates with interindividual different rates of lipid peroxidation. Biomedica and Biophysica Acta, 43: 755-763

Weiss V, 1986 From memory span and mental speed toward the quantum mechanics of intelligence. Personality and Individual Differences, 7: 737-749

Wuketits FM, 1986 Evolution as a cognition process: Towards an evolutionary epistemology. Biology and Philosophy, 1: 191-206

HUMAN BRAIN EVOLUTION:
I. EVOLUTION OF LANGUAGE CIRCUITS

Terrence W. Deacon
Biological Anthropology
Harvard University
Cambridge, MA 02138

The investigation of the neural basis for human language abilities has principally proceeded in neuropsychological and neurosurgical contexts with the study of brain damaged patients exhibiting aphasia, patients undergoing electrical stimulation mapping during neurosurgical procedures, and through the recent use of new in vivo imaging techniques of the working conscious brain. As powerful and illuminating as these approaches have been, there are still major gaps in our understanding of language systems of the brain and as yet no comprehensive model of language processing. One of the most serious gaps is the lack of circuit information regarding the neural connections of brain areas involved in linguistic processes. And despite the enormous progress in developing animal models for many other brain functions the uniqueness of linguistic abilities and uncertainties about their neural substrates have seemingly provided little hope of useful non-human models of language processes. However, more complete connectional data and informative non-human models are potentially available through a broader understanding of neuroanatomical homologies.

Neurobiological investigations comparing the morphogenesis, connectional patterns, and biochemistry of non-human brains are now in a position to begin to help in discerning some of the critical similarities and differences between human and non-human brains and provide clues and testable predictions concerning human brain circuitry, morphogenesis, and evolution. The following discussion will review some recent findings from axonal tracer research in monkeys and electrical stimulation studies in human patients that may help advance both the study of language processes in the brain and the study of human brain evolution.

Axonal tracer studies and language circuits

The development of enzymatic, immunological, fluorescent, and autoradiographic tracer techniques in the 1970s and 80s has radically changed the prospects for understanding the "wiring" of the nervous system. It is probably not overly

NATO ASI Series, Vol. G17
Intelligence and Evolutionary Biology
Edited by H. J. Jerison and I. Jerison

optimistic to predict that well before the turn of the century we will have compiled essentially complete maps of the neural connection patterns of the brains of some the most common experimental animals (e.g. rats, cats, and macaques).

Given the fundamental uniqueness of human language--its demands on the learning of skilled vocal sequences, its rule governed syntactic organization, and its semantic displacement of reference--it might be argued that the circuits underlying language processes in the human brain are the most likely of any brain circuits to have become reorganized during human evolution. This expectation is further supported by the apparent dichotomies between the neural control of naturalistic primate vocalization and speech (Myers, 1976; Jurgens and Ploog, 1976). Stimulation studies have located the principal vocalization circuits of monkeys within the limbic structures, medial diencephalon, and central midbrain, and along with lesion experiments have demonstrated that frontal motor cortex plays no necessary role (Jurgens et al., 1982). In contrast, limbic and diencephalic damage in humans seldom is associated with loss of language abilities, but damage to inferior frontal areas in humans including the areas surrounding the motor face area, or damage to posterior temporal and inferior parietal areas can result in severe speech and comprehension disturbances (Kolb and Whishaw, 1985).

Our present models of human language circuits come largely from inference from studies of brain-damage induced language loss (aphasia), and also from careful gross dissections of preserved human brains, especially in postmortem aphasia cases. From these lines of evidence crude circuit diagrams of the principal language areas and their critical inter- connections have been proposed. These have also been informed by so-called "disconnection" syndromes where a cortico-cortical connection is interrupted by damage but the involved cortical areas are presumably left intact (Geschwind, 1965).

The positions of language-specialized areas and their presumed interconnections are diagrammed in Figure 19.1 for the left hemisphere of the brain based on contemporary interpretations from language pathology. The location of Broca's area adjacent to the motor area (m1) for the face, tongue, and larynx and the location of Wernicke's area adjacent to the primary auditory area (a1) has suggested that Broca's area is the speech production area and Wernicke's area the speech comprehension area. A presumed reflex arc for language is completed by a connection between Broca's and Wernicke's areas, identified as the arcuate fasciculus. Damage to this fiber system is presumed to result in "conduction aphasia" in which both comprehension and production of speech are unaffected but the ability to repeate heard utterances is significantly impaired. In general this simplified model of language circuits fails to explain many more subtle and complex aspects of these aphasic syndromes: including syntactic comprehension deficits in Broca's aphasia, phonetic production deficits in Wernicke's aphasia, and anomia and phonetic production deficits in conduction aphasia. This simplified model of language circuitry is augmented by the inclusion of some additional areas and pathways. The angular gyrus, supra-

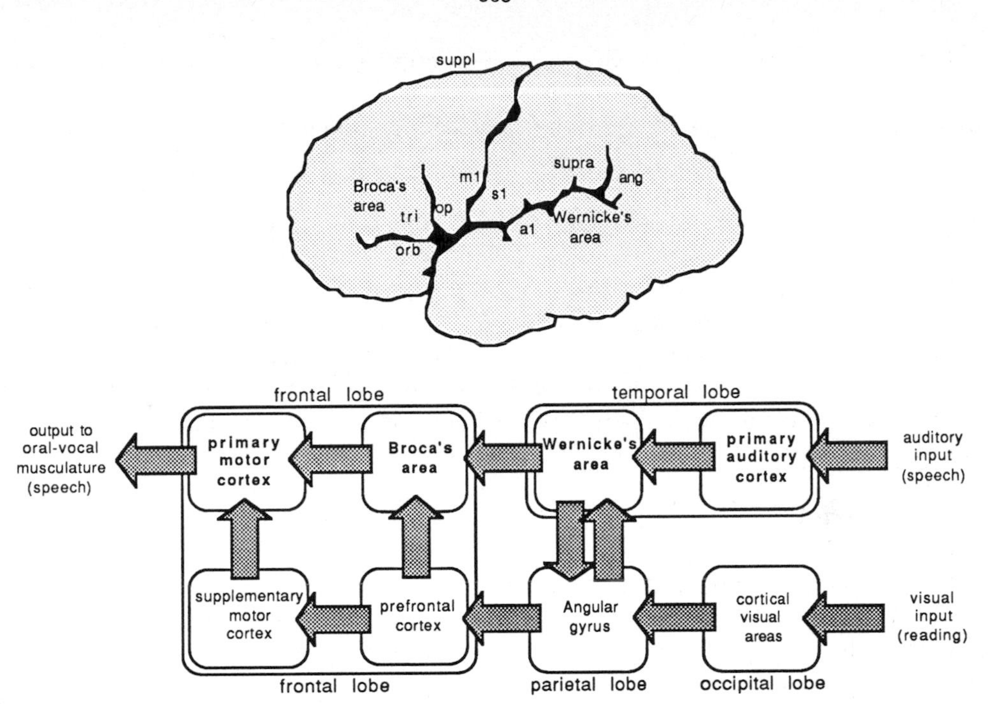

Figure 19.1 Human language areas and presumed major neural circuits. a. Major language areas: Broca's area (por = pars orbitalis; pt = pars triangularis; pop = pars opercularis), Wernicke's area (W), angular gyrus (ang), and the supramarginal gyrus (sm). b. An updated model of the classic connectionist view of human language circuits. The presumed major connections of the "speech reflex arc" as conceived by Carl Wernicke (1874) and his followers are depicted along the top series of arrows. Additionally, pathways presumed to be involved in reading, semantic analysis, sequential speech organization, and speech initiation are depicted along the bottom (based on a current introductory review by Kolb and Whishaw, 1985).

marginal gyrus and supplementary motor cortex have all been implicated in the study of aphasia: the angular gyrus for its role in reading and semantic retrieval; the supramarginal gyrus for its role in fluency and naming; and the supplementary motor cortex have all been implicated in the study of aphasia: the angular gyrus for its role in reading and semantic retrieval; the supramarginal gyrus for its role in fluency and naming; and the supplementary motor area for its role in initiating and sustaining spontaneous utterances. The prefrontal cortex anterior to Broca's area is not directly associated with any particular language disorder (aspontaneity of speech and a loss of "word fluency"--spontaneous word-list production--are typical language symptoms) but is included because of its presumed role in higher cognitive functions and as a relay for multimodal information.

Using both architectonic criteria (cell and axon layering patterns) and topographic criteria (relative surface position with respect to other cortical regions) it is possible to identifiy the position of morphological homologues to the

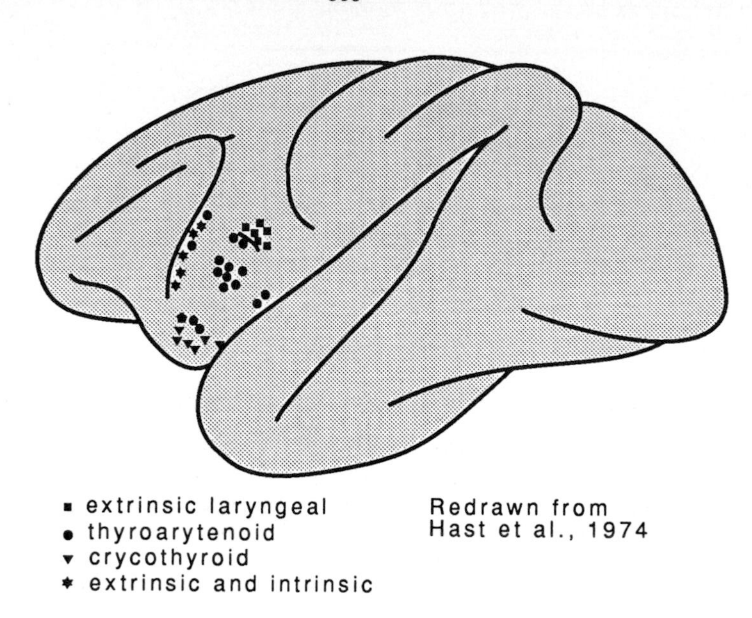

Figure 19.2 Laryngeal motor control area in the inferior frontal cortex of the macaque brain.

major human language areas in monkey brains. On these purely anatomic grounds Wernicke's area has been homologized to an area called Tpt and adjacent areas in the monkey temporal lobe and the different sectors of Broca's area have been homologized to areas surrounding and within the banks of the inferior branch of the arcuate sulcus (Braak, 1980; Galaburda and Sanides, 1980; Galaburda and Pandya, 1982; Deacon, 1984). Additionally, studies utilizing electrical stimulation, recording, or lesion techniques have demonstrated some functional homologies. For example, stimulation studies in the frontal cortex of monkeys demonstrate that the area bordering and behind the inferior branch of the arcuate sulcus is involved in controlling facial, oral, and laryngeal musculature (Vogt and Vogt, 1919; Jurgens, 1974; Hast and Milojevic, 1966; Hast et al., 1974; Jurgens and von Cramon, 1982--see Figure 19.2). In the temporal region cells in the dorsal bank of the superior temporal sulcus are "tuned" to the distinction between self-produced and other-produced conspecific vocalizations (Hupfer et al., 1977; Manley and Muller-Preus, 1978), the area posterior and adjacent to the left primary auditory cortex has been found to be crucial to sound sequence discrimination (Dewson, et al., 1969; Dewson and Burlingame, 1975; Dewson, 1977), and there appears to be left hemispheric specialization (at least in Japanese macaques) for discriminating conspecific vocalizations (Petersen et al., 1978; Beecher et al., 1979).

These perisylvian areas in monkeys have also been studied using axon tracing techniques to develop detailed "maps" of the connection patterns of each,

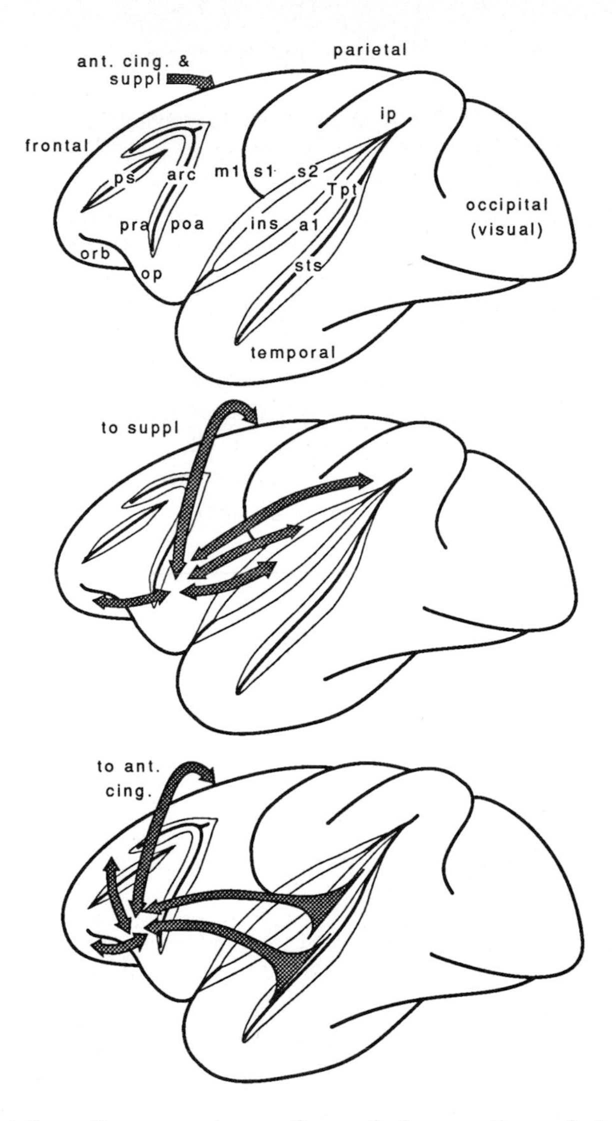

Figure 19.3 Depiction of some major cortico-cortical connections of the macaque inferior arcuate region, based on a series of tracer studies by Deacon (1982; 1984). The top figure identifies some of the major cortical areas involved. The middle figure shows projections from inferior parietal cortex (ip), supplementary sensory area (s2), insula (ins), supplementary motor area (suppl), and orbital frontal cortex (orb) converging within the postarcuate area (poa) in the monkey brain. This area is deemed roughly homologous to the pars opercularis of the human brain (see Figure 1). The bottom figure shows projections from posterior auditory cortex (Tpt), auditory association cortex within the dorsal bank of the superior temporal cortex (sts), anterior cingulate cortex, prefrontal cortex, and orbital frontal cortex (orb) converging within the prearcuate area (pra). This area is roughly deemed homologous to the pars triangularis of Broca's area. Other areas indicated include primary somatic cortex (s1), primary auditory cortex (a1), the opercular region of the inferior frontal lobe of the monkey (deemed roughly homologous to the pars orbitalis of Broca's area), the principal sulcus (ps), and the arcuate sulcus (as).

including both afferent and efferent connections with cortical and subcortical areas (Deacon, 1982; 1984). The results demonstrate the complexity of the interconnections among these areas, as well as the complexity of their subcortical connections. Figure 19.3 diagrams some major types of cortical connections of the Broca's area homologue in the monkey brain. It is located along the posterior and anterior banks of the inferior branch of the arcuate sulcus. Its cytoarchitectonic organization has been compared to that of areas 44 and 45 in the human brain and the physiology and connection patterns corroborate this comparison. This area is subdivided into sectors that are interconnected with a variety of auditory areas, perisylvian somatic sensory areas, insular cortex, ventral motor cortex, supplementary motor cortex, prefrontal cortex, and anterior cingulate cortex. This convergence of multimodal afferents is also characteristic of the frontal eye field, the periarcuate area bordering the convexity of the arcuate sulcus just dorsal to this inferior arcuate area. The two periarcuate areas are distinguished by the predominance of visual afferents within the convexity as compared to auditory afferents inferiorly.

There are many parallels between the inferior arcuate areas and the frontal eye field. The frontal eye field receives converging visual and hand-arm-face somatic information (Barbus and Mesulam, 1981), while the inferior areas receive converging auditory and face-tongue-larynx somatic information. These parallels may also apply to function. The frontal eye field has been so named because of its role in controlling consciously directed gaze and eye movements associated with shifts of attention (Ferrier, 1874; Bizzi and Schiller, 1970). It also plays a critical role in conditional learning (Petrides, 1987). A conditional learning task is one in which one of two alternative behavioral responses is signalled by one of two alternative stimuli. More specifically Petrides (1987) has shown that in the vicinity of the frontal eye field the prearcuate area is most critical when the different visual stimuli are used to cue different alternative behaviors, while the adjacent postarcuate area is most critical only when the visual stimuli are used to cue either the immediate or delayed performance of the same behavior (go/no-go paradigm). After postarcuate damage the reduction of somesthetic distinction is most disrupting, while after prearcuate damage the nature of the response is less important while the conditional association of alternative visual stimuli and alternative behaviors is most important. Almost no physiological or neuropsychological experiments with monkeys have been performed in the inferior periarcuate area. But because of the topographic, architectonic and connectional resemblance between these two periarcuate regions it is very likely that the inferior area performs parallel functions, but with respect to auditory stimuli and sequential facial-oral- vocal behaviors instead of visual stimuli and eye-hand behaviors.

The tracer data are inevitably much richer and more complete than predictions for human connectivity. It can immediately be discerned that all of the major pathways presumed to link language areas in humans are predicted by monkey tracer data. This includes especially connections presumed to run through the arcuate fasciculus linking Wernicke's area to Broca's area (some also demonstrated by Galaburda and Pandya, 1982). What is far more interesting, however, is that the monkey data predict that connections involved in language circuits are far more complex than previous theories have suggested. These findings suggest that the language areas are composed of heterogeneous functionally discrete subregions interconnected so as to comprise distinct language subsystems. Areas of the frontal lobes primarily receiving oral-laryngeal information are adjacent to, but distinct from, areas primarily receiving auditory information (see Figure 19.3). Although they are interconnected and probably functionally integrated, focal damage to one and not the other should result in distinctly different deficits.

Also there are a number of parallel efferent circuits likely playing complementary roles in the production of speech (see Figure 19.4a and 19.4b). Besides direct projections from primary motor area (m1) to motor nuclei in the brain stem there are parallel premotor (pm) and postarcuate (poa) projections to pontine nuclei (pont), prearcuate (pra) projections to the dorsal tegmentum and deep superior colliculus areas (dta), supplementary motor (sup) projections to pontine nuclei, and anterior cingulate (ac) projections to the dorsal tegmentum and central gray area (cg) of the midbrain. The pontine nuclei, dorsal tegmentum, deep superior colliculus, and central gray each indirectly project to motor nuclei of the brainstem, so these represent indirect pathways for speech.

Figure 19.4a depicts the efferent circuits shown to be the substrates of primate vocalizations (Jurgens, 1979) while Figure 19.4b depicts the efferent circuits demonstrated to originate from the primate homologues to perisylvian language areas. These two systems of descending connections comprise parallel efferent pathways for oral-vocal activity. The limbic-midbrain circuit is somewhat more indirect and predominantly involved with control of respiratory and laryngeal muscles, while the neocortical system is predominantly involved with control of the muscles of the face, jaw, and tongue. The motor programs for movements controlled by the limbic-midbrain system are represented in deep midbrain areas and remains intact after cortical damage, while the motor programming of the neocortical system depends on the integrity of neocortical areas. The tracer data indicate that at the cortical level these two parallel descending systems are directly interconnected. Different sectors of the Broca's area homologue in the arcuate area are reciprocally connected with the anterior cingulate vocalization areas. These interconnections provide a substrate for the interaction between primitive vocalization systems and the more complex speech control systems. In human speech the anterior cingulate cortex may play a major role in arousal and activation of laryngeal and respiratory function, while more lateral neocortical areas superimpose complex manipulation of oral musculature and modulation of laryngeal movements upon this generalized base. The mutism and aspontaneity of speech that often result from anterior cingulate and adjacent supplementary motor damage in humans (Freemon, 1971; Arnesi and Botez, 1961; Masdeau, Schone and Funklestein, 1978) thus may be reflections of the disturbance of this more ancient portion of the speech system. In this sense the apparent neural dichotomy between primate calls and human speech is more a difference in the number of levels of cortical processing involved and the range of muscle systems

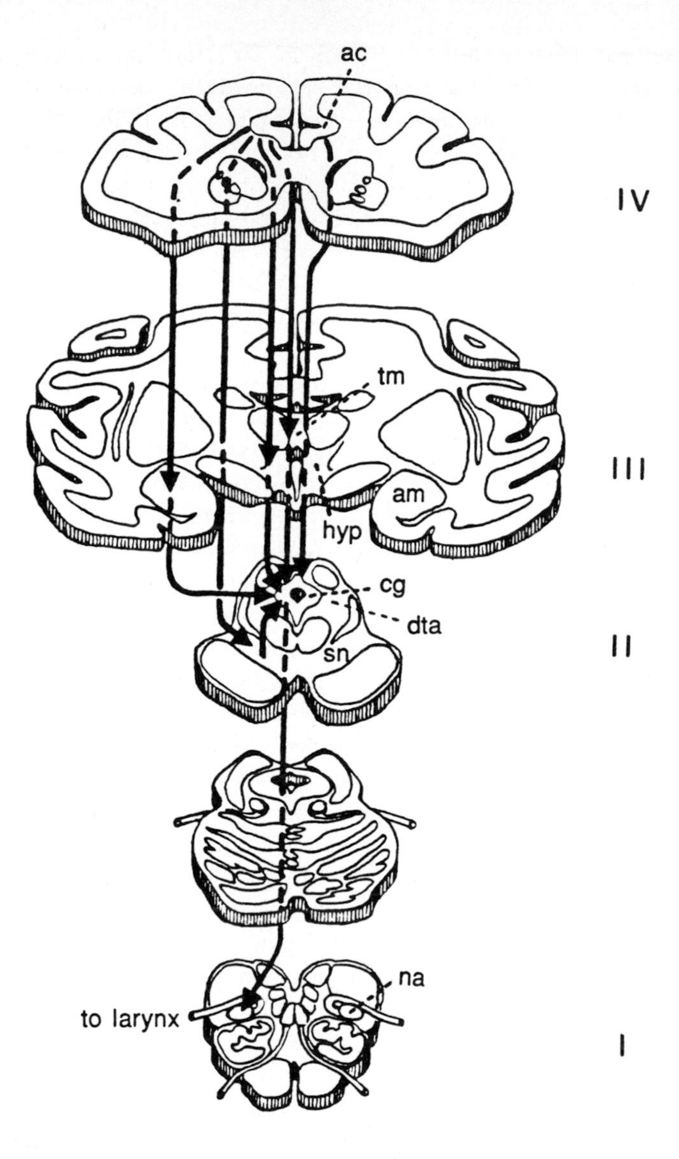

Figure 19.4a Efferent vocalization pathways demonstrated by electrical stimulation and tracer studies in monkeys. Four levels are indicated beginning with (I) laryngeal motoneurons in the nucleus ambiguus (na) and more distributed brainstem systems controlling respiration (not shown), controlled by projections from (II) the midbrain central gray (cg) and adjacent dorsal tegmental area (dta). Structures at this level are likely responsible for the basic motor program of each distinct call type. These midbrain areas receive projections from (III) the substantia nigra (sn), hypothalamus (hyp), medial thalamus (tm), and amygdala (am), and from (IV) limbic cortex, particularly the anterior cingulate cortex (ac) and hippocampus not shown. Anterior cingulate areas also project directly to level II midbrain structures.

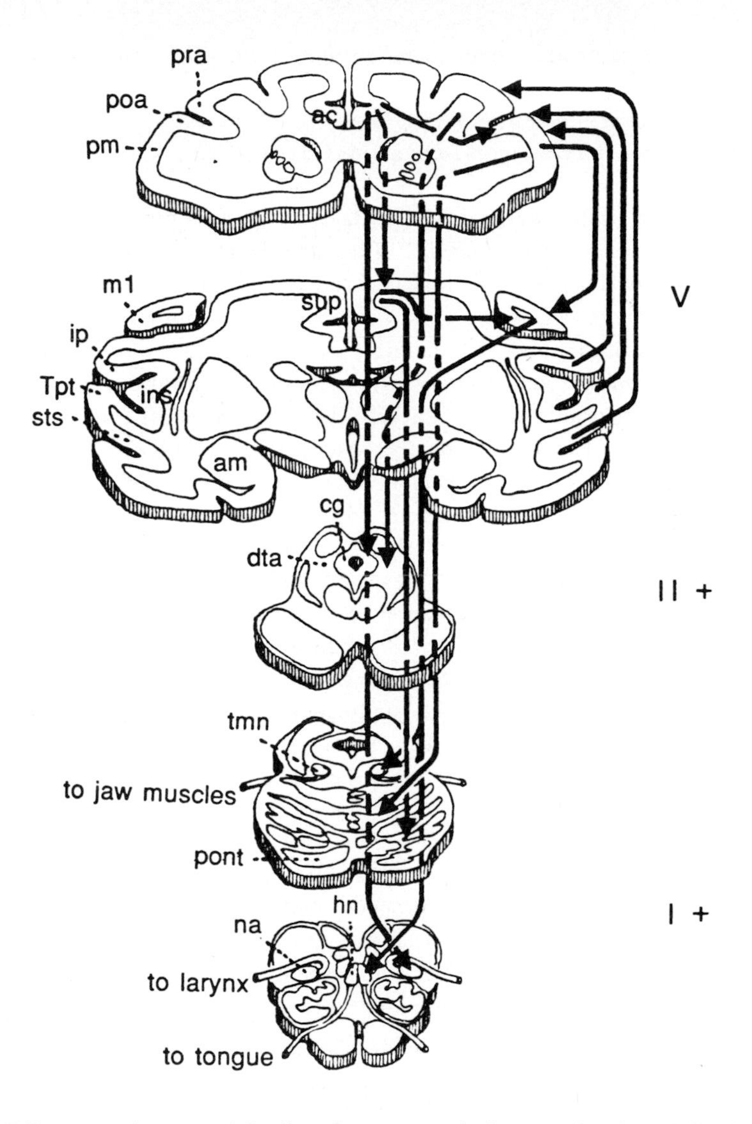

Figure 19.4b Efferent pathways originating from neocortical areas implicated in the control of speech. In addition to the four levels distinguished for the limbic-midbrain vocalization circuits, a fifth neocortical level is distinguished with respect to the control of speech. Neocortical motor (m1), premotor (pm), and supplementary motor (sup) areas probably project directly to most brainstem motor nuclei controlling oral-facial musculature (tmn=trigeminal motor nucleus; hn=hypoglossal nucleus), and possibly to the nucleus ambiguus (na) (in humans) as well as to pontine nuclei (pont). Pre-arcuate and post-arcuate areas project to pm and to pontine (pont), dorsal tegmental (dta), and deep tectal areas. The pre-arcuate area (pra) receives input from temporal auditory areas (e.g. Tpt and sts) and the post-arcuate area (poa) receives input from somatic face-mouth-larynx areas (ins=insular cortex; ip=inferior parietal area. These direct and indirect efferent connections demonstrate considerable parallelism and possible redundancy in the oral-vocal control system. They also demonstrate the parallel organization of limbic-midbrain vocalization circuits (see Figure 19.4a) and oral-facial language circuits.

that must be coordinated, rather than a total shift of vocal functions from limbic and midbrain to neocortical areas.

Comparison with human cortical stimulation data

Are there human data that might substantiate predictions about language circuits derived from these monkey studies? The data from aphasia patients can only hint at possible subdivisions of the language areas. Wernicke's aphasics, for example, can exhibit a range of symptoms variously involving problems analyzing (and spontaneously organizing) phonemic patterning of words and difficulties associating correctly articulated or perceived nouns or verbs with their appropriate interpretations. These two components of the deficit can independently vary, and are typically assumed to involve portions of Wernicke's area either closer to or more distant from the primary auditory area (though not including it) depending on the degree of the phonemic deficit. Similarly, Broca's aphasics exhibit a wide range of deficits. Broca's aphasia is most often defined in terms of an articulatory deficit, presumed to be dissociable from simple motor paralysis (anarthria) of the oral-vocal musculature (Mohr et al., 1978). These patients may, for example, be able to produce a variety of oral-facial gestures and yet not be able to arrange these in a simple sequence (Kolb and Milner, 1981). Recently, considerable data have indicated that Broca's aphasics also have comprehension deficits as well, that are specific to the analysis of grammatical relations and comprehension of small grammatical function words (Goodglass, 1976). These different types of deficit may also vary in severity independently, although in both aphasias absolute size of the lesion is the best predictor of the severity of any of these deficits. In addition to these classic aphasic syndromes anomic disturbances (inability to produce the name of a familiar object, though often with an intact ability to describe and define it) may be associated with aphasias or occur as isolated deficits. Placement of lesions in anomias tend to be in perisylvian areas near or overlapping presumed language areas in temporal, parietal, or frontal sites. These too may represent specialized sectors of large complex language areas.

Since the pioneering work of Wilder Penfield in the 1940s and 50s a new source of evidence has become available that may provide better resolution of the possible functional subdivision of language areas (see Penfield and Roberts, 1959). The use of electrical stimulation to identify critical brain areas in awake locally anesthetized patients during neurosurgery allows analysis of more specific linguistic functions in very precisely localized (though few) cortical sites. In a series of recent studies George Ojemann and his colleagues have also employed this stimulation technique coupled with a battery of language tasks that are more systematic and specific than those tested by Penfield. The result is that they have provided a technique that can be compared from patient to patient, and which is highly specific for dissociable language functions (see Ojemann, 1983 for a review). Some of the results of these electrical stimulation studies are depicted in Figure 19.5.

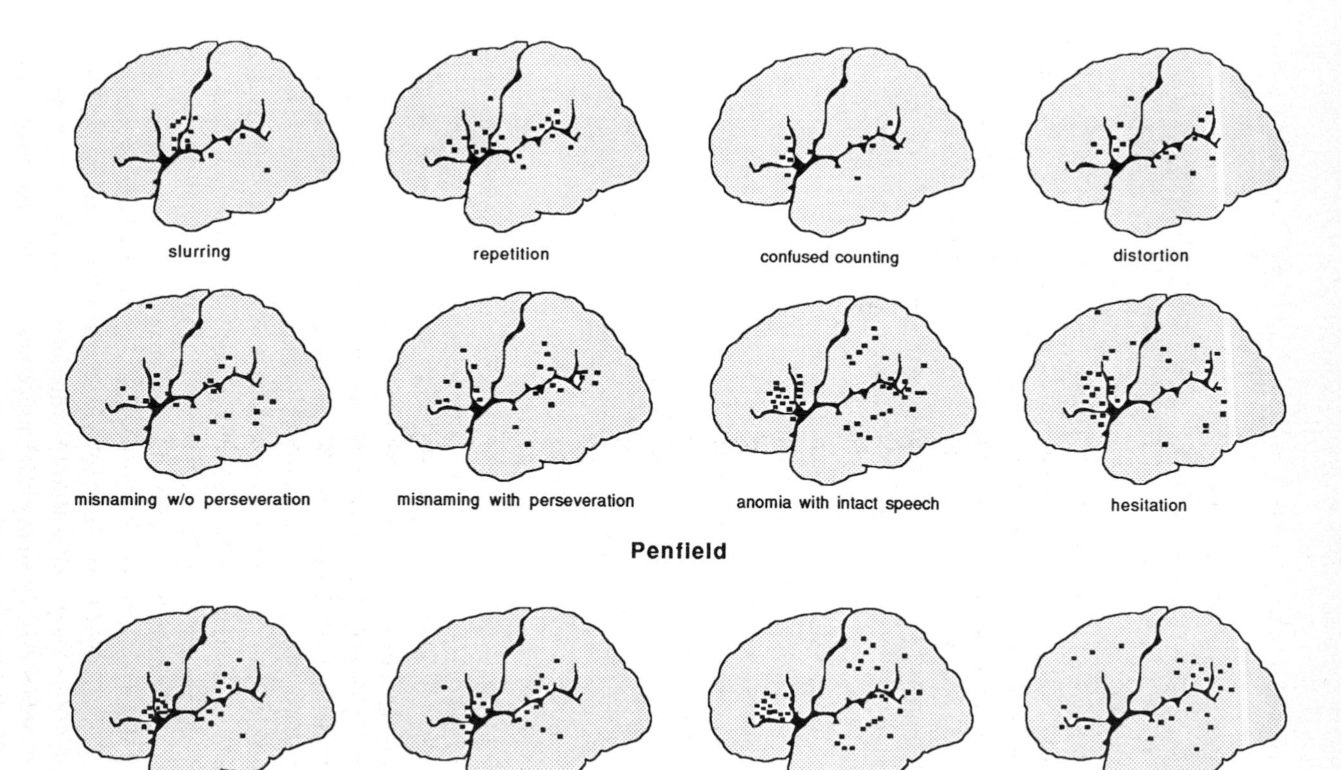

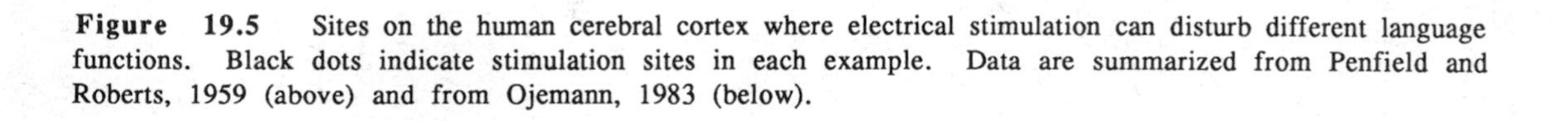

Figure 19.5 Sites on the human cerebral cortex where electrical stimulation can disturb different language functions. Black dots indicate stimulation sites in each example. Data are summarized from Penfield and Roberts, 1959 (above) and from Ojemann, 1983 (below).

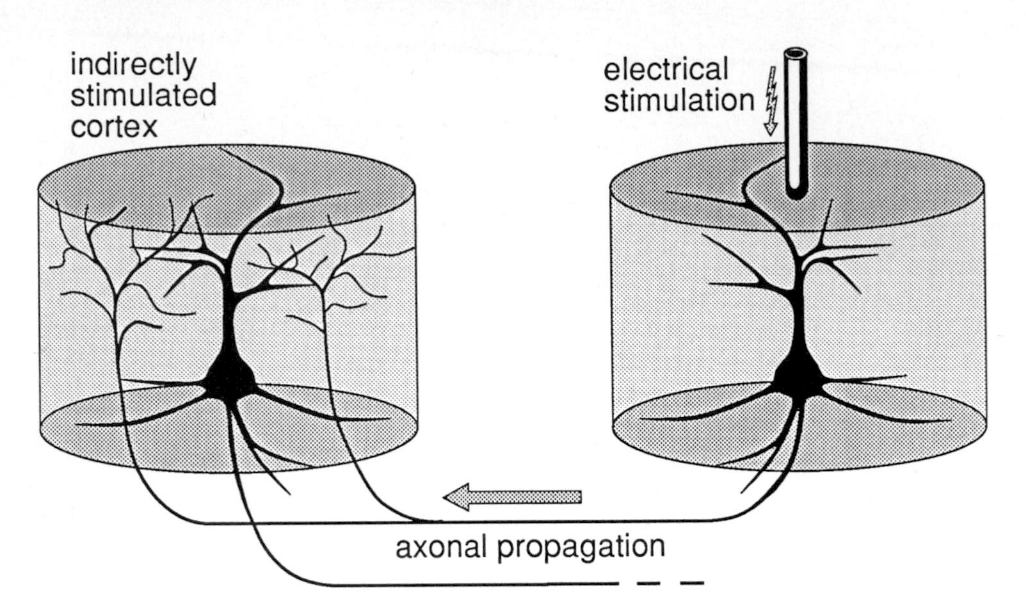

Figure 19.6 Diagram of the means by which localized electrical stimulation of the cerebral cortex might be propagated to areas in direct connection with the stimulated site. Stimulation applied to the region of cortex on the left excites neurons to produce essentially random action potentials that are transmitted to the cortical areas to which its axon projects thereby disrupting activity there.

These findings suggest that each of the classic language areas are divisible into discrete functional areas, and that a wide array of cortical sites are involved to some extent even in the simple language processes tested in these limiting conditions. They also demonstrate a geometric pattern that corresponds to differences in function. Areas subserving common functions in frontal, temporal, and parietal areas are arranged in a roughly concentric pattern extending out from the sylvian fissure. For example, stimulation in all three lobes close to the sylvian fissure and the central sulcus produces direct interference with speech movements (slurring), while more anteriorly in inferior frontal cortex, more posteriorly in parietal cortex, and more laterally in temporal cortex stimulation tends to interrupt both the ability to sequence oro-facial movements and the ability to make phonemic perceptual distinctions. Slightly further out in all three lobes (and not overlapping at any point the sites for sequencing and phoneme identification) electrical stimulation disturbs counting, reading, repetition, naming, and sometimes grammatical processes. Furthest from the sylvian fissure (and again not overlapping other sites) are areas in each lobe where stimulation produces hesitation or loss of spontaneity (Penfield and Roberts, 1959) and disturbs verbal short-term memory (Ojemann, 1983). It is interesting that considerable variability in position and absolute size of language areas is noted from patient to patient, even with respect to fairly predictable morphological landmarks (Ojemann and Whitaker, 1978; Ojemann, 1979).

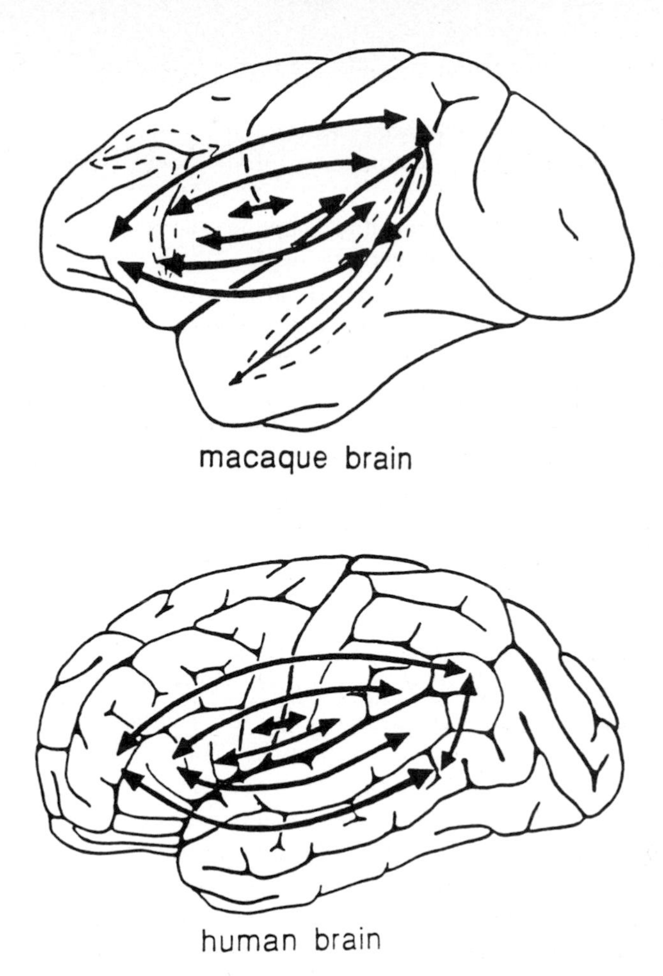

Figure 19.7 Concentric pattern of cortical connections linking monkey perisylvian areas and presumably linking human perisylvian language areas.

These electrical stimulation studies present neuropsychologists with an apparent paradox. There are mirror-image patterns of function identified in each perisylvian area and yet lesions in these different regions produce very different aphasic syndromes. The paradox is resolved by recognizing that electrical stimulation does not just have local effects. Unlike lesions that eliminate portions of the language circuits, electrical stimulation superimposes a blast of "neural noise" at a particular point within a complex circuit. Consequently, it may have disruptive effects throughout that circuit. Lesions, on the other hand, interrupt some part of a complex circuit but likely do not block function of other areas involved in that system. Thus, the fact that stimulation of certain sites in anterior and posterior areas can block the same function suggests that these sites are interconnected within the same close knit functional circuit, i.e. that direct connections link the areas in which these sites are found. Stimulating one site also results in the stimulation of other directly connected sites (diagrammed in Figure 19.7), blocking function of the entire circuit. This connectional stimulation effect has in fact been utilized in conjunction with the metabolic tracer 2-deoxyglucose to demonstrate distributed functional circuits in experimental animals.

In this light, electrical stimulation provides a new and interesting tool for investigating human neural connectivity with higher resolution than previously available, with the added advantage of also providing functional information. With this in mind we can reverse the logic: knowing the connectivity of cortical areas it should be possible to predict which areas of cortex will yield the same results when stimulated. Using the patterns of connections between perisylvian cortical areas that have been determined for the monkey brain we can make predictions concerning patterns of correlated stimulation effects in human brains. To the extent that these are concordant it is reasonable to imply connectional homology, and to the extent that they are not we have evidence for a divergence ofthe perisylvian circuits of human and monkey brains. Figure 19.7 summarizes the pattern of cortico-cortical connections linking perisylvian language areas in humans as compared to monkeys based on anatomical data in monkeys and inferred from stimulation data in humans. It is significant that connections in monkeys predict the spatial pattern of functionally correlated areas in temporal, parietal, and frontal lobes of human brains. Because of the cellular level of resolution monkey connection patterns can also be used predict functional correlations in the human brain not yet dissociated in aphasia cases nor observed with the limited functions that have been tested during electrical stimulation. Such predictions can help focus future research and behavioral testing.

Conclusions

The similarities between neural circuits of monkey brains and indirect data concerning functional connections in human brains suggest that language functions have recruited cortical circuits that evolved for very different purposes in our primate ancestry. The resolving power of techniques for studying the connection patterns of the human brain remains far below that for non-human species. As a result we cannot entirely rule out the possibility of rewiring at some smaller scale of cortical language circuits during human evolution. On the other hand, we can feel fairly confident that the major outlines of these circuits have been retained from their more general anthropoid origins. Homology is not identity, but it is more than just similarity. The common ancestry of these human and monkey circuits suggests that they likely share deep functional similarities as well, despite the fact that only in humans are they the substrates for language. For this reason a fuller understanding of their non-linguistic functions in non-human primate brains may well amplify our understanding of language processes.

The relative conservatism of connectivity patterns among homologous brain regions in species that use these areas for such different functions appears to be characteristic of vertebrate brain evolution in general. Tracer studies being pursued in a wide range of vertebrate species including fish, amphibians, reptiles, birds, and mammals are only now beginning to demonstrate the remarkable extent of connectional conservatism in the evolution of central nervous systems (e.g. Ebbesson, 1980; Deacon, et al., 1983; Reiner, Brauth, and Karten, 1984). Evolutionary changes in the patterns of neural connectivity appear to be highly constrained by some ancient neuro-embryological organizing principles (Ebbesson, 1984). These studies suggest that quantitative changes in particular populations of neurons or their projections may be far more common than the novel rerouting of neural connections or the denovo creation of new brain structures in vertebrate brain evolution.

At this time we can only guess at whether there is some as yet undiscovered, subtle but crucial difference in the connectivity of language circuitry that is responsible for our special language abilities or whether these can be accounted for by other sorts of differences between human and non-human primate brains (specifically quantitative differences; see Deacon, this volume). We can however, rule out some of the more radical hypotheses such as the existence of a species-specific "language organ" (Chomsky, 1972; 1975) or the denovo appearance of speech areas in the brains of early hominids (Geschwind, 1964; Lancaster, 1968; Holloway, 1975; Passingham, 1981; Tobias, 1981; Falk, 1982; 1983). Although there is an immense discontinuity between linguistic and non-linguistic communication, there probably is no comparable dishomology of neural connectional organization that underlies this difference. Identification of an anatomical homology to major language circuits may have the additional benefit of providing an animal model for the investigation of the neural processes that underly language processes. Although these monkey circuits are not adequate in some important ways to subserve language functions, those functions they do subserve are likely also homologous in important ways. Understanding what functional attributes of these circuits were recruited by language adaptations will provide a deeper understanding of both the origins and functional organization of the language process.

Unfortunately for the study of human brain evolution these connectional findings beg more questions than they answer. If monkey brains (and by implication, ape brains as well) have all the basic circuits necessary for language, why are they seemingly unable to acquire vocal skills, and at best capable of only very rudimentary symbol learning? Although comparative connectional analyses are still too crude to entirely rule out species differences in perisylvian (or possibly subcortical) connection patterns distinguishing humans from other primates, the present findings suggest that non-connectional features, such as quantitative factors, may provide the insights necessary to explain how these circuits became capable of supporting a uniquely human communicative faculty.

References

Arnesi C and Botez MI (1961) Speech disturbances caused by tumor of the supplementary motor area Acta Psychiatrica Scandinavica 36:279

Beecher MD, Petersen MR, Zoloth SR, Moody DB, and Stebbins WC (1979) Perception of conspecific vocalizations by Japanese macaques. Evidence for selective attention and neural lateralization Brain Behav Evol 16:443-460

Bizzi E and Schiller PH (1970) Single unit activity in the frontal eye fields of unanesthetized monkeys during eye and head movement. Exp Brain Res 10:151-158

Braak H (1980) Architectonics of the Human Telencephalic Cortex. Berlin-Heidelberg-New York: Springer-Verlag

Chomsky N (1972) Language and Mind. New York: Harcourt Brace Javonovich

Chomsky N (1975) Reflections on Language. New York: Pantheon

Deacon TW (1984) Connections of the inferior periarcuate area in the brain of *Macaca fascicularis*: An experimental and comparative investigation of language circuitry and its evolution. PhD Thesis Harvard University

Deacon TW, Rosenberg P, Eckert MK, and Shank C (1982) Afferent connections of the primate inferior arcuate cortex. Neurosci Abstr 8:2682

Deacon TW, Eichenbaum H, Rosenberg P, and Eckmann K (1983) Afferent connections of the perirhinal cortex in the rat. J Comp Neurol 220:168-190

Dewson JH III (1977) Preliminary evidence of hemispheric asymmetry of auditory function in monkeys. In S Harnard R W Doty L Goldstein I Jaynes and G Krauthamer (eds) Lateralization in the Nervous System New York: Academic Press pp 63-74

Dewson JH and Burlingame AC (1975) Auditory Discrimination and recall in monkeys. Science 187:267-268

Dewson JH III, Pribram KH, and Lynch JC (1969) Effects of ablations of temporal cortex upon speech sound discrimination in the monkey. Exp Neurol 24:579-591

Ebbesson SO (1980) The parcellation theory and its relation to interspecific variability in brain organization evolutionary and ontogenetic development and neuronal plasticity. Cell Tissue Res 213:179-212

Ebbesson SO (1984) Evolution and ontogeny of neural circuits. Behav Brain Sciences 7:321-366

Falk D (1982) Mapping fossil endocasts. In E Armstrong and D Falk (eds), Primate Brain Evolution. Plenum Press: New York pp 217-226

Falk D (1983) Cerebral cortices of East African early hominids. Science 221:1072-1074

Ferrier D (1874) The localization of function in the brain. Proc R Soc London B 22:229-232

Franzen EA and Myers RE (1973) Neural control of social behavior: prefrontal and anterior temporal cortex. Neuropsychologia 11:141-157

Freemon FR (1971) Akinetic mutism and bilateral anterior cerebral artery occlusions. J Neurol Neurosurg Psychiat 34:693-698

Galaburda AM and Pandya DN (1982) Role of architectonics and connections in the study of primate brain evolution. In E Armstrong and D Falk (eds), Primate Brain Evolution. Plenum Press: New York pp 203-217

Galaburda AM and Sanides F (1980) Cytoarchitectonic organization of the human auditory cortex. J Comp Neurol 190:597-610

Geschwind N (1964) Development of brain and evolution of language. Georgetown University Monograph Series on Language and Linguistics 17:155-170

Geschwind N (1965) Disconnexion syndromes in animals and man. Brain 88: 237-294 585-644

Goodglass H (1976) Agrammatism. Studies in Neurolinguistics 1:237-260

Hast MH and Milojevic B (1966) The response of the vocal folds to electrical stimulation of the inferior frontal cortex of the squirrel monkey. Acta Oto-laryng 61:197-204

Hast MH, Fischer JM, Wetzel AB, and Thompson VE (1974) Cortical motor representation of the laryngeal muscles in *Macaca mulatta*. Brain Res 73: 229-240

Holloway RL (1975) 43rd James Arthur Lecture on the evolution of the Human Brain: The role of human social behavior in the evolution of the brain. Am Mus Nat Hist NY

Hupfer K, Jurgens U, and Ploog D (1977) The effects of superior temporal lesions on the recognition of species-specific calls in the squirrel monkey. Exp Brain Res 30:75-87

Jurgens U (1974) Elicitability of vocalization from the cortical larynx area. Brain Res 81:564-566

Jurgens U (1979) Neural control of vocalization in non-human primates. In HD Steklis and MJ Raleigh (eds), Neurobiology of Social Communication in Primates. New York: Academic Press pp 11-44

Jurgens U (1982) Afferents to the cortical larynx area in the monkey. Brain Res 239:377-89

Jurgens U (1983) Afferent fibers to the cingular vocalization region in the squirrel monkey. Exp Neurol 80:395-409

Jurgens U and Cramon D von (1982) On the role of the anterior cingulate cortex in phonation. A case report. Brain Lang 15:234-248

Jurgens U, Kirzinger A, and Cramon D von (1982) The effects of deep- reaching lesions in the cortical face area on phonation. A combined case report and experimental monkey study. Cortex 18:125-139

Jurgens U and Ploog D (1976) Zur Evolution der Stimme. Arch Psychiat Nervenkr 222:117-237

Kolb B and Milner B (1981) Performance of complex arm and facial movements after focal brain lesions. Neuropsychologia 19:505-514

Kolb B and Whishaw IQ (1985) Fundamentals of Human Neuropsychology, Second Edition New York: W H Freeman

Lancaster JP (1968) Primate communication systems and the emergence of human language. In PC Jay (ed), Primates: Studies in Adaptation and Variability. New York: Holt Rhinehart and Winston; pp 439-457

Manley JA and Muller-Preuss P (1978) Response variability of auditory cortex cells in the squirrel monkey to constant acoustic stimuli. Exp Brain Res 32:171-180

Mohr JP, Pessin MS, Finkelstein HH, Duncan GW, and Davis KR (1978) Broca aphasia: Pathologic and clinical aspects. Neurology 28:311-324

Myers RE (1976) Comparative neurology of vocalization and speech: Proof of a dichotomy. In SR Harnad, HD Steklis, and J Lancaster (eds), Origins and Evolution of Language and Speech. Annals N Y Acad Sci 280:745-757

Ojemann GA (1979) Individual variability in cortical localization of language. J Neurosurg 50:164-9

Ojemann GA (1983) Brain organization for language from the perspective of electrical stimulation mapping. Behav Brain Sciences 2:189-230

Ojemann GA and Whitaker HA (1978) Language localization and variability. Brain Lang 6:239-260

Passingham RE (1981) Broca's area and the origins of human vocal skill. Phil Trans R Soc Lond B 292:167-175

Penfield W and Roberts L (1959) Speech and Brain Mechanisms. Princeton: Princeton University Press

Petersen MR, Beecher MD, Zoloth SR, Moody DB, and Stebbins WC (1978) Neural lateralization of species-specific vocalizations by Japanese macaques (*Macaca fuscata*). Science 202:324-327

Petrides M (1987) Conditional learning and the primate frontal cortex. In E Perecman (ed), The Frontal Lobes Revisited. New York: The IRBN Press pp 91-108

Reiner A, Brauth SE, and Karten H (1984) Evolution of the amniote basal ganglia. Trends in Neurosci, Sept pp 320-325

Robinson BW (1967) Vocalization evoked from the forebrain in *Macaca mulatta*.. Physiol Behav 2:345-354

Vogt C and Vogt O (1919) Allgemeinere Ergebnisse unserer Hirnforschung. J Psychol Neurol 25:277-462

HUMAN BRAIN EVOLUTION:
II. EMBRYOLOGY AND BRAIN ALLOMETRY.

Terrence W. Deacon
Biological Anthropology
Harvard University,
Cambridge, MA 02138

The unusually large size of the human brain with respect to the human body as compared to other mammals is a dominating fact in the study of human evolution. No account of the similarities and differences between human and other primate brains is complete without accounting for this quantitative fact. The intuitive correlation between this statistic, *encephalization*, and the seemingly astronomical gulf between human and non-human intellectual capabilities has led many to argue that this disproportionate size in itself reflects the fundamental adaptation in human brain evolution (Dubois, 1898; 1924; von Bonin, 1952; Jerison, 1973; Van Valen, 1974; Gould, 1975, Passingham, 1975a).

With increased size comes increased neuron number, increased number of neuronal connections, and presumably, increased processing capacity (Jerison, 1973; 1977; 1979). With increasing size of the brain and body there is also likely an increase in the information processing demands that the body places upon the brain. The question with respect to encephalization is how to calculate the number of "extra neurons" in the brain, i.e. those that can be said to be available in addition to those minimally necessary for purely somatic information processing, and thus available for "higher order" mental processes. Alternatively, one might phrase this in terms of neurons available in addition to the mean number utilized for purely somatic processing in a representative sample of species. Lacking some independent means for assessing the minimally necessary number of neurons for basic somatic functions, we must settle for an the operationally defined calculation with respect to the mean trends observed among an appropriate reference group of species (i.e. all mammals, or just primates with respect to humans). But even this simple statistic hides a number of exceedingly complex questions concerning (1) the nature of the mean trend with respect to which this statistic is determined; (2) the actual distribution of the "extra neurons" and how their number and distribution is determined; and most critically, (3) the functional significance of there being more neurons than expected in one part of the brain as opposed to another, or of extra neurons being somehow evenly distributed throughout brain structures. The following discussion will focus primarily on the second of these questions and how this

NATO ASI Series, Vol. G17
Intelligence and Evolutionary Biology
Edited by H. J. Jerison and I. Jerison

answer may inform investigations of functional significance.

The widespread acceptance of the centrality of relative brain size in human evolution has meant that attention has largely focused on encephalization itself as a fundamental unit of analysis. This underlying assumption has given rise to studies endeavoring to determine its accurate calculation (Jerison, 1973; Hofman, 1982a; 1982b), the evolutionary time course of its disproportional development in the hominid lineage (Tobias, 1975; McHenry, 1976; Hofman, 1983), ontogenetic factors that may determine it (Martin, 1983; Hofman, 1983; 1984), energetic factors that may constrain it (Martin, 1982; Armstrong, 1982; Hofman, 1983), feeding and social adaptations that may be associated with it (Clutton-Brock and Harvey, 1980; Harvey, Clutton-Brock, and Mace, 1980; Mace, Harvey, and Clutton-Brock, 1981), and the assessment of the information processing capacity provided by it (Jerison, 1979; Hofman, 1985). If human encephalization should turn out not to be a single simple additive phenomenon, then all theories that assume that it is would require critical re-evaluation.

Critics of the encephalization hypothesis for human brain evolution have argued that the correlation between encephalization and intelligence is not corroborated by actual comparative intelligence tests (MacPhail, 1982), and that internal reorganization of the brain resulting from selection for specialized "modular" functions may be more fundamental, producing an overall enlargement of the brain as a secondary effect of the enlargement of a few selected brain structures (Holloway, 1976; 1982). But beyond suggesting general reorganization (Holloway, 1966; 1983), addition of specific new structures (e.g. Passingham, 1981; Tobias, 1979; Falk, 1983), or selective enlargement of individual structures (e.g. Armstrong, 1982; 1986) no comprehensive alternative model has emerged.

There are purely anatomical reasons for thinking that the encephalization argument is at the very least an over-simplification. Because of the precise connectional relationships that exist between peripheral and central structures it is unlikely that the human brain could have disproportionately enlarged with respect to the body without internal reorganization as well. There are probably few neurons within the brain that are more than 4 or 5 synapses separated from peripheral receptor or effector cells. Association cortex in the classic sense (i.e. lacking subcortical afferents or efferents) is a misnomer since all neocortical areas have direct thalamic input and map sensory or motor modalities (Diamond, 1979; 1982). Finally, considerable experimental evidence has accumulated to suggest that the sizes of central neural populations and sensory-topic maps are influenced by the sizes of primary peripheral afferent sources or efferent targets (Hollyday and Hamburger, 1976; Rubel, Smith, and Miller, 1976; Sohal, 1976; D'Amato and Hicks, 1978; Goldman, 1978; Goldman and Galkin, 1978; Katz and Lasek, 1978; Merzenich, Kaas, and Nelson, 1980; Goldman-Rakic, 1981; Finlay and Slattery, 1983). Absolute brain and body size in mammals is the fundamental determinant of differences in quantitative structural organization within the brain (Sacher, 1970). It should also follow that a substantial deviation from the trend of relative brain and body proportions is correlated with a deviation from trends in the

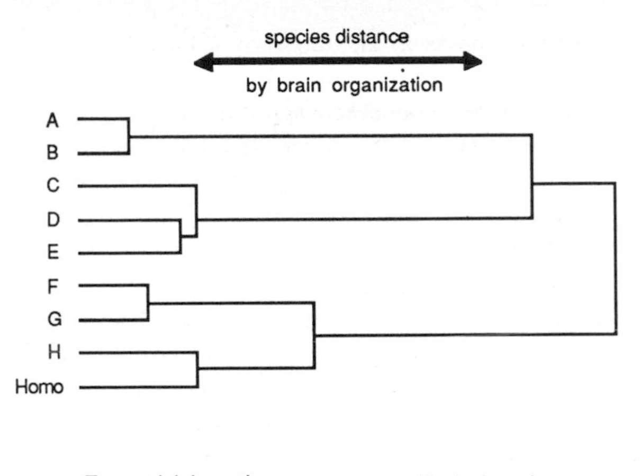

Figure 20.1 Figure from Bauchot (1982) showing the dendrogram derived from his analysis of structural similarities of the brains of primates and insectivores. Although successfully reflecting phylogenetic relationships between most major genera it incorrectly associates humans with talapoin and cebus monkeys (humans are not closely clustered with these except with respect to all other major groups). Chimpanzees and gorillas are closely clustered with other old world monkeys and are not particularly closely clustered with each other within this group.

proportions of brain structures with respect to one another. The possibility of such a correlation between internal organization and encephalization has not been carefully investigated.

Support for this hypothesis comes from studies of the patterns of quantitative organization in different species. Two independent studies have investigated the possibility of using correlations among the volumes of different brain structures to predict phylogenetic relationships between species. Though demonstrating that quantitative brain organization has some correlation with phylogenetic trends, their findings also point out the atypicality of the quantitative organization of the human brain. The significance of the anomalous position of the human brain with respect to the brains of related species in these analyses is not carefully considered in either study, and it appears to have gone unnoticed by subsequent investigators. Douglas and Marcellus (1975) compared eight major brain regions to the medulla (to correct for overall size). Their analysis resulted in a dendrogram in which the human brain is clustered closest to the brains of the spider monkey (*Ateles*) the wooly monkey (*Lagothrix*). In a similar study Bauchot (1982) analyzed multiple correlations between encephalization factors for 22 major divisions of the brain (corrected for brain/body size allometry assuming an exponent of 2/3). This analysis places the human brain in a clade closest to the brains of cebus and talapoin monkeys, all three of which are only distantly linked with other new and old world monkeys (see Figure 20.1). The majority of other species in these analyses were clustered appropriately. Although each study attempted to correct for differences in encephalization, which should have made the human brain more comparable to other species, neither classed humans with old world monkeys and apes. Bauchot's results are particularly interesting because *Homo* is linked most closely with the two species of anthropoid primates who exhibit the next highest encephalization values.

The fact that deviation from the allometric trend of brain and body size proportions has some predictive value for the clustering of species with respect to internal brain structure, even after the effect of encephalization has been taken into account, suggests that brain/body disproportion may have some direct organizational effect on the internal structure of the brain. This extrinsic allometric influence on internal brain organization is particularly important for the analysis of human brain evolution where such a brain/body disproportion is extreme.

There have been numerous analyses of interspecific trends of quantitative brain organization in primates (e.g. Stephan and Andy, 1969; Sacher, 1970; Gould, 1975) and in human-primate comparisons (Passingham, 1973; 1975b; 1979b; Passingham and Ettlinger, 1973). However, there are a number of embryological considerations that have not previously been appreciated that have a major bearing on the interpretation of these data. Statistical problems with some of these studies have also escaped attention, and have led to a number of confusions concerning human brain evolution. Both will be investigated here. Two claims deriving from such confusions are (1) that the human brain has enlarged in such a way that its internal structural proportions have changed along typical primate trends (e.g. Jerison, 1973; Passingham, 1975a and b; 1979a); and (2) that there are some structures within the brain that have become enlarged or reduced in isolation from the general systemic trend exhibited by the rest of the brain (e.g. Armstrong, 1982). The first interpretation is often cited to support arguments for an increase in general intelligence, while the second is often cited to support theories of specialized, modular adaptations of specific brain areas in hominid evolution.

The part/whole problem

The major statistical problem I wish to address can be characterized as the "part/whole fallacy" in allometric analysis. This fallacy results from comparing some statistic of a body part with some statistic of the whole body (or some other larger unit) and not correcting for the fact that the part is contained in the whole. Examples of this fallacy appear without notice (and often without significant consequence) in nearly every study of comparative brain size and its correlates, as well as in numerous other allometric studies. However, it has been particularly troublesome for the study of human brain organization. For example, in his statistical analyses of brain structure volumes with respect to total brain size

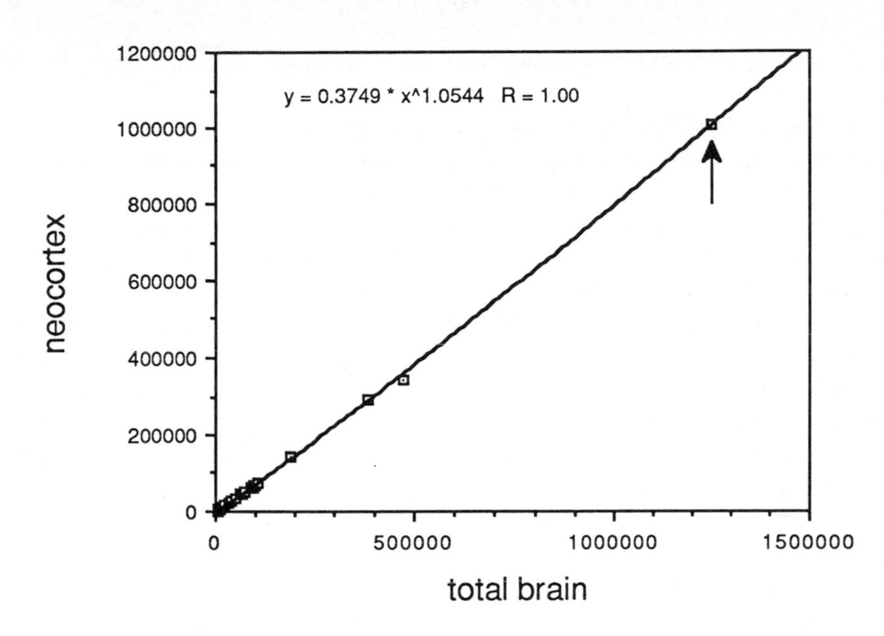

Figure 20.2 Plot of the relationship of neocortical volume to total brain size among 26 anthropoid species including humans (see arrow). Data from Stephan, Frahm, and Baron, 1981. The line represents a least squares regression best fit curve based on logarithmically transformed data. The equation for this line and correlation are listed at the top.

Passingham (1975b) provides a demonstration that total neocortical volume (including underlying white matter) in the human brain is not significantly greater than predicted from primate data for a brain of human proportions (Figure 20.2). This presumably would suggest that the enlargement of the human brain cannot be attributed to disproportionate cortical expansion in excess of normal primate trends, and more generally that human brain expansion followed patterns predictable from other primate brains. This finding (also implicit in other analyses, e.g. Sacher, 1970; Gould, 1975; Jerison, 1979; Hofman, 1985) has been uncritically accepted and incorporated into a number of evolutionary theories, despite a central methodological error.

The problem is that the cerebral cortex and underlying white matter comprise most of the total brain size, with the relative percentage increasing with brain size to nearly 80% in the human brain. Comparing neocortical size to total brain size thus results in a part/whole analysis in which one variable in a bivariate relationship is largely comprised of the the other! Although part/whole bias inevitably contaminates all brain/body and brain structure/total brain analyses it can be ignored (with caution depending on the purpose of the analysis) in most cases where the part is a very small fraction of the whole. Nonetheless, the effect of such bias, even if subtle, is an important bio-statistical problem that warrants careful consideration in any allometric analysis.

As the part becomes an increasingly larger portion of the whole the best fit line is correspondingly biased toward isometry and measures of individual deviations become increasingly insensitive. An example of a fairly trivial effect can be found in brain/body allometry itself. The relative percentage of brain to body ranges from high values approaching 4% in small mammals to values far below 1% in large mammals (in humans the brain is 2% of body size, a proportion more typical of small mammals). Consequently, the low end of the brain/body allometric curve is subtly biased toward isometry. In the case of the cerebral cortex or even just the neocortex, the part in question is not only a major fraction of the whole brain, but is a relatively larger fraction in large brains as opposed to small brains. So the bias is amplified in the case of humans whose brains are 3 times larger than the next closest primates.

This problem is evident in contradictory statistics of forebrain structural deviation in the human brain when analyzed this way (Deacon, 1984). Although the total volume of the human telencephalon, cerebral cortex, or neocortex is well predicted by primate scaling trends with respect to brain size, the diencephalon and basal ganglia are considerably smaller than predicted (see Figure 20.8). One component of the telencephalon (cerebral cortex) cannot be appropriate in size with respect to the whole telencephalon while the remainder (diencephalon and corpus striatum) is too small with respect to the whole. The bias hidden in this calculation has in fact underestimated the positive deviation of the volume of cerebral cortex from the expected trend while only minimally distorted values for the smaller brain structures.

The statistically appropriate approach to this problem is to compare measurements from non-overlapping structures in the body, or from a structure and the whole body less the value from that structure (Huxley, 1932). The uncorrected use of part/whole statistics should be considered only as a useful shortcut to comparing all parts with each other and for emphasizing size-linked trends. In cases where very small fractions of the whole are involved the bias is effectively negligible and so the subtraction of the part from the whole becomes a trivial exercise (Figures 20.8-20.10 take advantage of this in order to graphically present statistics for a large number of brain structures with respect to the whole). Accurate estimation of the deviations of brain structure volumes in the human brain from the general trends within primates requires treating each structure as an independent variable with respect to each other non-overlapping structure. This does not necessarily throw us back to piecemeal analysis of every relationship between every possible structural subdivision of the brain. Fortunately, there is extensive intercorrelation of the allometric trends of many structures (Sacher, 1970; Gould, 1975; Bauchot, 1982). In addition it will turn out that human deviations from these trends also exhibit systematic patterns.

Turning to an analysis of systematic size trends in the relationships between brain structures, the question with respect to the human brain becomes whether or not the entire brain is enlarged in an undifferentiated or systemic fashion or if the extra volume of the human brain is unevenly distributed, with a few major regions accounting for the total brain enlargement. If the enlargement is undifferentiated there are two plausible null hypotheses: (1) the whole brain is

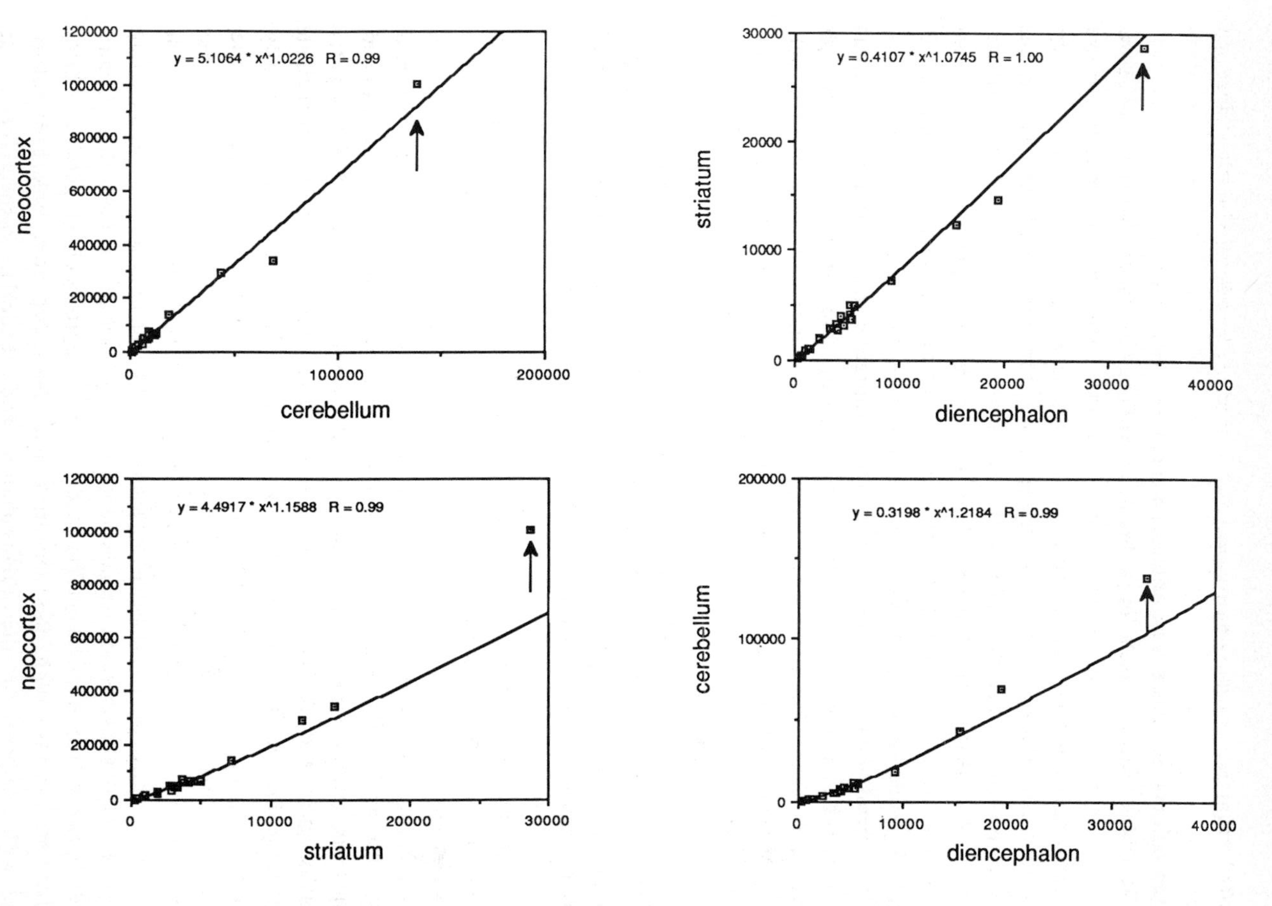

Figure 20.3 Plots of the allometric relationships between (a) total neocortex and cerebellum, (b) total corpus striatum and total diencephalon, (c) total neocortex and striatum, and (d) total cerebellum and diencephalon showing a dichotomous scaling relationship between the cortical and nuclear structures (data and curve fitting as in Figure 20.2).

isometrically enlarged from some ancestral value (e.g. approximating that of modern chimpanzees) (see Gould, 1975); or (2) the whole brain has enlarged along the expected allometric trend for primate brains (but not bodies) (see Passingham, 1975b). The first would result in a brain in which the ratios between the parts would be the same as for the chimpanzee brain (the operationally defined ancestral condition), although all values would be enlarged by the same scalar value. In this case human and chimpanzee values for any structural comparisons should be connected by a line with isometric slope, irrespective of the general primate trend for that comparison. The second should produce no apparent deviations from the primate trend for internal brain organization with respect to brain size. In either case there would be consistency among all comparisons. If on the other hand there is internal allometric reorganization within the human brain the separate comparisons should not all be consistent. The deviant structures should stand out from the primate trend in pair-wise comparisons with non-deviant structures (though not with other co-varying structures) while other comparisons should not exhibit deviations.

When part/whole bias is eliminated in this way, the analysis of human deviations produces a more complicated picture. Initial analysis argues against either null hypothesis for undifferentiated enlargement. In the following analyses data are taken from the extensive work of Stephan, Frahm, and Baron (1981), the compiled lists of Blinkov and Glezer (1968), and from a number of other sources for particular structures. Best fit curves are computed using log-transformed data by the least squares regression method, then presented on linear scales to depict absolute deviation from primate trends. In each graph the human value is indicated by an arrow.

The comparison of major structures such as the total neocortex and cerebellar cortex (including underlying white matter) indicates that the human relationships fall closely along the primate trends (Figure 20.3a). Similarly, the comparison of the total diencephalon and corpus striatum (Figure 20.3b) indicate that these structures also have enlarged with respect to each other as primate trends would have predicted. But comparison of either cortical structure to either nuclear structure (Figures 20.3c and 20.3d) indicates that the relationship between these pairs of brain structures in humans markedly deviates from the primate trend. A similar dichotomous pattern is seen in other comparisons as well. The scaling relationship between the total neocortex and total diencephalon (Figure 20.4a) and between the mesencephalon and medulla oblongata (Figure 20.4b) is also deviant in humans, with the neocortex and mesencephalon enlarged out of proportion with respect to the diencephalon and medulla, respectively. In contrast the scaling relationship between the cerebellum and mesencephalon follows the primate trend for those structures (Figure 20.4c).

We must conclude from these initial results that the volumetric scaling relationships between brain structures in the human brain reflect a departure from some otherwise highly correlated primate size trends. It appears that structures within the human brain can be grouped according to correlated enlargement patterns. Comparison of structures within such a correlated group

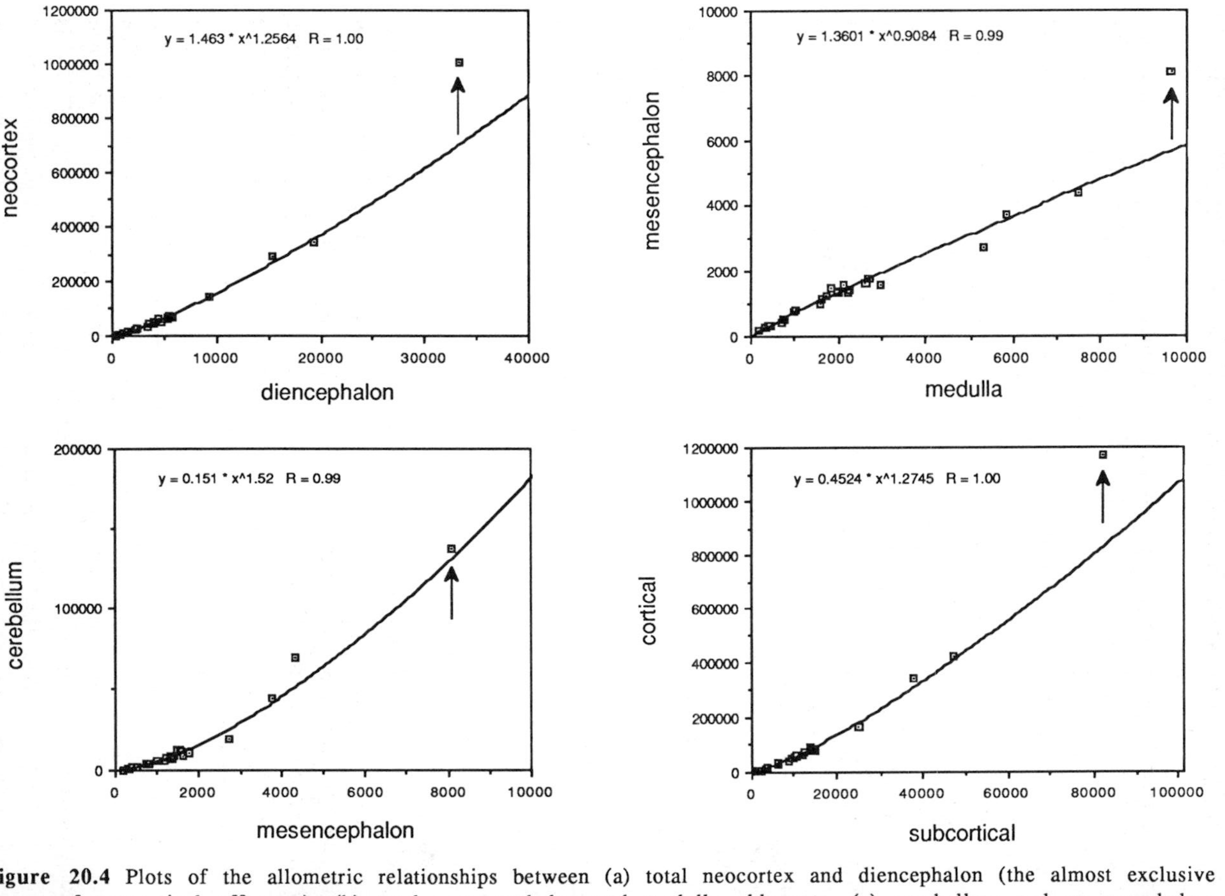

Figure 20.4 Plots of the allometric relationships between (a) total neocortex and diencephalon (the almost exclusive source of neocortical afferents), (b) total mesencephalon and medulla oblongata, (c) cerebellum and mesencephalon, and (d) the total of all cortical and all subcortical brain structures (data and curve fitting as in figure 2). Note the close scaling of the cerebellum to the mesencephalon. The latter structure is the route by which the cerebral cortex and cerebellum are interconnected. However, even if this white matter is ignored it appears that the tectum and tegmentum of the mesencephalon are themselves enlarged structures (see Figure 20.9).

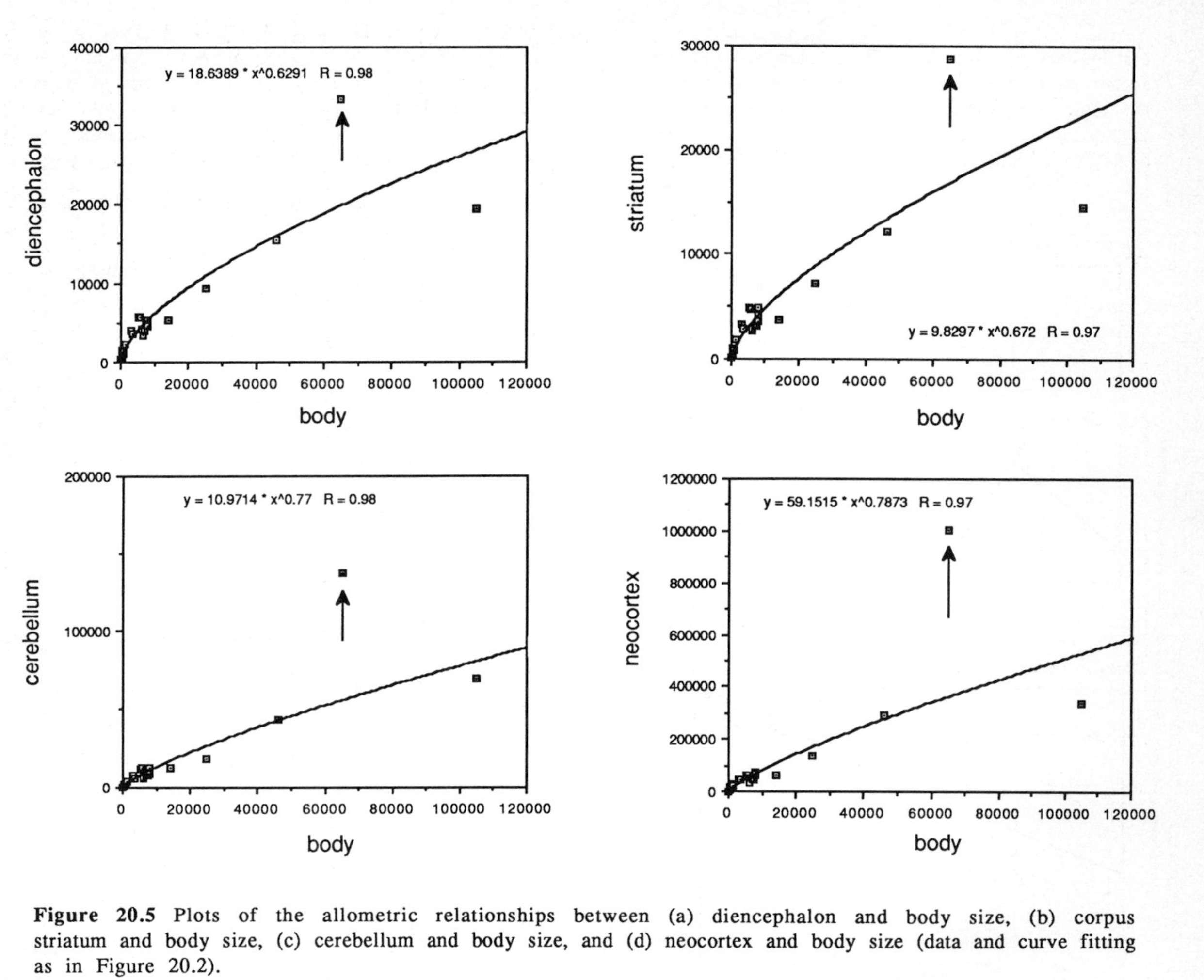

Figure 20.5 Plots of the allometric relationships between (a) diencephalon and body size, (b) corpus striatum and body size, (c) cerebellum and body size, and (d) neocortex and body size (data and curve fitting as in Figure 20.2).

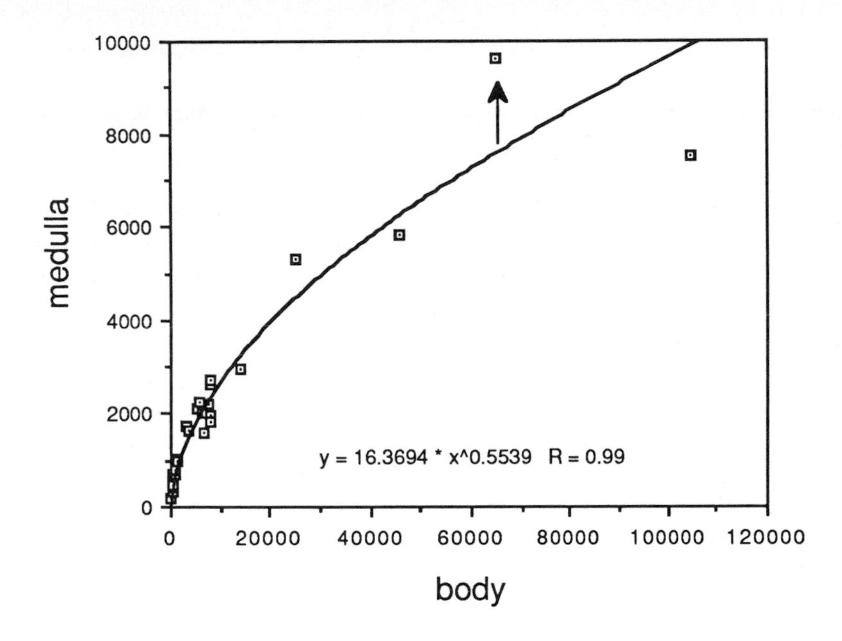

Figure 20.6 Plot of the allometric relationship between medulla oblongata (brain stem) and body size (data and curve fitting as in Figure 20.2).

exhibits a typical primate pattern while comparison of structures from different groups exhibits a divergence from primate trends. It appears that these groups of structures comprise a series of three or four grades of relative enlargement. In general (in descending order), the cerebral cortex, hippocampus, cerebellum and mesencephalon exhibit higher values and the striatum, diencephalon and medulla oblongata exhibit lower values with respect to one another. This roughly follows a simple structural dichotomy. Two broad classes of brain structures, roughly identified with cortical as opposed to nuclear structures, correspond to the higher as opposed to lower scaling structures. In Figure 20.4d all cortical structures are summed together and all subcortical structures are summed together. The human deviation from the primate trend demonstrates the generality of this dichotomy.

To discern which among these sets of structures are responsible for the departure from the primate trend (i.e. reduction of nuclear structures versus enlargment of cortical structures) we need to appeal to an appropriate external reference value. With respect to total body size (where brain structures are insignificant fractions of the whole) nearly all structures deviate in the positive direction from predictions from primate trends (see Figure 20.5; note the position of *Gorilla* below and to the right of the trend in all cases, indicating its deviantly large body size). The deviations of the cortical structures are far more extreme than those of the nuclear structures (which, although enlarged, do not reach statistical significance in this analysis). The one major human brain structure that is best predicted by body size: the medulla oblongata, or brain stem (Figure

20.6). This is an important clue to human brain reorganization since it reflects the fact that this major somatic input-output structure is more closely constrained by body proportions than other major brain structures (the implications will be examined in the next section).

Contrary to the conclusions of Passingham (1975b) and others these findings indicate that the human neocortex and the cerebellum do not scale in size along typical primate trends with respect to other structures as overall brain size has increased. Nor are these allometrically deviant human brain structure volumes explainable as isometric enlargements from chimpanzee proportions, since cortical to nuclear comparisons strongly deviate from isometric predictions from chimpanzee values (for example see Figures 20.3c and 20.3d; note that the x and y axes are not to the same scales and that both regression curves have exponents [=log/log slope] that are greater than 1.00). More generally then, human brain expansion as a whole cannot be accurately be explained by any undifferentiated enlargement model. Nor can it be explained with respect to phylogenetic trends linking it with our closest relatives the great apes.

To appreciate the nature of the influence of both encephalization (i.e. brain/body disproportion) and disproportional neural populations in different brain structures on human brain organization it is first necessary to take into consideration the ontogenetic processes that partition the developing brain into distinct growth fields. Following this a more extensive body of quantitative brain data can be integrated into a systematic model of human brain reorganization with respect to other primates. Quantitative brain data are still in many ways too incomplete and inadequate to reach any definitive statistical conclusions. But they are sufficient to help sketch the outlines of a plausible pattern of ontogenetic events which can guide both further quantitative and developmental neurobiological approaches to this question.

Embryological considerations

Not taking the phasic nature of embryological growth processes into account can lead to a number of problems when interpreting interspecific trends (Huxley, 1932). One major source of difficulty stems from a tendency to treat comparative adult data, such as brain-to-body relationships or brain-structure-to-whole-brain relationships, in a hypostatic fashion rather than as the results of independent dynamic growth processes (Gould, 1975). During ontogenesis the sizes of different body structures covary in systematic ways with different trajectories at different growth phases. Many structures are only distinguishable in later stages of development and derive from common origins at some earlier stage. Processes which determine the adult size of some structure may include cell proliferation, cell growth, cell death and atrophy, and sub-differentiation of structures which have previously attained final size. Knowing which of these processes are involved and in which order they progress in ontogenesis can direct statistical approaches to questions that are functionally meaningful, and can provide important clues for interpreting the results. Ignoring these effects and treating all measurements and statistics as equivalent can lead to superfluous associations and spurious correlations that are biologically meaningless.

Quantitative studies of neuron number and density have demonstrated that brain size and total neuron population size are strongly correlated (e.g. Shariff, 1953; Tower, 1954; Rockel, Hiorns, and Powell, 1980). The relationship is not linear since neuron density decreases with increasing size, and mean neuron volume and size diversity also increase (Hofman, 1985). Since nearly all neurons in mammals are non-mitotic from an early point in gestation, ultimate adult brain size is largely determined early in embryological development (neglecting the effects of cell death--considered more carefully below) with remaining growth primarily determined by glial mitosis and cell growth. This does not however mean that the ultimate adult size of any particular brain structure is also determined by the end of the mitotic phase of embryological development. It is significant that the size of a particular structure or functional subdivision is not determined by factors intrinsic to any particular embryonic population of cells thereby destined to constitute it.

Both intrinsic and extrinsic factors must be considered. In general, the size of any particular brain structure is determined by a three stage process involving independent mechanisms. The first phase is a *mitotic neurogenesis phase* during which the total potential population of neurons is determined within major brain regions by a limited period of mitotic division and migration of protoneuronal cells. Once its last mitotic division is completed a neuron will migrate (in some regions) to its final position, and enter a second phase where it begins to send out neuronal processes, including a growing axon, and where it interacts with invading axons from other areas seeking to make synaptic contacts. Functional circuits are not established until after neurogenesis is largely completed in a region. A glial mitosis phase follows neuron mitosis and also has a significant effect on overall adult brain size. However, the very precise predictability of neuron density within a brain region with respect to overall brain size (Rockel, Hiorns and Powell, 1980) suggests that glial populations are matched to neuronal populations within a region and so do not play an independent role in size determination (except in pathological cases).

The second phase can be called the *connection phase* during which major axonal connections are first established. Once cells have finished mitosis and migration afferents from other areas begin to establish synaptic contact with them. It appears that, although there is genetically determined selectivity of target sites for axons within the developing nervous system, it is of a fairly low level of precision compared to the specificity of connections in the adult. During this phase axons establish many more synaptic contacts in far more areas than will survive into adulthood (Koppel and Innocenti, 1983). This over-exuberant and somewhat non-specific production of axonal connections sets up the third and final phase.

The third phase can be called the *parcellation phase* during which interaxonal competition plays a prominent role in subdividing major structures into distinct subregions. This phase may include considerable programmed cell death in some areas although generally most of the competition results in the trimming back of over-exuberant axon collaterals while their cells of origin are not eliminated (Goldman-Rakic, 1981; O'Leary et al., 1981; Finlay and Slattery, 1983; Koppel and Innocenti, 1983; Rakic et al., 1986). Local competition for synaptic space appears largely determined by two factors: the size of the source pool of these afferents and their level of neural activity in correlation with other afferents. This process also determines efferent cell populations to the extent that axonal elimination determines which cells will and will not have an efferent output to a particular target area (thus affecting the morphologies of these cells). The functional specificity and architectonic organization of the various subdivisions of the thalamus and cerebral cortex, for example, are probably in large part determined by this parcellation process.

The implications of this three stage process are important when considering quantitative brain data. The ultimate size of a particular structure is determined, in part, systemically. In other words, the size of the major afferent source structures and efferent target structures plays a major role in determining the size of a functional area with respect to neighboring areas. Because parcellation can only subdivide existing cell populations, but cannot influence their total size (except where cell death plays a major role, and the evidence is against this within the mammalian forebrain; see O'Leary, et al., 1981; Koppel and Innocenti, 1983; Rakic, et al., 1986), these factors are probably of little importance in determining the size of the entire cerebral cortex, cerebellum, or diencephalon, but are likely of major importance in the determination of the size of the many subdivisions of each. The major neurogenetic fields, such as total neocortical cell population, define the absolute size of an arena of competition between afferent and efferent axons. Within this constraint the axonal competition determines the relative sizes of their subdivisions (e.g. cortical fields or nuclei).

These considerations suggest that we approach data about brain structure volumes with a multi-leveled analysis: first determining the allometry of major neurogenetic fields within the brain and then, in the context of these constraints, analyzing the partitioning of these structures with regard to their afferent and efferent associations. Consequently, it is not possible (nor would it be informative) to correlate every adult brain structure with some particular embryological neurogenetic field. However, many of the major structures of the brain provide close approximations: these include the total cerebral cortex; the total basal ganglia; the total diencephalon; and the cerebellar cortex, among others. Earlier it was noted that the cerebellar and cerebral cortices are larger than expected with respect to the diencephalon and striatum. This dichotomy between cortical and nuclear structures holds for the comparison of the total of all cortical structures with respect to total of all nuclear structures and to many other comparisons as well (Figure 20.6), including cerebellar cortex vs. cerebellar nuclei, and limbic cortex vs. hypothalamus (not shown; see Figures 20.8-10).

Besides their surface sheet versus deep solid structure it would appear that these two classes of divergently enlarged structures within the human brain share little else in common. Each class includes some of the most recent and most archaic

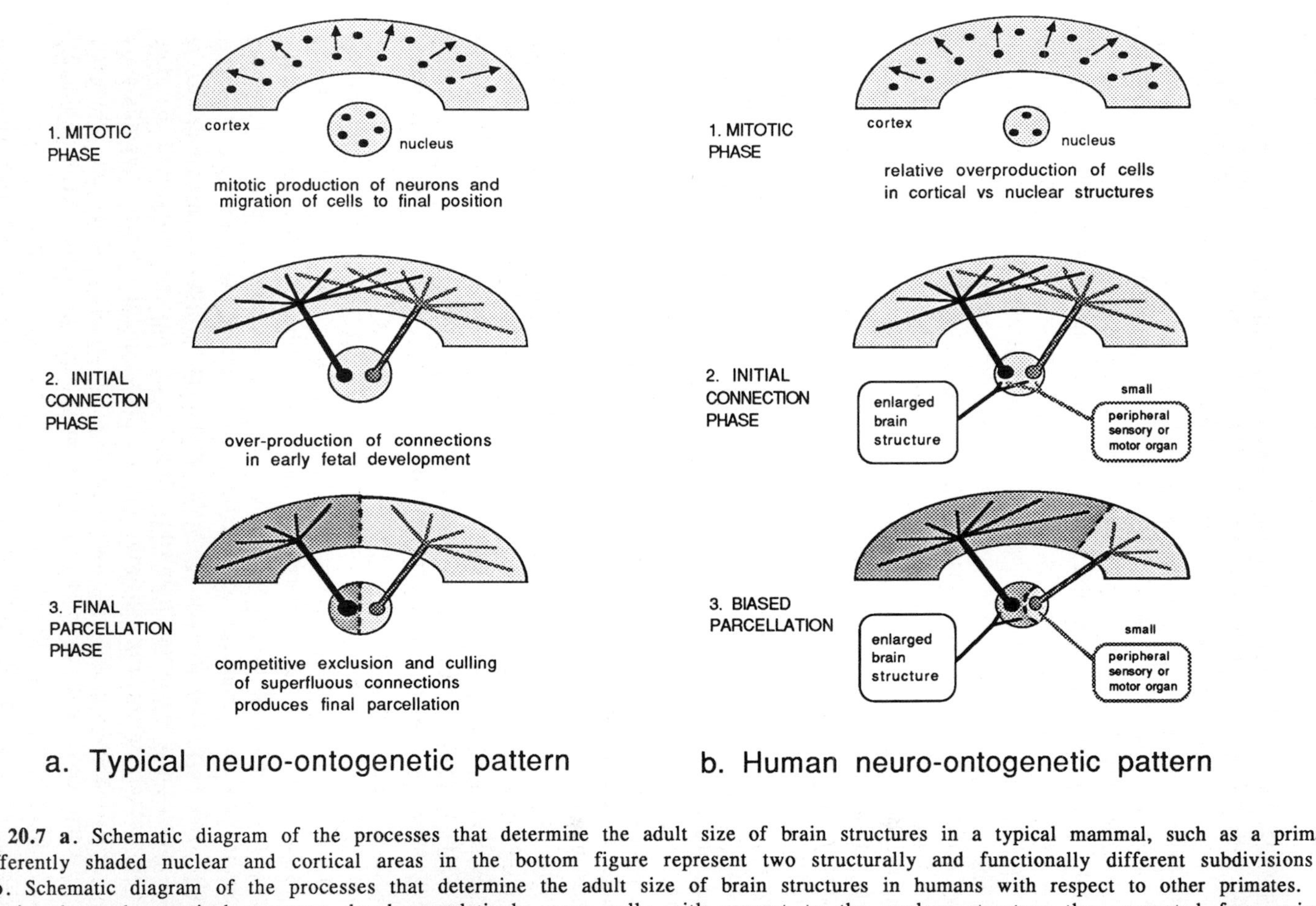

Figure 20.7 **a**. Schematic diagram of the processes that determine the adult size of brain structures in a typical mammal, such as a primate. The differently shaded nuclear and cortical areas in the bottom figure represent two structurally and functionally different subdivisions of each. **b**. Schematic diagram of the processes that determine the adult size of brain structures in humans with respect to other primates. In the mitotic phase the cortical structure develops relatively more cells with respect to the nuclear structure than expected from primate ontogenetic trends. With the added disproportion between small peripheral sensory or motor organs and an enlarged cortical area the final parcellation phase of development is biased so that structures most directly linked with the periphery do not recruit as much neural space as structures that are more centrally connected.

brain structures, and includes all modalities and functional specializations. However, attention to embryological considerations suggests a simpler and more ontogenetically basic similarity. This cortical/nuclear dichotomy coincides with a dichotomy in another ontogenetic feature: neural migratory patterns. All the "enlarged" structures (including neocortex, limbic cortex, paleocortex, septum-diagonal band, cerebellar cortex, tectum and tegmentum) are laminated surface structures whose neurons migrate into place by passing through or across previously laid down neurons (Jacobson, 1982). This pattern has been called "inside-out" migration, although this term does not capture the full variety of migratory patterns. In contrast, the "non-enlarged" structures are all deep gray matter structures in which neurogenesis is not accompanied by migration in the same sense. Neurons "born" later simply displace previously produced cells away from the mitotic zone and are subsequently displaced by the generation of later cells (reviewed in Deacon, 1984). Somewhat mixed or intermediate patterns are found in the basal ganglia, amygdala, and the dentate gyrus of the hippocampus.

It is hypothesized that during human evolution an entire embryological class of proto-neurons characterized by the "inside-out" migratory pattern (and likely some common mitotic regulatory mechanisms or receptor sites associated with it) may have slightly elongated their period of mitotic cycling with respect to other classes of neurons. Whatever the causes of cellular overproduction in these structures they must have their effect at this early stage of neurogenesis in order to produce a correlated deviation of size in such diverse brain structures.

This may also point to a possible correlate with the divergent brain and body size relationship in humans. Although it is often tacitly assumed that brain size is some function of body size (e.g. allometrically constrained by metabolic capacity or by the size of effector and receptor systems) the actual physiological determination of this size relationship in a growing animal is likely determined by the brain. The growth trajectories of normal developing animals (and humans) predict quite accurately the target adult body size, so this relationship is determined early in development. The somatic growth process is largely regulated by pituitary hormones and by their direct and indirect effects via other hormone systems. The pituitary hormones are in turn regulated by specialized nuclei of the hypothalamus. A major nuclear structure of the brain, the human hypothalamus is not among the class of "enlarged" structures. Since hypothalamic cellular populations are essentially determined by the end of the mitotic phase of neuron production in the first trimester of gestation, a correlated determination of target adult body size may also be set by brain structures within the hypothalamus at that time (subsequent deviation from this target size may be influenced by diet, disease, or other mitigating growth factors not under direct hypothalamic control). An explanation for the human body's failure to enlarge in allometric fashion to keep pace with an enlarged brain may also be found in the bimodal distribution of neural over- production.

Figures 20.8-10 list deviations of human brain structure volumes and some relevant peripheral organs from values predicted from anthropoid primate trends with respect to brain size (Figures from Deacon, 1984; all predictions are based on

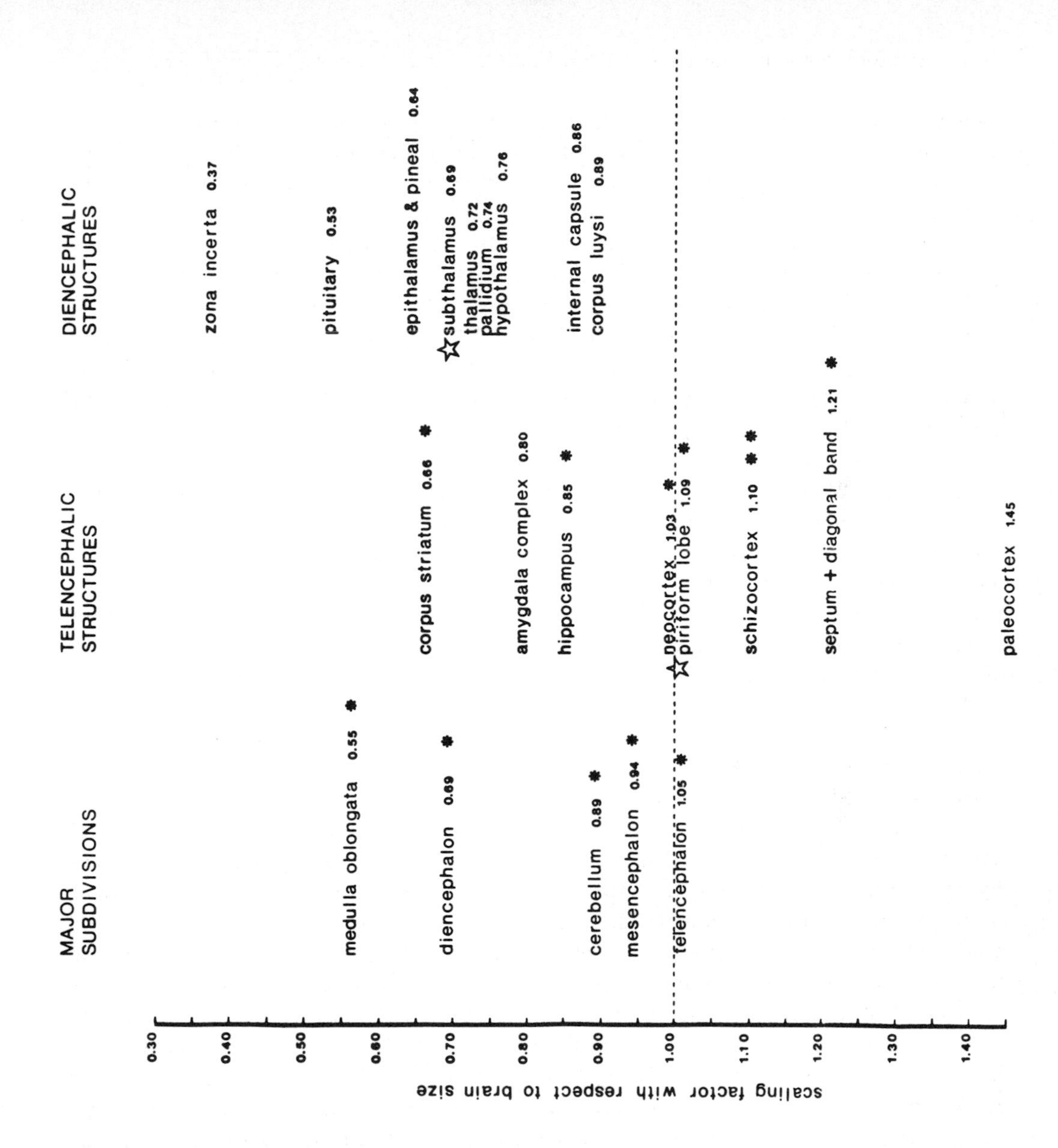

Figures 20.8, 20.9, and 20.10 Charts of the size deviations of major human brain areas from predictions based on anthropoid primate trends with respect to total brain size (calculation is by log transformed least squares regression not including human values; brain data are from Stephan, et al., 1981; Blinkov and Glezer, 1968; and Armstrong, 1979, 1980a, and 1980b; eye data are from Schultz, 1940; muscle mass data are from Grand, 1977; and Zihlman, 1979; and the nasal aperture data are from a personal communication from Michael Billig, Harvard University). The dashed line indicates size predicted from human brain size, asterisks indicate 0.05 significance with respect to either brain or body size predictions (calculated with human points included), and stars indicate the position of the total of a whole group taken together. Because of part/whole bias the values for telencephalon, cerebellum, and neocortex are misleading (see text). Reprinted from Deacon, 1984.

THALAMIC NUCLEI	BASAL GANGLIA	NEOCORTICAL SUBDIVISIONS
lateral geniculate 0.37	zona incerta 0.37	area 4 0.35
(lateral geniculate 0.49 *)	putamen 0.48	area 17 0.60
pregeniculate 0.54	globus pallidus 0.55	area 6 0.77
intralaminar 0.58	☆	inferior parietal 0.91
pulvinar 0.61	caudate nucleus 0.84	☆area 52 1.02
lateralis 0.66	corpus luysi 0.88	area 42 1.06
reticularis 0.72	substantia nigra 1.01	area 22 1.17
☆ centromedian + parafascicularis 0.73		areas 41 & 41/42 1.27
limitans 0.91		prefrontal 2.02 *
medial geniculate 0.98		
mediodorsalis 1.07		
cucularis 1.25		
anterior dorsalis 1.43		
anterior principalis 1.81		
dorsal superficialis 1.95		

Scale: 0.30, 0.40, 0.50, 0.60, 0.70, 0.80, 0.90, 1.00, 1.10, 1 20, 1.30, 1.40, 1.50, 1.60, 1.70, 1.80, 1.90, 2.00

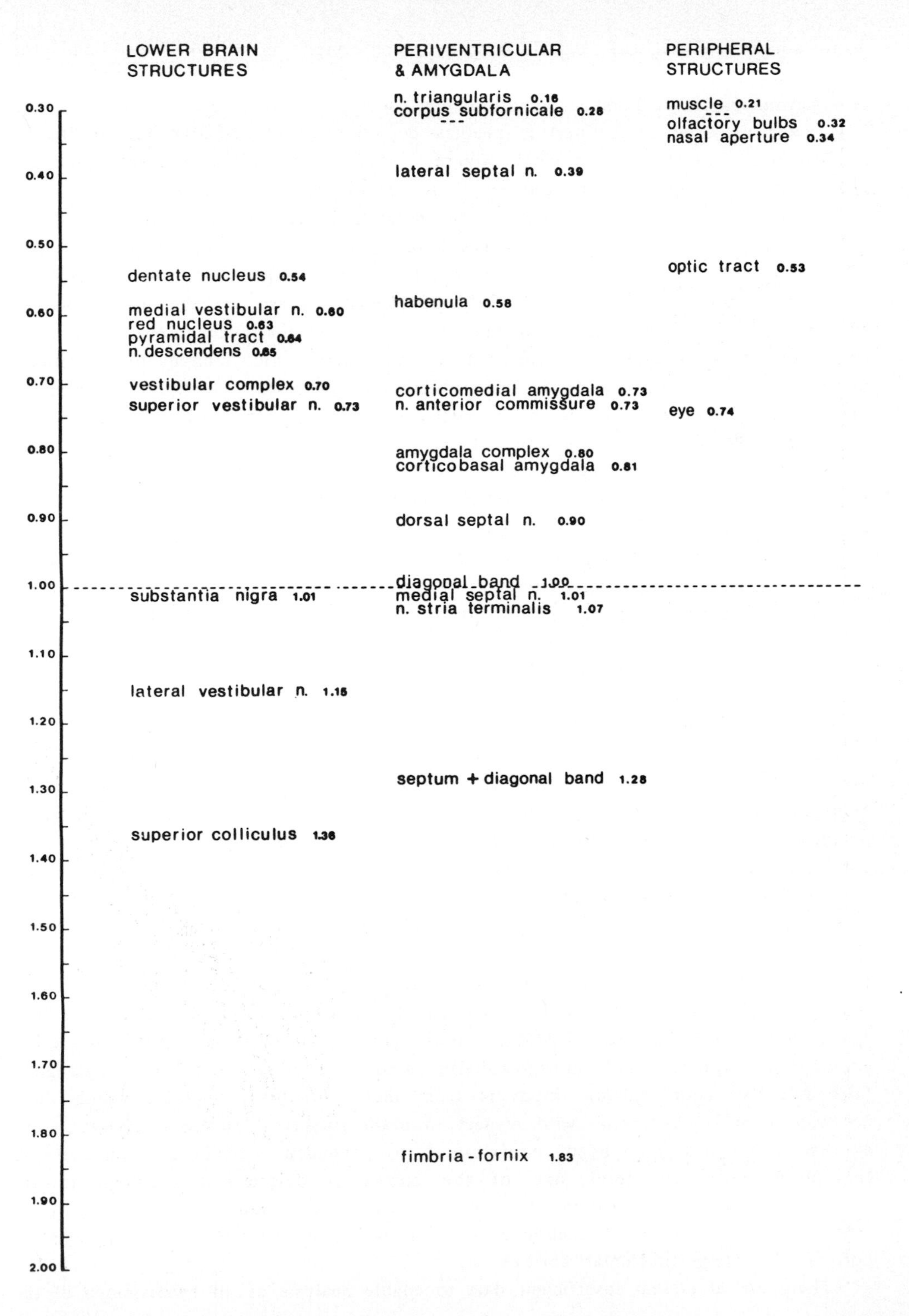

LOWER BRAIN STRUCTURES
PERIVENTRICULAR & AMYGDALA
PERIPHERAL STRUCTURES
n. triangularis 0.16
corpus subfornicale 0.28
muscle 0.21
olfactory bulbs 0.32
nasal aperture 0.34
lateral septal n. 0.39
dentate nucleus 0.54
optic tract 0.53
habenula 0.58
medial vestibular n. 0.60
red nucleus 0.63
pyramidal tract 0.64
n. descendens 0.65
vestibular complex 0.70
superior vestibular n. 0.73
corticomedial amygdala 0.73
n. anterior commissure 0.73
eye 0.74
amygdala complex 0.80
corticobasal amygdala 0.81
dorsal septal n. 0.90
diagonal band 1.00
substantia nigra 1.01
medial septal n. 1.01
n. stria terminalis 1.07
lateral vestibular n. 1.15
septum + diagonal band 1.28
superior colliculus 1.36
fimbria - fornix 1.83
0.30
0.40
0.50
0.60
0.70
0.80
0.90
1.00
1.10
1.20
1.30
1.40
1.50
1.60
1.70
1.80
1.90
2.00

least squares fitting of logarithmically transformed anthropoid data and the human deviations from this prediction are presented in linear terms). Although this approach underestimates deviations for some of the larger structures listed (e.g. neocortex and cerebellum) it enables depiction of the relative proportions of a great many structures in a single figure. Most of the values in this figure are taken from samples that are too small for significance testing and so must be taken as preliminary findings only. In general, it can be seen that the major cortical and nuclear structures all reflect the general dichotomy of relative size. In contrast, it appears that within a major nuclear complex or cortical structure different subdivisions reflect a diversity of trends. These are the more complex systemic effects of the parcellation process. The bimodally disproportionate sizes of different interconnected structures in the embryonic human brain should produce a complicated biasing of the parcellation processes. Reduced or enlarged afferent axon populations should respectively have a competitive advantage or disadvantage in the parcellation of structures at all levels (Hollyday and Hamburger, 1976; Katz and Lassek, 1978; Finlay and Slattery, 1983). This relationship is schematically depicted in Figure 20.11.

Quantitative data on structural subdivisions of the thalamus and cerebral cortex support this analysis (see Figures 20.8-10). For example, thalamic nuclei receiving primary afferents from an enlarged source (e.g. anterior thalamic nuclei with afferents from limbic cortex) are relatively larger than expected, while those receiving primary afferents from a peripheral or non-enlarged source (e.g. the lateral geniculate complex with afferents from the eye) are relatively smaller than expected (which is the case, see Armstrong, 1979, 1980a; 1980b; 1982; Passingham, 1979; Deacon, 1984). Also cortical areas with direct representation of peripheral (bodysize-limited) systems (e.g. the primary visual areas) should be more constrained than areas primarily linked with other enlarged brain structures (e.g. prefrontal cortex). Figure 20.12 summarizes some of these cortical trends. Striate cortex in the human brain is only 60% of the size predicted from primate trends. It is only one synapse removed from its peripheral representation and is smaller than predicted on the basis of brain size but not for body size (Passingham, 1979; Deacon, 1984). Prefrontal cortex on the other hand may be as much as 200% larger than predicted from brain size, and as much as 600% larger than predicted from body size (based on calculations from two independent sources of data from Brodmann, 1909 and Kononva, 1962; see Deacon, 1984; and also Passingham, 1975b). It is most intricately connected with other neocortical, limbic, and mesencephalic areas--all of which are "enlarged" (although its major thalamic input is from the mediodorsal nucleus, which does not appear to be "enlarged" with respect to brain size--see Armstrong, 1980a). As a result, it has not been constrained by reduced peripheral connections. Inevitably, since the total size of the cortex is determined previous to its parcellation, when some functional systems come to occupy less territory (e.g. striate cortex) others must occupy more than would be predicted on the basis of cortical size (e.g. prefrontal cortex).

There are at present insufficient data to enable analysis of all subdivisions of the

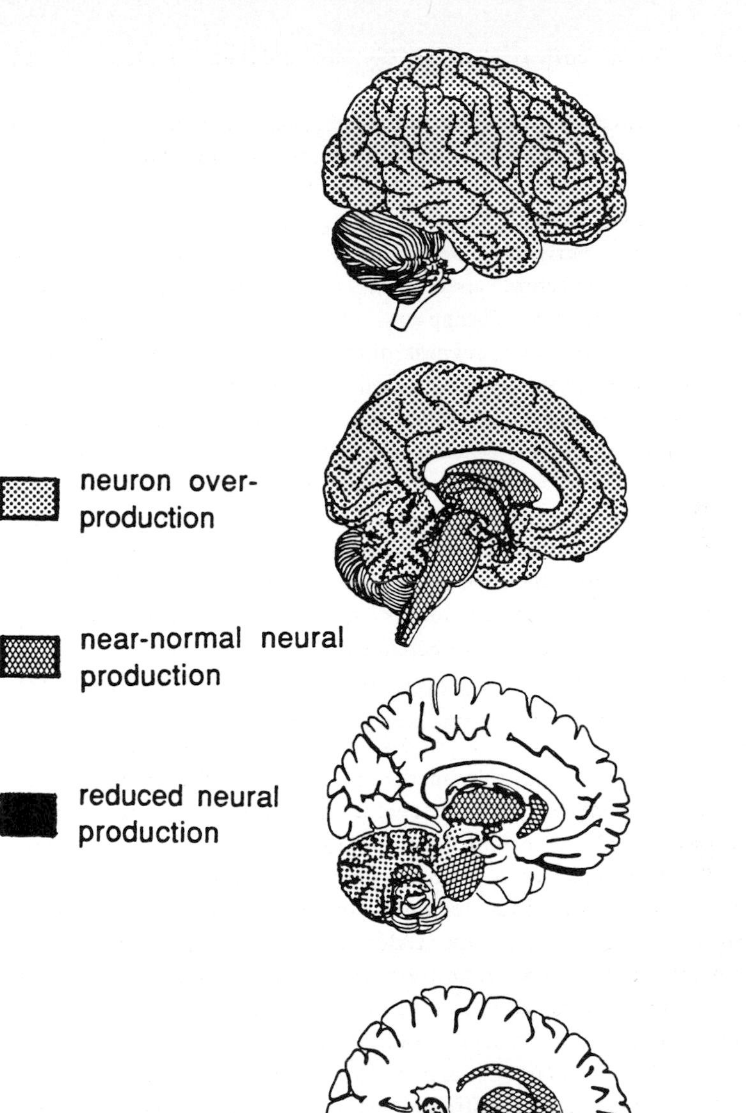

Figure 20.11 Summary diagram showing the general distribution of neuron production in human embryogenesis. Cerebral cortex, cerebellar cortex, and mesencephalon are presumed to exhibit cell over-production with respect to primate trends. Thalamus, hypothalamus, striatum, pallidum, cerebellar nuclei, and brainstem are only slightly larger than predicted from body size and so are assumed to be close to the primate trends. These structures with "near normal" neural production may be slight enlarged with respect to body size because of afferents they receive from "enlarged" brain structures. Only the olfactory bulbs are significantly smaller than expected with respect to total body size.

thalamus or cerebral cortex (limited by the number of species and number of structures represented) much less other subcortical structures. More comprehensive testing of this model must await future data. Hopefully, new computer-assisted image analysis techniques will overcome both the interpretive and labor intensive hindrances to this research. The model presented here predicts that primary motor cortex and probably somatic sensory cortex are constrained by their peripheral representation, while it is unclear whether this should be true of auditory areas which are separated from their peripheral receptor organs by numerous synapses. Preliminary data presented here support these predictions and further suggest that even the secondary and tertiary belt areas in each modality may be constrained to the same degree as primary areas.

There is much to learn regarding the quantitative features of the parcellation process in mammal brains in general, and still more to be discovered with respect to the uniqueness of the human brain. Although many of the quantitative results discussed here must be considered provisional and incomplete, the ontogenetic approach to these data cannot be overlooked. Body size, brain size and the internal organization of the brain are determined with respect to each other during embryogenesis. It is no longer defensible to treat brain/body proportions and internal brain organization as unrelated.

Discussion

One important implication of this model of human brain evolution is that changes of brain organization during human evolution should correlate with the degree of brain and body size disproportion. Since this latter relationship can be determined from fossil evidence it is possible to make estimates of the structural organization of various extinct hominid brains. Such estimates would not, for example, support claims based on endocast analysis that australopithecine brains are significantly reorganized with respect to other apes (e.g. Holloway, 1983; however, this does not rule out purely morphological reorganization. For example, Holloway has noted a reorganization of the "gestalt" of the fissurization of the australopithecine brain. This might result from the more ventral position of the foramen magnum in australopithecines as compared to great apes, rather than from internal reorganization). But what hypotheses about natural selection for brain traits are consistent with these preliminary data?

The non-uniform, non-allometric enlargement of the human brain could be seen as evidence bearing on the debate between those who argue that encephalization itself is the central morphological adaptation in response to selection for intelligence, and those who argue that increased size is secondary to mosaic reorganization in response to selection on distinct "modular" functions of the brain. Both simple forms of these arguments are contradicted.

The "extra neurons" in the human brain have been included in some systems and not others, and those with added size are distributed throughout the brain, not just in so-called "higher" centers. Although the overall increase in neural population above the primate trend can in some general sense be equated with

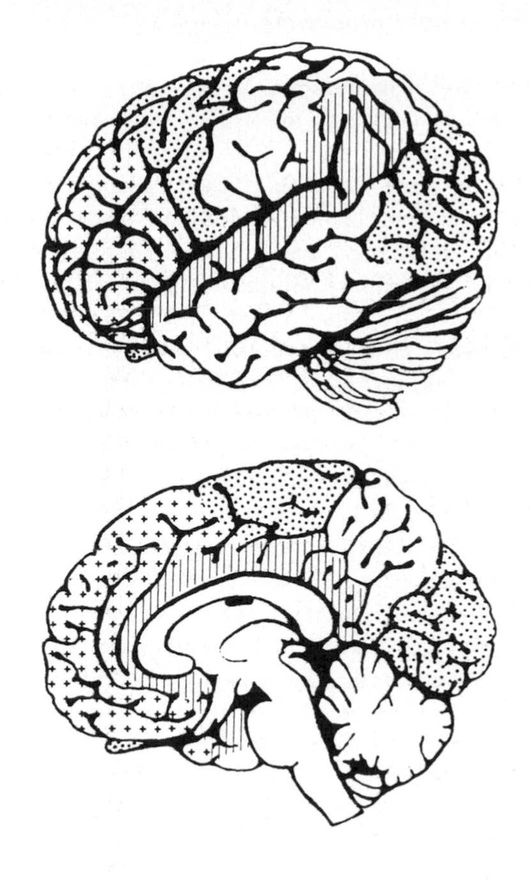

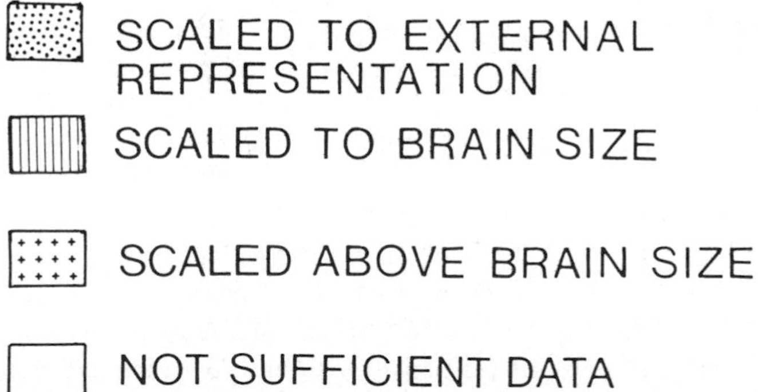

Figure 20.12 Summary diagram showing the relative sizes of cerebral cortical subdivisions with respect to parcellation constraints. Only data for prefrontal, limbic, and primary visual cortex are statistically significant, values for auditory, inferior parietal, and motor cortex are provisional, because they are based on small sample sizes.

increased information processing capacity, the non-equivalent spatial distribution of the "extra neurons" argues against this being the fundamental adaptation behind this reorganization. The only way I can see of salvaging a general intelligence model is to argue that these many disproportions of parcellation are secondary pleiotropic effects (paraptations) of the cortical-nuclear disproportion, and that this basic disproportion is the embryological means by which brain and body disproportion itself has been achieved. In this interpretation, the relatively small body determined by the non-enlarged hypothalamus makes a smaller information processing demand upon the enlarged brain.

The alternative mosaic evolution theories do not fare much better. Embryological considerations argue a priori against the possibility of independent and isolated enlargement of specific brain structures in response to natural selection. But even if we grant this possibility, the data would require that each of the distinct enlarged cortical and subcortical areas represents an independent adaptation. This piecemeal approach runs into particular difficulty creating adaptational individual scenarios with respect to both the number of different structures with deviant sizes and the variety of their functions. Mosaic evolution theories may be salvageable in the context of embryological considerations. They must, however, abandon the implicit assumption that a distinct adaptation is independently represented by each enlarged brain structure, and correspondingly they must recognize that enlargement of a particular brain structure may not indicate selection for the functions of that structure (e.g. Armstrong, 1982). Selection for enlargement of one structure may produce a range of correlated size disproportions (paraptations) in other structures via their embryological coupling. In this interpretation natural selection can be said to act selectively on modular functions but only one among a number of mosaic effects may be the focus of the principal adaptation. Because there is not likely to be a one-to-one correlation between structural sizes and functional adaptations, evolutionary interpretations must take into account organizational changes throughout the brain and not just focus on individual structures out of context.

In summary, both approaches must be radically qualified in the context of the embryological considerations raised here. Neither the undifferentiated addition of extra neurons nor the specific addition of extra neurons to particular functionally distinct structures appears to characterize the reorganization of the human brain. The rephrasals of these theoretical positions presented above manage to salvage the adaptationist positions of each at the expense of their descriptions of central nervous system mechanisms. Unfortunately, data concerning brain reorganization cannot alone distinguish between adaptational scenarios for human brain evolution. They can only elucidate the constraints within which such speculations must be framed. Within these constraints we have only the rule of parsimony to guide us.

Conclusions

It is my own speculation that the indirect prefrontal enlargement is most significant for understanding the selectional factors that reorganized and enlarged hominid brains. The prefrontal region appears the most enlarged of all major structures. Functional evidence from monkeys and humans suggests that this prefrontal expansion is probably correlated with the evolution of language. This association is suggested by the central role of medial frontal and anterior limbic areas in primate vocal, facial, and gestural communication; the role of lateral prefrontal areas in conditional learning and motor sequence organization (particularly sequencing oral- facial movements by inferior frontal areas); and the connections of inferior prefrontal areas with auditory areas, somatic mouth-tongue-larynx areas, anterior limbic vocalization areas, and motor mouth-tongue-larynx areas (e.g. Broca's area; see Deacon, this volume). Evidence from electrical stimulation studies in human subjects (Ojemann, 1979; Ojemann, 1983) indicates that the amount of cortex recruited for language processes is inversely proportional to the verbal IQ of the subject. This also may be interpreted to mean that the more difficult the language task the larger the distribution of areas recruited in its performance. In evolutionary terms, the difficulty of utilizing sequential organization and other grammatical features of language may have given individuals with more prefrontal cortex available for recruitment a selective advantage, despite any other quantitative disruptions of human brain organization that were paraptations of this disproportion.

Studies of connectional patterns in non-human primate brains that are homologous to human language circuits (Deacon, this volume) have suggested that the evolution of human language abilities did not result from major connectional reorganization. In contrast, allometric analysis of human brain structures demonstrates a pattern of quantitative reorganization that directly involves regions with known contributions to language functions. Together these findings provide a new basis for theories of human brain evolution with respect to the evolution of language. The focal point for this analysis now becomes the changes in the relative numbers of cells and projections comprising the neural pathways involved in the interrelation of auditory, facial, lingual and laryngeal functions. The preceding analysis suggests that the prefrontal area within which these connections converge (loosely identified as Broca's area) has become considerably enlarged with respect to other cortical and subcortical structures. As a result, efferent projections from this area are far more numerous with respect to target areas than in non-human primates. Some of these target areas include motor cortex, Wernicke's area, the anterior cingulate vocalization center, and brainstem nuclei indirectly invoved in vocalization.

Although at present we can only guess at the effects of such a shift in proportions of connections, it seems plausible to assume that, in humans, the functions of this enlarged area have taken on a more dominating role within these circuits than in primates. These proportional changes, occuring without major connectional changes, appear to be responsible for the appearance of a qualitatively unique adaptation: language. It is instructuve to compare this to the changes in relative size and shape of limb bones that makes possible such diverse forms of locomotion as swimming, walking, grasping and climbing, and flying in vertebrates, despite the fact that there has been no modification of the basic topology of bone connectivity. Quantitative changes in numbers of cells and connections, but not in the specificity of connections of the hominid brain established a new functional relationship among the parts that was without precedent as an adaptation, but was not without anatomical homology.

On the basis of these findings I suggest the following tentative evolutionary scenario: The earliest development of symbolic communication (in whatever form--probably multimodal) approximately two million years ago provided significant (unspecified) reproductive advantage to those most facile with its acquisition and use and so established powerful selection for these abilities. Most significant of these were selection for enhanced learning of complex conditional associations and enhanced oral-vocal motor-sequencing skills. Since prefrontal areas, and most particularly the posterior inferior region destined to become Broca's area, play a significant role in these functions, increased information processing capacity of these areas was selected for. This relative increase of functional capacity (and perhaps functional dominance of these processes over other competing processes) occurred in brains in which these areas occupied a relatively and absolutely larger proportion of the cerebral cortex than in a typical large ape brain. However, the ontogenetic process by which prefrontal areas acquired proportionately more cortical space is indirect (though perhaps the most direct mechanism when one considers embryological constraints) and involves many other areas than just prefrontal cortex. On the basis of volumetric analyses of brain structures I have proposed an ontogenetic mechanism for this change that does not invoke a highly specialized genetic basis. Rather, an essentially bimodal distribution of neuron overproduction throughout the brain is responsible for a cascade of secondary, system by system parcellation processes, one of which is relative prefrontal expansion. The overall expansion of brain volume with respect to body size is considered secondary to this effect. The evolution of language is thus considered the principal correlate of human brain evolution, implicated in the beginning stages of relative brain enlargement two million years ago. Relative brain enlargement is correspondingly the major fossil indicator of brain reorganization for language. It suggests that neural specialization for language began with *Homo habilis* and essentially reached the modern condition in neanderthal and archaic *Homo sapiens* specimens with modern brain-body proportions.

References

Armstrong E (1979) A quantitative comparison of the hominoid thalamus, I: Specific sensory relay nuclei. Am J Phys Anthrop 51:365-82

Armstrong E (1980a) A quantitative comparison of the hominoid thalamus, II: Limbic nuclei, anterior principalis and lateralis dorsalis. Am J Phys Anthrop 52:43-54

Armstrong E (1980b) A quantitative comparison of the hominoid thalamus, III: A motor substrate--the ventro-lateral complex. Am J Phys Anthrop 52:405-19

Armstrong E (1982) Mosaic evolution in the primate brain: Differences and similarities in the hominoid thalamus. In E Armstrong and D Falk (eds), Primate Brain Evolution. New York: Plenum Press, pp. 131-161

Bauchot R (1982) Brain organization and taxonomic relationships in insectivora and primates. In E Armstrong and D Falk (eds), Primate Brain Evolution. New York: Plenum Press, pp. 163-175

Blinkov SM (1950) Cited in Blinkov SM and Glezer II (1968) The Human Brain in Figures and Tables. New York: Plenum Press

Blinkov SM and Glezer II (1968) The Human Brain in Figures and Tables. New York: Plenum Press

Blinkov SM and Zvorykin VP (1950) Dimensions of the auditory cortex and the medial geniculate body in man and monkeys. Cited in SM Blinkov and II Glezer, The Human Brain in Figures and Tables. New York: Plenum Press

Bonin G von (1937) Brain-weight and body-weight in mammals. J Gen Psych 16:379-389

Bonin G von (1941) Sidelights on cerebral evolution: brain size of lower vertebrates and degree of cortical folding. J Gen et Psychol 25:273-282

Bonin G von (1952) Notes on cortical evolution. A M A Archives of Neurology and Psychiatry 672:135-144

Brodmann K (1909) Vergleichende Lokalisationslehre der Grosshirnrinde in ihren prinzipien dargestellt auf Grund der Zellenbaus. Leipzig: JA Barth

Brody H (1955) Organization of the cerebral cortex. III. A study of aging in the human cerebral cortex. J Comp Neurol 102:511-556

Clutton-Brock TH and Harvey PH (1980) Primates, brains and ecology. J Zool Lond 190:309-323

Cowan WM (1973) Neuronal death as a regulative mechanism in the control of cell number in the nervous system. In M Rockstein (ed), Development and Aging in the Nervous System. New York: Academic Press, pp. 119-141

Cragg BG (1967) Changes in visual cortex on first exposure of rats to light: Effect on synaptic dimensions. Nature 215:251-253

D'Amato CJ and Hicks SP (1978) Normal development and post-traumatic plasticity of cortico-spinal neurons in rats. Exp Neurol 60:557-569

Deacon TW (1984) Connections of the inferior periarcuate area in the brain of *Macaca fascicularis*: An experimental and comparative investigation of language circuitry and its evolution. Ph.D. Thesis, Harvard University

Diamond IT (1979) The subdivisions of neocortex: A proposal to revise the traditional view of sensory, motor and association areas. In JM Sprague and JN Epstein (eds), Prograss in Psychobiology and Physiological Psychology (Vol 8). New York: Academic Press

Diamond IT (1982) The functional significance of architectonic subdivisions of the cortex: Lashley's criticism of the traditional view. In J Orbach (ed), Neuropsychology after Lashley. London: Lawrence Earlbaum Assoc.

Douglas RJ and Marcellus D (1975) The ascent of man: Deductions based on a multivariate analysis of the brain. Brain Behav Evol 11:179-213

Dubois E (1897) Sur le rapport du poids de l'encephale avec la grandeur du corps chez les mammiferes. Bulletins de la Societe d' Anthropologie de Paris 8:337-376

Dubois E (1898) Uber die Abhangigkeit des Hirngewichtes von der Korpergrosse beim Menschen Arch Anthrop 25:423-441

Dubois E (1913) On the relation between the quantity of brain and the size of the body in vertebrates. Verh Kon Akad Wetenschappen Amsterdam 16:647

Dubois E (1923) Phylogenetic and ontogenetic increase of the volume of the brain in the vertebrata. Proc Kon Akad Wetenschappen Amsterdam 25:235-255

Ebbesson SO (1980) The parcellation theory and its relation to interspecific variability in brain organization, evolutionary and ontogenetic development, andneuronal plasticity. Cell Tissue Res 213:179-212

Finlay BL and Slattery K (1983) Local differences in the amount of early cell death in neocortex predict adult local specializations. Science 219:1349-51

Goldman PS (1978) Neuronal plasticity in primate telencephalon: anomalous projections induced by prenatal removal of frontal cortex. Science 202:768-70

Goldman PS and Galkin TW (1978) Prenatal removal of frontal association cortex in the fetal rhesus monkey: anatomical and functional consequences in postnatal life. Brain Res 152:451-85

Goldman-Rakic PS (1981) Development and plasticity of primate frontal association cortex. In O Schmidt, FG Worden, and SG Dennis (eds), The Organization of the Cerebral Cortex. Cambridge: MIT Press

Gould SJ (1975) Allometry in primates with emphasis of scaling and the evolution of the brain. Contributions to Primatology 5:244-292

Grand TI (1977) Body weight its relation to tissue composition, segment distribution, and motor function: Part 1 interspecific comparision. Am J Phys Anthrop 47:211-241

Harvey P, Clutton-Brock TH, and Mace GM (1980) Brain size and ecology in small mammals and primates. Proc Natl Acad Sci 77:4387-4389

Hofman MA (1982a) Encephalization in mammals in relation to size of the cerebral cortex. Brain Behav Evol 20:84-96.

Hofman MA (1982b) A two-component theory of encephalization in mammals. J Theor Biol 99:571-584

Hofman MA (1983a) Energy metabolism, brain size and longevity in mammals. Q Rev Biol 58:495-512

Hofman MA (1983b) Evolution of brain size in neonatal and adult placental mammals: a theoretical approach. J Theor Biol 105:317-332

Hofman MA (1984) Energy metabolism and relative brain size in human neonates from single and multiple gestations. An allometric study. Biol Neonate 45:157-164

Hofman MA (1985) Neuronal correlates of corticalization in mammals: a theory. J Theor Biol 112:77-95

Holloway RL (1976) Paleoneurological evidence for language origins. In SR Harnad, HD Steklis and J Lancaster (eds), Origins and Evolution of Language and Speech, Annals N Y Acad Sci 280:330-348

Holloway RL (1979) Brain size, allometry, and reorganization: Toward a synthesis. In M Hahn, C Jensen, and B Dudek (eds), Development and Evolution of Brain Size. New York: Academic Press, pp.59-88

Holloway RL (1983) Cerebral brain endocast pattern of *Australopithecus afarensis*. Nature 303:420-422

Holloway RL and Post DG (1982) The relativity of relative brain measures and hominid mosaic evolution. In E Armstrong and D Falk (eds), Primate Brain Evolution. New York: Plenum

Hollyday M and Hamburger V (1976) Reduction of the naturally occuring motor neuron loss by enlargement of the periphery. J Comp Neurol 170:311-320

Huxley J (1932) Problems of Relative Growth. London: Methuen

Jacobson M (1978) Developmental Neurobiology (Second edition). New York: Holt, Rhinehart, and Winston

Jerison HJ (1973) Evolution of the Brain and Intelligence. New York: Academic Press

Jerison HJ (1977) The theory of encephalization. Ann NY Acad Sci 299:146-160

Jerison HJ (1979) The evolution of diversity in brain size. In MHahn, B Dudek, and C Jensen (eds), Development and Evolution of Brain Size: Behavioral Implications. New York: Academic Press, pp. 29-57

Jerison HJ (1982) Allometry, brain size, cortical surface, and convolutedness. In E Armstrong and D Falk (eds), Primate Brain Evolution. New York: Plenum, pp. 77-84

Katz MJ and Lasek RJ (1978) Evolution of the nervous system: Role of ontogenetic mechanisms in evolution of matching populations. Proc Nat Acad Sci USA 75:1349-52

Kelahan AM, Ray RH, Carson LV, and Doetsch GS (1980) Functional organization of racoon somatosensory cortex: Effects of early peripheral injury. Soc Neurosci Abstracts 10

Kononova EP (1962) The Frontal Region of the Brain. Leningrad. Cited in SM Blinkov and II Glezer, The Human Brain in Figures and Tables. New York: Plenum (1968)

Koppel H and Innocenti GM (1983) Is there a genuine exuberancy of callosal projections in development? A quantitative electron- microscopic study in the cat. Neurosci Lett 41:33-40

Lu SM, Schmechel DE, and Lin C-S (1983) Transplantation between neonatal visual and somatosensory cortex in pigmented rats. Soc Neurosci Abstracts 9:112.5

Mace GM, Harvey P, and Clutton-Brock TH (1981) Brain size and ecology in small mammals. J Zool Lond 193:333-354

Martin RD (1983) Human brain evolution in an ecological context. Fifty- second James Arthur Lecture on the Evolution of the Human Brain. American Mus Nat Hist, New York

Merzenich MM, Kaas JH, Nelson RJ, Wall J, Sur M, and Felleman DJ (1980) Progressive topographic reorganization of representations of the hand within areas 3b and 1 of monkeys following median nerve section. Soc Neurosci Abstracts 10:222.1

Ojemann GA (1979) Individual variability in cortical localization of language. J Neurosurg 50:164-9

Ojemann GA (1983) Brain organization for language from the perspective of electrical stimulation mapping. Behav Brain Sciences 2:189-230

O'Leary D, Stanfield B, and Cowan W (1981) Evidence that the early postnatal restriction of cells of origin of the callosal projection is due to the elimination of axonal collaterals rather than to the death of neurons. Developmental Brain Res 1:607-17

Passingham RE (1973) Anatomical differences between the cortex of man and other primates. Brain Behav Evol 7:337-359

Passingham RE (1975a) The brain and intelligence. Brain Behav Evol 11:1-15

Passingham RE (1975b) Changes in the size and organization of the brain in man and his ancestors. Brain Behav Evol 11:73-90

Passingham RE (1979a) Brain size and intelligence in man. Brain Behav Evol 16:253-270

Passingham RE (1979b) Specialization in the language areas. In HD Steklis and MJ Raliegh (eds), Neurobiology of Social Communication in Primates. New York: Academic Press

Passingham RE and Ettlinger G (1973) A comparison of cortical functions in man and other primates. Brain Behav Evol 7:337-359

Perry VH and Cowey A (1982) A sensitive period for ganglionic cell degeneration and the formation of abberant retino-fugal connections following tectal lesions in rats. Neuroscience 7:583-594

Rakic P, Bourgeois J-P, Eckenhoff MF, Zecevic N, and Goldman-Rakic PS (1986) Concurrent overproduction of synapses in diverse regions of the primate cerebral cortex. Science 232:232-234

Rockel AJ, Hiorns RW, and Powell TPS (1980) The basic uniformity in structure of the neocortex. Brain 103:221

Rubel EW, Smith DJ, and Miller LC (1976) Organization and development of brainstem auditory nuclei of the chicken: Ontogeny of n. magnocellularis and n. laminaris. J Comp Neurol 166:469-490

Sacher GA (1970) Allometric and factorial analysis of brain structure in insectivores and primates. In CR Novack and W Montagna (eds), The Primate Brain: Advances in Primatology. New York: Appleton, pp. 245-287

Schultz AH (1940) The size of the orbit and of the eye in primates. Am J Phys Anthro 26:389-408

Shariff GA (1953) Cell counts in the primate cerebral cortex. J Comp Neurol 98:381-400

Sohal GS (1976) An experimental study of cell death in the developing trochlear nucleus. Exp Neurol 51:684-698

Stephan H and Andy OJ (1969) Quantitative comparative neuroanatomy of primates: an attempt at a phylogenetic interpretation. In Comparative and Evolutionary Aspects of the Vertebrate Central Nervous System, Annals NY Acad Sci 167:370-387

Stephan H, Frahm H, and Baron G (1981) New and revised data on volumes of brain structures in insectivores and primates. Folia Primatol 35:1-29

Tobias PV (1980) The anatomy of hominization. In EA Vidreo (ed), Progress in Clinical and Biological Research, Vol. 59B. Advances in the Morphology of Cells and Tissues. New York: Alan R. Liss, pp. 101-110

Tobias PV (1981) The emergence of man in Africa and beyond. Philos Trans R Soc Lond B Biol Sci 292:43-56

Tower DB (1954) Structural and functional organization of the mammalian cerebral cortex. The correlation of neurone density with brain size. Cortical density in the finwhale with a note on the cortical neurone density in the Indian elephant. J Comp Neurol 101:19-53

Zihlman AL (1979) Differences in body weight composition of Pygmy Chimpanzees *Pan paniscus* and common chimpanzees *Pan troglodytes*. Am J Phys Anthro 50:496

THE FUNCTION OF INFORMATION PROCESSING IN NATURE

J.M.H. Vossen
Department of Comparative and Physiological Psychology
University of Nijmegen PO Box 9104
6500 HE Nijmegen
The Netherlands

The processing of information I want to discuss concerns only one part of the general processing of information that takes place in the organism. It relates to information processing in which psychological functions like perception, learning, thinking, memory, retrieval, and action are involved. These functions determine behavioral relationships of the organism to its environment. In accordance with Piaget (1976) behavior is defined as all action directed by organisms towards the external environment in order to change conditions therein or to change their own situation in relation to the environment.

From the point of view of psychology and ethology, information processing refers to the use, the acquisition, the storage, and the retrieval of "units of information" (Hirsh, 1974). According to Hirsh, "items of information are assumed to consist of descriptions of an environmental situation or stimulus, an operation upon that situation having specific consequences or action, and the consequences of the action" (p. 423). A stimulus, an action with respect to that stimulus, and the consequences of that action constitute the three components of every item of information. A specific item of information can develop by maturation, it can be modified by learning, or it can result from learning.

What is the function of items of information? The answer to that question must be looked for in the third element of the item of information: the consequences of the action. The consequences of the action can be studied from two different points of view: a biological one and a psychological one.

In biology the consequences of the action for the organism are studied with special reference to the biological function of the consequences. The question then is whether the consequences of the action contribute to the maintenance or survival of the genetic material that directly or indirectly has to do with the unit of information. The preservation of the genetic material can imply - but it is not necessarily so in all cases - the preservation or improvement of the adaptation of the individual to its environment, and, as a result of that, the preservation of the individual. It is because of the consequences of the action within the unit of information that the principle of natural selection can operate on the prerequisites of that unit of information.

NATO ASI Series, Vol. G17
Intelligence and Evolutionary Biology
Edited by H.J. Jerison and I. Jerison

In psychology the consequences of the action for the organism are translated in the principles of reinforcer and reinforcement. These principles, that underlie the development of units of information, are elaborated in the various theories of learning, which in substance refer to the paradigms of classical conditioning and of operant conditioning or instrumental learning.

The Psychological Approach to the Consequences of the Action

In theories of classical conditioning the consequences of the action for the organism usually do not play a significant role. The unconditional reflex is considered to be functional or not. The conditional reflex develops as a result of the contiguity of conditional stimulus and unconditional stimulus; generally the conditional reaction is not considered to be of functional significance. Usually the development of the conditional reflex is not associated with the consequences of the conditional reaction for the organism.

Although large differences exist with respect to the speed of classical conditioning of various reflexes, these differences are not taken into account in explanatory models or theories of classical conditioning. Experiments on classical conditioning show that some systems are easily conditioned and other systems only very slowly or not at all.

The transformation of a stimulus into a conditional stimulus is not exclusively produced by the contiguity of that stimulus and an unconditional stimulus. The results of many experiments make clear that the unconditional stimulus sets limits to the characteristics and form of the neutral stimulus, if that stimulus is to be transformed into a conditional stimulus. A nice example can be found in Testa's (1975) experiment in rats. In that experiment two unconditional stimuli were used: a square wave air blast UCS originating from the floor of the cage or a pulsed air blast originating from the ceiling of the cage. Moreover, two neutral stimuli were introduced: a square wave visual CS located at the floor of the cage or a pulsed visual CS located at the ceiling of the cage. All four combinations were presented in a conditioned suppression paradigm. In those combinations in which neutral stimuli and unconditional stimuli were similar with respect to presentation and location in the cage, classical conditioning occurred at a faster rate than in the other combinations.

Particularly relevant in this respect are data concerning classical conditioning of drug responses. Administration of a drug often functions as an unconditional stimulus; the effect of that drug is considered to be the unconditional reaction. Stimuli offered in contiguity with the unconditional drug stimulus eventually are transformed into conditional stimuli. However, contrary to the usual course of classical conditioning, these conditional stimuli do not elicit a conditional reaction that is similar to the unconditional reaction. On the contrary: the conditional reaction is the reverse of the unconditional reaction. Administration of insulin produces hypoglycaemia, presentation of a conditional stimulus associated with the insulin injection produces hyperglycaemia. Another example: administration of ethanol leads to hypothermia, the conditional stimulus associated with ethanol administration, to hyperthermia. A survey of these compensatory conditional reactions is given by Siegel (1979). The existence of compensatory conditional reactions renders theories incomplete in which the accent lies on contiguity and contingency only.

Modern theories of classical conditioning should pay attention to the function of information processing, to the consequences of actions for the organism, if they are to progress.

Other theories of learning in psychology have to do with operant conditioning or instrumental learning. In these theories room is made for the consequences of the action for the organism - even a lot of room. The consequences of the action are described in terms of reinforcer and reinforcement. The reinforcer strengthens or weakens the association between the stimulus and the response; it changes the frequencies of behaviors that precede the reinforcer. The reinforcer functions like the cement of the unit of information; it binds the components of the unit of information together. It is the *conditio sine qua non* for the development of associations, for learning.

The association between stimulus and action is realized by means of the reinforcer. Without reinforcer stimulus and response have no relation to each other. The relationship of stimulus and response to the reinforcer is completely determined by the situation in which learning takes place. The connection between stimulus, response and reinforcer within the context of learning was formulated for the first time by Thorndike (1913) in his empirical and theoretical law of effect.

It is very difficult to find experimental evidence for the theoretical neutrality of stimulus, response and reinforcer with respect to each other. In the last fifteen years a growing number of experiments demonstrate that large differences exist with respect to the speed, with which associations between particular stimuli and particular responses develop under the influence of specific reinforcers (Seligman, 1970). Under the concept "Constraints on Learning" Shettleworth (1972) and, later LoLordo (1979) brought together experiments that show that 1) between some stimuli and some responses associations develop more easily than between other stimuli and other responses (stimulus-response interactions), and 2) the effectivity of reinforcers is co-determined by the characteristics of the stimuli in the learning situation (stimulus-reinforcer interactions), and 3) the effectivity of reinforcers is co-determined by the characteristics of the responses in the learning situation (response-reinforcer interactions).

An example of a stimulus-reinforcer interaction can be found in an experiment by Garcia and Koelling (1966). In this experiment thirsty rats learned a passive avoidance task that had to do with drinking behavior. While drinking, these animals were confronted with a complex stimulus which contained both a taste element and an audiovisual element. After the drinking session half of the animals were made sick by means of x-radiation; the other half of the animals received an electrical shock. Confronted afterwards with a situation in which they had to drink in the presence of particular elements of the original complex stimulus, those animals that had been made sick avoided water that was flavored with the taste they had experienced before. These animals did not avoid water when the drinking of it was accompanied by the audiovisual stimulus they had experienced before. On the other hand, animals which had received an electrical shock avoided water when the drinking of it was accompanied by the audiovisual stimulus, but they did not avoid water that was flavored with the taste experienced previously. Clearly, the *reinforcer* sickness has to do with the taste stimulus rather than with the audiovisual stimulus; the reinforcer *pain*, on the contrary, has to do with the audiovisual stimulus rather than with the taste stimulus.

The Biological Approach to the Consequences of the Action.

Garcia and Koelling offer an interpretation for their data, which holds out a prospect of a more biologically oriented approach of information processing. They state: "Natural selection may have favored mechanisms which associate gustatory and olfactory cues with internal discomfort, since the chemical receptors sample the materials soon to be incorporated into the internal environment." (p. 124). Implicitly in their interpretation they discuss the consequences of the action for the organism and the biological function of the action. Natural selection only can have played a role if the association between gustatory stimuli and passive avoidance of water (or food) increased the reproductive success of individuals. Moreover, the mechanisms as intended by Garcia and Koelling must have a genetic base: effects of natural selection on the phenotype are mediated by the genotype.

Against the background of these reflections the concept of "unit of information" - the unit consisting of stimulus, action and consequences of that action - has to be reconsidered. If stimulus-response associations exist that have not been produced by learning, these stimulus-response associations must have a genetic base. The consequences of a response must have advanced the survival of the genetic material that underlies the stimulus-response association, and, in the first instance, by keeping alive the carrier of that genetic material - the organism.

The assumption is justified that the genetic base of units of information expresses itself in neural structures, which constitute the association between stimulus and action. Plasticity of these neural structures can be increased by natural selection. If as a result of this plasticity stimulus-response associations are modified in such a way that the adaptation of the individual organism to its environment is advanced, the genetic material producing this plasticity will enhance its probability of survival. The plasticity manifests itself in learning, which theoretically is conceived as strengthening or weakening associations between stimuli and responses or between stimuli mutually.

It seems reasonable to assume that the plasticity as a result of natural selection is not equally great in all neural structures which underlie units of information. Dependent upon the characteristics of the environment, plasticity in some structures will have a greater impact on survival chances of the organism than plasticity in other structures. This will influence the formation of new units of information starting from old ones.

In line with this thought the nature of the plasticity of neural structures does not have to be identical for each structure. On the other hand, the similarity of nerve cells which underlie neural structures should limit the number of alternative forms of plasticity. Local theories concerning the formation of units of information should precede more general theories concerning information processing.

Let us return once more to the definition of "unit of information." Psychologically the element of consequences of the action has to do with the concepts of reinforcer and reinforcement; biologically the element has to do with the concepts of function and natural selection. Are both concepts - reinforcer and function - related? In the domain of operant conditioning and instrumental learning it is the reinforcer which produces the association between stimulus and response. The question whether or not the consequences of the action elicited by the stimulus have implications with respect to the adaptation of the organism is basically not a relevant one. If a stimulus, a response, and reinforcer are in the right place and at the right time, the formation of an association is - from the point of view of learning theory - inevitable, regardless of an eventual improvement, maintenance, or deterioration of the adaptation of the organism, produced by that association. However, it has to be expected that the biological function of units of information that are produced limits the learning-theoretical non-commitment. The capacity of the reinforcer to produce associations has its limits where systematically units of information are produced that decrease or even eliminate chances of survival of the genetic material which has to do with those units of information. The consequences of the action in terms of reinforcement are exceeded by the consequences of the action in terms of biological function.

Are the concepts of reinforcer and function also related in the domain of classical conditioning? Earlier we have seen that generally speaking no functional meaning is attached to the conditional reaction. The question is whether or not this is correct. In a recent article Karen Hollis (1982) presents an ethological analysis of experiments on classical conditioning. Departing from the concept of biological function she reaches an interesting hypothesis, which she labels "the prefiguring hypothesis." In her opinion an essential feature of the conditional reaction is the anticipatory nature of that reaction. The conditional stimulus signals to the organism the imminence of a species-specific stimulus. The conditional reaction is the preparatory reaction of the organism to the species-specific stimulus. Sometimes the conditional reaction is similar to the unconditional reaction; sometimes it is opposite to the unconditional reaction, as is the case in compensatory reaction.

Whether or not classical conditioning takes place, whatever stimulus becomes conditional, and whatever form the conditional reaction takes ultimately depends on its selective advantage for the organism. To put it in another way, it depends on the consequences of the action. The psychological approach to information processing needs to be placed within the context of a biological approach, in which the accent is on the function of information processing. An example may clarify this statement.

In the spring in their breeding area males of the Japanese quail (Coturnix coturnix Japonica) create a territory which they fiercely defend against other males. Females are lured with a particular call. As soon as a female settles in the territory, the male virtually immediately starts courting her. In reaction to that the female adopts the mating posture. For mating itself, the cooperation of the female is not necessary. In such a situation, in which there is a strong competition among males with respect to obtaining females, it is of the utmost importance to the male that he start mating as soon as a female is available.

Farris (1967), who studied this species, succeeded in classical conditioning the courtship behavior of the male. Classical conditioning resulted in a substantial reduction of latency time to the start of courtship behavior: some males even started mating immediately if a female was available. Under natural conditions reduction of the period of courtship behavior results in an increase of mating chances, and consequently, an increase in the chances of a successful reproduction. This implies a selective advantage in the genetic material that underlies classical conditioning of the courtship system.

Evolutionary Implications

In the study of the evolutionary history of an organ or a capability two questions are important: (1) what is the origin of the organ or the capability, and (2) which modification preceded which modification? The question concerning the origin of an organ or capability is ultimately the question concerning the phylogeny of the species in which that organ or capability is studied (cladogenesis). The question concerning the sequence of modifications is the question concerning trends in the development of the organ or capability (anagenesis) (for a further elaboration, see: Gottlieb, 1984). Cladogenesis of information processing capability is difficult or even impossible to study because that capability does not fossilize; the same is true for the neural structures, which play an essential role in information processing. The way information processing capability has developed in a phyletic line is hardly accessible for research.

As to the anagenesis of information processing capability it is different. Careful study of the way in which information is processed in various species and within species in various behavior systems makes it possible to indicate trends in the development of information processing capability. On the basis of a very extensive analysis of research on learning in many species in the American as well as in the Russian tradition, Razran (1971) reaches the conclusion that in the development of learning, i.e. in the formation and modification of units of information, a trend runs from habituation and sensitization to classical conditioning and reinforcement learning, from there to sensory preconditioning, configuring and educative learning and from there to symbosemic, sememic and logicemic thinking. Razran bases this sequence on the following premises: "(1) Higher levels of learning, arising from lower levels as antecedents, should bring into being some new forms and laws of learning manifestations. (2) Lower levels of learning should continue as subsystems within higher levels so that higher-level learning is normally a resultant of higher and lower learning. (3) Higher-level learning should be more efficient in organism-environment interaction than lower-level learning, but lower-level learning should be more universal and less disruptable by outward stimuli and inward organismic states. (4) Normally, higher-level learning should control lower-level learning because of its greater efficiency; yet under certain conditions lower-level learning may predominate because of greater universality and less disruptability. (5) Interactions between higher and lower levels of learning will be either synergic or antagonistic" (p.23-24).

I would like to pay some attention to one of the transitions mentioned by Razran, namely, the transition from classical conditioning to configuring. One of the paradigms in which the difference between both levels becomes clear runs as follows: The organism is confronted with two elementary stimuli A and B (for instance, a flashing light and a tone) and their compound (A+B). Presentation of the elementary stimuli and the compound are in a random sequence. If the stimuli A and B are presented separately, they never are followed by the unconditional stimulus; the compound (A+B), on the contrary, is always followed by the unconditional stimulus. The question is whether in this paradigm the elementary stimuli A and B as well as their compound (A+B) are going to elicit a conditional reaction. Configuring is said to occur if the compound (A+B) elicits the conditional reaction and the elementary stimuli A and B, presented separately, do not.

An analysis of experiments on a number of species with respect to the problem of configuring brings Razran to the suggestion that fish and amphibians probably can not configure, but birds and mammals can. In fish and amphibians the conditional reaction elicited by the compound seems to be the result of the conditional reactions elicited by each of the two elementary stimuli. In birds and mammals the compound appears to be different from the sum of the two elementary stimuli: the compound elicits the conditional reaction; the elementary stimuli presented separately, do not.

Research with respect to the transition from simple instrumental learning to configuring suggests a similar dividing line between fish and amphibians on the one hand, and birds and mammals on the other hand. In birds and mammals a compound discriminative stimulus is something else than the sum of the two elementary discriminative stimuli out of which it is composed; in fish and amphibians the compound is equal to the sum of the elementary stimuli. According to Wickelgren (1979) configuring is the manifestation of a more general development from an exclusively horizontal associative memory to an associative memory which is vertical as well as horizontal. The verticalization is produced by means of a chunking process: free neurons are going to represent the compound. While, for instance, in simple classical conditioning the compound stimulus is represented by a horizontal association between the neurons which represent the elementary stimuli, in configuring a free neuron (or a group of free neurons) is going to represent the compound via vertical associations with the neurons representing the elementary stimuli. Wickelgren suggests that in this chunking process the hippocampal (limbic) arousal system plays an essential role.

In neodarwinian evolution theory the environment often is looked upon as the selector and the genetic material as what finally is selected, although via the detour of the phenotype. This approach may be correct with respect to the genetic material which determines phenotypic characteristics of structures and processes in plants and, perhaps, in animals. However, if one has to do with genetic material which is related to behavior, one should proceed carefully. The evolution of behavior has its complications, which make the study of the evolution of behavior far more interesting than the evolution of, for instance, form and color features.

The unit of information discussed before, which consists of stimulus, action, and consequences of that action, can be described as a series of events in this sequence. First, there is a stimulus; to that the organism reacts with an action; that action has specific consequences. If the unit of information has its origin in learning, the consequences of the action will have strengthened the association between the stimulus and the action as a positive reinforcer. If the consequences of the action promote the survival of the genetic material that has to do with the unit of information, there will be selective pressure in favor of the survival of the prerequisites for the unit of information. This description may be correct as far as it concerns the behavior of lower animals, but it is too simplistic when it concerns higher organisms.

According to Wickelgren the phase in the anagenesis of information processing capability, which we discussed before in relation to the phenomenon of configuring, is also characterized by the development of expectancy learning. Organisms in which that phase of anagenesis is present appear to hold expectations with respect to the consequences of their actions; these expectations can develop in the course of learning. Eventually, these expectations become manifest by means of contrast effects. Powers (1978) suggests "that the inputs to an organism are affected not only by extraneous events but possibly by the organism's own action (p. 422). Organisms can develop into goal-directed systems, i.e. systems in which the guiding principle is control of input.

From the point of view of evolution theory this development, in which consequences of action are going to be evaluated against expected consequences, has interesting implications. Because of the availability not only of consequences of action but also of expected consequences of action, and because of eventual discrepancies between consequences and expected consequences, correction of action becomes possible. As a result of that, selection pressure must shift from the unit of information itself to the norm-value ("Sollwert"), against which the consequences of the action in the unit of information are evaluated. To the extent that this occurs, the role of the environment as selector will be limited. The organism is going to take the initiative in selecting the environment and by doing so, in a way it becomes director of evolution instead of solely being directed by evolution.

I would like to illustrate this thought with the example of Darwin's finches on the Galapagos Islands, which once were only one species probably originating in Latin America, but nowadays are several species, each one with its own ecological niche. The specific ecological niche goes together with a particular form of the beak, which must enable the animals to obtain their species-specific food. Imagine that in the past finches looking for food, and perhaps guided by genetically determined food preferences having a high degree of intraspecific variability, arrived at a particular ecological niche. In order to obtain food certain skills were required. For some animals that skill had to do with cracking a hard shell, for other animals it had to do with catching insects on leaves, for yet other animals it had to do with pecking out insects out of the bark of trees. The degree to which the skill became successful was also dependent on the form of the beak. The initial preference and the skill related to that preference therefore started to function as a selector with respect to the form characteristics of the beak. Here is Darwinian (and - it has to be emphasized - not Lamarckian) evolution, therefore, with the organism's own preference and behavior, and not the external environment, as a selector.

In this respect Popper (1973,1974) presents an interesting hypothesis, which he labels the hypothesis of the genetic dualism. This hypothesis originates in the question of why random mutations do not lead to random development, or, to put it in another way, why anagenesis is possible. From where does the "goal-directedness" come, which is so striking in the evolution of organs, limbs, form characteristics, etc.? Popper considers it meaningful to distinguish - at least in higher organisms - between a behavior-controlling part (especially the nervous system) and an executive part (like the sense organs, limbs etc.). Within the behavior-controlling part he wants to distinguish between the preference structure, which controls the behavior of the organism, and the skill structure. He assumes that each one of these parts - preference structure, skill structure, and executive structure - is under genetic control. The question then is whether mutations in any one of these three groups of genes are equally relevant. That question Popper answers in the negative. A mutation in the genes that underlie the executive structure does not make sense if the executive structure, changed by that mutation, cannot be used adequately. On the other hand, mutations in genes that determine the preference structure eventually will lead to a selection pressure on the skill structure and via the skill structure to a selection pressure on the executive structure. Popper compares this situation with that of an airplane guided by an automatic pilot. Mutations in the airplane that do not go together with adequate mutations in the automatic pilot will always, or very often, be disadvantageous, resulting in the destruction of the airplane and the automatic pilot. Mutations in the automatic pilot are, so to speak, held in check by properties of the airplane. Given, however, a specific mutation in the automatic pilot, particular mutations of the airplane will do better than others. This will result in a gradual evolution of the airplane in the direction of the capacities or peculiarities of the automatic pilot.

Popper summarizes his hypothesis in an elegant way in his intellectual autobiography: "At first sight Darwinism (as opposed to Lamarckism) does not seem to attribute any evolutionary effect to the adaptive behavioural innovations (preferences, wishes, choices) of the individual organism. This impression, however, is superficial. Every behavioural innovation by the individual organism changes the relation between that organism and its environment: it amounts to the adoption of or even to the creation by the organism of a new ecological niche. But a new ecological niche means a new set of selection pressures, selecting for the chosen niche. Thus the organism, by its actions and preferences, partly selects the selection pressures which will act upon it and its descendants. Thus it may actively influence the course which evolution will adopt. The adoption of a new way of acting, or of a new expectation (or "theory"), is like breaking a new evolutionary path. And the difference between Darwinism and Lamarckism is not one between luck and cunning, as Samuel Butler suggested: we do not reject cunning in opting for Darwin and selection" (p. 180).

By associating the concept of "function" with the concept of information processing, the concept of "function" gets a meaning which is different from the classical biological one. On the one hand, the function of information processing capacity refers to the contribution of that capability to the survival of the genetic material that pertains to that capability. In that sense, the concept of function refers to the passive adaptive reaction of the organism and its genetic contents to a dangerous environment. On the other hand, the function of the information processing capability can be described in terms of the expression of preference structures by means of behavior. In that sense the concept of function refers to the active adaptive action of the organism to a dangerous environment.

Psychology as biology of behavior can in a very particular and essential way contribute to the further theoretical development of neo-Darwinian evolution theory.

References

Farris HE (1967) Classical conditioning of courting behavior in the Japanese Quail (Coturnix coturnix japonica). Journal for the Experimental Analysis of Behavior 10:213-217

Garcia J and Koelling RA (1966) Relation of cue to consequence in Avoidance Learning. Psychonomic Science 4:123-124

Gottlieb G (1984) Evolutionary trends and evolutionary origins: relevance to theory in Comparative Psychology. Psychological Review 91:448-456

Hirsh R (1974) The hippocampus and contextual retrieval of information from memory: a theory. Behavioral Biology 12:421-444

Hollis KL (1982) Pavlovian conditioning of signal-centered action patterns and autonomic behavior: a biological analysis of function. In: Rosenblatt JS, Hinde RA, Beer C, Busnel MC (eds) Advances in the study of behavior, vol 12, Academic Press, New York, p 1-64

LoLordo VM (1979) Constraints on Learning. In: Bitterman ME, LoLordo VM, Overmier JB, Rashotte ME Animal Learning: survey and analysis. Plenum Press, New York, p 473-504

Piaget J (1976) Le comportement moteur de l'evolution. Gallimard, Paris

Popper KR (1973) Objective Knowledge: an evolutionary approach. Oxford University Press, London

Popper KR (1974) Unended Quest. Fontana/Collins, London

Powers WT (1978) Quantitative analysis of purposive systems: some spadework at the foundations of scientific psychology. Psychological Review 85:417-435

Razran G (1971) Mind in Evolution: an East-West synthesis of Learned Behavior and Cognition. Houghton Mifflin Company, Boston

Seligman MEP (1970) On the generality of the laws of learning. Psychological Review 77:406-418

Shettleworth SJ (1972) Constraints on Learning. In: Lehrman DS, Hinde RA, Shaw E (eds) Advances in the study of behavior, vol 4, Academic Press, New York, p 1-68

Siegel S (1979) The role of conditioning in drug tolerance and addiction. In: Keehn JD (ed) Psychopathology in animals: research and clinical implications, Academic Press, New York, p 143-168

Testa TJ (1975) Effects of similarity of location and temporal intensity pattern of conditioned and unconditioned stimuli on the acquisition of conditioned suppression in rats. Journal of Experimental Psychology: Animal Behavior processes 104:114-121

Thorndike EL (1913) Educational psychology, Vol II. The psychology of learning. Teacher's College, Columbia University, New York

Wickelgren WA (1979) Chunking and consolidation: a theoretical synthesis of semantic networks, configuring in conditioning, S-R versus cognitive learning, normal forgetting, the amnestic syndrome, and the hippocampal arousal system. Psychological Review 86:44-60

A NEUROPHYSIOLOGICAL BASIS FOR THE HERITABILITY OF VERTEBRATE INTELLIGENCE

T. Edward Reed
Departments of Zoology and Anthropology University of Toronto
Toronto, Ontario, M5S 1A1, Canada

For this paper "intelligence" is assumed to include, but not to be limited to, "problem-solving ability." For a given problem, the quicker its correct solution, the higher, on the average, is the intelligence of the solver. The problems may range from the simplest possible, e.g., comparing the lengths of two lines, to complex analogical puzzles, and on to difficult "real-life" situations (which have solutions). This usage of "intelligence" recognizes that there is no unanimity on its definition, and that most investigators believe that it comprises several, or many, more-or-less distinct mental abilities. I will focus on problem-solving ability as one important, practical, and commonsense component of intelligence. It is "practical" because we can devise simple tests for it among contemporary humans, and "commonsense" because this is the aspect -- rapid problem solving -- that should have been important in vertebrate evolution, i.e., in getting enough to eat, avoiding being eaten, and reproducing.

The fact and the degree of the genetic determination -- heritability -- of intelligence are also central to this presentation. "Heritability" here is the proportion of the variation (variance) of a trait which is determined by heredity. (More precisely, here we want the heritability due to additive gene action -- the effect of heredity which is transmitted from parent to offspring: Falconer, 1981). There are three distinct kinds of evidence for the heritability of intelligence: (1) indirect evidence from vertebrate evolution, (2) direct studies on contemporary humans, and (3) an a priori argument from the effect of genetic diversity (among individuals of a species) on the speed of information processing. The evolutionary evidence is the general increase in brain size (after correcting for body size) in the phylogenetic progression from fishes to mammals and, in the mammals, from monotremes to primates, and in the primates, to Homo sapiens. This argument accepts the data and views of Jerison (1973, 1982) that brain size, as corrected for body size -- encephalization --, is correlated with and evidence for a level of intelligence. The contrary view of Macphail (1982), that intelligence (as measured, primarily, by learning tests on modern vertebrates) did not increase, is rejected. (It should be noted that in most of the learning tests he cites, speed of learning was not measured). It is further assumed that the increase in intelligence is due to natural selection. It should have been advantageous for a given individual in a vertebrate species to be more intelligent, in a problem-solving sense (e.g., in quickly finding

NATO ASI Series, Vol. G17
Intelligence and Evolutionary Biology
Edited by H.J. Jerison and I. Jerison

food and mates and avoiding dangerous situations), than other individuals. In order for there to be such selection, intelligence must necessarily, to some degree, be heritable.

Evidence for the heritability of intelligence of modern humans is vast (e.g., Scarr & Carter-Saltzman, 1982). Here I note only that the largest family study of intelligence, the Hawaii Family Study of Cognition (De Fries et al., 1979), which used 15 different intelligence tests on parents and offspring (830 families of Americans of European ancestry, 305 families of Americans of Japanese ancestry), found significant regressions of mid-child (mean of the children of a family) on mid-parent (mean of the two parents) for each test. The regressions for the overall mean (first principal component) were 0.62 and 0.43 for AEA and AJA, respectively. The regressions for Progressive Matrices, a non-verbal test of pattern similarities, a kind of analogical problem-solving, were 0.60 and 0.28, respectively. If there were no effect of environment on human intelligence -- which is not the case -- these regressions would be the heritabilities for these tests. To see the effect of environment on family correlations in intelligence, we can examine results of the exhaustive survey (111 studies) of Bouchard and McGue (1981). Monozygous twins reared apart (100% common heredity, very little common environment), have a weighted mean correlation (3 studies) of 0.72, not much below the corresponding value of 0.86 for MZ twins reared together (34 studies). An adopting parent and his/her adopted children share no common heredity but do have a common environment. Six adoption studies give a weighted correlation of 0.19 (significantly above zero) but this is below the value of 0.42 (32 studies) for a parent and his/her biological offspring. Common environment clearly affects familial correlations in IQ scores but not overwhelmingly so. The heritability (for additive genes) of intelligence in Western Caucasoids is usually considered to be around 0.5 (Teasdale & Owen, 1984; T.J. Bouchard, pers. comm.).

A theoretical a priori argument for the heritability of intelligence appears to have been proposed first by Reed (1984); this argument used human examples. It begins with the great amount of human genetic diversity; half or more of structural gene loci are polymorphic (having two or more common alleles -- genes -- at that locus) (Nei & Roychoudhury, 1982). This genetic diversity produces marked genetic differences -- reflected in protein (and enzyme) diversity -- among individuals within an ethnic (racial) group. A second level of genetic control, in the amount of a given protein, also operates. For example, the levels of enzymes involved in the synthesis and metabolism of neurotransmitters are genetically determined (Winter et al., 1978; Weinshilboum, 1979). This genetic variation in kind and amount of proteins and enzymes must affect the speed of transmission of nerve impulses along nerve fibers (dendrites and axons) and across synapses between nerve cells. This prediction is borne out by studies in mice of peripheral nerve conduction velocity (Hegmann et al., 1973; Reed, 1983; T.E. Reed, unpubl.), which show significant additive heritability of conduction velocity. This argument also predicts that a similar genetic determination occurs in the brain, affecting the speed of impulse transmission -- which must affect the speed of information processing -- and therefore having consequences for intelligence. Because of similar biochemistry and genetic diversity, this argument should apply to vertebrates in general.

The question, therefore, arises as to exactly what is there about variation in intelligence which is heritable. It was suggested by Reed (1984) that speed of information processing (SIP) could fill this role; more accurately, variations in SIP, which are heritable as discussed above, can account for an important fraction of the variations in intelligence. Note that this focus on variation, which presently is unavoidable, by-passes the very large question of what intelligence actually is. Although the SIP involved in cognition cannot now be studied with any generality, one specific type of SIP, reaction time (RT), has been extensively investigated, both for itself and for its relation to human intelligence. After considering RT and this relation, it will be the goal of this paper to dissect RT into its major components, as quantitatively as possible, and to show that two of these components play a major part in information processing speed in the brain and are also very likely heritable.

Reaction Time

The concept of RT -- the time from first seeing or hearing a stimulus to the time of the first response -- goes back more than 160 years. Over a century ago, Francis Galton suggested that RT is related to intelligence. Many investigators since then have studied RT and its relation to intelligence; this history has been reviewed by Jensen (1982). I will follow his account and approach below.

As measured by Jensen and his collaborators, RT is the time, in milliseconds (msec), from the turning on of a light stimulus to the time the subject removes his index finger from a "home" button (in order to press a button by the light). The light and buttons are standard distances apart on a test apparatus; times are measured electronically. Simple RT is measured when there is only one light to watch; choice RT occurs when there are two or more lights to watch so that a decision must be made as to which button must be pressed after removing the finger from the home button. On this apparatus, with a standard test paradigm (with warning signal and repeated trials), the mean simple RT for normal subjects is about 300 msec while the mean choice RT (with 8 lights to watch) is about 350 msec (Jensen, 1982, Fig. 10). Other RT paradigms, with more complex procedures, may take up to four times longer (Vernon & Jensen, 1984). These various RT paradigms have consistently shown an inverse relation between RT and intelligence, more intelligent persons usually having shorter RTs (Jensen, 1982; Vernon, 1985). The correlations with intelligence (IQ) of single RT tests typically are in the range of -0.2 to -0.5 while multiple correlation coefficients for several different RT tests with IQ may be higher, -0.6 to -0.7 (Vernon, 1985). More complex RT tests (with long mean RTs) are usually more highly correlated with IQ than are simple RT tests (Vernon, 1985).

The above results make it very likely that variation in intelligence, as measured by various standard IQ tests, is to some degree a simple consequence of variation in speed of information processing (SIP) as measured by RT scores.

We can now begin to partition the time required for a simple RT task into its major components. For convenience and explicitness we may use 300 msec as a mean RT, this being the minimum mean time of the various RT tests considered above. The "input time", from eye to the visual cortex of the brain, is about 85 msec (Kaufman

& Williamson, 1986, p.96) while the "output time", from motor cortex to finger tip, is about 20 msec (conduction time from motor cortex to wrist is 17 msec (Rossini et al., 1986, p.78) and wrist to index finger tip is about 3 msec). This leaves about 200 msec for total processing time in the various regions of the brain involved in RT response. This figure includes times within the various cerebral cortices (including visual, association, pre-motor, motor) and sub-cortical nuclei, and connection times between these regions.

Cortical Conduction Times

If there were a single "path" followed by the stimulus, from input to output, one might, in theory, trace it, noting the time spent in each region. Since this is not possible, the path (or paths) and times not being known (Martin, 1984, p.277; Gevins, 1986, p.424), we can try something much simpler: estimate, to order of magnitude, the input and output times of a typical mammalian cortical neuron which may be in the "path." We may neglect synaptic delay between axons and dendrites in the following rough calculations because this time is very probably less than one msec (e.g., Hubbard et al., 1969, p.164). The input time of a cortical neuron is defined here as the time from the first synaptic input to its dendrites to the time (after summation of the many excitatory post-synaptic potentials) of the beginning axonal action potential. This time, the dendritic conduction time (DCT), in an experimental environment, can vary greatly, from 2 to 100 msec, depending on the strength of the excitation (Connors et al., 1982, Figs. 3 and 4). For a physiological stimulus 10 msec seems a reasonable rough mean value for the DCT. This is approximately the time for a flashing visual stimulus to cause a visual cortical neuron to reach a maximal firing rate (Martin, 1984, p.262) and is also a rough mean DCT for the experimentally stimulated olfactory bulb (Freeman, 1976, Fig. 4A).

The cortical neuron output time, which is the axonal conduction time (ACT) of the action potential to the dendrites of other neurons, is less well-known but its order of magnitude can be estimated. Based on the diameter of cortical axons, 0.5 - 1 um (Feldman, 1984, p.161), the conduction velocity of action potentials is thought to be 1 - 10 m/sec (Martin, 1984, p.272). The mean distance to the dendrites of local neurons is thought to be about 0.5 mm (Peters, 1985, Fig. 9) (although lengths up to 8 mm are known: Feldman, 1984, p.166) so that the local ACT is usually 0.5 - 0.05 msec. Pyramidal cells, which are about 70% of the cortical neurons, in addition to sending axons to local neurons, usually send axons to more distant neurons, either in the same cortical region or in different regions. This is well-established for the different visual areas of the cerebral cortex (Rosenquist, 1985, p.103). It is relevant to note here that these connections between different visual areas (in the same cerebral hemisphere) are reciprocal (in both directions). There are further reciprocal connections between visual areas of different hemispheres (Rosenquist, 1985, p.108). These distant axonal paths might vary from, say, 2 to 100 mm in length. If a mean distant axonal length is about 20 mm, say, and there is a mean conduction velocity of 5 m/sec, the mean distant ACT is of the order of 4 msec [O(4 msec)]. Since a given pyramidal cell is usually simultaneously communicating with both local and distant cortical neurons, the distant ACT time, O(4 msec), is the one to compare with the DCT, O(10 msec).

These calculations, although very rough, indicate that the dendritic conduction time of a typical cortical neuron is of the same order of magnitude as its axonal conduction time. This conclusion is not universally acknowledged (e.g., Sutherland, 1986; he presents no data for assuming that most brain processing time is in ACT). If this is correct, we should consider both conduction processes in the cortex in order to understand how these two neurophysiological parameters may relate to the heritability of intelligence.

Reasons Why Cortical Conduction Times Should Be Heritable

In the Introduction reasons were given for believing that the extensive genetic variability found in almost all vertebrate species, together with the genetic determination of proteins and enzymes involved in nerve impulse transmission, should produce appreciable heritability of nerve conduction velocity (CV). This expectation was realized in each of three studies of peripheral CV in mice. The heritabilities (due to additive gene action, parent to offspring) are about 0.2. There is good reason to believe that the corresponding heritabilities in humans would be appreciably greater because humans are very probably more genetically variable than the mice used (HS strain; Reed, 1977).

The theoretical reasons discussed above for the existence of heritability of CV should apply equally to peripheral axons and to axons of the central nervous system (CNS), in particular to cortical axons. The biochemical and biophysical mechanisms for propagation of action potentials are the same in both systems. There are no data on CNS CV to test this expectation but the demonstration of CV heritabilities in the PNS leads to a strong expectation of CV heritability in the CNS.

Dendritic conduction differs from axonal conduction in being passive (without an inward Na+ current to produce the self-sustaining action potential) and in taking place in an unmyelinated nerve fiber. Each active synapse on a dendrite of a cortical neuron produces a small post-synaptic potential (PSP) which is passively conducted along the dendrite, decreasing in magnitude with distance travelled. If a large number of such PSPs reach the cell body, the temporally and spatially summated potential may be sufficient to produce an action potential in the axon. The speed of dendritic conduction, at a given site, will depend on the cross-sectional area of the dendrite and the chemical composition of the dendrite. This composition is directly under genetic control, both in kind and amount of constituents. The area also, perhaps less directly, should be under some genetic control. For example, the cross-sectional areas of caudal nerves of a line of mice selected for increased CV were larger than the areas of a line selected for low CV (Hegmann, 1979, p.168). Genetic diversity, producing protein diversity, will produce variations in dendritic composition and size and, consequently, produce some genetic determination of dendritic conduction velocity. There are no data to test this expectation directly but there are good indirect data which support it, namely twin studies of electroencephalogram (EEG) patterns. EEG waves are summated dendritic currents (Freeman, 1981, p.574). The resting wave patterns (amplitude and frequency) are under strong genetic control; identical (monozygous) twins, which have exactly the same genes, have very similar patterns while fraternal (dizygous) twins, each of whom shares half of the other's genes, usually have quite different patterns (Propping, 1977).

Conclusions

The partitioning of simple visual reaction time into input time, output time, and brain processing time (BPT), shows that BPT is about two-thirds of the total RT. For choice or more complex RTs BPT increases and so does the proportion. BPT can be divided into axonal conduction time and dendritic conduction time. These two times are within an order of magnitude of each other so that both are important components of BPT.

On strong theoretical grounds, stemming from genetic diversity and genetic control of protein synthesis, both ACT and DCT and their sum, BPT, are expected to be heritable. The expectation for ACT is supported by the demonstration of heritability of peripheral conduction velocity while the expectation for DCT is supported by the heritability of EEG patterns. Since RT is mainly composed of BPT, it should also be heritable. One small study of RT (McGue et al., 1984) in twins reared apart indicates that it is indeed heritable.

Since RT is negatively correlated with intelligence (IQ score), BPT and its components, ACT and DCT, should also be negatively correlated. ACT and DCT, consequently, are two neurophysiological parameters which probably account for an important part of the variation in IQ and also are very likely to be under appreciable genetic control. For these reasons it appears that we are now beginning to see the neurophysiological basis for the heritability of intelligence. This new understanding should have implications in a number of disciplines.

Acknowledgements

For comments and discussion I thank Professors O.-J. Grusser, T.J. Bouchard, M.C. Diamond, W.J. Freeman, R.A. Nicoll, E.W. Larsen, and S.J. Shettleworth. This research was supported by the Natural Sciences and Engineering Research Council of Canada.

References

Bouchard T J & McGue M 1981 Familial studies of intelligence: A review. Science 212: 1055-1059

Connors B W, Gutnick M J & Prince D A 1982 Electrophysiological properties of neocortical neurons in vitro. Journal of Neurophysiology 48: 1302-1320

De Fries J C, Johnson R C, Kuse A R, McClearn G E, Polovina J, Vandenberg S G & Wilson J R 1979 Family resemblance for specific cognitive abilities. Behavior Genetics 9: 23-43

Falconer D S 1981 Introduction to Quantitative Genetics, 2nd ed , p 113 Longman, London and New York

Feldman M L 1984 Morphology of the neocortical pyramidal neuron. In (A Peters & E G Jones, eds) Cerebral Cortex, Vol 1 Cellular Components of the Cerebral Cortex, pp 123-200 Plenum, New York

Freeman W J 1976 Quantitative patterns of integrated neural ability. In (J C Fentress, ed) Simpler Networks and Behavior, pp 280-296 Sinauer Association, Sunderland, Massachusetts

Freeman W J 1981 Physiological hypotheses of perception. Perspectives in Biology and Medicine 24: 561-592

Gevins A S 1986 Quantitative human neurophysiology. In (H J Hannay, ed) Experimental Techniques in Human Neuropsychology, pp 419-456 Oxford University Press New York.

Hegmann J P, White J E & Kater S B 1973 Physiological function and behavioral genetics. II Quantitative genetic analysis of conduction velocity of caudal nerves of the mouse, Mus musculus. Behavior Genetics 3: 121-131

Hegmann J P 1979 A gene-imposed nervous system difference influencing behavioral covariance Behavior Genetics 9: 165-175

Hubbard J I, Llinas R & Quastel D M J 1969 Electrophysiological Analysis of Synaptic Transmission. Edward Arnold, London

Jensen A R 1982 Reaction time and psychometric g. In (H J Eysenck ed) A Model for Intelligence, pp 93-132 Springer-Verlag, New York

Jerison H J 1973 Evolution of the Brain and Intelligence. Academic Press, New York

Jerison H J 1982 The evolution of biological intelligence. In (R J Sternberg, ed) Handbook of Human Intelligence, pp 723-791 Cambridge University Press, New York

Kaufman L & Williamson S J 1986 The neuromagnetic field In (R Q Cracco & I Bodis-Wollner, eds) Evoked Potentials. Frontiers of Clinical Neuroscience, Vol 3, pp 85-98 Alan R Liss, New York

Macphail E M 1982 Brain and Intelligence in Vertebrates. Oxford University Press, Oxford

Martin K A C 1984 Neural circuits in cat striate cortex. In (E G Jones & A Peters, eds) Cerebral Cortex, Vol 2 Functional Properties of Cortical Cells, pp 241-284 Plenum, New York

McGue M, Bouchard T J, Lykken D T & Feuer D 1984 Information processing abilities in twins reared apart. Intelligence 8: 239-258

Nei M & Roychoudhury A K 1982 Genetic relationship and evolution of human races. In (M K Hecht, B Wallace & C T Prance, eds) Evolutionary Biology, Vol 14, pp 1-59 Plenum, New York

Peters A 1985 The visual cortex of the rat. In (A Peters & E G Jones, eds) Cerebral Cortex, Vol 3 Visual Cortex, pp 19-80 Plenum, New York

Propping P 1977 Genetic control of ethanol action on the central nervous system. Human Genetics 35: 309-334

Reed T E 1977 Three heritable responses to alcohol in a heterogeneously mated mouse strain: Inferences for humans. Journal of Studies on Alcohol 38: 618-632

Reed T E 1983 Nerve conduction velocity in mice: A new method with results and analysis of variation. Behavior Genetics 13: 257-265

Reed T E 1984 Mechanism for the heritability of intelligence. Nature 311: 417

Rosenquist A C 1985 Connections of visual cortical areas in the cat. In (A Peters & E G Jones, eds) Cerebral Cortex, Vol 3 Visual Cortex, pp 81-117 Plenum, New York

Rossini P M, Marciani M G, Caramia M, Hassan N F & Cracco R Q 1986 Transcutaneous stimulation of motor cerebral cortex and spine: Noninvasive evaluation of central efferent transmission in normal subjects and patients with multiple sclerosis. In (R Q Cracco & I Bodis-Wollner, eds) Evoked Potentials Frontiers of Clinical Neuroscience, Vol 3, pp 76-84 Alan R Liss, New York

Scarr S & Carter-Saltzman L 1982 Genetics and intelligence. In (R J Sternberg, ed) Handbook of Human Intelligence, pp 792-896 Cambridge University Press, New York

Sutherland S 1986 Parallel distributed processing. Nature 323: 486

Teasdale T W & Owen D R 1984 Heredity and familial environment in intelligence and educational level -- a sibling study. Nature 309: 620-622

Vernon P A & Jensen A R 1984 Individual and group differences in intelligence and speed of information processing. Personality and Individual Differences 5: 411-423

Vernon P A 1985 Individual differences in general cognitive ability. In (LC Hartlage, ed) The Neuropsychology of Individual Differences, pp 125-150. Plenum, New York

Weinshilboum R M 1979 Catecholamine biochemical genetics in human populations In (X O Breakfield, ed) Neurogenetics: Genetic Approaches to the Nervous System, pp 257-282 Elsevier, New York

Winter H, Herschel M, Propping P, Friedl W & Vogel F 1978 A twin study on three enzymes (DBH, COMT, MAO) of catecholamine metabolism. Psychopharmacology 57: 63-69

BRAIN, MIND AND REALITY: AN EVOLUTIONARY APPROACH TO BIOLOGICAL INTELLIGENCE

Michel A. Hofman
Netherlands Institute for Brain Research
Meiberdreef 33
1105 AZ Amsterdam ZO, The Netherlands

Biological Intelligence: A Hypothesis

Organisms are faced during their lives with an immense variety of problems, ranging from purely physical ones, such as changes in climate or geomorphic disturbances, to organism-specific problems related to food supply, predation, homeostasis, reproduction, etc. In order to enhance their chances of survival, living organisms have to find adequate solutions for the problems with which they are confronted, for any of them could easily be fatal. Problem solving, in other words, is an essential dynamic survival mechanism, evolved to cope with disturbances in the ecological equilibrium. It can therefore be looked upon as an adaptive capacity enabling organisms to adjust themselves to one another and to their physical environment.

The kind of problems with which organisms are confronted, however, and their relative significance, varies from one species to another, according to the ecological niche or adaptive zone that it occupies. These specific environmental challenges form the selection pressures that have given rise to the evolution of species-specific neural mechanisms and action patterns. According to this view, phototropic effects in plants and avoidance reactions in protozoans, as well as the linguistic skills of modern man, must be seen as species-specific mechanisms evolved in accordance with specific environmental demands. This means that all organisms, plants as well as animals, including human beings, are problem solvers, and that the problem solving capacity of a species or population reflects the temporal and spatial complexity of its environment. Consequently, the ability to solve problems will manifest itself in all those situations in which subjects are required to respond adequately to novel objects and changing circumstances, as well as in situations in which successful adaptation involves the detection of an appropriate response to regularities in the external world or the formation of rules and hypotheses (Popper & Eccles, 1977; Popper, 1982; Macphail, 1982; Griffin, 1984). Environmental adaptation, therefore, can be considered to be the primary function of problem solving in that it serves as a pre-eminent mechanism for survival.

The organism's adaptability, however, is but one aspect of its fitness. Free-moving organisms, for example, can actively explore their environment, and thus ge-

NATO ASI Series, Vol. G17
Intelligence and Evolutionary Biology
Edited by H.J. Jerison and I. Jerison

nerate new selection forces which can modify the structures involved. Mayr (1982, p. 612) even argues that "many if not most acquisitions of new structures in the course of evolution can be ascribed to selection forces by new acquired behaviors." This implies that in higher organisms, such as vertebrates, behaviour rather than environmental change is the major driving force for evolution at the organismal level. However, this does not detract from the fact that all organisms, whether they are simple reflex automata or active and complex explorers, are above all concerned with keeping track of their local spatio-temporal environment, as part of their struggle for existence. Since sensory information processing and the ability to model reality (or certain parts of it) are essential components in this process, our idea of problem solving seems to correspond reasonably well to the notion of intelligence. However, the common use of the term "intelligence" applies not only to processes involving complex information processing (perception) but includes operations of the "mind"' as well (see e.g., Macphail, 1982, 1985; Dennett, 1983; Griffin, 1984, 1985). This means that if no thoughts, intentions, expectations and the like can reasonably be attributed to a group of organisms, they are considered to be creatures lacking in intelligence. In order to avoid such subjective criteria, intelligence in the present essay is defined as the problem solving capacity of a species. It implies that intelligence is not a quality restricted to the functional domain of complex neural structures, but applies even to unicellular organisms. It is therefore fitting to speak of *biological* (instead of animal) intelligence, which at once distinguishes it from the problem solving capacity of machines, also known as *artificial* intelligence.

Although each organism is equipped with an execution potential which enables it to cope with a variety of problems in a specified environment, there are, of course, tremendous differences in the problem solving capacity among living beings, and thus in biological intelligence. Most of these differences are connected with the functional characteristics of the organism's "perceiving and executing apparatus". A coelenterate, for instance, with its diffuse nerve fiber network, has an action pattern which is of quite a different order of magnitude as compared to the rich behavioural repertoire of birds and mammals, with their highly evolved central nervous system. Therefore, it is only meaningful to compare the intellectual capacities of species when they have certain basic features in common. The mammalian brain can be considered to be such a structure in that it is a multimodal integration system composed of analogue neural units and connections. In these highly organized animals, information from the external world passes through three distinct stages or systems: a sensory transducer system, a perceptual input system (or systems) and finally a central cognitive system. During this "journey" the otherwise overwhelming amount of sensory information is selected, analyzed, integrated and stored in accord with the species' attention and its interests. This means that the picture that an animal has of its external world depends on (i) the quality of its sense organs, (ii) its information processing capacity and (iii) its informational and emotional states of mind. Consequently, "world pictures" of animals must perforce differ from each other, and can be looked upon as highly individual representations of the external world. Hence it is appropriate to speak of species-specific perceptual worlds (Jerison, 1973, 1979) or models of reality (Riedl, 1980; Plotkin, 1982; Wuketits,

1982, 1986). The idea of a species-specific model of reality corresponds, to a certain extent, to Kant's assumption that the world as we know it is our interpretation of the observable facts in the light of theories that we ourselves invent (see Popper, 1958). This means that models of reality, at least those of higher vertebrates, are related to both the external and the internal worlds and that we assume "the existence of some knowledge in the form of dispositions and expectations" (Popper, 1972, p. 71). The more complete and reliable these knowledge-based specific models are, the better the chance of survival. They enable the animal to make better predictions, especially predictions relating to features or situations which do not occur in stereotyped patterns.

If we now assume that biological intelligence in higher organisms is the product of processes of complex sensory information processing and mental faculties, responsible for the planning, execution and evaluation of intelligent behaviour, variations among species in intelligence must in principle be observable.

Brain and Intelligence

Before attempting to determine the underlying neural mechanisms of intelligence, we should have in mind a specific biological entity towards which to direct our attention. Conceiving biological intelligence as the capacity of an organism to construct an adequate model of reality implies that the spectrum of inquiry may range from the sensory receptor system to behaviour in its broadest sense (that is to say, overt activity as well as internal homeostatic action). Usually, however, valid comparisons at the extremes of the spectrum i.e., at the level of sense organs and complex behaviour patterns, respectively, are difficult to make in view of the very great sensory-motor differences that exist among species (see e.g., Passingham, 1981; Macphail, 1982). How to compare in mammals, for example, the sensory capacity of diurnal monkeys with stereoscopic vision with that of nocturnal echolocating bats, or the learning ability of terrestrial shrews with that of marine dolphins? Differences in intelligence may in fact be uncorrelated with *measurable* differences in overt behaviour, nor are such differences implicit in many learning situations, since both activities depend on the behavioural potential of the organism as well as on its internal state (attention, motivation, etc.; a starving rat, for instance, is probably not more intelligent than a satiated one!).

To avoid these formal problems, one should instead investigate the neural structure, where both perception and instruction take place, that is to say, the central nervous system (CNS) itself. A further argument for studying the CNS in relation to intelligence is that it is the neural substratum where the external world is interpreted and modelled, where concepts are formed and hypotheses tested, in short, where the physical world interacts with the mind. These properties make the CNS, and especially the brain an excellent, if not the most suitable, system for seeking the neural basis of biological intelligence. Since the primary function of the brain is to adequately interact with the external world, brain function can be most readily characterized by the manner in which the brain senses the physical environment and how it responds to it by generating motor actions. From experimental and theoretical studies it has become evident that the brain is a distributed parallel processor where most of the sensory information is analyzed in parallel involving large neuronal networks (Freeman, 1975; Ballard, 1986). The principle of parallel processing implies that activities of ordered sets of nerve cells can be considered to be mathematical vectors. An important aspect of this vector approach is that it focuses on the explanation of brain function in terms of neuronal networks, i.e., mass action, and that it is therefore compatible with the modular organization of the brain.

Despite these major developments to explain brain function in terms of mathematical vectors (or tensors), the attempts have so far been confined to sensori-motor operations (Pellionisz, 1986), whereas no current theory successfully relates higher brain functions to details of the underlying neural structure. This is hardly surprising in view of the enormous functional complexity of the mammalian brain. Instead, one may ask a more general question, one which is related to the evolution, development and function of neural information processing: Does the mammalian brain operate according to a general mechanism or principle in processing sensory signals and stored information, despite the manifold differences in brain subsystems and their interconnections? After all, brains must be able to process extremely complex information, but they must also have a simple enough underlying organization to have evolved by natural selection (Glassman, 1985). If the neuron can be regarded as the "atomic" unit of function in the nervous system, then the "molecular" unit of information processing is in a way akin to the module. In particular, the mammalian cerebral cortex has been found to be uniformly organized and to be composed of such modular units interacting over fairly short distances (Mountcastle, 1978; Szentagothai, 1978). It appears that the module for information processing in the cortex consists of a functional neuronal network having the capability of quite sophisticated spatial-temporal firing patterns. Their widespread occurrence, furthermore, qualifies them to be considered as fundamental building blocks in neural evolution (for reviews see Hodos, 1982; Hofman, 1985).

These processing units operate as prewired neural assemblies where individual neurons "are configured to execute fairly complex transactions analogous to an integrated circuit (i.c.) chip" (Russell, 1979, p. 151). It has become evident that modules integrate at higher levels of information processing as a result of the hierarchical organization of the brain, thus enabling the system to combine dissimilar views of the world. It implies that if we seek the neural basis of biological intelligence, we can hardly localize it in a specific region of the brain, but must suppose it to involve all those regions through whose activity, ideas, memories and emotions are experienced, and plans for action entertained. Perhaps this would even encompass the whole of the brain! In accordance with this view is the idea that a large brain is indicative of a high degree of biological intelligence. Gross brain size, however, is an insufficient (although necessary) variable if we wish to understand biological intelligence (Holloway, 1983), as can be deduced from the problem solving abilities of small-brained mammals. Jerison, therefore, developed a theorem in which he relates biological intelligence to the doctrine of encephalization (Jerison, 1973, 1983, 1985). The central idea upon which this theorem is based presumes that the nature of the evolution of intelligence is an aspect of the evolution of encephalization, that is, the evolution of the information processing capacity of the brain, beyond that required for routine bodily functions. Consequently, the residual processing capacity can be looked upon as the phenotypic expression of intelligence (see Jerison, this volume). Similar hypotheses have been formulated with respect to the mammalian cortex, in which the degree of corticalization is thought to be indicative of differences in intelligence among species (Hofman, 1982, 1984).

Recently, new light has been thrown on this matter by linking the modular concept to the idea of corticalization, in an attempt to explain the augmented information processing capacity of the mammalian cerebral cortex (Hofman, 1985). According to this theory, the neural processing units or modules of the cortex are wired together so as to form complex processing and distribution units, having spatial dimensions depending on the species' degree of corticalization. As a result, the structural complexity of these processing units increases with the evolutionary expansion of the cerebral cortex, and with that, their functional capacity. Analogue organizational principles are known from computer technology, where the achievements of complex systems are found to depend on emergent mechanisms (cf. species-specific neural processing units) rather than on the quantitative addition of units and interconnections with the same properties as found in simpler systems.

The conclusion can be reached that the *complexity* of neural processing units and the richness of the intermodular processes are at least as important to the information processing capacity of the cortex as the *number* of modules is. This means that the complexity of the neural circuitry of the cerebral cortex, in addition to corticalization, can be looked upon as a neural correlate of biological intelligence in mammals.

Evolution and Intelligence

Considering biological intelligence as the problem solving capacity of an organism makes it possible to speak of degrees of intelligence, and of its evolution from amoeba to man. But what does it mean precisely when one says that species differ in intelligence, or that vertebrates are in general more intelligent than invertebrates? It means that there are differences in the abilities of organisms to perceive and interpret the physical world. Biological intelligence can thus be conceived as to reflect the temporal and spatial complexity of the species' niche, without referring, however, to the kinds of situations organisms encounter in everyday life. It is, in fact, a measure of capacity, independent of the way the capacity is used, and it may be treated as a trait for "anagenetic" rather than "cladistic" analysis (Gould, 1976; Jerison, 1985). It implies that when distantly related species are comparable in their problem solving capacity, we should consider the species to be comparable in biological intelligence. Yet the near equality in intelligence may be based upon radically different adaptations. Since neural mechanisms and action patterns evolve in the contexts of the environments in which they are effective, and since species never occupy identical niches, "many and various intelligences (in the plural) must have evolved in conjunction with evolving environments" (Jerison, 1985, p. 29). In theory, each ecological niche requires its own degree of biological intelligence. That means that specific neural and sensorimotor adaptations always occur in relation to particular environments. A striking example is the mammalian brain, where the evolutionary changes in the balance of the sensory systems are the result of the adaptive radiation of species into many different ecological niches (Russell, 1979; Macphail, 1982). These sensory systems, like any other biological feature, could evolve as a result of natural selection, because any subject that forms inadequate representations of outside reality will be doomed by natural selection.

In this view, cognitive systems and emotional phenomena can also be considered to be the result of interactions between genetic aptitude and natural environment, as they have a number of biologically useful functions. One is to keep track of the individual's whereabouts in the world by constructing a schematic model of reality (Popper, 1982; Wuketits, 1986). It is evident that the mind, as an emergent property of sufficiently complex living systems, has its evolutionary history like any other trait that increases adaptation to the environment, and that its functions have increased with the evolution from lower to higher organisms (Popper, 1972). Although adaptation of an organism to its environment is the chief process directing biological evolution, with the evolution of intelligence organisms became more and more independent of their environments by modifying the environments according to their needs. This process culminated in the evolution of mankind, which can be understood only as a result of the interaction of two kinds of evolution, the biological and the cultural (Ayala, 1986).

Cultural evolution, however, being the emergent result of the evolution of mind, cannot dispense with biological preconditions; it builds on biological facts and faculties (Vollmer, 1984; Wuketits, 1986). Though cultural evolution indeed presupposes biological evolution, it is not fully explicable in terms of theories and methods of the latter, or as Wuketits (1986, p. 199) puts it, "Cultural evolution . . . has transgressed organic evolution and shows a certain autonomy". The special status of cultural heredity can be derived from the fact that most cultural innovations are devised precisely in order to meet the environmental challenges or to improve our models of reality, whereas biological evolution has a mindless, random character. It is appropriate therefore to distinguish adaptations to the environment due to cultural selection from those that take place by the selection of genotypes. Cultural inheritance, since it is based on the transmission of information through direct communication, and through books, the arts and the media makes it possible for "a new scientific discovery or technical achievement . . . [to] be transmitted to the whole of mankind in less than one generation" (Ayala, 1986, p. 253).

It is evident that the role of human language in the transmission of knowledge is extremely important, even so prominent and pervasive that "it is not possible to estimate human general intellectual capacity independent of linguistic capacity" (Macphail, 1982, p. 328). Its manifestations and, in particular, those of its newly acquired functions - description and argumentation - is the most peculiar phenomenon in human problem solving. While animals can communicate by expressing their inner state by means of their behaviour, and by signalling to conspecifics, (e.g., in case of danger), man is the only creature that is able to make true and false statements, and to produce valid and invalid arguments (see Popper, 1968; Popper and Eccles, 1977).

If we assume that part of the basis of human speech is inherited in the DNA then changes in the brain that permit the advantageous supplement of language acquisition to perception and communication would have had obvious selective advantages throughout the period of hominid evolution. We may conceive human language, therefore, as a superorganic form of adaptation, evolved not only as a cognitive adaptation contributing to the knowledge of reality of each individual, but also as a means of sharing and, even more importantly, influencing states of mind among conspecifics. Indeed, because of language, human beings not only are able to construct individual representations of the external world, but also can contribute to and learn from *collective* models of reality, that is, the cumulative experience of the whole of mankind. With its cognitive and linguistic skills *Homo sapiens* tries to know its world and even exerts itself to the utmost to control it. It is obvious that by virtue of language, human beings tend to have highly organized informational states of mind, and, consequently, are excellent problem solvers. But although knowledge of reality may be a necessary condition for survival, it is surely not enough. The degree of intelligence reached by a species does not determine the propensity of its reproductive success. This may be inferred from the indiscriminate elimination of millions of species through the eras, from ammonites to australopithecines. It means that though adaptability increases with the evolution of biological intelligence environmental catastrophes can always be fatal to a species. But not only external factors can threaten the existence of organisms; *Homo sapiens*, despite its impressive intellectual capacities, might in the end become the victim of its own mind by, paradoxically, creating problems that it is then unable to solve.

Concluding Remarks

I suggest, with Karl Popper, that all organisms are constantly engaged, day and night, in solving problems. Of course, most living creatures are unconscious problem solvers. Even human beings are not always conscious of the problems they are trying to solve. Being aware of these problem situations or not, living organisms must have fitting and relevant models of their specific environments in order to enhance their chances of survival. Consequently, the problem solving capacity of a species is assumed to reflect the temporal and spatial complexity of its ecological niche. My thesis is that biological intelligence can be considered to be a correlate of the problem solving capacity of a species, manifesting itself in the complexity of the species' model of reality. Though this view has its heuristic value, Jerison already pointed out that "it obviously adds significantly to the difficulties of a scientific analysis requiring objective tests" (1985, p. 29). This is true, in particular, when comparing levels of biological intelligence, because it may be impossible, even in principle, to equate the environments used in testing different species. Instead, one should try to find objective and quantitative measures of biological intelligence by studying, for example, the underlying neural substratum of intelligence among congeneric species, as has been done for mammals. In comparing levels of intelligence, however, we have to keep in mind that it is meaningless to speak of "worse" or "better" problem solvers, since the evolutionary history of "extant" species is, by definition, the story of successful problem solvers: species only differ in their problem solving *capacity*.

With the evolution of sensory systems as adaptations to specialized environments, the capacity to process large amounts of sensory information increased and, with that, the power to create more complex physical realities. The processing of large amounts of information originating from the various sense organs, and the construction of complex models of reality require a neural system that selects, integrates, stores and models: in other words, a system with mindlike properties that enables the organism to make sense of an otherwise chaotic world. But once we allow mindlike properties to come in, such as motivation, emotion, preference and anticipation, we must concede that it is not only the hostile environment which plays an organizing or designing role in the evolution of biological intelligence, but also the active search of an organism for a new ecological niche, a new mode of living (Popper, 1982). Since the mind, prehuman and human, takes a most active part in evolution and especially in its own evolution, hominization and the evolution of our linguistic world may have begun as a cultural adaptation to new ecological niches. As Marshack (1984, p. 503) puts it: "During hominization the increasing potential for open and variable behaviors was probably the crucial and significant adaptive mode being developed neurologically and morphologically to interface with a seasonally variable mosaic ecology" (see also Marshack, 1985). The process probably began at the time of hominid divergence as part of "the cognitive and manipulative adaptation" to what was in essence a more complex physical reality. It seems that with the evolution of biological intelligence in Homo sapiens the behavioural adaptive capacity of the organic world has reached its pinnacle.

Acknowledgements

I owe a special debt of gratitude to Michael A. Corner and Raymond H. Hofman with whom I have discussed some of the ideas presented in this essay.

References

Ayala FJ (1986) Booknotes. Biology and Philosophy 1:249-262

Ballard DH (1986) Cortical connections and parallel processing: Structure and function. The Behavioral and Brain Sciences 9:67-120

Dennett DC (1983) Intentional systems in cognitive ethology: The "Panglossian paradigm" defended. The Behavioral and Brain Sciences 6: 343-390

Freeman WJ (1975) Mass action in the nervous system. Academic Press, New York

Glassman RB (1985) Parsimony in neural representations: generalization of a model of spatial orientation ability. Physiological Psychology 13: 43-47

Gould SJ (1976) Grades and clades revisited. In: Masterton RB, Hodos W, Jerison HJ (eds) Evolution, Brain and Behavior: persistent problems. Erlbaum, Hillsdale, New Jersey, p 115

Griffin DR (1984) Animal thinking. Harvard Univ. Press, Cambridge, Massachusetts

Griffin DR (1985) The cognitive dimension of animal communication. Fortschritte der Zoologie 31: 471-482

Hodos W (1982) Some perspectives on the evolution of intelligence and the brain. In: Griffin DR (ed) Animal mind - human mind. Springer, Berlin Heidelberg New York, p 33

Hofman MA (1982) A two-component theory of encephalization in mammals. Journal of theoretical Biology 99: 571-584

Hofman MA (1984) Towards a general theory of encephalization. An allometric study on the evolution and morphogenesis of the mammalian brain. Academic thesis, Amsterdam

Hofman MA (1985) Neuronal correlates of corticalization in mammals: a theory. Journal of theoretical Biology 112: 77-95

Holloway RL (1983) Human brain evolution: a search for units, models and synthesis. Canadian Journal of Anthropology 3: 215-230

Jerison HJ (1973) Evolution of the brain and intelligence. Academic Press, New York

Jerison HJ (1979) The evolution of diversity in brain size. In: Hahn ME, Jensen C, Dudek B (eds) Development and evolution of brain size. Academic Press, New York, 29

Jerison HJ (1985) Animal intelligence as encephalization. Philosophical Transactions of the Royal Society, London B 308: 21-35

Macphail EM (1982) Brain and intelligence in vertebrates. Clarendon, Oxford

Macphail EM (1985) Vertebrate intelligence; the null hypothesis. Philosophical Transactions of the Royal Society, London B308:37-51

Marshack A (1984) The ecology and brain of two-handed bipedalism: An analytic, cognitive and evolutionary assessment. In: Roitblat HL, Bever TG, Terrace HS (eds) Animal cognition. Erlbaum, Hillsdale, New Jersey, p 491

Marshack A (1985) Hierarchical evolution of the human capacity: The paleolithic evidence (54th James Arthur lecture) American Museum of Natural History, New York

Mayr E (1982) The growth of biological thought. Diversity, evolution, and inheritance. Belknap Press, Cambridge, Massachusetts

Mountcastle VB (1978) An organizing principle for cerebral function: the unit module and the distributed system. In: Edelman GM, Mountcastle VB (eds) The mindful brain. MIT Press, Cambridge, Massachusetts, p 7

Passingham RE (1981) Primate specialization in brain and intelligence. Symposium of the Zoological Society, London 46:361-388

Pellionisz AJ (1986) Tensor network theory of the central nervous system and sensorimotor modeling. In: Palm G, Aertsen A (eds) Brain theory. Springer, Berlin Heidelberg New York (to be published)

Plotkin HC (1982) Evolutionary epistemology and evolutionary theory. In: Plotkin HC (ed) Learning, development, and culture: Essays in evolutionary epistemology. Wiley, Chichester, p. 3

Popper KR (1958) On the status of science and of metaphysics. Ratio 1: 97-115

Popper KR (1968) Conjectures and refutations. The growth of scientific knowledge. Routledge and Kegan Paul, London

Popper KR (1972) Objective knowledge. An evolutionary approach. Clarendon, Oxford

Popper KR (1982) The place of mind in nature. In: Elvee RQ (ed) Mind in nature. Harper and Row, San Francisco, p. 31

Popper KR, Eccles JC (1977) The self and its brain. Springer, Berlin Heidelberg New York

Riedl R (1980) Biologie der Erkenntnis: Die Stammesgeschichtlichen Grundlagen der Vernunft. P. Parey, Berlin

Russell IS (1979) Brain size and intelligence: A comparative perspective. In: Oakley DA, Plotkin HC (eds) Brain, behaviour and evolution. Methuen, London, p 126

Szentagothai J (1978) The neuron network of the cerebral cortex: A functional interpretation. Proceedings of the Royal Society, London B201: 219-248

Vollmer A (1984) Mesocosm and objective knowledge - On problems solved by evolu tionary epistemology. In: Wuketits FM (ed) Concepts and approaches in evolutionary epistemology: Towards an evolutionary theory of knowledge. D. Reidel, Dordrecht Boston, p 69

Wuketits FM (1982) Grundriss der Evolutionstheorie. Wissenschaftliche Buchgesell schaft, Darmstadt

Wuketits FM (1986) Evolution as a cognition process: Towards an evolutionary epistemology. Biology and Philosophy 1: 191-206

THE EVOLUTIONARY BIOLOGY OF INTELLIGENCE: AFTERTHOUGHTS

Harry J. Jerison
Department of Psychiatry and Biobehavioral Sciences
University of California, Los Angeles, Medical School
Los Angeles CA 90024 USA

Our Institute was organized to provide a two-week course of advanced study designed to cover a field as best we could within its time limits. My job was to recruit the main lecturers and a few of the participants. Other participants were self-selected according to their interests. Although it was not designed to be a conference, the Institute worked in many ways like a conference or symposium. This will be evident from these published proceedings. Rather than present the material to be covered as if it could be described by a syllabus about which there is broad agreement, I recognized that the subject matter is still in flux. It was important to present sometimes conflicting points of view and to encourage rather than avoid controversy. The title of the Institute was "The Evolutionary Biology of Intelligence," but a more modest title was chosen for this publication, acknowledging that there is as yet too little evolutionary biology of intelligence that is of the same grandeur as other areas of evolutionary analysis. There are intersections between the fields, however, as indicated in our revised title, and this Institute has been able to cover many of them.

My plan was to emphasize, in order, evolutionary biology, neurobiology, analytic issues (philosophy and artificial intelligence) and, finally, behavioral data from ethology and psychology. The order reflects my judgment of what people interested in intelligence know and ought to know about its biological foundations, with the least widely known disciplines given the highest priority.

Reprise

One unusual pleasure was hearing contributions on new and important research by participants whose contributions had not been arranged in advance as tutorial material. Terry Deacon's outstanding analysis of human brain evolution requires two chapters for a full report. Ian Deary's excellent review of research on human intelligence and Hans Peter Lipp's discussion of his work on brain-behavior genetics are other high points. Evolutionists will appreciate my delight with the presentation by Dieter Kruska of integrated data on the effect of domestication upon relative brain size. Here I could truly say, "Never since Darwin." Darwin's only contribution to evolutionary neurobiology, I think, was his observation that the brains of domestic

NATO ASI Series, Vol. G17
Intelligence and Evolutionary Biology
Edited by H.J. Jerison and I. Jerison

rabbits were smaller than those of their wild contemporaries. The same was recognized as true for laboratory versus wild rats, but Kruska's work is the first systematic analysis of this problem. I have some trivial quarrels with his methods of analysis, which are essentially those used by Stephan and his colleagues (see Kruska for citations), but the results are easily transformed to those produced by other methods. The important thing is to have the data.

In the area of evolutionary biology in the strictest sense Martin Pickford's tutorial review of the fossil evidence on primate evolution may be the most up-to-date available. To the extent that it varies from other accounts, I am persuaded that Pickford's position is superior. It was my privilege as editor to participate in his development of the intriguing hypothesis that the transition between prosimians and simians involved a specialized morphological trait. He conjectures that the enclosure of the eye's orbit in bone made eye movements more independent of jaw movements in anthropoid primates compared to their lemur-like relatives, and enhanced their capacities to use visual information. This is a new hypothesis, which suggests several experimental tests for vision researchers. There are interesting parallels between this conjecture and Philip Lieberman's analysis of the development of the voice-box area in humans to be functionally independent of constraints associated with eating and drinking.

Paul Harvey reviewed allometry, an evolutionary subject that is especially close to my personal interests. His contribution raised many questions that deserve more discussion, and I will spend a bit of time in a later section on some of these. This topic must be analyzed with a mathematical language, but once the "words" are mastered there is real pleasure in playing with the ideas.

Another contribution that compels comment is in the area of the relationship between brain and intelligence as reviewed by William Hodos. There are difficulties at many levels and much more misunderstanding and disagreement than I had expected. Also presented in a later section, my comments have to be fairly detailed: a kind of debate with Hodos, which I hope will contribute to a greater understanding of the issues and to some appreciation about how people who share much of the same information can reach rather different conclusions.

Deacon's contributions are also in the sphere of neurobiology, and Lipp's paper provides a genetic perspective. Michel Hofman chose to present a theoretical discussion of biological intelligence and evolution, although he is best known for his analysis of morphological encephalization. Aldo Fasolo and his co-author, G. Malacarne, writing on homologies in the nervous system, also surprised me with the depth of their concern with philosophical issues. All of our contributors included outstanding bibliographies, but I recommend theirs in particular for its unusual breadth.

Still under this heading, Jerre Levy's chapter is especially important. The functional asymmetry of the human brain has been known for more than a century, but the discovery of structural asymmetries, and even more, the demonstration of both structural and functional asymmetries in other species came much later. Levy, who pioneered evolutionary consideration of this issue, updates her analysis in this volume. Hers is also the most general discussion of animal language. The topic is treated in several other contributions (Lieberman, Poli, and notably Schusterman), but for comprehensive recent coverage see Weiskrantz, 1985, as well as Savage-Rumbaugh's 1986 monograph cited by Levy.

Schusterman and Gisiner's contribution requires special comment as well. These authors presented a detailed analysis on techniques of research with sea lions and dolphins with emphasis on new and unfamiliar data on marine mammals. Their chapter provides enormous detail on how one approaches the problem. It should alert serious students to difficult methodological problems that arise in this area. The authors devote considerable space to controversial issues concerning work by L. M. Herman on bottlenosed dolphins (see Schusterman and Gisener for citations). Since Herman was unable to attend the Institute, I asked him for comments, from which I will quote later in this chapter.

Philip Lieberman's analysis of the evolutionary biology of language is classic. His contribution here is a precis of his important, recently published, book on the subject. Another neurobiological contribution is Reed's, who tackles the thorny subject of heritability of intelligence in terms of reaction time and conduction velocity.

Michael Ruse and Paul Thompson, our philosopher-lecturers, covered the philosophical foundations of the analysis of evolution and intelligence in enough detail to lead readers into the literature. The problem is often one of disagreements on the nature and substance of the definition of intelligence. I am almost certain that this is what keeps Hodos and me from agreeing more completely.

Margaret Boden, a most lucid writer on computer science, turns out to have been trained in philosophy and psychology. From under her various hats Boden has given us a uniquely understandable introduction at an advanced level to the problem of applying research on artificial intelligence to biological intelligence. Her evaluations and descriptions provide reliable entries to recent developments in AI and the latest from the cognitive frontiers of parallel-distributed-processing (PDP). Look for Rumelhart in her bibliography.

Is Henry Plotkin after all another philosopher in disguise? Plotkin's essay is intriguing in its recognition of the interaction of the observer with the observed in natural phenomena, comparable to that noted in popular accounts of recent developments in physical theories. A genuinely original contribution.

In his discussion of information processing in nature Vossen profitably draws on Karl Popper, the arrogant intellectual Dutch uncle of evolutionary epistemology, which was Plotkin's theme. Popper may be the most important philosopher of science for evolutionists, though sometimes more for ill than for good. His notions about the importance of falsifiability in scientific theories have sent evolutionists, and particularly developers of cladistic doctrine (see Cracraft and Eldredge, 1979), scurrying for scientific virtue and purity. I recommend Popper's (1979) essays on "Clouds and Clocks" and "Evolution and the Tree of Knowledge" in addition to the works cited by Vossen. Michel Hofman has also mentioned some of Popper's contributions relevant for our symposium. The autobiography (Popper, 1986), from which Vossen quotes, clarifies some of his extreme criticisms of evolutionary theory.

In my introductory chapter I remarked that learning has been overemphasized in the analysis of intelligence. I was, nevertheless, both impressed and surprised by Bitterman's remarkable discoveries about the performance of honeybees in situations like those he has analyzed so successfully in vertebrate species. Despite the excellent reviews by Bitterman as well as Poli and Csanyi, I am not convinced that the major phenomena treated in the analysis of learning are not properly considered as cognitive phenomena: "What does an animal know about its world?" I am not ready to accept Poli's equation of intelligence with learning capability. I can, however, accept experiments on learning ability as controlled tests that can be related to an animal's knowledge of its world. In that sense, the literature on comparative learning abilities is as relevant for my somewhat idiosyncratic approach to our problem as for the more conventional approaches that emphasize learning ability.

And now for more controversial comments, first on allometry, because problems raised by Harvey also affect the other two chapters; next on Schusterman and Gisiner, primarily with respect to the controversy between them and Herman; and lastly on Hodos.

On Allometry

The point about Paul Harvey's chapter that concerns me is not the "discovery" that 3/4 rather than 2/3 is the true mammalian allometric exponent for brain-body relations but what one means by a true exponent. Harvey's position is essentially that the exponent must be determined empirically by statistical curve-fitting. This implies that the average for the available sample arises from the fundamental generating process that produces an allometric relationship and that deviations from allometry that occur both in the sample and the population that it represents are fundamentally random. Furthermore, Harvey implies that theoretical analyses were suggested by the empirically determined exponents, so that when the 2/3 exponent was discovered empirically this led to interest in surface-volume relations, whereas, following its correction to 3/4, interest shifted to the Kleiber (1947) function in which such a relationship was found between metabolic rate and body size.

This account of the relation between empirical data and theory is not quite right for brain-body analysis, although it is correct for the later formalization of allometric analysis initiated by Huxley's (1932) monograph on relative growth. It is true that the relationship to metabolic rate was emphasized recently when the 3/4 exponent was discovered, but the brain-body problem was seen as a surface-volume problem when it was first addressed quantitatively (Brandt, 1867). The first derivation of a formula that has since been called allometric was by Snell (1891), who recognized that surface-volume relations were fundamental for the organization of the nervous system. It was for that reason that I have used an exponent of 2/3, an exact number, and have tried to correct editors when they changed it to 0.67, a number with two significant figures, in printing my work.

Actually, if the true theoretical exponent is 2/3 the empirical exponent should be about 3/4, for an interesting reason. The theoretical exponent is based on the simplifying assumption that all mammalian brains are identical in all respects except size and that external sensory and motor information is received on an extended (abstractly defined) peripheral sensory surface that is estimated by the body surface. The peripheral sensory surface is then viewed as a map that is projected and repeated in the brain proper, and the brain's size is determined primarily by the sum of the surfaces required for the maps. The basic surface may be thought of as retina, basilar membrane, skin, etc., the *area* of which is estimated from body weight or volume. For dimensional reasons that body measure must be raised to the 2/3 power to convert it into a surface, but the other constants must always be determined empirically (see Jerison, 1977).

The shift from a theoretical 2/3 exponent to an expected empirical exponent of 3/4 arises because biological surfaces are really thin sheets that are 3-dimensional and not 2-dimensional. Of the "sheets," the most extensive in the mammalian brain is the neocortex, and although the simplifying assumption in the theory is that projections from the abstract body to the abstract brain are true surface-to-surface mappings, much of the real world mapping is to and within the neocortex, the average thickness of which across species varies approximately with the 1/6 power of brain size. One must go through a bit of dimensional algebra at this point, but the result of that exercise is that an errorless empirical brain-body exponent would be the sum: $2/3 + (1/6)^{2/3} = 2/3 + 1/9 = 7/9$, or about 0.78. This would be the case if the thickness exponent were exactly 1/6. Since the thickness exponent is empirical and has not been carefully determined, there is uncertainty in the determination of the empirical brain-body exponent, but only if thickness were completely independent of brain size would the empirical exponent be approximately 2/3. My theoretical approach to brain-body allometry, which is based on a dimensional analysis that assumes the brain's size to be determined by mappings from the body and within the brain, should always have predicted empirical exponents closer to 3/4 than 2/3 (See Jerison, 1985, for more details.)

I have gone through this exposition to illustrate my view of one correct use of regression analysis of the type that Harvey described, namely to determine agreement with theoretical expectations. Empirically determined allometric exponents can be used directly as theoretical values only if the process that generates one's data results from forces comparable to the ones that generate a regression model. In other words, there must exist a fundamental equation that determined the regression variance, and there must be random generators determining the residuals. This is probably never true in living systems, which are determined by many variables. However, when a major variable is represented by the allometric exponent (the "regression" effect), the remaining variables may produce such varied and different effects that their combined activity appears haphazard. Their combined effect may, therefore, be estimated by a linear regression model as a single "residual" effect, that is, as if they were random deviations from the regression. The "theory" for using regression approaches to determine fundamental allometric parameters is, therefore, the same as the theory of errors. The approach is not "objective" (i.e., theory-free). It simply uses a curve fitting procedure that has become so conventional that its theoretical basis is usually overlooked. There is plenty of justification for this tactic, but I prefer a theory that is closer to the biology of the problem. A few years ago I published a comparison between a conventional allometric analysis of within-genus data on cercopithecus monkeys in which the empirically determined exponent was about 1/3, and I compared this result with curve-fitting according to an equally parsimonious dimensional (biological) theory (Jerison, 1982). The curves generated by the biological theory and by the allometric analysis were almost indistinguishable.

I have never suggested an evolutionary model or narrative that would indicate why a particular amount of processing capacity would ever evolve. My model has been, essentially, that within, e.g., a mammalian body plan, if a species evolved to a particular body size a certain amount of brain would have to develop to control that body. The theoretical analysis of allometry was with respect to the requirements of mapping that would force a 2/3 exponent, and I now recognize that the fact that the maps are three dimensional sheets would force a 3/4 exponent. There are other constraints that can be examined, such as the energetic and ecological constraints discussed by Harvey. Nevertheless, whatever the effects of such constraints there will be a correlated effect upon the mapping functions and information-processing capacity. Information processing capacity may be less a constraint than a response to the effects of constraints. If metabolic rate limits brain size in species, it limits information processing capacity. If ecological requirements establish constraints on natural selection for brain size because of the energetics of brain tissue these, too, eventually limit processing capacity. If situations arise in which there are unusual selection pressures for additional processing capacity, then the effect would be not on the allometric function but on a displacement of the function, in other words, on encephalization rather than allometry. Harvey's main concern seems to be with the evolutionary significance of allometric relations. Mine has been with the evolutionary role of encephalization.

To quantify encephalization as a residual it is first necessary to determine the appropriate allometric (= regression) brain-body relationship. I have usually preferred to treat the issue "nonparametrically" by imagining a brain-body space in which various species fall and by considering the relationships among species in terms of their position in that space. This kind of analysis has been performed only pictorially with convex polygons drawn about arrays of data-points (Figure 24.1, below). It avoids the curve-fitting problems that remain to be resolved in allometric analysis, but it loses the power of numerical methods. Harvey appears to be on the road to discovering correct allometric analyses on which to base a numerical analysis of encephalization.

On Problems with Marine Mammals

An issue in allometry and encephalization arose in the contribution by Schusterman and Gisener when they reported in a footnote that the encephalization quotient (EQ) in the bottlenose dolphin *(Tursiops)* was 2.8. I added to their footnote to clarify the computation, because other samples and parameters give different EQ's. Using data on two other samples of the same species (Jerison, 1978), I calculated EQ's of 4.1 for one sample and 3.4 for the other. The expected value in "average" mammals is 1.0, which is approximately the value for the California sea lion.

Schusterman and Gisiner's footnote may give a wrong impression of the place of bottlenose dolphins among the cetaceans and among other mammals. In my sample, the bottlenose dolphin had the highest EQ among the cetaceans, and in the sample cited by Schusterman and Gisiner they had the second highest, with the highest value occurring in the Pacific Striped Dolphin (*Lagenorhynchus obliquidens*). In indicating that *Tursiops* was in the lower half of the range of cetaceans, Schusterman and Gisiner were evidently referring to the absolute measurements of EQ, which are too unstable to be used in this way. Even the difference between the dolphin EQ and the human EQ (approximately 5.8 when calculated this way) is too unstable to suggest greater encephalization in humans than in dolphins. The problem is with the EQ measurement, which is, of course, no more that the residual from a regression.

I am using the Schusterman and Gisiner footnote to make a point about the analysis of encephalization. Numerical measurements (including some that I have suggested) involve assumptions that can be hard to justify. My preferred strategy for analyzing encephalization is to "quantify" graphically either with convex polygons or simply by graphing and labelling points in order to differentiate groups. In Figure 24.1, I have copied a graph that is relevant for the Schusterman and Gisiner footnote, in which comparisons are primarily with convex polygons. The similarities between the dolphin species and *Homo sapiens* in their location in "brain-body space" in this graph are obvious. Harvey's Figures 12.5, 12.6, 12.7 and 12.9 illustrate

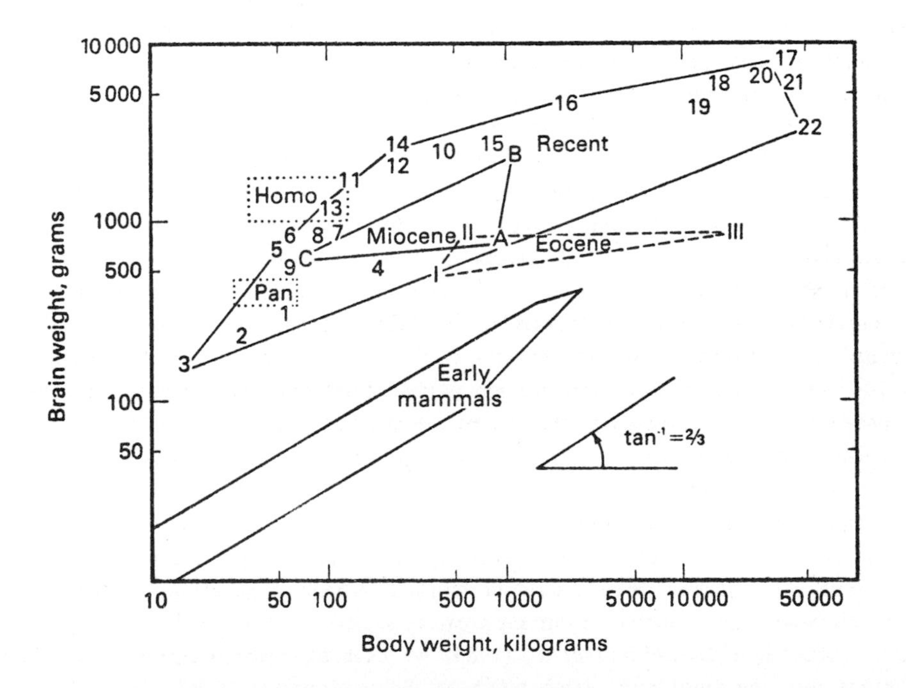

Figure 24.1 Brain-body relations in cetaceans, early mammals, humans ("Homo") and chimpanzee ("Pan"). Numbers are cetacean genera: American *Tursiops truncatus* (11); Mediterranean *Tursiops truncatus* (12); *Lagenorhynchus spp.* (13). Polygons labeled Miocene and Eocene are each based on three fossil cetaceans from those epochs, illustrating some encephalization (upward displacement of the polygon) between the Eocene and the Miocene, but no clear change after the 18 million years old strata from which the Miocene specimens were obtained. [Graph from Jerison (1978), where there is a more complete description of the data.]

the use of encoded data points on a graph to allow equally clear non-numerical comparisons. The conclusion, generally, is that it is inappropriate to rely on numerical indexes, based on incompletely understood formulae, when good alternative unbiased methods are available.

To turn to the substantive issues on language research in marine mammals that are presented by Schusterman and Gisiner, these contributors had developed their procedure to be comparable with earlier analyses by L. M. Herman of the University of Hawaii (see references in Schusterman and Gisiner). Their evaluation of Herman's work was surprisingly negative, and to give a balanced report on this evidently controversial topic (on which I am not expert), I will quote from comments provided by Herman at my request. Herman's complete comments might have been a 12-page contribution to this book had he been able to accept my invitation to lecture at our Institute.

As I indicated in the footnote to the Schusterman and Gisiner text, Herman believes that his most complete response to the criticisms discussed by Schusterman and Gisiner is in his 1987 review, cited by Schusterman and Gisiner. The excerpts from his additional comments that I quote now should clarify the issues in this very important topic that extends our understanding of the cognitive capacities of living mammals. The citations are to publications in the list of refences in Schusterman and Gisiner. The excerpts from Herman's critique are as follows:

Linguistic Terms

Schusterman and Gisiner take their cues from Premack (1986) in protesting that our interpretation of our dolphin's performance is "in terms which are linguistically loaded" and "gratuitous" (p. 325). They follow Premack in objecting to our use of such words as "grammar," "sentence," "lexical," "direct object," " indirect object" and "semantic proposition."

There are several inter-related issues here: (a) how we actually used the stated linguistic terms; (b) when such terms are used does it imply an equivalence in meaning to that occurring when those same terms are used in reference to humans? (c) whether the use of such terms is unique to our animal language enterprise; and (d) whether linguistic terms can be used productively in domains other than human language, or whether they should be banned in any analyses or discussions of animal cognitive or communicative skills.

To begin, in Herman et al. (1984) we carefully stated our definitions of the linguistic terms cited by Schusterman and Gisiner. A word was defined as "a unique semantic entity," a sentence as "a sequence of words that expressed a unique semantic proposition," and language as "the lexical component and the set of syntactic rules that governed the construction and interpretation of sentences in the acoustic or gestural form" (p. 135). Grammar was used in a more general sense to refer to the set of syntactic rules found within a language. . . . We stated that ["indirect object" and "direct object"] were used for convenience to indicate the referents of the words that occupy those syntactic slots and not to describe the grammatical functions of those words (Herman et al., 1984, p. 146). Thus, the terms were used restrictively, as defined. While our use was generally consistent with the use of these terms in human linguistic analyses, nowhere do we imply that our use of such terms carries all of the baggage associated with those terms by linguists dealing with human language. Just the opposite. . . . For example, in reference to our glosses of the sentences of the artificial languages we stated: "The English glosses for the instructions [to the dolphins] are provided only as an aid to the reader. They are not intended to suggest that the dolphin understood the sentence in the sense of the gloss" (Herman et al., 1984, p. 134).

The use of . . . linguistic terms . . . is common within the publications of researchers into animal language capabilities. . . . The banishment of mental terms by the behaviorist movement of the 1920's through 1950's resulted in a great constriction in experiment and theory that retarded the understanding of complex behaviors. Morgan's canon was over-applied and animals were denied mental experiences. . . . Broadbent [in *Perception and Communication*, Pergamon, 1958] argued eloquently for the use of cybernetic language (information processing language) over the then traditional S-R terminology. A major component of his argument was that the information processing approach leads us to consider once again processes within the mind of the organism, and provides a structure within which to view those mental processes. Language, as a cognitive process, is a legitimate, broad domain for study, and its study in animals should not be encumbered by proscriptions against the use of common terms "reserved" for humans. . . .

Methods and Data Interpretations

Relational sentences. . . . Schusterman and Gisiner (p. 340) seem to imply that we were not cognizant of the effect of the category of the goal object on performance until very recently. In fact, very early we discussed with Schusterman our experiments that attempted to make sense of this phenomenon. One of these experiments was described in Herman (1986). Basically, I view the problem as arising from a coding process, in which less interference in memory occurs if objects are categorically different than if they are from the same category. . . . An additional problem for the animal is coding conflict: two codes from the same category may be more mutually interfering than codes from different categories. In this sense, two transportable objects, one of which is the goal object, would be expected to produce greater difficulty than one transportable and one nontransportable object.

Rules and syntax. Syntactic understanding (of word-order rules) by the dolphins . . . was shown through responses to semantic contrasts in reversible sentences, to syntactically anomalous sentences, to structurally novel sentences, to sentences in which later words modified the interpretation of earlier words, to interrogative versus imperative sentence forms, and through variations in modifier placement. Schusterman and Gisiner, considering only a restricted subset of our studies, propose that the behavior of our dolphin Ake can be explained by two simple operational rules (p. 348). In fact, the rules [they] stated . . . are ad hoc, simple re-statements of the responses that the dolphin must take . . . [and] are inadequate to account for the all of the cases enumerated above. . . .

[Schusterman and Gisiner] take issue with our interpretation of how modifiers are attached to object names. We stated a precedence rule in which the modifier term is attached to the object names it precedes rather than follows. Their argument, however, is not theoretical, but methodological, stating "if the sign sequence BALL, LEFT PIPE is given there must necessarily be only one ball, but two pipes in the pool. How could the animal express an incorrect assignment of the LEFT modifier sign to BALL, if there were not two balls from which to choose?" (p. 342). This claim presupposes that the dolphin keeps an accurate tally of the type and number of objects in the tank, a claim which is clearly false. . . .

Referential quality of a word. In focusing on the ability to exchange sign and referent, Schusterman and Gisiner . . .state the following about Ake's performance: "Are the signs and their referents interchangeable as are human words and the things they signify? The answer is no" (p. 325). In fact, the answer is more complicated, is probably yes, and is decidedly yes for the dolphin Phoenix specialized in the acoustic language. . . .For example, we might show Phoenix a frisbee (in air) followed by a sound meaning OVER. The real frisbee was substituted for the sound FRISBEE. In either case -- whether shown a frisbee or hearing a sound denoting a frisbee--Phoenix made the same response: She jumped over a frisbee that was floating in her tank. Thus, referent was exchanged for symbol. . . .

Error probabilities. . . . In Herman et al. (1984) we outlined several alternative models, all of which yielded above-chance probabilities, and chose a conservative one. Schusterman and Gisiner add yet another model, for calculating error probabilities in relational sentences, but it is a model for the probability of correctly choosing the goal object rather than for executing the sentence as a whole correctly. While it is important to evaluate goal-object errors, the theme of our work was on sentence processing abilities and our statistics mirrored that theme.

In conclusion, Schusterman and Gisiner have shown two things. One, of utmost importance, is the high degree of replicability of our findings when using the same techniques we applied to the dolphins. Replicability of this type has always been questionable in the ape language work emphasizing language production rather than language comprehension, and is therefore most welcome in the present case. The second thing is that the cognitive bases for performance in these languagelike tasks (and Schusterman and Gisiner acknowledge that the task utilizes a simplified language, which they call a pidgin) may be fairly widespread among mammals. (L. M. Herman, personal communication.)

Anatomy and Intelligence

My deepest disappointment at the Institute was the extent to which my good friend and colleague, William Hodos, and I were not able to agree on fundamental issues. Although this may have been a simple failure of communication, or differences about inadequately analyzed quantitative data, I believe that we have real differences in our views on localization of function as well as on the nature of intelligence. We also have philosophical differences, which we might have resolved in discussion meetings at the Institute had they been as apparent then as they are now. Among the outstanding activities at the Institute, unfortunately unpublishable because we lacked the recording facilities, were the philosophical rump sessions that Michael Ruse organized at which everything imaginable was discussed.

My differences with Hodos may begin with the meaning of the principle of proper mass. In his lecture at the Institute, Hodos presented this as pretty much the same as Lashley's old idea of mass action, that the brain must be seen as working as a whole rather than according to its parts. He assured me when I questioned him that he had checked his understanding (= misunderstanding) with other neuroscientists, and that they interpreted proper mass as he did. We were able to correct the error in personal discussion, and at his urging, I restate the principle now to try to correct an apparently common misconception. Beyond the fact that both refer to quantities of tissue, proper mass has nothing to do with mass action. Hodos states the principle of proper mass well on p. 94 of his contribution, although his statement might be expanded to indicate that the principle is primarily one of brain organization and not of intelligence: If a species specializes more in one function than in another (e.g., vision more than hearing), then it will have relatively more brain tissue in the systems controlling its more specialized function than in those controlling its less specialized function. The comparisons with other species should be with closely related species in which the specializations are different.

Although Hodos stated the principle correctly, there still seems to be some confusion about its interpretation. Hodos's suggested test of the principle, which applied it to distantly rather than closely related species (p. 94), reveals some of this. It would hardly be appropriate, for example, to compare the absolute sizes of the optic lobes in birds and fish as tests of proper mass. One need simply make absolute measurements to determine that there is more neural machinery in these lobes (homologous to the superior colliculus) in any bird compared with any bony fish. Obviously, both fish and birds may use visual information; the data on the optic lobes suggest only that the repertoire of visual activities associated with this structure in birds is either broader or more flexible but certainly very different from that of fish. I would not compare a trout with an owl and expect larger optic lobes in a trout because of its use of vision and the owl's unusual nocturnal specialization. The entire trout brain would fit inside an owl's optic lobes. On the other hand, one might compare the relative amounts of optic and auditory neural control machinery within each of the two species, with the expectation that the balance would be toward the auditory in the owl and toward the optic in the trout. This is a trivial comparison because on other grounds we know that central neural auditory representation in all lower vertebrates is limited.

Some residual misunderstanding may also be reflected in Hodos's suggestion that the approach by Welker and by Radinsky (p. 98) is an alternative to "global" views of the brain and of intelligence. That is true, but it might also be mentioned that their approach relies on the proper mass principle, which I think of as complementary to global principles. It is inappropriate to put global and localized views in opposition in this way. For example, in comparisons among primates the amount of neocortex is proportional to gross brain size; a single equation correctly estimates neocortex from brain size in all primates including humans. Yet the absolute size of human neocortex is so much larger than that in other primates that to relate an analysis of parts of the neocortex in, e.g., rhesus monkeys to that of humans would require some control for the differences in gross brain size. The "proper mass" of primary visual cortex appears to be about the same in the two species, whereas that for forebrain is clearly different. Deacon's contribution to this volume, which refines this approach, illustrates this very elegantly (Figure 20.12).

A bit of history on how I came to state the principle may help clear things up. When I was writing my 1973 book on brain evolution I was looking for a succinct statement of one of the more familiar generalizations in comparative neurobiology, that the size of a brain system was proportional to its importance. Wally Welker, an old friend from University of Chicago days, had recently (1963 and 1964) published his classic papers with Seidenstein, Campos, Pubols, and Johnson. Was there a good word or catch phrase to describe what they had found? Welker couldn't think of any so I invented the phrase, "proper mass." Among other things, the idea epitomizes Welker's work on procyonids -- that the raccoon, a paw specialist, has an expanded "paw area" in its somatosensory neocortex, whereas the coatimundi, which explores with its nose and vibrissae, has an expanded "nose area." The principle is evident in the relatively enlarged auditory cortex in echolocating mammals, and it is even more dramatic in the grossly visible neural specializations among fish (Jerison, 1973, Fig. 5.6). Bottom fish that use tactile and other data from barbels have expanded bulbar facial nuclei; surface fish relying on vision have expanded optic lobes; mormyrids and other electric fish have expanded cerebellum, the target organ for information from the electric organs.

The principle of proper mass is fundamental, but a special reason for emphasizing it in my book was to keep people who were naive about neurophysiology and neuroanatomy from thinking that gross brain size (which was the measure that I was developing) was enough to understand the brain. I doubt that people unfamiliar with brain research got my message, but I am afraid that some first rate neuroscientists also missed it. Since they know that the parts of the brain are important (and usually know quite a bit about their structure and function), some of them have difficulties with analyses of the whole brain. They may consider global analysis as contrary to normal science in neurobiology in which the principle of localization of function is so well established. There is increasing recognition of whole-brain effects, however, because many brain systems are "distributed systems" (Mountcastle, 1978). For the analysis of such systems the concept of localization has to be refined, and the refinement will require some procedures that are like those used in the analysis of allometry and encephalization.

If we wish to apply the proper mass principle to distributed systems how should we determine their size? The answer is with a statistical knife rather than a surgical or anatomical knife. The analysis of encephalization is anatomy with a statistical knife, dividing the brain's mass into fractions according to a mathematical procedure. The first cut is into an allometric and encephalization fraction. But the analysis can be carried much further. Deacon shows one direction of the cutting; I have recently made cuts in another way to determine the extent of neocorticalization in fossil animals (Jerison, in press). There is still anatomy to be done, because the procedure at present usually involves the measurement of parts of the brain and relating them to one another in a way that enables one to apply a proper mass principle. Without using the words, "proper mass," Stephan and his colleagues have done this for years (Stephan et al., 1981; Baron et al., 1983), and these are all developments that begin with a global approach.

Hodos's proposal that one can partition a brain into visceral, somatic, and intellectual components may have some heuristic appeal, but, as he makes clear, these are not located in completely different places in the brain. If the medulla is a visceral center it is also a somatic center involved in analysis and transmission of sensory and motor information. If frontal neocortex is an intellectual center, it is also a visceral center and a somatic center. I do not believe that the analysis of intelligence can proceed as if it has no significant visceral and somatic dimensions. And I take Hodos's use of this almost Aristotelian vocabulary as intended primarily to indicate weaknesses in the global approach rather than as proposing those as serious categories for partitioning either brain or behavior.

For a complete critique of Hodos's position I would have to write a chapter at least as long as his. To avoid this, I conclude with some remarks about the four assumptions that Hodos (p. 93) thinks must be true to justify my approach to the relation between brain and intelligence. I do not accept Hodos's judgment that the assumptions he lists are fundamental for analyses like mine, nor have I had to assume that any of them is correct. (My only assumption in this vein has been that information processing capacity is proportional to brain size, and I have been at some pains to justify this in mammals [see Jerison, 1985, for a review], the only vertebrates for which appropriate data are available.) But since Hodos's criticisms are organized around a discussion of these assumptions, it is probably most sensible to devote the remainder of this comment to them.

1. Is there a proportionality among the parts of the brain? The answer is yes to a remarkable extent in the mammalian brain. Analyses like Sacher's, which Hodos cites, show this strongly. (I did a few analyses like these myself, at about the same time as Sacher, and published them mainly in my 1973 book; see also Jerison [1979].) It is true that with respect to their relative size, the olfactory bulbs and a few brain structures associated with them seem to follow different rules than

other parts of the brain, but for most other brain structures, the only way to demonstrate much independence among the sizes of the parts of the mammalian brain is to "control" the general size effect statistically. Even after controlling for size, a 3- or 4-factor model accounts for almost all of the variation among structures in the mammalian brain. The best data showing these effects are from primates and insectivores (Stephan et al., 1981), but to the extent that other data have become available (e.g., from bats) they are consistent with most features of the primate-insectivore picture. In the primate-insectivore data about 80 per cent of the variance in the whole sample is "explained" by overall size, about 12 per cent is explained by olfactory bulbs and related structures, and the remaining 8 per cent is all that remains to be explained by "other factors" including error of measurement. The greatest deviation from a general mammalian pattern like this may be in cetaceans (Garey and Leuba, 1986), yet even here it is appropriate to consider the similarity between humans and dolphins in encephalization (Figure 24.1) as indicating similarities in residual information processing capacities.

1a. How about comparisons across vertebrate classes? Nobody has done the quantitative neuroanatomy to my knowledge, but we should note that comparisons have to be among homologous structures. Cerebral cortex (certainly neocortex which accounts for about half the brain) is a uniquely mammalian adaptation, but as Hodos points out, avian forebrain probably includes tissue homologous to neocortex. If we are to compare gross sizes of parts of the brain we probably have to compare telencephalon, diencephalon, etc. Here we should be prepared for surprises if we quantify rather than rely on impressions. The "gigantic" cerebellum of mormyrids may be no larger than that of a hummingbird, because the whole mormyrid brain is so small (see Jerison, 1973, Fig. 5.6).

1b. What is to be gained from such comparisons? A clearer understanding of the relationships among the parts of the brain and of whole-part relationships. This is especially true if the analysis can be extended to the cellular and neurochemical level. The information-processing capacities of brains and their parts are analyzed on the assumption that cells are efficiently (and similarly) packed in brains of all vertebrates. This is a good working assumption, but it is certainly not exactly true. It may be seriously wrong for lower vertebrates compared with birds and mammals, and that is all the more reason to correct it. Systematic differences among groups could be determined and provide correct estimations of processing capacities of the parts of (and thus of the whole) brain in different species. This is one direction for the development of the "global" approach to the brain.

2. Must there be general agreement about what intelligence is? Of course not. Whether one likes it or not, working definitions are operational definitions, and their metaphysical baggage can be ignored. This is true for all scientific entities. These entities (mass, molecules, membranes, medullas, etc.) are what people make them by the way they use them. Hodos knows that my own definition in terms of encephalization, which is not completely tongue-in-cheek, is not universally accepted, and he even suggests (p.102) that there is circularity in my views. But that does not prevent me from writing as if mine were the one true definition, nor from trying to convince my colleagues that it is the correct definition.

I reject Hodos's tactic of trying to find a definition in the work of students of human intelligence. I would argue (agreeing in this instance with Macphail and many others) that from an evolutionary perspective, the intelligence that interests IQ testers evolved with the evolution of species-typical human language and with those functions of the "nonlanguage" hemisphere of the brain that complement language (see Levy in Chapter 10). Hodos's list of abilities derived from this approach (p. 100) is phrased in terms that should be translated with great care to the domain of animal behavior. They are human terms meaningful in human contexts even if translations to an animal domain are possible.

Although definitions are important and deserve discussion, the discussions can be awful time-wasters for scientific work. Any genuine definition, even mine, can be used very easily. If my definition is accepted, only those behavioral differences that are related to encephalization can be related to biological intelligence. That is how definitions work. It is a matter of philosophy, not science. The only questions about a definition are whether it is self consistent and whether it is interesting. (My definition is, of course, in terms of the information-processing capacity that can be estimated from measured encephalization, and the major scientific difficulties are with the latter measurement.) As Hodos indicates there are few definitions that compete with mine, but a comparative psychology of intelligence can be developed about any definition, including intuitions that serve as definitions. This seemingly anarchic situation raises no real problems, because as a consistent body of knowledge is developed it will become clearer that some definitions are more useful and comprehensive than others, and these will survive by a kind of natural selection.

3. Is intelligence unitary, like body weight, etc.? If intelligence is defined by encephalization, or by processing capacity without specifying how the capacity is used, then it cannot be unitary. I said as much in my introductory chapter (p. 11): "The evolutionary message about intelligence, like the message about so many other dimensions in biology, is a message about pluralism and diversity, about the variety of intelligences in the biological world." I am obviously not alone in this view. Margaret Boden introduces her chapter with the following quotable segment (p. 45): "[T]here is no such thing as biological intelligence. There are, however, biological *intelligences*." We might note, of course, that body weight is not unitary either, and its utility as a measure is perhaps a good model for the utility of intelligence(s) as a concept.

4. Is intelligence the same in different taxonomic groups? Since there is no single "intelligence" the question would have to be restated: are there aspects of the intelligence(s) of different taxonomic groups that are, in some sense, the same? We have a very fine answer to that question in Bitterman's chapter and its remarkable demonstration of convergence between honeybees and various vertebrate species in behavioral capacities.

When Hodos accepted my invitation to participate in our Institute he sent me a draft of Hodos (1982). I was concerned that we agreed so much that we might misrepresent the extent to which there is consensus. It may be just as well that now we appear to disagree as much as we do, because the issues that separate us deserve debate. His response to these comments is appended to this chapter, following the list of references.

There were other interesting debates at the Institute which, unfortunately, were not recorded in publishable form. The amount of consensus and of disagreement among us on evolutionary biology of intelligence was proper for an active scientific field. There is a body of information and plenty of theory. Good theory is being developed by cognitive scientists concerned with animal mind, as mentioned by Hodos and also by Boden, and equally impressive theory is appearing in the evolutionary literature as game theory, optimal forager theory, and as mathematical evolutionary biology and ecology. We saw parts of this development as well as some supporting data at the Institute (e.g., in Paul Harvey's work), but we haven't exhausted the subject. Nor have we integrated enough of our data, especially in the area of learning, with sophisticated modern evolutionary theories. There is plenty left to do.

References

Baron G, Frahm HD, Bhatnagar KP, Stephan H (1983) Comparison of brain structure volumes in Insectivora and Primates. III. Main olfactory bulb (MOB). Journal für Hirnforschung, 24:551-558

Brandt A (1867) Sur le rapport du poids du cerveau à celui du corps chez différents animaux. Bull Soc impér Naturalistes Moscou 40 (III-IV), 525-543

Cracraft J, Eldredge N (Eds.) (1979) Phylogenetic analysis and paleontology. New York, Columbia University Press

Garey LJ, Leuba G (1986) A quantitative study of neuronal and glial numerical density in the visual cortex of the bottlenose dolphin: Evidence for a specialized subarea and changes with age. Journal of Comparative Neurology, 247:491-496

Hodos W (1982) Some perspectives on the evolution of intelligence and the brain. In: Griffin, DR (ed) Animal Mind - Human Mind. Springer Berlin

Huxley JS (1932) Problems of Relative Growth. London, Allen Unwin

Jerison HJ (1973) Evolution of the Brain and Intelligence. New York, Academic Press

Jerison HJ (1977) The theory of encephalization. Annals of the New York Academy of Sciences 299:146-160

Jerison HJ (1978) Brain and intelligence in whales. In Frost, S (Ed) Whales and Whaling, Vol 2. 159-197. Canberra, Australian Govt Publ Service

Jerison HJ (1979) The evolution of diversity in brain size. In Hahn, M. E., Jensen, C., & Dudek, B. C. (eds.). Development and Evolution of Brain Size: Behavioral Implications. pp. 29-57. New York, Academic Press

Jerison HJ (1982) The evolution of biological intelligence. In Sternberg, R. J. (ed) Handbook of Human Intelligence. 723-791. New York London, Cambridge Univ. Press

Jerison HJ (1985) Issues in brain evolution. Oxford Surveys in Evolutionary Biology, 2:102-134

Jerison HJ (in press) Fossil evidence on the evolution of the neocortex. In Jones, EG and Peters, A (eds) Cerebral Cortex, Vol. 8. New York, Plenum

Kleiber M (1947) Body size and metabolic rate. Physiological Reviews 27:511-541

Mountcastle VB (1978) An organizing principle for cerebral function: The unit module and the distributed system. In Edelman GM, Mountcastle VB. The Mindful Brain. 7-50. Cambridge, Mass.: MIT Press

Popper KR (1979) Objective Knowledge: An Evolutionary Approach. (Rev Ed). Oxford, Oxford University Press

Popper KR (1986) Unended Quest: An Intellectual Autobiography. (With Postscript and Updated Bibliography). Glasgow, William Collins Sons

Snell O (1891) Die Abhängigkeit des Hirngewichtes von dem Körpergewicht und den geistigen Fähigkeiten. Arch. Psychiat. Nervenkr. 23:436-446

Stephan H, Frahm H, Baron G (1981) New and revised data on volumes of brain structures in insectivores and primates. Folia Primatologica, 35:1-29.

Weiskrantz L (ed) (1985) Animal Intelligence. Clarendon Press, Oxford

Some Comments on "Afterthoughts: Anatomy and Intelligence" (By William Hodos)

My friend and colleague, Harry Jerison, has taken issue with a number of the points raised in my chapter in this volume. I believe that we actually disagree on relatively few issues in comparative psychology and brain evolution. I summarize our disagreements here and then comment on specific points that Jerison raised.

Having surveyed the history of the concept of intelligence as used by writers on human and animal behavior, I am convinced that they use the term to refer to some cognitive process or processes, whether discussing human or animal behavior. I object to definitions of intelligence such as "total information processing capability" because I feel that they include too many things that are not cognitive. For example, many behavioral phenomena are affected by the length of the day. The brain processes information about day-length, which affects the pineal, the pituitary and the endocrine system as a whole; these endocrine changes have consequences for animal behavior, such as premigratory restlessness, nest building, courtship, hibernation, etc. I do not believe, however, that any of these behaviors represent cognitive processes any more than the food-seeking behavior that results from reduced food intake is a cognitive process. To be sure cognitive processes may be involved in getting food; but the food seeking itself is a reflection of a motivational state that developed as a result of the brain processing information about the levels of glucose and lipids in the blood. I also would not disagree with the notion that the extent and speed of information processing are aspects of intelligence; they are, in my opinion, only two of a much wider set of specific abilities that together make up intelligence as a global property.

Jerison's definition of encephalization is a good one and I see no problem in it. My point is that it should be applied only to the intellectual brain if one is attempting to correlate intelligence and brain size.

In my chapter, I raise questions about the utility of using global brain measures as an estimate of the size of brain constituents. When I first looked at Sacher's data, I felt that his observations indicated that such estimates were quite reasonable. I was especially impressed with the very high correlation between neocortex surface and total brain size. On the other hand, I was concerned by the considerably lower correlation between the volume of the olfactory bulb and brain size. The extent of covariation between brain size and the size of the olfactory bulb would depend very much on whether the animals were microsmatic or macrosmatic. This led me to think about other variations in brain structures within mammals and across classes.

Another possible problem with global indices of the brain is the extent to which they may fail to take into account the columnar organization and microcircuitry that, by common consent -- at least for the moment -- are the hallmarks of sophisticated processing. These organizational systems are so condensed and so neatly packed that they might not be greatly reflected in total brain weight or volume and yet they very likely play a significant role in intelligence, by anyone's definition. While, the high correlation between cortical surface and brain size in mammals suggests that columnar organization is being taken into account in this class, I still have serious questions about non-mammalian brains.

In my chapter, I do not take a position against general intelligence or global measures of intelligence. My point was to caution that general intelligence cannot be measured with a single test; rather, it only emerges as a statistical entity from a correlation matrix based on the results of a diversity of tests of specific abilities. To administer one or two tests of specific abilities to animals and then assume that we know something about their general intelligence is a grave error, in my opinion. Moreover, few animal experimenters are willing to give batteries of tests to animals for reasons of technical difficulty and the near impossibility of finding financial support for something that sounds very pedestrian, but in fact is the only way to answer the general-intelligence question. My suggestion is that [we] face up to the reality of the situation and deal with the specific abilities and not make strong inferences about general intelligence. But if we do that, does it make sense to do so in the context of global brain indices? I would not be surprised at all to learn that global intelligence correlated highly with global brain indicators; but since we cannot measure global intelligence in animals, are we on safe ground with global brain indices?

I have difficulty with Jerison's definition of biological intelligence as a tool to understand the relationship between the size of the brain and the amount of intelligence because this definition logically locks the independent variable (brain size) and the dependent variable (intelligence) together. It does not permit the one to float freely so that we can determine whether it is related to the other.

In my comments on the findings of Welker and Radinsky, I was not taking issue with Jerison's principle of proper mass; indeed their findings are excellent examples of the proper-mass principle. My purpose was to build my case for local brain indicators as measures of specific intellectual abilities. I certainly did not intend to suggest the the global and specific intelligences are somehow mutually exclusive. The global process or processes are built up from the specific processes.

I disagree with Jerison's use of the term "whole-brain effects." Perhaps a better term would be "major-system effects" or "multiple-system effects." Jerison uses the example of the frontal lobe to challenge the usefulness of my suggested parcelling of the brain into visceral, somatic and intellectual because the frontal lobe has been implicated in all three of these functions. Within the frontal lobe, however, the somatic areas are distinct from the visceral and from the intellectual;

i.e., the parcelling can be accomplished, but not at the level of the frontal lobe as a whole.

At the beginning of my chapter, I challenged four assumptions about the relationship between brain size and intelligence. Jerison has taken issue with my challenges and the following are my responses:

We do not disagree about the proportionality of brain constituent parts to total brain size in mammals. But there are many instances in non-mammals in which the proportionality argument would fail. For example, I would be quite surprised [if] the size of the giganto-cerebellum were proportional to total brain size in mormyrid fishes. These animals have a cerebellum that is related to their use of electroreception. Another illustration may be seen in the vagal and facial lobes of carps and catfishes. These lobes vary not in proportion to brain size, but in proportion to the fishes' use of gustation. But even within mammals, compare a tree shrew's optic lobes to those of an insectivore of equal body size -- they would be proportionately larger because the tree shrew is an arboreal, diurnal animal.

I have no doubt that, as a general rule, the assumption of proportionality holds for those animals close to the regression line. But when animals lie well above the line because of these specialized structures, should one conclude greater intelligence? An information-processing definition of intelligence would suggest an affirmative conclusion. A cognitive definition of intelligence would suggest that some increase in cognitive ability would be represented by the increased sophistication of the sensory modality. Indeed, an application of the proper-mass principle leads us to conclude that if a brain region has a lot of specialized neural machinery, the animal must be doing something very interesting with it. My point here is that an increase in overall brain size due largely to hypertrophy of a single specialized region may not contribute as much to global intelligence as an increase in brain size proportional in all parts. An analysis based on specific regions and specific abilities would reveal these differences more clearly than a global analysis.

I do not believe that it is seriously wrong for "lower vertebrates" (by which I assume Jerison means anamniotes) to be compared with mammals. Indeed, when the comparisons are made, at least in the area of learning, where most of the data are, the similarities are more striking than the differences. This is certainly the case in simple associative learning. Differences in complex learning generally have revealed differences between amniotes and anamniotes. But even here, phenomena that once were thought to be demonstrable only in birds and mammals have been reported in fishes. Indeed, in this volume, Bitterman has reported an impressive array of complex associative learning skills in honeybees -- animals with considerably fewer neurons than in any mammal or bird.

I cannot agree with Jerison's assertions about the lack of necessity for clear definitions of terms. Without some general agreement on definitions, different investigators will be studying different things but will call them the same thing. The potential for confusion is serious. If Jerison called his approach "the correlation between global information processing and global brain indices," I would be perfectly satisfied. It is the equation of total information-processing capacity with intelligence to which I object. Perhaps the specific abilities ought not to be called intelligence either but merely specific abilities. The term intelligence could then be reserved for global intelligence -- whenever we are able to measure it in animals.

Jerison raised the issue of language and human intelligence. I stated in my chapter that I specifically wished to avoid dealing with the language-intelligence question, but it is appropriate to review it now. Jerison and I agree that much of human intellectual ability comes from the formulation of problems linguistically. In my list of attributes of intelligence that could be considered in studies of animal behavior, I listed only processes that I felt were relatively free from linguistic bias. My position is that intelligence is a human concept, and when we look for intelligence in animals it is to find features equivalent to human intelligence. I have written elsewhere about the dangers of this approach. Bias is introduced because the tests are made by humans based on traits that they value. We should look at what animals use in their own environments. But we will find it very hard to get away from the notion that being intelligent means doing something clever. My list of specific abilities aims to do that bearing in mind the caveats just raised.

I hope that readers will conclude, as I have, that Jerison and I are not very far apart on most of the major issues in the area of the brain and intelligence. We differ mainly on questions of definition and the relative weights to be given to global versus local indices of brain size. Even here, we seem to differ more in emphasis than in substance. In any case, we have had a useful exchange of views and perhaps have revealed additional facets of these issues.

AUTHOR INDEX

Abbott ME, 110, 117
Abelson AR, 354, 359
Abplanalp P, 222-226
Abramson CI , 265, 267, 268, 272
Acher R, 131, 136
Ackley DG, 60, 68
Adler JE, 137
Aertsen A, 445
Agnati LF, 134-136, 138
Alberch P, 140
Alcorn, 357
Alkon DL, 272
Allen E, 37, 43
Alloway TM, 284, 295-296
Altbacher V, 316
Ammon D, 271, 272
Amsel A, 259,272
Anderson JA, 69, 146, 155
Anderson JR, 293, 302, 315
Anderson M, 360
Andersson K, 138
Andreone C, 138
Andy OJ, 229, 386, 414
Angermeier WF, 280, 293
Annett M, 116-117
Apfelbach K, 214, 226
Arbib MA, 304, 308, 315
Archer J, 302, 315
Arentzen R, 129, 141
Ariens Kappers CU, 94, 97, 104, 215, 227
Armstrong E, 120, 136-137, 140, 196, 207-209, 384, 386, 390, 399, 402, 406, 409, 41
Arnesi C, 369, 378
Arnold SJ, 123, 136
Asaka A, 353, 361
Atlan H, 135-136
Ayala FJ, 14, 15, 30, 31, 442, 444

Baettge EE, 129, 139
Baettig K, 110, 117
Bakeman R, 168, 172
Baldino F, 129, 141
Baldwin, 75
Ballard DH, 440, 444
Ballinger JC, 276
Bargiello TA, 141
Barney HL, 151, 152
Baron G, 229, 387, 390, 414, 459, 462-463
Barrett MD, 155
Barrington EJW, 82, 90
Bartlett I, 317
Bastian J, 321, 349
Bates E, 154-155
Bateson P, 123, 136
Batteau DW, 321, 322, 322, 349
Battistini N, 138
Bauchot R, 95, 97, 104, 217, 226, 229, 385-386, 388, 409
Baum SR, 156
Beach FA, 281, 293
Beatty J, 37, 43
Beck LF, 355, 360
Becker HC, 274
Beecher MD, 164, 170, 172, 366, 378, 381
Beer RA, 426
Behrend ER, 262, 272
Belyaev DK, 221, 226
Benedito MAC, 110, 118
Bennett PM, 201-206, 209, 210, 215, 227
Benson DA, 163, 173
Benyamina M, 137
Bergeron R, 208-209
Beritashvili IS, 301, 315
Berlucchi G, 163, 173
Bernstein M, 362
Bever TG, 99, 101, 103, 106, 445
Bhatnagar KP, 462
Binford LR, 194, 196

Bingham E, 357-358, 361
Bishop J, 361
Bitterman ME, 84-90, 102-105, 119, 137, 251-276, 280, 294, 296, 426, 449-450, 461
Bizzi E, 368, 378
Black IB, 131, 136
Black M, 17, 30
Blake DV, 70
Blinkhorn SF, 356, 360
Blinkov SM, 390, 399, 409, 413
Block N, 29, 30
Boakes RA, 283, 291, 294
Bobrow DG, 58, 68
Bochenski I, 23, 31
Bock WJ, 121, 137
Boden MA, 45, 49-50, 68, 449, 461-462
Bodis-Wollner I, 435, 436
Boesch C, 194, 196
Boesch H, 194, 196
Bogen JE, 159, 170
Boice R, 223-224, 226
Bolles RC, 282, 284, 294
Bonin G von, 383, 409
Boring E, 351, 352, 356, 358, 360
Botez MI, 369, 378
Bouchard TJ, 430, 434-435
Bourgeois J-P, 414
Bouton, ME, 257, 273
Boyd R, 40, 44
Braak H, 366, 378
Braden W, 356, 361
Bradford MR Jr, 126-127, 140
Bradie M, 90
Bradshaw JL, 157, 160, 165, 170
Braithwaite RB, 36, 43
Brand CR, 357-358, 360
Brandon R, 25, 31
Brandt A, 450, 462
Brauth SE, 140, 377, 381
Braveman NS, 284, 286, 294
Breland K, 283, 294
Breland M, 283, 294
Brenner S, 135
Briedermann L, 226
Broadbent D., 455
Broca P, 159, 170
Brodmann K, 402, 409
Brody H, 409
Bronson RT, 217, 226
Bronstein P, 284, 286, 294
Brooks R, 353, 360
Brookshire KH, 103, 105, 119, 137
Brown PL, 260, 273, 286, 294
Bullock D, 262, 273
Bullock TH, 119-120, 137
Bunge M, 145, 155
Burian R, 25, 31, 37-38, 43, 140
Burkhard B, 341, 349
Burlingame AC, 366, 378
Burns A, 353, 360
Burt C, 351, 353, 356, 360-361
Busnel MC, 426

Cade WH, 123, 137
Cain DP, 163, 170
Callebaut, 74, 90-91
Calt KJ, 139
Camarda R, 163, 173
Campbell CBG, 124, 137, 140, 280, 294-295
Campbell DT, 14, 31, 74-75, 78, 90, 313, 315, 353, 360
Campbell ME, 104-105
Campos GB, 98, 107, 458
Canavan AGM, 351, 360
Cantley L, 129, 137
Caplan AL, 37, 43
Caramia M, 436
Carew TJ, 272, 273
Carey N, 354, 360
Carlson NR, 96, 105
Carnap R, 36, 43
Carson LV, 412
Carter-Saltzman L, 25, 32, 430, 436
Cartmill M, 189, 196
Caryl PG, 356-358, 360
Castellucci VF, 140
Cattell JM, 352, 356, 360
Cattell RB, 352, 356, 360
Changeux JP, 133, 137, 144, 155
Charniak E, 50, 64, 68

Chase TN 353,360
Chatilllon M, 156
Chau Kun HO, 196
Checke S, 274
Cherfas J, 129, 137
Chi Je G, 163, 170
Chomsky N, 17-18, 31, 48, 143-145, 155, 377, 378
Church RM, 289, 294
Clark BT, 141
Clarke R, 163, 173
Clemente CD, 227
Cloak FT Jr, 303, 315
Cloud P, 15, 31
Clutton-Brock TH, 176, 197, 200, 204, 210, 384, 410-411, 413
Colbert KK, 351, 361
Cole-Harding S, 118
Collett TS, 48, 68
Collier E, 141
Connors BW, 432, 434
Conway BE, 362
Cook PE, 253, 274
Corkin S, 102, 106
Corning WC, 252, 273, 275, 276
Corruccini RS, 198
Couvillon PA, 252, 253, 254, 255, 256, 257, 258
Cowan W, 410, 413-414
Cowey A, 414
Cracco RQ, 435-436
Cracraft J, 449, 462
Cragg BG, 410
Craik KJW, 300, 315
Cramon D Von, 366, 380
Cranach von, 137
Creelman, 358
Crelin ES, 148, 155-156
Cronbach LJ, 105
Crosby E, 94, 97, 104,227
Croze H, 302, 315
Csanyi V, 299, 302, 309-310, 312-317, 450
Culebras A, 163, 171
Cumming S, 358, 361
Cunningham CL, 257, 275
Danger JM, 137
Daniels N, 29, 31
Darlington PI, 313, 316
Davis H, 288, 289, 294
Davis KR, 380
Davis RE, 137, 141
Davis RT, 188, 196
Davis R, 52, 69
Dawkins R, 3-5, 10-11, 26, 31, 304, 308, 313, 316
De Fries JC, 118, 430, 434
De Groot J, 227
De Valois RL, 188, 196
De Waal F, 23, 31
Deacon TW, 133, 152, 363, 366-368, 377-378, 383, 388-389, 398-399, 402, 407, 410, 447-459
Deary IJ, 351, 353-354, 357-360, 447
Delius JD, 101, 105
Delson E, 176, 178, 180-181, 198
Demski LS, 96, 105
Denes G, 160, 171
Denis-Donini S, 132, 137
Dennett DC, 76, 90, 301, 316, 438, 444
Dessi-Fulgheri, 136-137
Deutsch-Klein N, 291, 296
Dewson JH III, 164, 170-171, 366, 378
Diamond IT, 222, 226, 384, 410
DiChiro G, 353, 360
Dickinson A, 259, 274
Dilger WC, 123, 137
Dobzhansky Th, 15, 24, 31
Doetsch GS, 412
Dohlinow P, 197
Doka A, 316
Domjan M, 284, 294, 307, 316, 341, 349
Donnelly M, 141
Dooling EC, 163, 170
Douglas RJ, 385, 410
Dragoin WB, 284, 297
Dreyfus HL, 65, 69, 136
Driscoll P, 110, 117
Dubois E, 1, 11, 95, 97, 105, 213, 215, 226, 383, 410
Dudek BC, 110, 117, 445, 462
Dufort RM, 262, 274

Dumont JPC, 271, 274
Duncan GW, 380
Duncker H-R, 229
Dunn G, 351, 360
Durlach PJ, 257, 258, 274, 275
Dyal JA, 273, 275, 276
D'Amato CJ, 384, 410
D'Este L, 138
D'Udine B, 123, 136

Ebbesson SOE, 126, 137, 140, 377-379, 410
Ebert PD, 110, 117
Ebinger P, 214, 216-218, 221-223, 226-227, 229
Eccles JC, 437, 442, 446
Echteler SM, 126, 137
Eckenhoff MF, 414
Eckert MK, 378
Eckmann K, 378
Edelman GE, 11, 132, 137, 313, 316, 360, 445, 462
Edinger L, 115, 117
Edwards CA, 290, 294, 297
Edwards C, 357-358, 360, 362
Egan V, 357-358, 360
Eibl-Eibesfeldt I, 122-123, 137, 308, 316
Eichenbaum H, 378
Eisenberg JF, 204, 210, 321, 349
Eisenstadt M, 69
Elcock EW, 69
Eldredge N, 78, 90, 449, 462
Elvee RQ, 445
Emery OB, 153, 155
Engen E, 153, 155
Engen T, 153, 155
Epstein R, 290-294, 301, 312, 316
Erber J, 252, 274
Ertl JP, 356
Ervin FR, 316
Erwin VG, 118
Espenkotter E, 214, 217, 227
Estenoz M, 132, 137
Ettlinger G, 98, 106, 386, 414
Evans FG, 126, 137
Evarts EV, 302, 316
Eysenck HJ, 351-353, 356, 360-361, 435

Fahlmann SE, 62, 69
Fairbanks RG, 184-186, 197
Falconer DS, 429, 434
Falk D, 187, 192, 198, 377, 379
Fantino E, 284, 294
Farabegoli C, 138
Farley J, 272
Farris HE, 422, 426
Fasolo A, 119, 127-128, 131, 136, 138, 448
Fedio P, 353, 360
Feirtag M, 96, 106, 133, 140
Feldman ML, 432, 435
Felleman DJ, 413
Felsenstein J, 123, 138
Fentress JC, 435
Ferguson W, 190-191, 196
Fernandes DM, 289, 294
Ferrier D, 368, 379
Feuer D, 435
Fietz A, 273
Fikes RE, 69
Finger T, 94, 106
Finkelstein HH, 380
Finlay BL, 384, 397, 402, 411
Fischer CJ, 217, 227-229
Fischer JM, 379
Fitzmaurice G, 358, 360
Flaherty CF, 270, 271, 274
Flamm L, 171
Fleagle J, 181, 186, 196
Fleischer G, 229
Fodor J, 143-144, 155
Folkins JW, 150, 155
Folstein MF, 353, 361
Folstein SE, 353, 361
Forshaw D, 360
Foster NL, 353, 360
Fouts D, 320, 349
Fouts RS, 166, 171, 320, 349
Fowler H, 101, 105
Frahm HD, 229, 387, 390, 414, 462-463
Franco L, 161-162, 171

Franzen EA, 379
Franzini C, 163, 173
Franzoni MF, 138
Fraser IC, 356, 360
Frearson, 356
Freeman WJ, 432-435, 440, 444
Freemon FR, 369, 379
Frick H, 214, 217, 227
Friedl W, 436
Frisby J, 49, 70
Frisch K von, 8, 11, 252, 274
Fromm, 151
Frost GJ, 193, 196
Frost S, 462
Fuller JL, 110, 117, 315, 316
Funt BV, 61-63, 69
Fuxe K, 134-136, 138

Gaertner I, 291, 294
Galaburda AM, 366, 369, 379
Galanter E, 308, 317
Galef BG Jr, 307, 316
Galkin TW, 384, 411
Gallistel CR, 302, 316
Gallup GG, 288, 292-293, 295, 297, 302, 316
Galton F, 354, 361, 431
Gamzu E, 286, 295
Gans C, 87, 91
Garber RI, 140
Garcia E, 283-286, 295
Garcia J, 306, 316, 419-420, 426
Gardner BT, 146, 155, 166, 171, 321, 323, 349
Gardner H, 160, 171, 173
Gardner RA, 146, 155, 166, 171, 321, 323, 349
Garey LJ, 460, 462
Gaudino G, 131, 138-139
Gazzaniga MS, 159-160, 162, 171
Gegenbaur, 125
Gehring WJ, 140
Gelperin A, 271, 275
Gerli M, 283, 296
Geschwind N, 163, 172, 364, 377, 379
Gevins AS, 432, 435
Giardino L, 138
Gibbon J, 286-287, 296
Gibson JJ, 54, 58, 69
Gillan DJ, 23, 31, 86, 90
Gilles FH, 163, 170
Gingerich PD, 178, 180, 185, 196-197
Ginsberg M, 23, 31
Gisiner R, 449-450, 452-457
Gittleman JL, 204, 210, 215-216, 227
Glassman RB, 313, 317, 440, 444
Glendon FM, 291, 296
Glezer II, 390, 399, 409, 413
Glick SD, 157, 171
Goelet P, 140
Goldberg ME, 101, 105
Goldman PS, 384, 411
Goldman-Rakic PS, 384, 397, 411, 414
Goldstein M, 138
Goodall G, 283, 294
Goodglass H, 372, 379
Goodwin BC, 132, 135, 138, 140
Gopnick A, 154-155
Goridis C, 132, 141
Gottesfeld JM, 141
Gottesman II, 118
Gottlieb G, 422, 426
Goudsblom J, 194, 197
Gould CG, 309, 317
Gould HJ, 150-152, 156
Gould JL, 309, 317
Gould SJ, 4, 11, 37, 43, 103, 105, 122, 136, 138, 200, 210, 271, 274, 352, 353, 361, 383, 386-388, 394, 411, 441, 444
Gowlett JAJ, 193, 197
Grand TI, 399, 411
Grastyan E, 302, 317
Grau JW, 269, 274
Gray K, 70
Green JD, 223, 227
Green S, 164, 169, 172
Griffin DR, 46, 69, 99, 101, 105, 119, 137-138, 288, 295, 301, 303, 317, 319, 349, 437-438, 444, 445, 462
Grimaldi R, 138
Grossmann KE, 259, 261, 274

Guglietta A, 139
Guilford, 352, 361
Gundlach H, 223, 227
Guntherschulze J, 227
Gustaffson 352, 354, 361
Guthrie DM, 96, 105
Gutnick MJ, 434
Guttman N, 274
Guy J, 137

Haberlandt K, 276
Hafez ESE, 223, 227
Haier RJ, 356, 361
Hailman JP, 121, 138
Hall WC, 222, 226
Hallam A, 20, 31
Hamburger V, 384, 402, 412
Hamby S, 160, 173
Hamilton CR, 163, 171, 173
Hamm A, 163, 173
Hannay HJ, 435
Hardy SM, 53, 69
Harfstrand A, 138
Harlow HF, 319, 349
Harris JWK, 197
Harris S, 156
Harrison T, 197
Hart PE, 69
Hartlage LC, 436
Hartline PH, 101, 105
Harvey PH, 176, 197, 199-206, 209-210, 215-216, 227, 228, 384, 410-411, 413, 448, 450-453, 462
Hassan NF, 436
Hassler R, 222, 227
Hast MH, 366, 379
Hayes PJ, 58-59, 60-61, 69-70
Hearnshaw L, 353, 361
Hearst E, 286-287, 295
Hebb DO, 302, 317
Heberer G, 228
Hecht MK, 435
Heffner HE, 164, 171
Heffner RS, 164, 171
Hegmann JP, 430, 433, 435
Heil J, 63, 69
Heilman KM, 160, 171, 173
Heimbuch IC, 148, 155-156
Hein A, 59, 69
Held R, 59, 69
Heller DP, 54, 69
Hempel CG, 21, 31, 36, 43
Hendrickson AE, 353, 356, 360-361
Hendrickson DE, 356, 361
Hennig W, 121, 138
Herman LM, 9, 11, 27, 31, 320, 322-327, 330-331, 333-338, 340-343, 449-450, 454-457
Herre W, 214, 227
Herrick CJ, 125-126, 138
Herrnstein RJ, 103, 105, 290, 295, 297
Herschel M, 436
Hess EH, 57, 69
Hick W, 355, 361
Hickie JP, 321, 350
Hicks SP, 384, 410
Hillyard SA, 160, 171
Hinde RA, 285, 294-297, 305, 317, 426
Hineline PN, 284, 295
Hinton GE, 60, 66, 68-69
Hiorns RW, 6, 11, 395, 414
Hirsh R, 417, 426
Hoban E, 320, 325, 349
Hobbs, 58, 69
Hodos W, 93, 103, 105, 119, 122, 124, 138, 280, 295, 440, 444-445, 448-450, 457-462
Hofman MA, 384, 395, 411, 437, 440-441, 444-445, 448-449
Hogan DE, 290, 297
Hokfelt T, 131, 138
Holland PC
Hollard VD, 101, 105
Hollis I, 320, 349
Hollis KL, 421, 426
Holloway RL, 176, 196-197, 377, 379, 440, 445 RW, 384, 404, 412
Hollyday M, 384, 402, 412
Holmes NK, 262, 274
Holst E von, 59, 69, 101, 105, 288, 294-295
Hooper FH, 351, 361

Hooper JO, 351, 361
Hopkins WD, 166-167, 172
Horn JL, 100, 105, 356, 362
Hubbard JL, 432, 435
Hubel DM, 305, 317
Huber F, 135, 138
Huber GC, 94, 97, 104, 227
Hull D, 36, 43
Hulse SH, 101, 103, 105
Humphrey NK, 302, 317, 318
Humphreys, LG 100, 105
Hupfer K, 366, 380
Hurwitz HMB, 289, 294
Hutton JT, 163, 172
Huxley JS, 75, 388, 394, 412, 450, 462
Huxley TH, 16, 31
Hyde JS, 110, 117

Ifune, 163, 171, 173
Imada H, 288-289, 295
Ingle DJ, 137-138
Innocenti GM, 395, 397, 413
Inouye E, 353, 361
Irons BJ, 139
Isard S, 145, 156
Ison JR, 253, 274
Itani J, 279, 295

Jackson FR, 112, 141
Jacobs WJ, 285, 296
Jacobshagen E, 125, 139
Jacobson M, 112, 117, 398, 412
Jagielo JA, 290, 294
Jakway JS, 227
James FC, 124, 139
James W, 75, 90
Jarman PJ, 123, 139
Jenkins HM, 260, 273, 286, 294-295
Jensen AR, 351-356, 361, 431, 435-436
Jerison HJ, 1, 3, 7, 11, 17, 27, 30-31, 90, 94-97, 99, 102, 104, 109, 117, 119, 138-139, 170-171, 176, 180-181, 191, 197, 201, 210, 211, 216, 227, 302, 317, 321, 349, 383-384, 386-387, 412, 429, 435, 438, 440-441, 443-445, 447, 451-453, 458-460, 462
Jerome H, 353, 362
Joh TH, 129, 139
Johanson D, 16, 17, 31
Johanson D, 190, 198
Johansson O, 138
John ER, 301, 317
Johnson JI, 124, 139, 458
Johnson RC, 434
Johnson TD, 303, 317
Johnson-Laird PN, 59-60, 70
Johnston TD, 87, 98, 307, 317
Jonakait GM, 137
Jones DL, 273, 274
Jones EG, 435-436, 462
Joosse J, 131, 139
Julesz B, 304, 317
Jungers WL, 210, 228
Jurgens U, 364, 366, 369, 380

Kaas JH, 384, 413
Kaiserman-Abramof R, 139
Kakihana R, 110, 117
Kalat JW, 286, 295
Kamin LJ, 13, 29, 31, 352-353, 361-362
Kampis Gy, 313, 316
Kandel ER, 140
Kant I, 20-21, 24, 27, 31-32
Kaplan PS, 287, 295
Karten HJ, 98, 105-106, 140, 377, 381
Kater SB, 435
Katz DF, 133, 137, 139
Katz HM, 102, 107
Katz MJ, 112, 117, 384, 402, 412
Kauffman S, 140
Kaufman L, 431, 435
Kawai M, 279, 295
Keehn JD, 426
Kelahan AM, 412
Kellogg WN, 321, 349
Kemper S, 153, 155
Kessel EL, 122, 139
Ketron JL, 362
Kicliter EE, 124, 126, 140
Kimbel W, 190, 198
Kimble GA, 274
Kimura D, 160, 171

King A, 223, 229
King BJ, 194, 197
King D, 283, 297
King J, 52, 69
Kirk KL, 262, 274
Kirsch JAW, 139
Kirshnit CE, 290, 294, 316
Kirzinger A, 380
Kitcher P, 14, 31
Kleiber M, 450, 462
Kling JW, 282, 295
Klosterhalfen S, 273
Knudsen EI, 101, 105
Koelling RA, 283-285, 295, 316, 419-420, 426
Kohler I, 69
Kohler W, 311-312, 317, 319, 349
Kolb B, 364-365, 372, 380
Kononova EP, 402 413
Koppel H, 395, 397, 413
Kraal PA, 284, 297
Krames L, 284, 295-296
Krechevsky I, 302, 317
Kremer EF, 268, 274
Krieger K, 319-320, 326-327, 340, 349-350
Kruijt JP, 308, 317
Kruska D, 211-222, 224, 226-229, 447-448
Kuczaj SA, 155
Kuhlenbeck H, 125, 139
Kuhn TS, 30, 31
Kunkel JG, 91
Kuroiwa A, 140
Kuse AR, 434
Kuwabara M, 262, 274
Kynette D, 153, 155

La Gamma EF, 137
Laitman JT, 148, 155-156
Lally M, 356-357, 361-362
Lancaster JP, 377, 380, 381
Land MF, 48, 68
Landahl HJ, 150-152, 156
Lande R, 4-5, 11, 140, 200, 202, 210
Lang TG, 321, 349, 350
Lanza RP, 290-294, 296, 316
Lapique L, 213, 228
Lasek RJ, 112, 117, 139, 384, 412
Lashley KS, 1, 11, 410
Lassek RJ, 402
Lawick-Goodall J, 87, 90
Lazarus LH, 130, 131, 138-139
Le Douarin NM, 132, 139
Leboulenger P, 137
Lehrman DS, 50, 52, 69, 426
Lejeune J, 353, 362
LeMay M, 163, 171-172
Lennel K, 23, 33
LeRoith D, 139
Lesniak MA, 141
Lettvin JY, 46, 69
Leuba G, 460, 462
Levitsky W, 163, 171
Levy J, 157, 160, 172, 448-449, 461
Levy-Agresti J, 161, 172
Lewin R, 135, 139
Lewontin RC, 13, 15, 29, 31, 37-38, 43, 77, 90, 352-353, 362
Lieberman PK, 18, 32, 143, 145, 149, 151-153, 156, 448-449
Lilly JC, 321, 323, 350
Lin C-S, 413
Ling PK, 171
Lipp H-P, 109, 112, 116-117, 133, 447-448
Litner JS, 267, 276
Llinas R, 435
Lloyd E, 37, 43
Lockardt RB, 281, 296
Lockhead J, 361
Lockwood MJ, 283, 294
Locurto CM, 286-287, 296
Logan CA, 284, 294
Logan FA, 276
LoLordo VM, 268, 275, 285, 294, 296
Longo N, 262, 274, 276
Longstreth LE, 355, 357, 361
Lorenz K, 29, 30, 32, 49-50, 52-53, 69-70, 75, 223, 228304, 317, 320, 350
Loumaye E, 131, 139
Lu SM, 413
Lubinski D, 291, 296

Lubow RE, 267, 274
Lucas GA, 287, 297
Lumsden CJ, 5, 11, 22, 29, 32, 39, 43
Lykken DT, 435
Lynch JC, 378

McCafferty M, 70
McCarthy J, 61, 70
McClearn GE, 110, 117, 434
McClelland JL, 48, 70
MacCorquodale K, 291, 296
McCrady E, 222, 228
McCulloch CE, 124, 139
McCullough WS, 46, 69
McDermott D, 50, 64, 68
McDonald K, 156, 166-167, 172, 323, 350
Mace GM, 199, 200, 204, 210, 384, 411, 413
Macedo H, 227
MacKay DM, 299, 301-303, 317
McFarland D, 99, 106
McGarry J, 141
McGinnis W, 140
McGonigle B, 289, 296
McGue M, 115, 118
McGue M, 430, 434-435
McHenry H, 197
McHugh PR, 353, 361
Mackenzie B, 357, 358, 361
Mackintosh NJ, 285, 287, 296, 353, 355-357, 359, 361
McNeill D, 147, 156
Macphail EM, 85-86, 90 103, 303, 317, 384, 429, 435, 437-439, 442, 445, 461
McPhillips SA, 273
Mahowald MW, 163, 172
Malacarne G, 119, 448
Malcolm N, 65, 70
Manley JA, 366, 380
Manning RW, 141
Mansi L, 353, 360
Marcellus D, 385, 410
Marciani MG, 436
Marean GC, 156
Markey KM, 137
Markey PR, 321, 322, 349
Marler PR, 164, 169, 172
Marquis DG, 95, 106
Marr DA, 54-55, 57, 60, 70
Marsh OC, 95, 106
Marshack A, 444-445
Martin KAC, 432, 435
Martin M, 65, 70
Martin RD, 122, 140, 175-176, 178, 180, 185, 197, 200-202, 207-210, 215-216, 228, 384, 413
Martire M, 138
Marzi CA, 163, 173
Mason IL, 212, 228
Masur J, 110, 118
Matolcsi J, 227
Matsuzawa T, 289, 296
Maturana HR, 46, 69
Mayhew J, 49, 70
Maynard Smith J, 3, 5, 11, 19, 32, 128, 140
Mayr E, 15, 32, 109, 118, 144, 156, 299, 317, 438, 445
Mech LD, 9, 11
Mellgren RL, 101, 103, 1O6
Meltzer B, 70
Meltzoff AN, 152, 154-156
Memmott J, 288-289, 294
Menzel EW, 303, 317
Menzel R, 252, 265, 270, 272-273
Merlo G, 138
Merzenich MM, 384, 413
Mesarovic MD, 301, 317
Meskill RH, 156
Mesulam MM, 160, 172
Michie D, 69
Miles HL, 320, 350
Milgram NW, 284, 295, 296
Mill JS, 21, 32
Millard WJ, 291, 296
Miller A, 59-60, 70
Miller GA, 145, 156, 305, 308, 317, 326, 350
Miller KG, 184-186, 197
Miller LC, 384, 414
Milner B, 372, 380
Milner JR, 141

Milner PM, 302, 317
Milojevic B, 366, 379
Minsky ML, 60, 70
Mischel T, 70
Mishkin M, 102, 106
Mobbs PG, 272, 275
Mohr JS, 372, 380
Montagna W, 229
Montague A, 37, 43
Montarolo PG, 132, 140
Moody DB, 164, 170, 172, 378, 381
Moore MK, 152, 156
Moore, RC 58, 69
Morgan CL, 278, 296, 299, 318
Morgan J, 140
Morita Y, 94, 106
Moriya M, 289,295
Morris, RJ 141
Moskowitz N, 95, 106
Motta M, 283, 296
Mountcastle VB, 6, 11, 440, 445, 459, 462
Muller-Preuss P, 366, 380
Murphy J, 23, 32
Myers RE, 9, 11, 364, 379, 381

Nadel L, 301, 318
Nageishi Y, 288, 295
Nagel E, 21, 32, 44
Nagel T, 66-68, 70
Napier J, 176, 181, 194, 197
Napier PH, 176, 181, 194, 197
Nauta WJH, 96, 98, 106, 115, 117-118, 133, 140
Nebes RD, 160, 162, 172
Nei M, 430, 435
Nelson RJ, 384, 413
Nettlebeck T, 356-359, 361-362
Nettleton NC, 157, 160, 165, 170
Niall HD, 129, 140
Nicholas DJ, 274
Nichols C, 97, 106
Nieuwenhuys R, 97, 106, 124, 140
Nillson N, 69
Noback CR, 95, 106, 126, 140, 229
Nolan JM, 358, 360
Nord HJ, 214, 217, 227
North AJ, 253, 275
Northcutt RG, 95, 97-98, 104, 106, 120-121, 124, 126-128, 132, 137, 140

Oakley DA, 11, 446
Oakley KP, 193, 197
Odling-Smee FJ, 40, 44, 74, 90-91, 299, 318
Ogren SO, 138
Ojemann GA, 372-374, 381, 407, 413
Oliverio A, 122, 140-141
Olivier A, 126, 140
Olton DS, 9, 11, 102, 106, 301, 318
Overmier JB, 294, 296, 426
Owen DR, 430, 436
O'Keefe J, 301, 318
O'Leary D, 397, 413
O'Shea T, 69-70

Pallie W, 163, 173
Palm G, 445
Pandya DN, 366, 369, 379
Pantin CFA
Papert S, 60, 70
Parent A, 126, 140
Park KS, 353, 361
Pask G, 315, 318

Pasnak R, 90
Passingham RE, 7, 11, 97-99, 104, 106, 377, 381, 383-384, 386-387, 394, 402, 413-414, 439, 445
Pattee HH, 313, 318
Patterson C 121, 140
Paul DB, 123, 140
Pavlov IP, 255-257, 263-265, 274-275
Pelletier G, 137
Pellionisz AJ, 440, 445
Penfield W, 372-374, 381
Perin CT, 253, 275
Perkins D, 361
Perry VH, 414
Pert A, 262, 275
Pessin MS, 380
Peters A, 432, 435-436, 462

Peters RP, 9, 11
Petersen DR, 118
Petersen MR, 164, 172, 366, 378, 381
Peterson GE, 151
Petitto LA, 323, 350
Petrides M, 368, 381
Pfungst O, 319, 350
Piaget J, 75, 77, 81, 91, 109, 118, 155, 351-353, 361, 417, 426
Piattelli-Palmarini M, 155
Pickford M, 175, 177, 180-181, 188, 197-198, 448
Picton WT, 302, 318
Pietrewicz AT, 87, 90
Pilbeam D, 16, 32
Pinker S, 144, 156
Pinxten R, 74, 90-91
Pisacreta R, 290, 296
Pitts WH, 46, 69
Ploog D, 364, 380
Plotkin HC, 4, 40, 44, 73-74, 90-91, 299, 318, 438, 445-446, 449
Poli MD, 119, 277, 283, 294, 296, 448, 450
Polovina J, 434
Pooley G, 87, 91
Popper KR, 75, 90, 110, 118, 360, 425-426, 437, 439, 442-444, 445-446, 499, 463
Porte A, 128, 141
Portmann A, 202, 205, 210
Post DG, 412
Potter B, 190, 198
Powell TPS, 6, 11, 395, 414
Power RJ, 51, 59, 63, 70
Powers WT, 424, 426
Prance CT, 435
Prato Previde E, 283, 296
Premack D, 9, 11, 23, 31, 33, 166, 46, 50-51, 57, 64, 70, 86, 90, 172, 290, 292-293, 296-297, 320-321, 323, 325, 350, 454
Pribram KH, 302, 308, 317, 318, 378
Price T, 276
Prince DA, 434
Pringle JWS, 313, 315, 318
Propping P, 434-436
Putnam H, 30, 32, 36, 44

Quastel DMJ, 435

Rachlin H, 284, 295
Radinsky L, 97, 98, 106, 176, 180, 198, 458
Raisman G, 141
Rak Y, 190, 198
Rakic P, 397, 414
Randich A, 268, 269, 275
Rao DC, 115, 118
Raphe, 128
Rashotte ME, 294, 296, 426
Raup D, 140
Ray RH, 412
Razran G, 422-423, 426
Reed TE, 429-431, 433, 435-436, 449
Reeke GN Jr, 353, 360
Reiher, 223, 228
Reiner A, 121, 140, 377, 381
Reiss S, 268, 275
Remane A, 121, 141
Rempe U, 213, 228
Renda T, 138
Rescorla RA
Revusky S, 84, 91
Reynolds GS, 283, 296
Reynolds W, 141
Richards DG, 320, 322-323, 349
Richerson PJ, 40, 44
Richter CP, 223, 228
Ridley M, 200, 210
Riedl R, 74, 91, 438, 446
Rizzolatti G, 163, 173
Roberts L, 372-374, 381
Robertson RM, 274
Robinson BW, 381
Robinson DL, 356, 361
Robinson DM, 101, 105
Robinson R, 222, 229
Rock I, 54, 70
Rockel AJ, 6, 11, 395, 414
Rohlf FJ, 200, 210
Rohrs, 214, 216-217, 219, 227-229

Roitblat HL, 99, 101, 103, 106, 320, 350, 445
Roith J, 130-131, 141
Romanes GJ, 277-278, 296
Rorty R, 13, 30, 32
Rose S, 13, 29, 31, 352-353, 362
Rosenberg A, 36, 44
Rosenberg P, 378
Rosenberger A, 180, 198
Rosenblatt F, 60, 70
Rosenblatt JJ, 294, 296, 297, 426
Rosenquist AC, 432, 436
Rosenthal R, 319, 350
Ross ED, 160, 172
Ross ME, 139
Ross RT, 269, 275
Rossini PM, 432, 436
Roth J, 130-131, 139, 141
Rothenbuhler WC, 123, 141
Roychoudhury AK, 430, 435
Rozin P, 284, 296
Rubel EW, 384, 414
Rubens AB, 163, 172
Rubert E, 166-167, 172
Rubin LC, 290, 294, 316
Ruby LM, 289, 296
Rudy JW, 271, 275
Rumbaugh DM, 146, 156, 166-167, 170, 172, 323, 350
Rumelhart DE, 48, 70, 449
Runciman WG, 18, 32
Ruse M, 13-17, 20-23, 25, 28-30, 32, 36, 39, 44, 449, 457
Russell IS, 440, 442, 446
Russell LH, 156
Rutishauser U, 132, 14

Sacerdoti, 51, 70
Sacher GA, 97, 104, 384, 386-388, 414, 459
Saether BE, 199, 210
Sahley CL, 252, 259, 265, 271, 272, 273, 275
Sahlins MD, 37, 44
Saidel WM, 126, 137
Salter W, 29, 33
Salthe SN, 78, 90
Samuelson RJ, 301, 318
Sanders GD, 252, 275
Sands SF, 290, 297
Sanides F, 366, 379
Santiago HC, 290, 297
Savage S, 168, 172
Savage-Rumbaugh S, 9, 64, 78, 146, 156, 166-167, 170, 172, 320, 323, 325, 350, 449
Scarr S, 25, 32, 430, 436
Schacher S, 140
Schafer S, 273
Schaffer EWP, 356
Scharrer E, 136, 141
Schiller CH, 69
Schiller PH, 368, 378
Schleidt WM, 122, 141
Schleifenbaum C, 214, 218, 229
Schmechel DE, 413
Schmitt FO, 131, 135, 141
Schneider GE, 222, 229
Schoenfeld D, 320, 349
Scholes R, 160, 171
Schott M, 214, 218, 221, 228
Schreiner L, 223, 229
Schucard DW, 356, 362
Schultz AH, 399, 414
Schultz W, 229
Schumacher M, 214-215, 218, 229
Schupack H, 156
Schusterman RJ, 11, 319-320, 326-327, 340, 346, 349-35, 448-450, 452-457
Schwartz A, 321, 350
Schwartz B, 286, 295
Schwegler H, 112, 117
Scruton R, 13, 32
Sebeok TA, 319, 350
Seidenberg MS, 323, 350
Seidenstein SS, 98, 107, 458
Seim E, 199, 210
Sejnowski TJ, 60, 68
Selfridge OG, 46, 70
Seligman MEP, 419, 426
Senut B, 191, 198
Sereno J, 154, 156

Sevcik RA, 166-167, 172
Shank C, 378
Shariff GA, 395, 414
Shaw E, 426
Sheffield FD, 260, 275
Shepherd M, 360
Shettleworth SJ, 87, 91, 285, 296-297, 419, 426
Shin HS, 129, 141
Shinoda A, 254, 275
Shuku H, 289, 295
Sidman M, 266, 275, 276
Siegel S, 419, 426
Sigurdson JE, 259, 261-262, 275
Silverman J, 171
Simmelhag VL, 315, 318
Simons E, 181, 196
Simpson GG, 121, 132, 141, 177, 180, 198, 271, 275
Sinet PM, 353, 362
Skinner BF, 260, 261, 275, 281, 291-294, 296-297, 315, 318
Slattery K, 384, 397, 402, 411
Sloman A, 48, 56, 61, 63, 64, 70
Slotnick BM, 102, 107
Smith DJ, 384, 414
Smith HAP, 321, 350
Smith J, 139
Smith R, 198
Smith UK, 70
Smith-Gill SJ, 122, 128, 141
Sneed J, 37, 44
Snell O, 450, 463
Sober E, 38, 44
Sohal GS, 384, 414
Sokal RR, 200, 210
Sokolov EN, 302, 318
Spearman C, 351, 354, 362
Sperry RW, 160-162, 171-173, 303, 318
Staal JF, 23, 33
Staddon JER, 315, 318
Stahnke A, 223, 229
Stanfield B, 413
Starck D, 215, 229
Starr J, 291, 296
Stebbins G, 31, 164, 170, 172, 378, 381
Steele Russell J, 99, 107
Stegmuller W, 37, 44
Stehlin HG, 186, 198
Stephan H, 95, 97, 104, 211, 214, 217-219, 226-229, 386-387, 390, 399, 414, 448-460, 462-463
Stephan M, 211, 229
Sternberg RJ, 29, 31-33, 100, 102, 105, 107, 144, 156, 352, 354, 362, 435-436, 462
Stevenson J, 282, 285, 297
Stevenson-Hinde J, 282, 285, 295-296, 305, 317
Stimmel DT, 253, 275
Stoeckel ME, 128, 141
Strasser E, 198
Stratton GM, 57, 70
Stuss DT, 302, 318
Suarez SD, 292, 297
Sudre J, 180, 186, 198
Sulston JE, 132, 141
Sumption LJ, 227
Suppe F, 37, 43-44
Suppes P, 37, 44
Sur M, 410, 413
Susman RL, 168, 172
Sussman GJ, 71
Sutcliff JG, 133, 141
Sutherland NS, 252, 276
Sutherland S, 433, 436
Swartzentruber, 273
Switzer RC, 139
Szalay F, 176, 178, 180-181, 198
Szentagothai J, 440, 446
Szeszulski, 357

Tauber ES, 54, 69
Taylor P, 21, 33, 358
Teasdale TW, 430, 436
Terrace HS, 99, 101, 106, 166-167, 172, 251, 275, 276, 286-287, 296, 323, 347, 349, 350, 445
Terzulo, 308, 318
Testa TJ, 418, 426
Teuber H-L, 102, 106
Thiede M, 214, 227

Thomas RK, 348-350
Thompson P, 35, 37-38, 44
Thompson VE, 379
Thompson WR, 110, 117, 315, 316
Thomson KS, 80, 91
Thoner J, 139
Thorndike EL, 251, 271, 276, 278-279, 281, 419, 427
Thurstone LL, 352, 362
Timberlake W, 283, 287, 297
Tinbergen N, 302, 318
Tobias PV, 384, 415
Toffano G, 138
Tolman EC, 254, 276, 281, 302, 318
Tonon MC, 137
Tower DB, 395, 415
Trevarthen C, 160, 171-172
Tucker DM, 160, 173
Tully T, 110, 118

Uexkull J von, 49, 57, 66, 71, 309, 318
Ullman S, 53-57, 60, 67, 71
Umilta C, 163, 173
Urcuioli PJ, 290, 297
Uttley AM, 70

Valenstein ES, 115, 118
Valentine JW, 15, 31, 33
Van Fraassen BC, 37, 44
Van Valen LM, 5, 11, 98-99, 107, 135, 141
Vandenberg SG, 434
Vandesande F, 138
Vaudry H, 137
Vermeire BA, 163, 171, 173
Vernon PA, 431, 436
Vickers D, 356, 362
Viviani P, 308, 318
Vogel F, 436
Vogt C, 366, 381
Vogt O, 366, 381
Vollmer A, 442, 446
Vossen JMH, 417, 449
Vrba E, 176, 178, 188, 198
Vreugdenhil, 357-358, 362

Wada JA, 163, 170, 173
Waddington CH, 88, 91, 116, 118
Wagner AR, 256, 268, 275, 276
Wahl G, 283, 297
Wahlsten D, 122, 141
Walker A, 187, 192, 198
Walker JJ, 272, 276
Wall J, 413
Wallace B, 435
Wallenberg A, 115, 117
Walsh, 357
Walton D, 197
Wang CC, 222, 229
Wapner W, 160, 173
Washburn S, 197
Watanabe M, 302, 318
Watkins, 151
Watson RT, 160, 171, 173
Wechsler AF, 160, 173
Weidemann W, 224, 229
Weinshilboum RM, 430, 436
Weiskrantz L, 68, 71, 95, 107, 449, 463
Weisman RG, 267, 276
Weiss V, 353, 362
Welker WI, 98, 107, 458
Wells PH, 252, 276
Wernicke C, 159, 173
Wetzel AB, 379
Whewell W, 21, 33
Whishaw IQ, 364-365, 380
Whitaker HA, 374, 381
White JE, 435
White JG, 132, 133, 141
White MJ, 166, 173
White T, 190, 198
Wickelgren WA, 423-424, 427
Wiesel TN, 305, 317
Wigger H, 219, 229
Wight, 357, 358
Wilcoxon HC, 284, 286, 297
Wilczynski W, 122, 141
Wiley EO, 141
Williams DR, 287, 297
Williams D, 356, 361
Williams GC, 16, 31, 33, 77, 82, 91
Williams H, 287, 297

Williamson SJ, 432, 435
Willows AOD, 273
Wills RH, 144-145, 156
Willson RJ, 356, 362
Wilson AC, 79, 91
Wilson DE, 204, 210
Wilson EO, 5, 11, 22-24, 28-29, 32-33, 39, 43-44
Wilson JR, 434
Wilson WE, 139
Winner E, 160, 173
Winter H, 430, 436
Wirz J, 140
Witelson SF, 163, 173
Witt K, 290, 296
Wolff P, 156
Wolfson B, 129, 141
Wolman B, 99, 105, 107
Wolpert L, 140
Wolz JP, 320, 322-323, 349
Wood B, 197
Wood FG, 320-321, 349, 350
Woodard WT, 251, 261, 264, 273, 274, 276
Woodruff G, 9, 11. 23, 31, 33, 50-51, 57, 70, 293, 296, 297
Woodward WT, 103, 104, 280, 294
Wright AA, 290, 297
Wuketits FM, 74, 91, 353, 362, 438, 442, 446
Wundrum IJ, 193, 198
Wyles JS, 88, 91

Xue ZG, 139

Yakalis G, 141
Yamagiwa J, 194, 198
Yeni-Komishian GH, 163, 173
Yerkes RM, 251, 276
Young MW, 141
Young RM, 52, 71

Zaidel E, 160, 169, 173
Zamboni G, 163, 173
Zangwill OL, 107
Zecevic N, 414
Zentall TR, 290, 294, 297
Zeuner FE, 212, 229
Zhou CF, 132, 141
Zihlman AL, 399, 415
Zimen E, 223, 229
Zoloth SR, 164, 170, 172, 378, 381
Zvorykin VP, 409

NATO ASI Series G

Vol. 1: **Numerical Taxonomy.** Edited by J. Felsenstein. 644 pages. 1983.

Vol. 2: **Immunotoxicology.** Edited by P.W. Mullen. 161 pages. 1984.

Vol. 3: **In Vitro Effects of Mineral Dusts.** Edited by E.G. Beck and J. Bignon. 548 pages. 1985.

Vol. 4: **Environmental Impact Assessment, Technology Assessment, and Risk Analysis.** Edited by V.T. Covello, J.L. Mumpower, P.J.M. Stallen, and V.R.R. Uppuluri. 1068 pages. 1985.

Vol. 5: **Genetic Differentiation and Dispersal in Plants.** Edited by P. Jacquard, G. Heim, and J. Antonovics. 452 pages. 1985.

Vol. 6: **Chemistry of Multiphase Atmospheric Systems.** Edited by W. Jaeschke. 773 pages. 1986.

Vol. 7: **The Role of Freshwater Outflow in Coastal Marine Ecosystems.** Edited by S. Skreslet. 453 pages. 1986.

Vol. 8: **Stratospheric Ozone Reduction, Solar Ultraviolet Radiation and Plant Life.** Edited by R.C. Worrest and M.M. Caldwell. 374 pages. 1986.

Vol. 9: **Strategies and Advanced Techniques for Marine Pollution Studies: Mediterranean Sea.** Edited by C.S. Giam and H.J.-M. Dou. 475 pages. 1986.

Vol. 10: **Urban Runoff Pollution.** Edited by H.C. Torno, J. Marsalek, and M. Desbordes. 893 pages. 1986.

Vol. 11: **Pest Control: Operations and Systems Analysis in Fruit Fly Management.** Edited by M. Mangel, J.R. Carey, and R.E. Plant. 465 pages. 1986.

Vol. 12: **Mediterranean Marine Avifauna: Population Studies and Conservation.** Edited by MEDMARAVIS and X. Monbailliu. 535 pages. 1986.

Vol. 13: **Taxonomy of Porifera from the N.E. Atlantic and Mediterranean Sea.** Edited by J. Vacelet and N. Boury-Esnault. 332 pages. 1987.

Vol. 14: **Developments in Numerical Ecology.** Edited by P. Legendre and L. Legendre. 585 pages. 1987.

Vol. 15: **Plant Response to Stress. Functional Analysis in Mediterranean Ecosystems.** Edited by J.D. Tenhunen, F.M. Catarino, O.L. Lange, and W.C. Oechel. 668 pages. 1987.

Vol. 16: **Effects of Atmospheric Pollutants on Forests, Wetlands and Agricultural Ecosystems.** Edited by T.C. Hutchinson and K.M. Meema. 652 pages. 1987.

Vol. 17: **Intelligence and Evolutionary Biology.** Edited by H.J. Jerison and I. Jerison. 481 pages. 1988.